GaN Power Devices for Efficient Power Conversion

GaN Power Devices for Efficient Power Conversion

Fourth Edition

Alex Lidow
Efficient Power Conversion Corporation
USA

Michael de Rooij
Efficient Power Conversion Corporation
USA

John Glaser
Efficient Power Conversion Corporation
USA

Alejandro Pozo
Efficient Power Conversion Corporation
USA

Shengke Zhang
Efficient Power Conversion Corporation
USA

Marco Palma
Efficient Power Conversion Corporation
Italy

David Reusch
Kilby Labs
Texas Instruments
USA

Johan Strydom
Kilby Labs
Texas Instruments
USA

This edition first published 2025
© 2025 John Wiley & Sons Ltd

Edition History
GaN Power Devices for Efficient Power Conversion, 1e, 2012, 2e, 2014, 3e, 2019 John Wiley & Sons Ltd

All rights reserved, including rights for text and data mining and training of artificial technologies or similar technologies. No part of this publication may be reproduced, stored in a retrieval system, or transmitted, in any form or by any means, electronic, mechanical, photocopying, recording or otherwise, except as permitted by law. Advice on how to obtain permission to reuse material from this title is available at http://www.wiley.com/go/permissions.

The right of Alex Lidow, Michael de Rooij, John Glaser, Alejandro Pozo, Shengke Zhang, Marco Palma, David Reusch, and Johan Strydom to be identified as the authors of this work has been asserted in accordance with law.

Registered Offices
John Wiley & Sons, Inc., 111 River Street, Hoboken, NJ 07030, USA
John Wiley & Sons Ltd, New Era House, 8 Oldlands Way, Bognor Regis, West Sussex, PO22 9NQ, UK

For details of our global editorial offices, customer services, and more information about Wiley products visit us at www.wiley.com.

Wiley also publishes its books in a variety of electronic formats and by print-on-demand. Some content that appears in standard print versions of this book may not be available in other formats.

Trademarks: Wiley and the Wiley logo are trademarks or registered trademarks of John Wiley & Sons, Inc. and/or its affiliates in the United States and other countries and may not be used without written permission. All other trademarks are the property of their respective owners. John Wiley & Sons, Inc. is not associated with any product or vendor mentioned in this book.

Limit of Liability/Disclaimer of Warranty
In view of ongoing research, equipment modifications, changes in governmental regulations, and the constant flow of information relating to the use of experimental reagents, equipment, and devices, the reader is urged to review and evaluate the information provided in the package insert or instructions for each chemical, piece of equipment, reagent, or device for, among other things, any changes in the instructions or indication of usage and for added warnings and precautions. While the publisher and authors have used their best efforts in preparing this work, they make no representations or warranties with respect to the accuracy or completeness of the contents of this work and specifically disclaim all warranties, including without limitation any implied warranties of merchantability or fitness for a particular purpose. No warranty may be created or extended by sales representatives, written sales materials or promotional statements for this work. The fact that an organization, website, or product is referred to in this work as a citation and/or potential source of further information does not mean that the publisher and authors endorse the information or services the organization, website, or product may provide or recommendations it may make. This work is sold with the understanding that the publisher is not engaged in rendering professional services. The advice and strategies contained herein may not be suitable for your situation. You should consult with a specialist where appropriate. Further, readers should be aware that websites listed in this work may have changed or disappeared between when this work was written and when it is read. Neither the publisher nor authors shall be liable for any loss of profit or any other commercial damages, including but not limited to special, incidental, consequential, or other damages.

Library of Congress Cataloging-in-Publication Data Applied for:

Hardback: 9781394286959

Cover Design: Wiley
Cover Image: © Efficient Power Conversion Corporation (EPC), epc-co.com.

Set in 10/12pt Warnock by Straive, Pondicherry, India
SKY10105772_051625

In memory of Eric Lidow,
the original power conversion pioneer.

Contents

Foreword   *xi*
Acknowledgments   *xiii*

1   GaN Technology Overview   *1*
1.1   Silicon Power MOSFETs: 1976–2010   *1*
1.2   The GaN Journey Begins   *2*
1.3   GaN and SiC Compared with Silicon   *2*
1.4   The Basic GaN Transistor Structure   *6*
1.5   Building a GaN HEMT Transistor   *11*
1.6   GaN Integrated Circuits   *15*
1.7   Summary   *21*
        References   *22*

2   GaN Transistor Electrical Characteristics   *25*
2.1   Introduction   *25*
2.2   Device Ratings   *25*
2.3   Gate Voltage   *30*
2.4   On-Resistance ($R_{DS(on)}$)   *31*
2.5   Threshold Voltage   *34*
2.6   Capacitance and Charge   *35*
2.7   Reverse Conduction   *38*
2.8   Thermal Characteristics   *40*
2.9   Summary   *42*
        References   *42*

3   Driving GaN Transistors   *45*
3.1   Introduction   *45*
3.2   Gate Drive Voltage   *47*
3.3   Gate Drive Resistance   *48*
3.4 dv/dt Considerations   *50*
3.5 di/dt Considerations   *53*
3.6   Bootstrapping and Floating Supplies   *56*
3.7   Transient Immunity   *59*
3.8   Gate Drivers and Controllers for Enhancement-Mode GaN Transistors   *61*
3.9   Cascode, Direct Drive, and Higher-Voltage Configurations   *61*
3.10   Using GaN Transistors with Drivers or Controllers Designed for Si MOSFETs   *67*
3.11   Driving GaN ICs   *68*

viii | *Contents*

3.12 Summary *69*
References *70*

4 Layout Considerations for GaN Transistor Circuits *75*
4.1 Introduction *75*
4.2 Origin of Parasitic Inductance *76*
4.3 Minimizing Common-Source Inductance *77*
4.4 Minimizing Power-Loop Inductance in a Half-Bridge Configuration *79*
4.5 Paralleling GaN Transistors *85*
4.6 Summary *93*
References *93*

5 GaN Reliability *95*
5.1 Introduction *95*
5.2 Getting Started with GaN Reliability *95*
5.3 Determining Wear-Out Mechanisms Using Test-to-Fail Methodology *95*
5.4 Using Test-to-Fail Results to Predict Device Lifetime in a System *98*
5.5 Wear-Out Mechanisms *99*
5.6 Mission-Specific Reliability Predictions *133*
5.7 Summary *150*
References *150*

6 Thermal Management of GaN Devices *155*
6.1 Introduction *155*
6.2 Thermal Equivalent Circuits *155*
6.3 Cooling Methods *160*
6.4 System-Level Thermal Overview: Single FET *163*
6.5 System-Level Thermal Analysis: Multiple FETs *176*
6.6 Experimental Thermal Examples *182*
6.7 Summary *191*
References *191*

7 Hard-Switching Topologies *195*
7.1 Introduction *195*
7.2 Hard-Switching Loss Analysis *196*
7.3 External Factors Impacting Hard-Switching Losses *217*
7.4 Frequency Impact on Magnetics *223*
7.5 Buck Converter Example *224*
7.6 Summary *245*
References *245*

8 Resonant and Soft-Switching Converters *249*
8.1 Introduction *249*
8.2 Resonant and Soft-Switching Techniques *249*
8.3 Key Device Parameters for Resonant and Soft-Switching Applications *254*
8.4 High-Frequency Resonant Bus Converter Example *261*
8.5 Summary *269*
References *271*

Contents | **ix**

9   RF Performance *273*

9.1   Introduction   *273*
9.2   Differences Between RF and Switching Transistors   *275*
9.3   RF Basics   *276*
9.4   RF Transistor Metrics   *277*
9.5   Amplifier Design Using Small-Signal *s*-Parameters   *284*
9.6   Amplifier Design Example   *285*
9.7   Summary   *292*
         References   *292*

10   DC–DC Power Conversion   *295*

10.1   Introduction   *295*
10.2   DC–DC Converter Examples   *295*
10.3   Summary   *317*
          References   *318*

11   Multilevel Converters   *321*

11.1   Introduction   *321*
11.2   Benefits of Multilevel Converters   *321*
11.3   Experimental Examples   *338*
11.4   Summary   *348*
          References   *348*

12   Class D Audio Amplifiers   *351*

12.1   Introduction   *351*
12.2   GaN Transistor Class D Audio Amplifier Example   *355*
12.3   Summary   *364*
          References   *364*

13   High Current Nanosecond Laser Drivers for Lidar   *367*

13.1   Introduction to Light Detection and Ranging (Lidar)   *367*
13.2   Pulsed Laser Driver Overview   *368*
13.3   Basic Design Process   *378*
13.4   Hardware Driver Design   *384*
13.5   Experimental Results   *388*
13.6   Additional Considerations for Laser Transmitter Design   *394*
13.7   Summary   *399*
          References   *399*

14   Motor Drives   *403*

14.1   Introduction   *403*
14.2   Motor Types   *403*
14.3   Inverter   *403*
14.4   Typical Applications   *404*
14.5   Voltage Source Inverters and Motor Control Basics   *404*
14.6   Field-Oriented Control Basics   *408*
14.7   Current Measurement Techniques   *410*
14.8   Power Dissipation in Motor and Inverter   *411*

x | *Contents*

14.9   Silicon Inverter Limitations   *412*
14.10   LC Filter Dissipation   *412*
14.11   Torque Sixth Harmonic Dissipation   *413*
14.12   GaN Advantage   *413*
14.13   GaN Switching Behavior   *413*
14.14   Dead Time Elimination Effect   *414*
14.15   PWM Frequency Increase Effect   *415*
14.16   Layout Considerations for Motor Drives   *420*
14.17   GaN Devices for Motor Applications   *421*
14.18   Application Examples   *421*
14.19   Summary   *430*
       References   *430*

15   **GaN Transistors and Integrated Circuits for Space Applications**   *433*
15.1   Introduction   *433*
15.2   Failure Mechanisms in Electronic Components Used in Space Applications   *433*
15.3   Standards for Radiation Exposure and Tolerance   *434*
15.4   Gamma Radiation   *434*
15.5   Neutron Radiation (Displacement Damage)   *437*
15.6   Single-Event Effects (SEE) Testing   *438*
15.7   Performance Comparison Between GaN Transistors and Rad-Hard Si MOSFETs   *440*
15.8   GaN Integrated Circuits   *441*
15.9   Summary   *445*
       References   *445*

16   **Replacing Silicon Power MOSFETs**   *449*
16.1   Introduction: GaN, Rapid Growth/Great Future   *449*
16.2   New Capabilities Enabled by GaN Devices   *449*
16.3   GaN Devices Are Easy to Use   *453*
16.4   GaN Cost Reduction over Time   *454*
16.5   GaN Devices Are Reliable   *454*
16.6   Future Direction of GaN Devices   *455*
16.7   Summary   *456*
       References   *456*

Appendix   Glossary of Terms   *459*
Index   *477*

Foreword

It is well established that the CMOS inverter and DRAM are the two basic building blocks of digital signal processing. Decades of improving inverter switching speed and memory density under Moore's Law has unearthed numerous applications that were previously unimaginable. Power processing is built upon two similar functional building blocks: power switches and energy storage devices, such as the inductor and capacitor. The push for higher switching frequencies has always been a major catalyst for performance improvement and size reduction.

Since its introduction in the mid-1970s, the power MOSFET, with its greater switching speed, has replaced the bipolar transistor. To date, the power MOSFET has been perfected up to its theoretical limit. Device switching losses can be reduced further with the help of soft-switching techniques. However, its gate-drive loss is still excessive, limiting the switching frequency to the low hundreds of kilohertz in most applications.

The recent introduction of GaN, with much improved figures of merit, opens the door for operating frequencies well into the megahertz range. A number of design examples are illustrated in this book and other literatures, citing impressive power density improvements by a factor of 5 or 10. However, I believe the potential contribution of GaN goes beyond the simple measures of efficiency and power density. GaN has the potential to have a profound impact on our design practice, including a possible paradigm shift.

Power electronics is interdisciplinary. The essential constituencies of a power electronics system include switches, energy storage devices, circuit topology, system packaging, electromagnetic interactions, thermal management, EMC/EMI, and manufacturing considerations. When the switching frequency is low, these various constituencies are loosely coupled. Current design practices address these issues in piecemeal fashion. When a system is designed for a much higher frequency, the components are arranged in proximity to minimize undesirable parasitics. This invariably leads to unwanted electromagnetic coupling and thermal interaction.

This increasing intricacy between components and circuits requires a more holistic approach, concurrently taking into account all electrical, mechanical, electromagnetic, and thermal considerations. Furthermore, all operations should be executed correctly, both spatially and temporally. These challenges would prompt circuit designers to pursue a more integrated approach. For power electronics, integration will take place at the functional level or the subsystem level whenever feasible and practical. These integrated modules will serve as the basic building blocks of further system integration. In this manner, customization can be achieved using standardized building blocks, much the same way as digital electronics systems. With the economy of scale in manufacturing, this will bring significant cost reduction in power electronics equipment and unearth numerous new applications previously precluded due to high cost.

GaN will create fertile ground for research and technology innovations for years to come. Dr. Alex Lidow mentions in this book that it took thirty years for power MOSFET to reach its current state of maturity. While GaN is still in an early stage of development, a few technical challenges require immediate attention. These issues are recognized by the authors and addressed in the book.

1) High dv/dt and high di/dt render most of the commercially available gate drive circuits unsuitable for GaN devices, especially for the high-side switch. Chapter 3 offers many important insights in the design of the gate drive circuit.
2) Device packaging and circuit layout are critical. The unwanted effects of parasitics should be contained. Soft-switching techniques can be very useful for this purpose. A number of important issues related to packaging and layout are addressed in detail in Chapters 4–6.
3) High-frequency magnetic design is critical. The choice of suitable magnetic materials becomes rather limited when the switching frequency goes beyond 2–3 MHz. Additionally, more creative high-frequency magnetics design practice should be explored. Several recent publications suggest design practices that defy the conventional wisdom and practice, yielding interesting results.
4) The impact of high frequency to EMI/EMC has yet to be explored.

Dr. Alex Lidow is a well-respected leader in the field. Alex has always been in the forefront of technology and a trendsetter. While serving as the CEO of IR, he initiated GaN development in the early 2000s. He also led the team in developing the first integrated DrMOS and DirectFET®, which are now commonly used in powering the new generation of microprocessors and many other applications.

This book is a gift to power electronics engineers. It offers a comprehensive view, from device physics, characteristics, and modeling to device and circuit layout considerations and gate drive design, with design considerations for both hard switching and soft switching. Additionally, it further illustrates the utilization of GaN in a wide range of emerging applications.

It is very gratifying to note that three of the authors of this book are from CPES, joining with Dr. Lidow in the effort to develop this new generation of wide-band-gap power switches – presumably a game-changing device with a scale of impact yet to be defined.

Dr. Fred C. Lee
Founder and Director Emeritus, Center for Power Electronics Systems
University Distinguished Professor Emeritus, Virginia Tech

Acknowledgments

The authors acknowledge the many exceptional contributions toward the content of this book from our colleagues.

Jianjun (Joe) Cao, Robert Beach, Alana Nakata, Yanping Ma, and Robert Strittmatter provided much of the technical foundation behind today's GaN transistors and integrated circuits. Adolfo Herrera made major contributions to Chapter 6 on thermal management. Steve Colino and Tiziano Morganti were the driving force behind our class D audio in Chapter 12. Tony Marini, Rob Strittmatter, and Max Zafrani provided much of the data used in Chapter 15 on space electronics.

The authors really do stand on the shoulders of these giants.

A special thank you is due to Joe Engle who, in addition to reviewing and editing all corners of this work, put all the logistics together to make it happen. Sometimes these logistics meant long continuous hours of editing, coupled with amazing diplomacy working with a wide spectrum of personalities. Jenny Somers, the lead graphic artist on this work, as well as many other GaN-related papers and application notes, deserves a medal of honor as well as an honorary degree in GaN for her creative and extremely accurate projection of scientific data into documentary communications.

A note of gratitude to the editors and staff at Wiley who were instrumental in undertaking a diligent review of the text and shepherding the book through the production process.

Finally, we thank Archie Huang and Sue Lin for believing in GaN from the beginning. Their vision and support will change the semiconductor industry forever.

Efficient Power Conversion Corporation
July 2024

Alex Lidow
Michael de Rooij
John Glaser
Alejandro Pozo
Shengke Zhang
Marco Palma
David Reusch
Johan Strydom

1

GaN Technology Overview

1.1 Silicon Power MOSFETs: 1976–2010

For over three decades, power management efficiency and cost have improved steadily as innovations in power Metal Oxide Silicon Field-Effect Transistor (MOSFET) structures, technology, and circuit topologies have kept pace with the growing need for electrical power in our daily lives. In the new millennium, however, the rate of improvement has slowed as the silicon power MOSFET asymptotically approaches its theoretical bounds.

Power MOSFETs first appeared in 1976 as alternatives to bipolar transistors. These majority-carrier devices were faster, more rugged, and had higher current gain than their minority-carrier counterparts (for a discussion of basic semiconductor physics, a good reference is [1]). As a result, switching power conversion became a commercial reality. Among the earliest high-volume consumers of power MOSFETs were AC–DC switching power supplies for early desktop computers, followed by variable-speed motor drives, fluorescent lights, DC–DC converters, and thousands of other applications that populate our daily lives.

One of the first power MOSFETs was the IRF100 from International Rectifier Corporation, introduced in November 1978. It boasted a 100 V drain–source breakdown voltage and a 0.1 Ω on-resistance ($R_{DS(on)}$), the benchmark of the era. With a die size over 40 mm^2 and a \$34 price tag, this product was not destined to supplant the venerable bipolar transistor immediately. Since then, several manufacturers have developed many generations of power MOSFETs. Benchmarks have been set, and subsequently surpassed, each year for 40-plus years. As of the date of this writing, the 100 V benchmark arguably is held by Infineon with the ISC022N10NM6. In comparison with the IRF100 MOSFET's resistivity figure of merit (4 Ω – mm^2), the ISC022N10NM6 has a figure of merit of 0.022 Ω – mm^2. That is almost at the theoretical limit for a silicon device [2].

There are still improvements to be made in power MOSFETs. For example, super-junction devices and IGBTs have achieved conductivity improvements beyond the theoretical limits of a simple vertical, majority-carrier MOSFET. These innovations may still continue for quite some time and certainly will be able to leverage the low-cost structure of the power MOSFET and the know-how of a well-educated base of designers who, after many years, have learned to squeeze every ounce of performance out of their power conversion circuits and systems.

GaN Power Devices for Efficient Power Conversion, Fourth Edition. Alex Lidow, Michael de Rooij, John Glaser, Alejandro Pozo, Shengke Zhang, Marco Palma, David Reusch, and Johan Strydom.
© 2025 John Wiley & Sons Ltd. Published 2025 by John Wiley & Sons Ltd.

1.2 The GaN Journey Begins

Gallium nitride (GaN) is called a wide bandgap (WBG) semiconductor due to the relatively large bonding energy of the atomic components in its crystal structure. Silicon carbide (SiC) is the other most common WBG semiconductor. GaN HEMT (High Electron Mobility Transistors) devices first appeared in about 2004 with depletion-mode radio frequency (RF) transistors made by Eudyna Corporation in Japan. Using GaN-on-silicon carbide substrates, Eudyna successfully produced transistors designed for the RF market [3]. The HEMT structure was based on the phenomenon first described in 1975 by Mimura et al. [4] and in 1991 by Khan et al. [5], which demonstrated the unusually high electron mobility described as a two-dimensional electron gas (2DEG) near the interface between an aluminum gallium nitride (AlGaN) and GaN heterostructure interface. Adapting this phenomenon to gallium nitride grown on silicon carbide, Eudyna was able to produce benchmark power gain in the multi-gigahertz frequency range. In 2005, Nitronex Corporation introduced the first depletion-mode RF HEMT device made with GaN grown on silicon wafers using their SIGANTIC® technology.

GaN RF transistors have continued to make inroads in RF applications as several other companies have entered the market. Acceptance outside of this application, however, has been limited by device cost as well as the inconvenience of depletion-mode operation (normally conducting and requires a negative voltage on the gate to turn the device off).

In June 2009, Efficient Power Conversion Corporation (EPC) introduced the first enhancement-mode GaN on silicon (eGaN®) field-effect transistors (FETs) designed specifically as power MOSFET replacements (since eGaN FETs do not require a negative voltage to be turned off). At the outset, these products were produced in high volume at low cost by using standard silicon manufacturing technology and facilities. Since then, Texas Instruments, VisIC, Cambridge GaN Devices, ST Microelectronics, Transphorm (now part of Renesas), GaN Systems (now part of Infineon), Panasonic, TSMC, Navitas, Innoscience, and Infineon, among others, have announced their intention to manufacture GaN transistors for the power conversion market.

The basic requirements for semiconductors used in power conversion are efficiency, reliability, controllability, and cost effectiveness. Without these attributes, a new device structure would not be economically viable. There have been many new structures and materials considered as a successor to silicon; some have been economic successes, others have seen limited or niche acceptance. In the next section, we will look at the comparison between silicon, SiC, and GaN as platform candidates to dominate the next generation of power transistors.

1.3 GaN and SiC Compared with Silicon

Silicon has been a dominant material for power management since the late 1950s. The advantages silicon had over earlier semiconductors, such as germanium or selenium, could be expressed in four key categories:

- silicon enabled new applications not possible with earlier materials
- silicon proved more reliable
- silicon was easier to use in many ways
- silicon devices cost less.

All of these advantages stemmed from the basic physical properties of silicon combined with a huge investment in manufacturing infrastructure and engineering. Let us look at some of

Table 1.1 Material properties of GaN, 4H-SiC, and Si [6].

Parameter		Silicon	GaN	SiC
Band gap (E_g)	eV	1.12	3.39	3.26
Critical field (E_{crit})	MV/cm	0.23	3.3	2.2
Electron mobility (μ_n)	cm^2/V·s	1400	1500	950
Permittivity (ε_r)		11.8	9	9.7
Thermal conductivity (λ)	W/cm·K	1.5	1.3	3.8

those basic properties and compare them with other successor candidates. Table 1.1 identifies five key electrical properties of three semiconductor materials contending for the power management market.

One way of translating these basic crystal parameters into a comparison of device performance is to calculate the best theoretical performance achievable for each of the three candidates. For power devices, there are many characteristics that matter in the variety of power conversion systems available today. Five of the most important are conduction efficiency (on-resistance), breakdown voltage, size, switching efficiency, and cost.

In the next section, the first four of the material characteristics in Table 1.1 will be reviewed, leading to the conclusion that both SiC [7] and GaN are capable of producing devices with superior on-resistance, breakdown voltage, and a smaller-sized transistor compared to silicon (Si). In Chapter 2, how these material characteristics translate into superior switching efficiency for a GaN transistor will be explored, and in Chapter 16, how a GaN transistor can also be produced at a lower cost than a silicon MOSFET of equivalent performance will be addressed.

1.3.1 Band Gap (E_g)

The band gap of a semiconductor is related to the strength of the chemical bonds between the atoms in the lattice. These stronger bonds mean that it is harder for an electron to jump from one site to the next. Among the many consequences are lower intrinsic leakage currents and higher operating temperatures for higher band gap semiconductors. Based on the data in Table 1.1, both GaN and SiC have higher band gaps than silicon.

1.3.2 Critical Field (E_{crit})

The stronger chemical bonds that cause the wider band gap also result in a higher critical electric field needed to initiate impact ionization, which results in avalanche breakdown. The voltage at which a device breaks down can be approximated with the formula:

$$V_{BR} = \frac{1}{2} w_{drift} \cdot E_{crit} \tag{1.1}$$

The breakdown voltage of a device (V_{BR}), therefore, is proportional to the width of the drift region (w_{drift}). In the case of SiC and GaN, the drift region can be 10 times smaller than in silicon for the same breakdown voltage. In order to support this electric field, there need to be carriers in the drift region that are depleted away at the point where the device reaches the critical field. This is where there is a huge gain in devices with high critical fields. The number of electrons (assuming an N-type semiconductor) between two terminals can be calculated using Poisson's equation:

$$q \cdot N_D = \varepsilon_o \cdot \varepsilon_r \cdot E_{crit}/w_{drift} \tag{1.2}$$

In this equation q is the charge of the electron (1.6×10^{-19} Coulombs), N_D is the total number of electrons in the volume, and ε_o is the permittivity of a vacuum measured in Farads per meter (8.854×10^{-12} F/m). ε_r is the relative permittivity of the crystal compared to a vacuum. In its simplest form under DC conditions, permittivity is the dielectric constant of the crystal.

Referring to Eq. (1.2), it can be seen that if the critical field of the crystal is 10 times higher, and from Eq. (1.1), the electrical terminals can be 10 times closer together. Therefore, the number of electrons, N_D, in the drift region can be 100 times greater, but only have one-tenth the distance to travel. This is the basis for the ability of GaN and SiC to outperform silicon in power conversion.

1.3.3 On-Resistance ($R_{DS(on)}$)

The theoretical on-resistance of a one square millimeter majority-carrier device (measured in ohms [$\Omega \cdot mm^2$]) is therefore

$$R_{DS(on)} = w_{Drift}/q \cdot \mu_n \cdot N_D \tag{1.3}$$

where μ_n is the mobility of electrons. Combining Eqs. (1.1)–(1.3) produces the following relationship between breakdown voltage and on-resistance:

$$R_{DS(on)} = 4 \cdot V^2_{BR}/\varepsilon_o \cdot \varepsilon_r \cdot \mu_r \cdot E^3_{Crit} \tag{1.4}$$

This equation can now be plotted as shown in Figure 1.1 for Si, SiC, and GaN. This plot is for an ideal structure. Real semiconductors are not always ideal structures and, therefore, it is always a challenge to achieve the theoretical limit. In the case of silicon MOSFETs, it took 30 years.

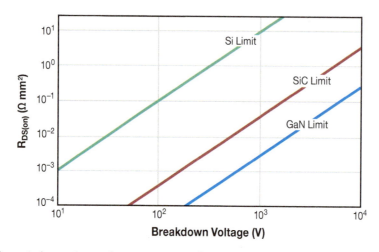

Figure 1.1 Theoretical on-resistance for a one square millimeter device versus blocking voltage capability for Si-, SiC-, and GaN-based power devices.

GaN Technology Overview | **5**

1.3.4 The Two-Dimensional Electron Gas

The natural structure of crystalline gallium nitride is a hexagonal structure named "wurtzite" is shown in Figure 1.2a, and the 4H-SiC structure is shown in Figure 1.2b. Because both structures are very chemically stable, they are mechanically robust and can withstand high temperatures without decomposition. The wurtzite crystal structure gives GaN piezoelectric properties that lead to its ability to achieve very high conductivity compared with either silicon or silicon carbide.

Piezoelectricity in GaN is predominantly caused by the displacement of charged elements in the crystal lattice. If the lattice is subjected to strain, the deformation will cause a miniscule shift in the atoms in the lattice that generate an electric field – the higher the strain, the greater the electric field. By growing a thin layer of AlGaN on top of a GaN crystal, a strain is created at the interface that induces a compensating two-dimensional electron gas (2DEG), as shown schematically in Figure 1.3 [8–10]. This 2DEG is used to efficiently conduct electrons when an electric field is applied across as in Figure 1.4.

This 2DEG is highly conductive, in part due to the confinement of the electrons to a very small region at the interface. This confinement increases the mobility of electrons from about 1000 cm^2/V·s in unstrained GaN to between 1500 and 2000 cm^2/V·s in the 2DEG region. The

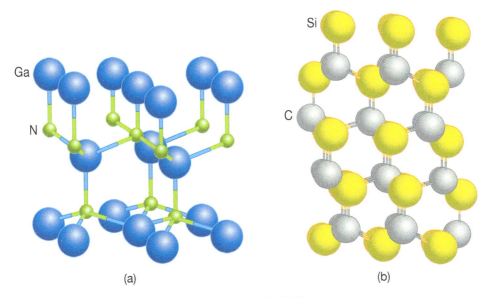

Figure 1.2 (a) Schematic of wurtzite GaN. (b) Schematic of 4H-SiC.

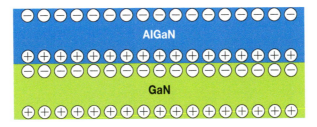

Figure 1.3 Simplified cross section of a GaN/AlGaN heterostructure showing the formation of a 2DEG due to the strain-induced polarization at the interface between the two materials.

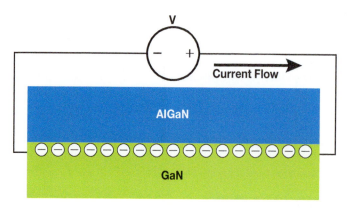

Figure 1.4 By applying a voltage to the 2DEG an electric current is induced in the crystal.

high concentration of electrons with very high mobility is the basis for the HEMT, the primary subject of this book.

1.4 The Basic GaN Transistor Structure

The basic depletion-mode GaN transistor structure is shown in Figure 1.5. As with any power FET, there are gate, source, and drain electrodes. The source and drain electrodes pierce through the top AlGaN layer to form an ohmic contact with the underlying 2DEG. This creates a short circuit between the source and the drain unless the 2DEG "pool" of electrons is depleted and the semi-insulating GaN crystal can block the flow of current. In order to deplete the 2DEG, a gate electrode is placed on top of the AlGaN layer. When a negative voltage relative to both drain and source electrodes is applied to the gate, the electrons in the 2DEG are depleted out of the device. This type of transistor is called a depletion-mode, or d-mode, HEMT.

There are two common ways to produce a d-mode HEMT device. The initial transistors introduced in 2004 had a Schottky gate electrode that was created by depositing a metal layer directly on top of the AlGaN. The Schottky barrier was formed using metals such as Ni-Au or Pt [11–13]. Depletion-mode devices have also been made using an insulating layer and metal gate similar to a MOSFET [14]. Both types are shown in Figure 1.6.

In power conversion applications, d-mode devices are inconvenient because, at the startup of a power converter, a negative bias must first be applied to the power devices. If this negative bias is not applied first, a short circuit will result, leading to catastrophic failure. An enhancement-mode (e-mode) device, on the other hand, would not suffer this limitation.

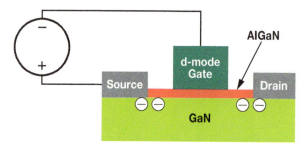

Figure 1.5 By applying a negative voltage to the gate of the device, the electrons in the 2DEG are depleted out of the device. This type of device is called a depletion-mode (d-mode) HEMT.

GaN Technology Overview | 7

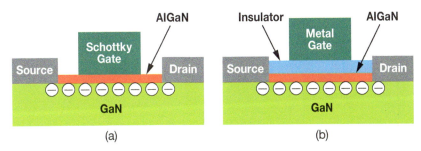

Figure 1.6 Cross section of a basic depletion-mode GaN HEMT with (a) Schottky gate, or (b) insulating gate.

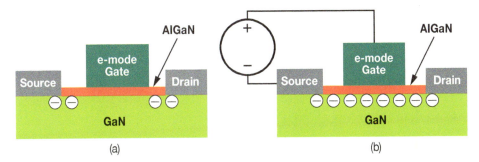

Figure 1.7 (a) An enhancement-mode (e-mode) device depletes the 2DEG with zero volts on the gate. (b) By applying a positive voltage to the gate, the electrons are attracted to the surface, re-establishing the 2DEG.

With zero bias on the gate, an e-mode device is OFF (Figure 1.7a) and will not conduct current until a positive voltage is applied to the gate, as illustrated in Figure 1.7b.

There are five popular structures that have been used to create enhancement-mode devices: recessed gate, implanted gate, pGaN gate, direct drive hybrid, and cascode hybrid.

1.4.1 Recessed Gate Enhancement-Mode Structure

The recessed gate structure has been discussed extensively in the literature [15] and is created by thinning the AlGaN barrier layer above the 2DEG (see Figure 1.8). By making the AlGaN barrier thinner, the amount of voltage generated by the piezoelectric field is reduced proportionally. When the voltage generated is less than the built-in voltage of the Schottky gate metal, the 2DEG is eliminated with zero bias on the gate. With positive bias, electrons are attracted to the AlGaN interface and complete the circuit between source and drain.

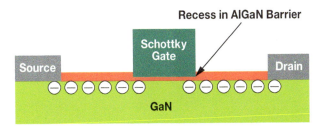

Figure 1.8 By etching away part of the AlGaN barrier layer a recessed gate e-mode transistor can be fabricated.

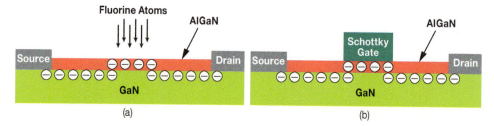

Figure 1.9 (a) By implanting fluorine atoms into the AlGaN barrier layer negative charges are trapped in the barrier. (b) A Schottky gate now can be used to reconstruct the 2DEG when a positive voltage is applied.

1.4.2 Implanted Gate Enhancement-Mode Structure

Shown in Figure 1.9a,b is a method for creating an enhancement-mode device by implanting fluorine atoms into the AlGaN barrier layer [16]. These fluorine atoms create a "trapped" negative charge in the AlGaN layer that depletes the 2DEG underneath. By adding a Schottky gate on top, an enhancement-mode HEMT is created.

1.4.3 pGaN Gate Enhancement-Mode Structure

The first enhancement-mode devices sold commercially had a positively charged (p-type) GaN layer grown on top of the AlGaN barrier (see Figure 1.10) [17]. The positive charges in this pGaN layer have a built-in voltage that is larger than the voltage generated by the piezoelectric effect, thus depleting the electrons in the 2DEG and creating an enhancement-mode structure [18]. The first mass production devices used a semi-insulating, or compensated gate technology [19, 20] that had the advantage of high gate impedance thus more closely imitating the characteristics of the Si MOSFET.

1.4.4 Hybrid Normally Off Structures

An alternative to building a single-chip enhancement-mode GaN transistor is to place an enhancement-mode silicon MOSFET in series with a depletion-mode HEMT device [21, 22]. Figure 1.11 shows two variations of these hybrid normally-off structures.

In the cascode circuit, shown in Figure 1.11a, the gate of the depletion-mode GaN HEMT is connected to the source of the enhancement-mode Si MOSFET. When the MOSFET is turned on with a positive voltage on the gate, the depletion-mode GaN transistor's gate voltage goes to near-zero volts and turns on as a result. Current can now pass through the depletion-mode GaN HEMT and the MOSFET, which is connected in series with the GaN HEMT. When the voltage on the MOS gate is removed, a negative voltage is created between the depletion-mode GaN transistor gate and its source electrode, turning the GaN device off.

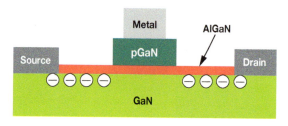

Figure 1.10 By growing a p-type GaN layer on top of the AlGaN, the 2DEG is depleted at zero volts on the gate.

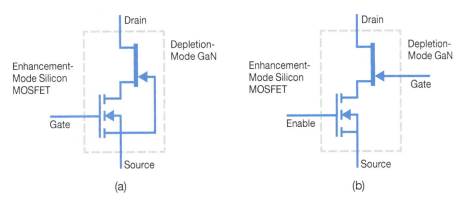

Figure 1.11 Schematic of low-voltage enhancement-mode silicon MOSFET in series with a depletion-mode GaN HEMT. (a) Cascode circuit and (b) enable/direct-drive circuit.

The second variation of this circuit has been described as an "enable circuit" or "direct drive circuit," shown in Figure 1.11 [23]. Here, the gate of the depletion-mode GaN HEMT is directly accessible to the external gate driver. The circuit has four terminals: gate, drain, source, and enable. Compared with the cascode configuration, this variation has the advantage of more direct control over the switching behavior of the GaN HEMT, but it also requires a more complex supporting circuit. Because the gate terminal directly drives the depletion-mode GaN HEMT, the gate drive circuit must provide voltage levels of about 0 V for turn-on and a negative voltage for turn-off (typically −12 to −14 V). The "enable" terminal is typically connected to the under-voltage lockout (UVLO) of the gate drive power supply so that the Si MOSFET is turned off when the gate drive circuit loses power. In contrast with the cascode variation, the Si MOSFET does not experience any switching during normal operation. When the low-voltage Si "enable switch" is off, gate node of the GaN HEMT can no longer be shorted to its source node to turn on, thereby functioning as a normally off device.

These types of solutions for an enhancement-mode GaN "system" work well when the GaN transistor has a relatively high on-resistance compared with the low-voltage (usually 30 V rated) silicon MOSFET. Since on-resistance increases with the device breakdown voltage, hybrid solutions are most effective when the GaN transistor is high voltage and the MOSFET is very low voltage. In Figure 1.12 is a chart showing the added on-resistance to the cascode

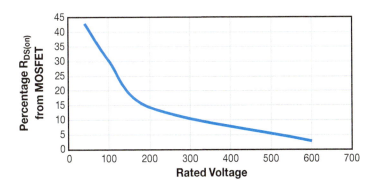

Figure 1.12 At a higher voltage rating the low-voltage MOSFET does not add significantly to the on-resistance of the cascode transistor system.

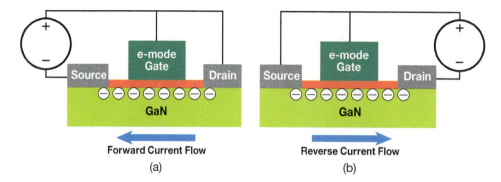

Figure 1.13 Enhancement-mode HEMT devices can conduct in either the forward (a) or reverse (b) direction.

circuit by the enhancement-mode silicon MOSFET. A 600 cascode device would only have about 3% added on-resistance due to the low-voltage MOSFET. Conversely, as the desired rated voltage goes down, and the on-resistance of the GaN transistor decreases, the MOSFET contribution becomes more significant. For this reason, hybrid solutions are only practical at voltages higher than 200 V.

1.4.5 Reverse Conduction in HEMT Transistors

Enhancement-mode GaN transistors can also conduct in the reverse direction. When current is forced into the source of an "off" device, such as the case of a synchronous rectifier during its dead time, a voltage drop is created from source to drain. When the drain voltage becomes lower than the gate voltage by at least $V_{GS(th)}$ (see Figure 1.13b), the 2DEG is again restored under the gate electrode and current can flow from source to drain. Because the enhancement-mode HEMT has no minority-carrier conduction, the device, operating similar to a diode, will turn off instantly when the forward bias is removed between the gate and drain electrodes. This characteristic is quite useful in certain power conversion circuits.

In the reverse direction, the cascode-connected transistor discussed in Section 1.4.4 conducts in the same way as an enhancement-mode GaN transistor, except that the diode of the MOSFET is conducting the reverse current, which then has to flow through the GaN device. The forward voltage drop of the MOSFET diode creates a slight positive voltage from gate to source in the HEMT which, therefore, is turned on in the forward direction. The HEMT on-resistance is added to the voltage drop of the MOSFET in this configuration. Unlike the enhancement-mode GaN transistor, the cascode-configured transistor does have a recovery time due to the injection of minority carriers in the silicon-based MOSFET.

1.4.6 Vertical GaN Transistors

Up to this point, the discussion has focused on GaN HEMT devices with a lateral conduction path for electrons through a 2DEG. An alternative topology is a vertical conduction device such as shown in Figure 1.14 [24]. This vertical conduction GaN device uses the depletion of the two pGaN regions to block the flow of electrons. The source of the electrons is the 2DEG at the AlGaN/GaN interface. By locating the drain electrode on the bottom of the device, the transistor can be made smaller while preserving the necessary standoff distance between two high-voltage terminals.

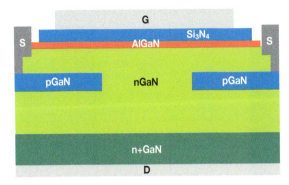

Figure 1.14 This vertical GaN transistor is turned off by applying negative voltage to the gate terminal such that the source-connected pGaN regions deplete the nGaN region under the gate and pinch off the flow of electrons.

This type of structure may be useful in applications requiring blocking voltages greater than 1000 V as well as fast switching speeds. The competition for the applications of these devices will come from high-voltage SiC transistors that have been in mass production for several years as of this writing. Before such a device can enter commercial markets, cost of single crystal GaN substrates needs to be lower.

1.5 Building a GaN HEMT Transistor

Building a GaN transistor starts with the process of growing the GaN/AlGaN heterostructure. There are four different starting bases, or substrates, that have been commonly used in fabricating GaN HEMT transistors: bulk gallium nitride crystal, sapphire (Al_2O_3), silicon carbide, and silicon.

1.5.1 Substrate Material Selection

The most obvious choice for a GaN device starting material would be a GaN crystal. The first attempts at growing GaN crystals were in the 1960s. Native defects from high nitrogen vacancy concentrations rendered these early attempts unusable for semiconductor device fabrication. Since then, progress has been made, and small-diameter, high-quality GaN crystals are becoming available, holding promise for use as a platform for active device fabrication.

Heteroepitaxy is a process whereby one type of crystal structure is grown on top of a different crystal. Because GaN crystals have not been readily available, there has been much work focused on growing GaN crystals on top of a more convenient platform such as sapphire, silicon carbide, silicon, or more recently Qromis Substrate Technology (QST™), which is an engineered substrate [22–24]. The starting point for trying to grow on a dissimilar crystal layer is to find a substrate with the appropriate physical properties.

Referring to Table 1.2, there are tradeoffs between any of the four listed choices for a substrate material. For example, sapphire (Al_2O_3) has a 16.1% mismatch to a GaN crystal lattice and has poor thermal conductivity. Thermal conductivity is especially important in transistors for power conversion because it generates a significant amount of heat flux during operation due to internal power dissipation. Silicon carbide (6H-SiC) substrate, on the other hand, has a reasonably good lattice match and excellent thermal conductivity. The

Table 1.2 Some key properties of Al_2O_3, SiC, Si, and QST [22, 24].

Substrate	Crystal plane	Lattice spacing (Å)	Lattice mismatch (%)	Relative thermal expansion ($10^{-5} \cdot K^{-1}$)	Thermal conductivity (W/cm · K)	Relative cost
Al_2O_3	(0001)	4.758	16.1	−1.9	0.42	Middle
6H-SiC	(0001)	3.08	3.5	1.4	3.8	Highest
Si	(111)	3.84	−17	3	1.5	Lowest
QST	(111)	3.84	−17	1.4	1.2	High

disadvantage comes from the cost of the starting crystal substrate, which can be up to 100 times the cost of a silicon substrate of the same diameter. Silicon is also not an ideal base for a GaN heteroepitaxial structure due to the lattice mismatch and the mismatch of thermal expansion coefficients. Silicon, however, is the least expensive material and there is a large and well-developed infrastructure to process devices on silicon substrates. QST substrates have a thin layer of (111) orientation Si on top of a polycrystalline core. The core material is engineered for a better match of the thermal expansion coefficient with that of GaN. The cost of QST substrates remains high compared with Si.

For the reasons cited above, SiC is commonly used for devices that require very high-power densities, such as linear radio frequency (RF) applications; silicon is used for devices in more cost-sensitive commercial applications such as DC–DC conversion, AC–DC conversion, class D audio amplifiers, and motion control.

1.5.2 Growing the Heteroepitaxy

There are several types of technologies that have been used to grow GaN on different substrates [25, 28–30]. The two most promising are Metal Organic Chemical Vapor Deposition (MOCVD) and Molecular Beam Epitaxy (MBE). MOCVD is faster and generally lower cost, whereas MBE is capable of more uniform layers with very abrupt transitions between layers. For GaN HEMT devices in power conversion applications, MOCVD is the dominant technology due to the cost advantages.

An MOCVD growth occurs in an inductively or radiantly heated reactor. A highly reactive precursor gas is introduced into the chamber where the gas is "cracked" by the hot substrate and reacts to form the desired compound. For GaN growth, the precursors are ammonia (NH_3) and trimethyl-gallium (TMG). For AlGaN growth, the precursors are trimethyl-aluminum (TMA) or triethyl-aluminum (TEA). In addition to the precursors, carrier gases such as H_2 and N_2 are used to enhance mixing and control the flow within the chamber. Temperatures in the range of 900–1100 °C are used for these growths.

A GaN heteroepitaxial structure involves at least four growth stages. Figure 1.15 illustrates this process. The starting material (i) of either SiC, QST, or Si is heated in the reaction chamber. A layer of AlN is then grown (ii) to create a seed layer that is hospitable to the AlGaN wurtzite crystal structure. An AlGaN "buffer layer" (iii) creates the transition to the GaN crystal (iv). Finally, the thin AlGaN barrier is grown on top of the GaN crystal to create the strain layer that induces the 2DEG formation.

Earlier in this chapter, several methods of making enhancement-mode GaN transistors were described. One method (pGaN enhancement-mode), illustrated in Figure 1.10, includes an additional GaN layer grown on top of the AlGaN barrier. This layer is most commonly doped

GaN Technology Overview

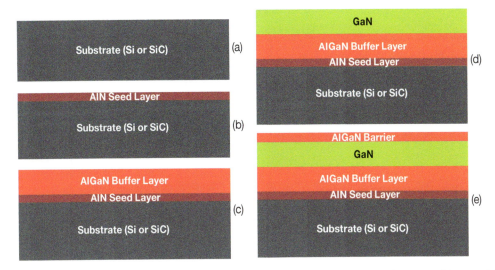

Figure 1.15 An illustration of the basic steps involved in creating a GaN heteroepitaxial structure (Not to scale). (a) Starting silicon substrate. (b) aluminum nitride (AlN) seed layer grown. (c) Various Al GaN layers grown to transition the lattice from AlN to GaN. (d) GaN layer grown. (e) AlGaN barrier layer grown. This layer creates strain in the underlying GaN layer that causes the 2DEG to develop.

Figure 1.16 An additional GaN layer, doped with p-type impurities, can be added to the heteroepitaxy process when producing an enhancement-mode device as illustrated in Figure 1.10.

with p-type impurities such as C, Mg, or Fe. A cross section of this heteroepitaxy structure is shown in Figure 1.16.

1.5.3 Processing the Wafer

Fabricating a HEMT transistor from a heteroepitaxial substrate can be accomplished in variety of sequential steps. One example of a simplified process for making an enhancement-mode HEMT with a pGaN gate is shown in Figure 1.17. A cross section of a completed device using this process is shown in Figure 1.18.

1.5.4 Making Electrical Connection to the Outside World

Following the device fabrication, provisions are needed to make the electrical connection to the electrodes of the device. There are two common ways of making connection to a power transistor: (i) attach bond wires or plate copper pillars between metal pads on the device and metal

14 *GaN Power Devices for Efficient Power Conversion*

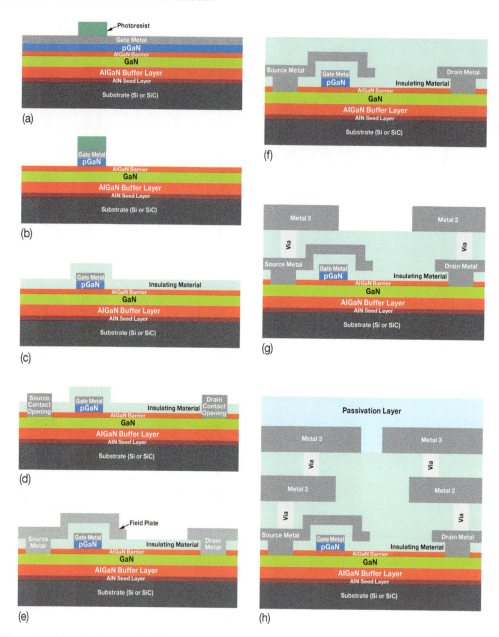

Figure 1.17 A typical process for fabricating an enhancement-mode GaN HEMT (not to scale) [27, 28]. The process steps are as follows: (a) Deposit gate metal and define gate pattern using photoresist as a protecting layer. (b) Etch the gate metal and pGaN crystal. (c) Deposit insulating material. (d) Create contact openings to source, drain, and gate (gate contact not pictured). (e) Deposit first aluminum metal layer and define metal pattern. (f) Deposit interlayer dielectric. (g) Cut vias between metal layers, form tungsten via plug, deposit and define second aluminum metal layer. (h) Deposit and define third aluminum layer and deposit final passivation layer.

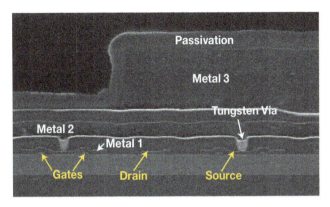

Figure 1.18 SEM micrograph of an enhancement-mode GaN HEMT made by Efficient Power Conversion Corporation [36–38]. *Source:* With permission of EPC.

posts in a plastic or ceramic package, or (ii) create contacts that can be soldered directly on the device while still in wafer form.

As explored further in Chapter 3, GaN transistors are able to switch very quickly and, therefore, are very sensitive to inductances in either the power loop or the gate–source loop. Wire bonds have a significant amount of inductance and limit the ultimate capability of the GaN device. Wire bonding also increases the possibility for poor bonds, which can reduce the reliability of the final product [31–33]. It is for this reason that the preferred method for making electrical connection is by soldering directly to contacts on the device either with a solder bump or bar that can mount directly to a PCB, or with copper pillars that can either be attached directly to a PCB or to a leadframe in a package. A common process for making these contacts that can be soldered is shown in Figure 1.19. The solder bars can either be a lead-tin (Pb-Sn) composition or a lead-free composition of silver-copper-tin (Ag-Cu-Sn). Following solder bar or ball formation, the completed wafer looks like the example in Figure 1.20.

The individual devices are singulated and the final chip-scale transistor may look like the example in Figure 1.21. This device is now ready to be soldered onto a printed circuit board (PCB) or onto a lead frame to be incorporated into a plastic molded package. In Figure 1.22 a GaN transistor is mounted onto a copper leadframe that is subsequently molded into a completed Power Quad Flat No-lead (PQFN) package such as shown in Figure 1.23.

1.6 GaN Integrated Circuits

Silicon-based power devices with rated voltages higher than about 20 V need to have vertical conduction paths due to the large separation required between source and drain terminals. Thanks to the higher critical electric field in GaN (see Section 1.3), GaN-on-Si devices can have a lateral conduction path while maintaining a very small size compared with Si-based MOSFETs. It is therefore straightforward to monolithically integrate multiple GaN-on-Si power transistors along with signal-level components. An early example of such single-chip integration is shown in Figure 1.24 [34].

This ability to integrate multiple functions monolithically holds the promise of producing a complete power conversion system on a single GaN-on-Si chip, a promise that, if realized, will significantly lower the cost and efficiency of power conversion.

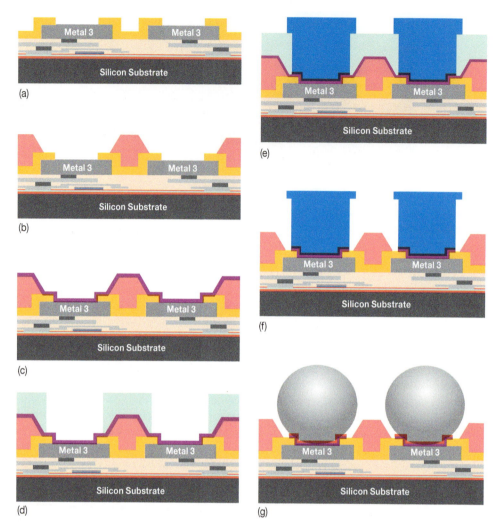

Figure 1.19 A typical process for creating solder bars on an enhancement-mode GaN HEMT (not to scale). The basic process steps are as follows: (a) The finished wafer with openings in the passivation. Metal layer 3 is partially exposed. (b) Photopolyimide is deposited and removed in the area where the solder is desired. (c) An under-bump metal is deposited to create an interface between the aluminum top metal and the solderable material. (d) Photoresist is used to define where the solder will be plated. (e) Copper and solder are plated in the opening. (f) The photoresist is removed, and the under-bump metal is etched. (g) The solderable metal is reflowed to form elongated solder bars.

The journey of integration has several steps, the first of which is to place multiple power devices on a single chip in a half-bridge configuration, as shown in Figure 1.25. This initial step is particularly compelling because a large percentage of power conversion applications revolve around this half-bridge topology. An early example of a monolithic half bridge is shown in Figure 1.26.

In Chapter 4 of this text, we will discuss the need for careful circuit layouts when using GaN devices. Circuit layouts need to have particularly low parasitic inductance. This is due to the extremely fast switching speed that, when combined with small amounts of inductance, and by

GaN Technology Overview | 17

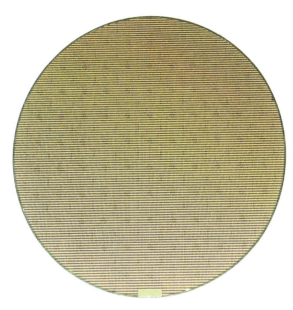

Figure 1.20 A completed 200 mm diameter enhancement-mode GaN HEMT wafer with approximately 30,000 individual power transistors. *Source:* With permission of EPC.

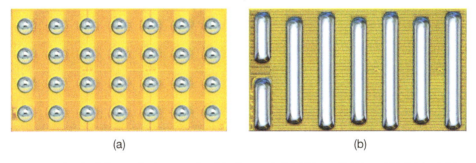

Figure 1.21 (a) finished chip-scale device with solder bars in Ball Grid Array (BGA), and (b) finished chip-scale device with solder bars in Land Grid Array (LGA). These devices each measure approximately 4 mm × 1.6 mm. *Source:* With permission of EPC.

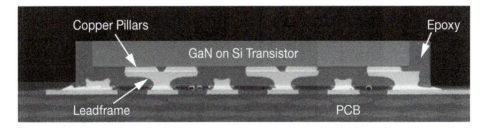

Figure 1.22 A GaN-on-Si transistor mounted inside an epoxy PQFN package. *Source:* With permission of EPC.

Figure 1.23 A finished packaged device in a PQFN format with exposed back surface for efficient heat extraction. This device measures approximately 5 mm × 3 mm. *Source:* With permission of EPC.

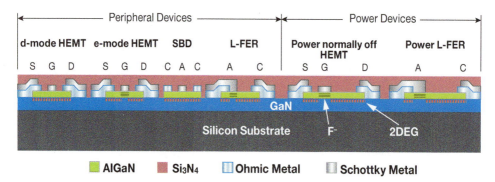

Figure 1.24 An early example of a GaN-on-Si integrated circuit including multiple power devices and signal level devices on the same chip. *Source:* Wong et al. [34]/with permission of IEEE/ Courtesy of Kevin J. Chen.

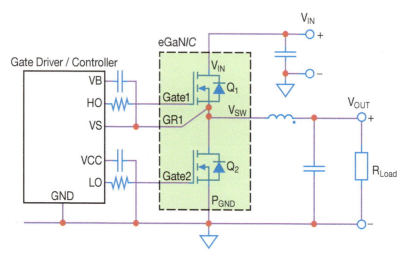

Figure 1.25 The half-bridge circuit is one of the most common circuit topologies used in power conversion. Shown here is a half bridge in a buck converter. The components inside the dotted lines (eGaN®IC) can be integrated monolithically.

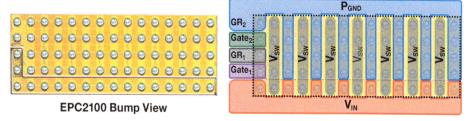

Figure 1.26 The first monolithic enhancement-mode GaN-on-Si half-bridge IC was the EPC2100 launched in 2014. This device measures 6 mm × 2.3 mm and has a BV_{DSS} = 30 V on both high-side and low-side transistors. *Source:* Ref. [39]/Efficient Power Conversion Corporation. On the left is a photo of the die and, on the right, are the pin assignments as per Figure 1.22.

virtue of the relationship between inductance (*L*), voltage (*V*), and fast changes in current (d*i*/d*t*) shown in Eq. (1.5), switching losses can be increased and large, unwanted voltage spikes can be generated.

$$V = L \cdot di/dt \tag{1.5}$$

A monolithic half bridge has several advantages; a small transistor with a large aspect ratio can reside next to a large transistor, the overall system area is reduced by eliminating the need for space on the PCB allocated to two discrete devices, and the parasitic inductance coming from the copper between discrete devices on the PCB is eliminated. This latter advantage can be seen more clearly in Figure 1.27. At higher output current in a simple buck converter (see Chapter 7), the parasitic power loop inductance becomes a key contributor to power losses. In Figure 1.27 a 12–1 V buck converter with two monolithic half bridges in parallel has an efficiency at 1 MHz and 40 A of about 87.5%, compared with about 85% for discrete devices of approximately the same size and on-resistance.

Following the production of a monolithic half bridge, the next logical step on the path to a GaN-on-Si power system on a chip is to add a driver to the power transistor. This driver can

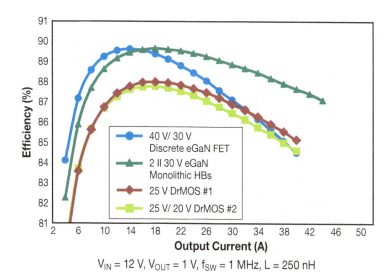

Figure 1.27 Total buck converter efficiency with the EPC2100 half bridge compared with discrete GaN transistors and discrete MOSFETs. *Source:* Reusch et al. [40]/with permission of IEEE.

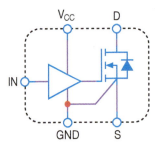

Figure 1.28 Block diagram of the driver circuit and the power transistor most commonly required in a power conversion system.

provide the interface to a logic device, such as a microcontroller, and can eliminate some of the sensitive design and layout requirements that are discussed in Chapters 3 and 4. In Figure 1.28 is a block diagram of the driver circuit that is most commonly needed in a GaN-based power conversion system. In Figure 1.29 is the chip-scale integrated circuit in a Ball Grid Array (BGA) format that implements the functions inside the dotted line in Figure 1.28.

As discussed in Chapters 3–16 of this book, most power conversion applications have at their core a half bridge in some form. This half bridge requires drivers and a level shifting function to drive and synchronize the high-side transistor. A basic block diagram of the half bridge with drivers and level shifting is shown in Figure 1.30, while Figure 1.31 is the chip-scale integrated circuit in an LGA format that implements the functions shown in Figure 1.30.

Once the building block of a half bridge with drivers and level shift has been developed, the door is open for producing a wide variety of integrated circuit variants, such as full-bridge converters and three-phase power stages. In addition, it is straightforward to add functions and features, such as analog or digital interfaces and controls.

One limitation to even greater integration of functions, however, is that the technical challenge of creating a complementary p-channel transistor to work with the enhancement-mode HEMT has yet to be met. As discussed in Section 1.2, GaN is a great host for a HEMT due to the ability to create a 2DEG. There have been numerous works demonstrating a two-dimensional hole gas (2DHG) [35], but overall hole mobility, typically less than 20 $cm^2/V\,s$,

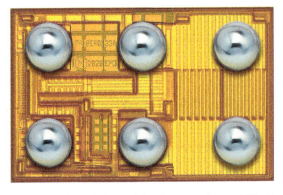

Figure 1.29 The EPC21601 integrated circuit has a driver integrated monolithically with the 30 V, 40 mΩ enhancement-mode transistor. This chip-scale device in a BGA format measures 1.5 mm × 1.0 mm and was designed for time-of flight (ToF) lidar. (See Chapter 13 for a detailed discussion of lidar.) *Source:* Ref. [41]/ Efficient Power Conversion Corporation.

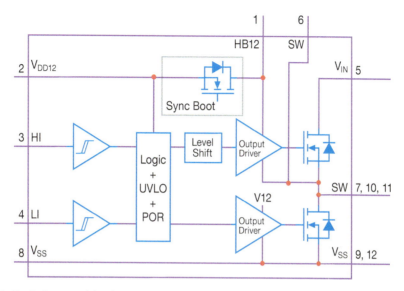

Figure 1.30 Block diagram of the driver circuits, level shift, and the power transistors most commonly required in a power conversion system.

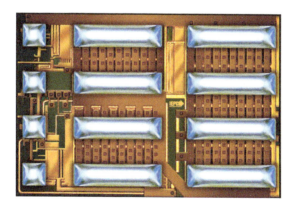

Figure 1.31 The EPC2152 integrated circuit has drivers and level shifting integrated in a single chip with the two 100 V enhancement-mode transistors in a half-bridge configuration. This device measures 3.9 mm × 2.6 mm and conducts 10 A at 1 MHz with 48 V_{IN} and 12 V_{OUT}. *Source:* Ref. [42]/Efficient Power Conversion Corporation.

is too poor to make a competitive complementary MOS (CMOS) circuit. Many other alternatives, including creating silicon-based p-channel devices in the silicon substrate either beneath, on top of, or beside a GaN epitaxial layer, have also been explored, but no commercial devices have been realized as of the date of this publication.

1.7 Summary

In this chapter, a platform for making switching power transistors using gallium nitride grown on top of a silicon substrate was introduced. Enhancement-mode transistors have in-circuit characteristics very similar to power MOSFETs, but with improved switching speed, lower on-resistance, and at a smaller size than their silicon predecessors. These new capabilities, married with high-density packaging, enable power conversion designers to reduce power losses, reduce system size, improve efficiency, and ultimately, reduce system costs.

Chapter 2 will connect these basic physical properties of GaN transistors to the electrical characteristics most important in designing power conversion systems. These electrical characteristics will be compared to state-of-the-art silicon MOSFETs to illustrate both the strong similarities and the subtle differences. In Chapters 3–15, these same electrical characteristics will be related to circuit and system performance in such a way as to give the designer the tools to get the maximum performance from GaN devices.

Finally, in Chapter 16 the "why, when, and how" that GaN transistors will displace MOSFETs will be discussed. Included is a discussion of cost trajectories, reliability, and technology directions for the years 2024–2030.

These are early years of a great new technology.

References

1 Sze, S.M. (1981). *Physics of Semiconductor Devices*, 2e. Hoboken, NJ: Wiley.

2 Baliga, B.J. (1996). *Power Semiconductor Devices*, 373. Boston, MA: PWS Publishing Company.

3 Mitani, E., Haematsu, H., Yokogawa, S., Nikaido, J., and Tateno, Y., Mass production of high voltage GaAs and GaN devices, *CS Mantech Conference*, Vancouver B.C., Canada (24–27 April 2006).

4 Mimura, T., Tokoyama, N., Kusakawa, H., Suyama, K., and Fukuta, M. (1979). GaAs MOSFET for low-power high-speed logic applications. *37th Device Research Conference*, University of Colorado, Boulder, CO, (25–27 June 1979).

5 Khan, M.A., Kuznia, J.N., and Olson, D.T. (1991). High electron mobility transistor based on a GaN-Al$_x$Ga$_{1-x}$N heterojunction. *Appl. Phys. Lett.* 65 (9): 1121–1123, 1129.

6 Kaminski, N., State of the art and the future of wide band-gap devices. *Proc. of the EPE Conference*, Barcelona, Spain (8–10 September 2009). NY: Institute of Electrical and Electronics Engineers

7 Baliga, B.J. (2005). *Silicon Carbide Power Devices*. Hackensack, NJ: World Scientific Publishing Co. Pte. Ltd.

8 Bykhovski, A., Gelmont, B., and Shur, M. (1993). The influence of the strain induced electric field on the charge distribution in GaNAlNGaN structure. *J. Appl. Phys.* 74: 6734–6739.

9 Yu, E., Sullivan, G., Asbeck, P. et al. (1997). Measurement of piezoelectrically induced charge in GaN/AlGaN heterostructure field-effect transistors. *Appl. Phys. Lett.* 71: 2794.

10 Asbeck, P., Yu, E., Lau, S. et al. (1997). Piezoelectric charge densities in AlGaN/GaN HFETs. *Electronic Lett.* 33: 1230.

11 Liu, Q.Z. and Lau, S.S. (1998). A review of the metal-GaN contact technology. *Solid State Electron.* 42 (5): 677–691.

12 Javorka, P., Alam, A., Wolter, M. et al. (2002). AlGaN/GaN HEMTs on (111) silicon substrates. *IEEE Electron Device Lett.* 23 (1): 4–6.

13 Liu, Q.Z., Yu, L.S., Lau, S.S. et al. (1997). Thermally stable PtSi Schottky contact on n-GaN. *Appl. Phys. Lett.* 70 (1): 1275–1277.

14 Kordoš, P., Heidelberger, G., Bernát, J. et al. (2005). High-power SiO$_2$/AlGaN/GaN metal-oxide-semiconductor heterostructure field-effect transistors. *Appl. Phys. Lett.* 87: 143501–143504.

15 Lanford, W.B., Tanaka, T., Otoki, Y., and Adesida, I. (2005). Recessed-gate enhancement-mode GaN HEMT with high threshold voltage. *Electron. Lett.* 41 (7): 449–450.

16 Cai, Y., Zhou, Y., Lau, K.M., and Chen, K.J. (2006). Control of threshold voltage of AlGaN/GaN HEMTs by fluoride-based plasma treatment: from depletion-mode to enhancement-mode. *IEEE Tran. on Electron Devices* 53 (9): 2207–2215.

17 Davis, S. (2010). Enhancement-mode GaN MOSFET delivers impressive performance. *Power Electron. Technol.* 36 (3): 10–13.

18 Hu, X., Simin, G., Yang, J. et al. (2000). Enhancement mode AlGaN/GaN HFET with selectively grown pn junction gate. *Electron. Lett.* 36 (8): 753–754.

19 Lidow, A., Beach, R., Cao, J., Nakata, A., and Zhao, G.Y. (2013). Compensated gate MISFET and method for fabricating the same. US Patent 8,350,294 B2, filed 8 April 2010 and issued 8 January 2013.

20 Sun, J., Mouhoubi, S., Silvestri, Z. et al. (2023). Gate characteristics of enhancement-mode fully depleted p-GaN HEMT. *IEEE Electron Dev. Lett.* 44 (12): 2015–2018.

21 Strite, S. and Morkoç, H. (1992). GaN, AlN, and InN: a review. *J. Vac. Sci. and Technol. B* 10 (4): 1237–1266.

22 Qromis, Inc (2017). Disruptive and validated solutions in substrate and device technologies: wideband semiconductor materials and devices [Brochure]. Santa Clara, CA. [Retrieved 15 May 2024]. www.qromis.com/.

23 Anderson, T.J., Koehler, A.D., Tadjer, M.J. et al. (2017). Electrothermal evaluation of thick GaN epitaxial layers and AlGaN/GaN high-electron-mobility transistors on large-area engineered substrates. *Appl. Phys. Express* 10: 126501.

24 Hsu, L.-H., Lai, Y.-Y., Tu, P.-T. et al. (2021). Development of GaN HEMTs fabricated on silicon, silicon-on-insulator, and engineered substrates and the heterogeneous integration. *Micromachines* 12: 1159.

25 Nakamura, S. (1991). GaN growth using GaN buffer layer. *Jpn. J. of Appl. Phys.* 30 (10A): L1705–L1707.

26 Brohlin, P.L., Ramadass, Y.K., and Kaya, C. (2018). Direct-drive configurations for GaN devices. Texas Instruments Application note, November 2018. http://www.ti.com/lit/wp/slpy008a/slpy008a.pdf.

27 Chowdhury, S., Swenson, B.L., and Mishra, U.K. (2008). Enhancement and depletion mode AlGaN/GaN CAVET with Mg-ion-implanted GaN as current blocking layer. *IEEE Electron Dev. Lett.* 29 (6): 543–545.

28 Nakamura, S., Iwasa, N., Senoh, M., and Mukai, T. Hole compensation mechanism of p-type GaN films. *J. Appl. Phys.* 31: 1258–1266.

29 Amano, H., Sawaki, N., Akasaki, I., and Toyoda, Y. (1986). Metalorganic vapor phase epitaxial growth of a high quality GaN film using an AlN buffer layer. *Appl. Phys. Lett.* 48: 353–355.

30 Hughes, W.C., Rowland, W.H. Jr., Johnson, M.A.L. et al. (1995). Molecular beam epitaxy growth and properties of GaN films on GaN/SiC substrates. *J. Vac. Sci. and Technol. B* 13 (4): 1571–1577.

31 Harman, G. (2010). *Wire Bonding in Microelectronics*, 3e. New York: McGraw-Hill Companies Inc.

32 Coucoulas, A. (1970). Compliant bonding. *Proceedings 1970 IEEE 20th Electronic Components Conference*, Washington, DC (May 1970), 380–389.

33 Heleine, T.L., Murcko, R.M., Wang C. (1991). A wire bond reliability model. *Proceedings of the 41st Electronic Components and Technology Conference*, Atlanta, GA (May 1991), 378–381.

34 Wong, K.Y., Chen, W.J., and Chen, K.J. (2009). Integrated voltage reference and comparator circuits for GaN smart power chip technology. *ISPSD 2009*, Barcelona, Spain (June 2009).

35 Nakajima, A., Sumida, Y., Dhyani, M.H. et al. (2010). *High density two-dimensional hole gas nduced by negative polarization at GaN/AlGaN heterointerface*, Published 10 December ©2010 The Japan Society of Applied Physics.

36 Lidow, A., Beach, R., Nakata, A., Cao, J., and Zhao, G.Y. (2013). Enhancement mode GaN HEMT device and method for fabricating the same. US Patent 8,404,508, 26 March 2013.

37 Lidow, A., Beach, R., Nakata, A., Cao, J., and Zhao, G.Y. (2013). Compensated gate MISFET and method for fabricating the same. US Patent No. 8,350,294, filed 4 April 2010 and issued 26 March 2013.

38 Lidow, A., Strydom, J., de Rooij, M., and Ma, Y. (2012). *GaN Transistors for Efficient Power Conversion*, 1e. El Segundo, CA: Power Conversion Publications.

39 Efficient Power Conversion Corporation (September 2014). EPC2100 – enhancement-mode power transistor half-bridge. EPC2100 datasheet. http://epc-co.com/epc/Portals/0/epc/documents/datasheets/epc2100_datasheet.pdf.

40 Reusch, D., Strydom, J., and Glaser, J. (2015). Improving high frequency DC-DC converter performance with monolithic half bridge GaN ICs. *Proceedings of the Seventh Annual IEEE Energy Conservation Congress and Expo*, Montreal, Canada, (September 2015).

41 Efficient Power Conversion Corporation (December 2022). EPC21601 – 30 V, 15 A integrated gate driver eGaN® IC. EPC21601 datasheet. http://epc-co.com/epc/Portals/0/epc/documents/datasheets/epc21601.pdf.

42 Efficient Power Conversion Corporation (March 2021). EPC2152–80 V, 15 A integrated ePower^{TM} stage IC. EPC2152 datasheet. https://epc-co.com/epc/products/gan-fets-and-ics/epc2152.

2

GaN Transistor Electrical Characteristics

2.1 Introduction

In this chapter, the basic physical properties of GaN transistors discussed in Chapter 1 will be connected to electrical characteristics that are important when developing power conversion circuits and systems. The key electrical parameters and ratings of a GaN transistor should give the designer most of the information necessary to design a system with predictable results. The most basic static parameters are blocking voltage, on-resistance, and threshold voltage. Furthermore, in order to understand how this device will work when switched on and off, intrinsic terminal capacitances and reverse-conduction characteristics need to be added.

These electrical characteristics will be compared to state-of-the-art silicon power MOSFETs in order to explore both their similarities and differences. Understanding these similarities and differences is fundamental to understanding the extent to which existing power conversion systems can be improved by GaN-based technologies.

2.2 Device Ratings

For a power-switching transistor, the primary device ratings are voltage, current, and temperature. This section provides analytical insight into the variance of these device parameters over an operating range determined by the end-use power conversion system.

2.2.1 Drain–Source Voltage

The rated breakdown voltage (BV_{DSS}) between the source and drain terminals of a GaN transistor is determined by several factors [1], including the fundamental breakdown electric field (E_{crit}) of GaN discussed in Chapter 1; the specific design of the device; the specifics of the heterostructure; internal insulating layers in the device structure above the gate, source, and drain electrodes; and the underlying substrate material properties. A semiconductor transistor will break down and conduct large amount of current when the critical electric field of any of the constituent materials is exceeded and may destroy itself in the process.

To visualize the important electric fields in a transistor, a good starting point is the structure illustrated in Figure 2.1. Applied from drain-to-source is a voltage such that the transistor is biased to block voltage. (The drain electrode is positive compared to the source electrode.) In

GaN Power Devices for Efficient Power Conversion, Fourth Edition. Alex Lidow, Michael de Rooij, John Glaser, Alejandro Pozo, Shengke Zhang, Marco Palma, David Reusch, and Johan Strydom.
© 2025 John Wiley & Sons Ltd. Published 2025 by John Wiley & Sons Ltd.

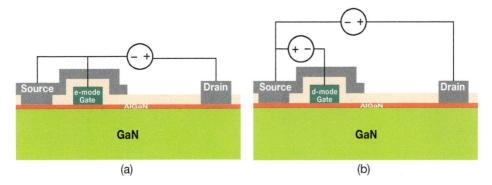

Figure 2.1 An illustration of (a) a basic enhancement-mode GaN transistor with reverse bias applied, and (b) a depletion-mode transistor with the gate turned off and reverse bias applied.

an enhancement-mode transistor, the channel is turned off by shorting the gate to the source. In a depletion-mode device, there would need to be a negative bias from the gate-to-source electrodes to keep the transistor from conducting current.

A simple two-dimensional analysis of the structure in Figure 2.1a, shown in Figure 2.2, illustrates the electric fields at any point in the device. Higher electric fields, where the contour lines are closest together, develop near the drain and gate electrodes. When the field at any location in the device structure exceeds E_{crit}, the device will break down and conduct current.

Breakdown can also occur between the metal busing layers in the device. For example, Figure 1.17 of Chapter 1 showed an enhancement-mode transistor with three levels of metal used to consolidate drain current and source current and bring them to the outside of the device where they can be connected into a circuit. When the device is blocking voltage and not conducting current, one of these layers may be connected to the source potential while an adjacent layer, or perhaps a higher or lower layer in the stack, might be at drain potential. If the E_{crit} of the dielectric material separating these two layers is exceeded, breakdown will occur. This can be prevented by either increasing the separation of the layers, or by switching to an insulating layer with a higher E_{crit}.

If a transistor device goes into breakdown, either from exceeding the E_{crit} of the semiconductor or of an insulating layer, the effect tends to be destructive. In the case where the insulating layers exceed their capacity for blocking voltage, a physical rupture of the dielectric material will develop. The closer the electric field approaches $E_{crit(insulator)}$, the sooner the

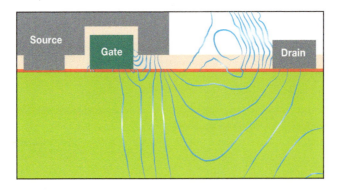

Figure 2.2 The device in Figure 2.1 showing the electric fields when voltage is applied from drain to source.

rupture will occur. This effect is called "time-dependent dielectric failure" and is discussed extensively in [2]. When the electric field of the GaN semiconductor layer exceeds $E_{crit(GaN)}$, a different mechanism causes device failure. When breakdown occurs in the GaN or AlGaN regions, the electrons generated can destroy the 2DEG, causing the device on-resistance to increase greatly [3].

When a transistor of any kind is in the blocking state, there is still a small amount of leakage current (I_{DSS}) that will flow between terminals. In the case of a transistor device, the leakage current can flow directly from the drain to the source electrodes, or indirectly from the drain to the gate electrodes and the drain to the substrate. The sum of these three leakage currents will be the total I_{DSS} measured between drain and source in a circuit.

These three components of leakage current are shown in Figure 2.3 and are measured on an enhancement-mode transistor with a breakdown voltage above 700 V. In this example, the substrate material is silicon (Si) and is connected to source potential. The current has been normalized to a 1 mm-wide gate structure.

When designing a power conversion system, I_{DSS} can become a significant source of power loss. For example, if a 100 V device has 100 μA I_{DSS} leakage current, the overall power dissipation due to the leakage current would be 10 mW. In certain applications requiring very low standby power, this amount of loss could become unacceptable.

The drain-to-source leakage current also varies with temperature. Figure 2.4 shows a family of curves from a sample of commercially available enhancement-mode transistors (EPC2204) with one-meter gate width, showing leakage current measured at various temperatures. Figure 2.4a traces an individual device at various temperatures as a function of V_{DS}. Figure 2.4b compares leakage current measurements of 13 different devices of the same part type plotted against the inverse temperature to illustrate device-to-device variation. The slope of these measurements for each device gives a consistent activation energy of $E_A = -0.4$ eV. This value of activation energy lies in between values reported in literature, a lower limit of −0.2 eV due to surface related traps [4, 5], and −0.99 eV due to a temperature-assisted tunneling mechanism [4].

Typically, commercial transistors are described using a datasheet. These datasheets vary from supplier to supplier and do not always follow consistent conventions for measurements

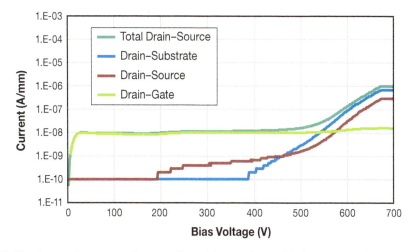

Figure 2.3 The three main sources of current that add up to the total leakage current between drain and source terminals: drain–gate leakage, drain–source leakage, and drain–substrate leakage.

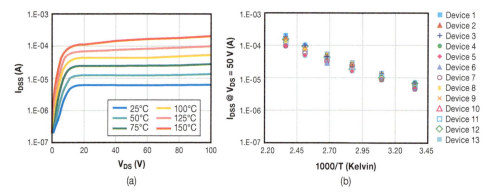

Figure 2.4 I_{DSS} over temperature for an EPC2204 enhancement-mode GaN transistor: (a) I_{DSS} versus temperature for an individual device, and (b) I_{DSS} measured at $V_{DS} = 50$ V for 13 different devices, showing device-to-device variation.

leading to the numbers listed in the document. However, for BV_{DSS} and I_{DSS}, there are several relevant sections of a datasheet that can be analyzed to deduce device behavior and compare parts from different manufacturers.

In Tables 2.1 and 2.2 are examples of data from two different manufacturers of GaN transistors, Efficient Power Conversion Corporation (EPC) and GaN Systems (now part of Infineon Technologies). Listed in Table 2.1 are the specifications for an enhancement-mode transistor from EPC with a nominal maximum drain–source voltage (V_{DS}) of 100 V, while in Table 2.2, the corresponding specifications for an enhancement-mode GaN transistor from GaN Systems with a nominal maximum V_{DS} of 650 V are outlined. In both cases, there is also a specification for a transient voltage capability above the maximum V_{DS}, which will be explained in Section 2.2.2. This transient capability means that, for short periods of time, the devices can withstand higher voltages than their respective rated maximums. In the case of a mature 100 V Si MOSFET shown in Table 2.3, there are no transient overvoltage ratings. Instead, an avalanche capability is specified, enabling the user to take the device into full drain–source breakdown with a certain amount of energy (specified in millijoules). MOSFET users rarely take advantage of this avalanche capability.

Table 2.1 Data from an Efficient Power Conversion EPC2204 datasheet showing sections relating to I_{DSS} and BV_{DSS}.

	Maximum ratings			
V_{DS}	Drain-to-source voltage (continuous)		100	V
	Drain-to-source voltage (1% duty cycle)		120	

	Static characteristics ($T_J = 25\,°C$, unless otherwise stated)					
	Parameter	Test conditions	Min.	Typical	Max.	Unit
BV_{DSS}	Drain-to-source voltage	$V_{GS} = 0$ V, $I_D = 300\,\mu A$	100	—	—	V
I_{DSS}	Drain–source leakage	$V_{GS} = 0$ V, $V_{DS} = 80$ V	—	40	200	µA

Source: Adapted from [6].

Table 2.2 Data from a GaN Systems datasheet showing sections relating to I_{DSS} and BV_{DSS}.

Parameter	Symbol	Min.	Typ.	Max.	Units	Conditions
Drain-to-source blocking voltage	BV_{DS}	650			V	$V_{GS} = 0\,V$, $I_{DSS} = 50\,\mu A$
Drain-to-source leakage current	I_{DSS}		2	50	μA	$V_{DS} = 650\,V$, $V_{GS} = 0\,V$, $T_J = 25\,°C$

Parameter	Symbol	Value	Unit
Drain-to-source voltage	V_{DS}	650	V
Drain-to-source voltage – transient[a]	$V_{DS\ (transient)}$	750	V

[a] For 1 μs.
Source: Adapted from [7].

Table 2.3 Data from an Infineon BSC060N10NS3G datasheet for a Si MOSFET showing sections relating to I_{DSS} and BV_{DSS} and avalanche energy.

Parameter	Symbol	Test conditions	Value			Unit
Maximum ratings ($T_J = 25\,°C$, unless otherwise specified)						
Avalanche energy, single pulse	E_{AS}	$I_D = 50\,A$, $R_{GS} = 25\,\Omega$	230			mJ
Static characteristics						
Drain–source breakdown voltage	$V_{(BR)DSS}$	$V_{GS} = 0\,V$, $I_D = 1\,mA$	100	—	—	V
Zero gate voltage drain current	I_{DSS}	$V_{DS} = 100\,V$, $V_{GS} = 0\,V$, $T_J = 25\,°C$	—	0.01	1	μA
		$V_{DS} = 100\,V$, $V_{GS} = 0\,V$, $T_J = 125\,°C$	—	10	100	μA

Source: Adapted from [8].

Tables 2.1 and 2.2 also list the static (DC) I_{DSS} characteristics for the respective devices. In each case, the test conditions are slightly different. The EPC2204 gives the I_{DSS} at 80 V and the I_D at BV_{DSS}, both at 25 °C, whereas the GS66508B lists the I_{DSS} at maximum V_{DS} at 25 °C. The silicon MOSFET, BSC060N10NS3 G from Infineon, has specifications for BV_{DSS} and I_{DSS} at yet a third set of test conditions. This again emphasizes the need for standardized operating condition specifications for similar devices.

2.2.2 Drain–Source Transient Overvoltage

Transient drain voltage overshoot is commonly observed in gallium nitride-based converters due to high slew rate and fast switching applications. A survey of transient overvoltage specification from a suite of GaN suppliers was conducted by JEDEC JC-70 committee and presented in JEP186 [9]. Most of the transient overvoltage specifications describe it as a device

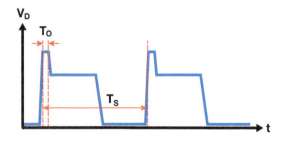

Figure 2.5 Illustration of the 1% overshoot duty cycle overvoltage specification. 1% is the ratio between T_O (overvoltage duration) and T_S (one switching period).

robustness indicator. In addition, many of them consider drain voltage overshoot as a single rare event or atypical occurrence. Hence, it is challenging for design engineers to effectively implement these specifications into their designs. Therefore, an application-driven and user-friendly repetitive transient off-state drain overvoltage specification on datasheets is important for the general adoption of GaN technology because of the absence of avalanche mechanisms in GaN HEMTs within normal operating conditions. In Table 2.1 is an example of a repetitive transient overvoltage specification that enables design engineers to account for small transients that may occur during each switching cycle [10].

For example, if a converter operates at 100 kHz, equivalent of 10 μs per switching period, the specification indicates that the GaN devices can withstand a repetitive 120 V overvoltage spike with a 100 ns duration in each switching cycle over 25 years of lifetime. This mathematical relation is demonstrated in Eq. (2.1) and further illustrated in Figure 2.5.

$$\text{Overshoot duty cycle} = \frac{120\% \text{ Overvoltage Duration at } 75°C \, (T_O)}{\text{Switching Period} \, (T_S)} \leq 1\% \quad (2.1)$$

where T_O is the overvoltage duration within each switching period and T_S is the switching period.

An alternative overvoltage specification can be found in Table 2.2 for a product made by GaN Systems, now part of Infineon. In this specification, the 650 V-rated GaN transistor is capable of withstanding 750 V for one microsecond.

MOSFETs do not have overvoltage specifications, but rather specify the device's ability to handle energy in avalanche breakdown. For example, in Table 2.3 there is a specification for the avalanche energy in a single pulse (E_{AS}). MOSFETs typically enter avalanche breakdown a few volts higher than their rated breakdown voltage. Specifying the avalanche capability therefore becomes a necessity. GaN transistors typically have a large margin between the rated maximum and the point at which the device goes into avalanche making it possible to specify significant overvoltage capability. Chapter 5 discusses the physics supporting this overvoltage capability.

2.3 Gate Voltage

As shown in Section 2.5, GaN transistors devices require a 5 V_{GS} to properly drive the devices, which leaves a small margin from the nominal drive voltage to the datasheet maximum specification, $V_{GS,Max}$, of 6 V as shown in Table 2.4. In the reverse direction, not commonly used in a typical power conversion circuit, the maximum gate voltage is specified as −4 V. After many years of mass adoption experience, gate failure is almost non-existent [11] due to the robust nature of GaN and the conservative gate ratios.

Table 2.4 Data from an Efficient Power Conversion EPC2204 datasheet showing sections relating to V_{GSmax}.

	Maximum ratings		
V_{GS}	Gate-to-source voltage	6	V
	Gate-to-source voltage	−4	

Source: Adapted from [6].

2.4 On-Resistance ($R_{DS(on)}$)

The on-resistance ($R_{DS(on)}$) of a transistor is the sum of all the resistance elements that make up the device, as shown in Figure 2.6. The source and drain metals have to connect to the 2DEG through the AlGaN barrier. This component of resistance is called the contact resistance (R_C). Electrons then flow in the 2DEG with a resistance R_{2DEG}. This resistance is proportional to the mobility of the electrons (μ_{2DEG}), the quantity of electrons created by the 2DEG (N_{2DEG}), the distance the electrons have to travel (L_{2DEG}), the width of the 2DEG (W_{2DEG}), and the universal charge constant, q ($1.6 \cdot 10^{-19}$ Coulombs). This resistance can be described by the following formula [12]:

$$R_{2DEG} = L_{2DEG}/(q \cdot \mu_{2DEG} \cdot N_{2DEG} \cdot W_{2DEG}) \tag{2.2}$$

As discussed in Chapter 1, the number of electrons in the 2DEG will depend on the amount of strain induced by the AlGaN barrier. However, the 2DEG could have a lower concentration beneath the gate than in the region between the gate and drain electrodes, depending on the type of gate (recessed gate, implanted MOS, Schottky, or pGaN), the particular process used, as well as the heterostructure deployed. It also depends on the voltage applied to the gate. A fully enhanced gate will have a higher electron concentration than a partially enhanced gate. A good approximation of the resistance of the transistor shown in Figure 2.6 can then be calculated as follows:

$$R_{HEMT} = 2 \cdot R_C + R_{2DEG} + R_{2DEG(Gate)} \tag{2.3}$$

Additional parasitic resistance ($R_{parasitic}$) can come in the form of metal resistance from the multiple metal buses that conduct the current from the individual source and drain electrodes to the terminals of the transistor. In a power conversion circuit, the conduction losses of the transistor are quite significant, and therefore the device is typically used either fully turned on

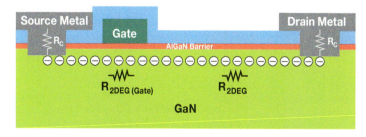

Figure 2.6 Cross section of a GaN transistor showing the major components of $R_{DS(on)}$.

(ohmic), or fully turned off. For this reason, a key parameter for any power transistor is the on-resistance, $R_{DS(on)}$, and can be defined as follows:

$$R_{DS(on)} = R_{HEMT \text{ (Fully Enhanced)}} + R_{parasitic} \quad (2.4)$$

Each of these components of $R_{DS(on)}$ will vary with temperature. The metal layers typically are made of combinations of copper and aluminum and have resistivity temperature coefficients in the range of $3.8 \cdot 10^{-3}/°C$ [13] for copper, to $3.9 \cdot 10^{-3}/°C$ [14] for aluminum. In contrast, a 2DEG's resistivity temperature coefficient is significantly higher, in the range of $1.3 \cdot 10^{-2}/°C$ [15, 16], and the contact resistance has a temperature coefficient in the range of $4.7 \cdot 10^{-3}/°C$ [15]. The transistor's $R_{DS(on)}$ as a function of temperature should be somewhere in between these numbers and can be approximated as:

$$R_{DS(on)}(T) = R_{parasitic}(T) + (2 \cdot R_C(T)) + (R_{2DEG} + R_{2DEG(Gate)})(T) \quad (2.5)$$

The overall temperature variation of $R_{DS(on)}$ will depend on the design of the device, and how much of the $R_{DS(on)}$ comes from 2DEG, contact resistance, or parasitic metal resistance. Devices in commercial use, however, typically demonstrate a variation of $R_{DS(on)}$ with temperature that is about the same as silicon MOSFETs, as shown in Figure 2.7. For devices designed for higher BV_{DSS}, the 2DEG will be a greater fraction of the total $R_{DS(on)}$; since the temperature coefficient of the 2DEG is higher than that of the parasitic elements and the contact resistance, the temperature coefficient will be higher.

Equation (2.2) breaks down the resistance of the 2DEG into two components, the 2DEG between the drain and gate, and the 2DEG under the gate. The resistance of the 2DEG under the gate changes from a very high value when the device is in the off state, to a very low value when the device is in the on state. This transition depends on the specific design of the gate and has a significant impact on device performance in a switching circuit.

In Figure 2.8, a graph of the $R_{DS(on)}$ of an enhancement-mode GaN transistor as a function of gate-to-source voltage for various drain currents and at various temperatures is shown. In this

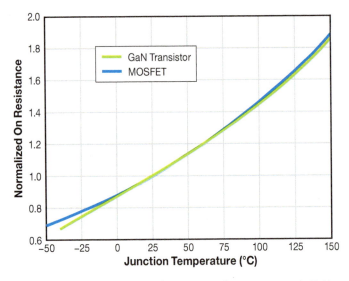

Figure 2.7 Normalized $R_{DS(on)}$ versus temperature for a 100 V enhancement-mode GaN transistor (EPC2204) compared with a Si power MOSFET with similar ratings [6, 8].

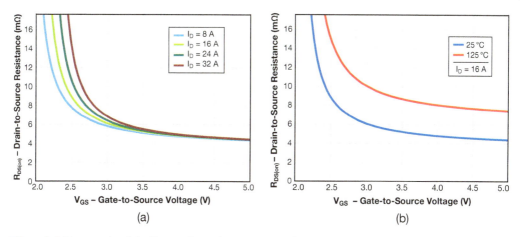

Figure 2.8 Two graphs of the $R_{DS(on)}$ of an enhancement-mode GaN transistor (EPC2204) as a function of gate voltage (a) for various drain currents and (b) at various temperatures. *Source:* [6]/Efficient Power Conversion Corporation.

example, the resistance of the gate 2DEG rapidly decreases until it is fully enhanced at about V_{GS} = 4.5 V. Beyond that voltage, there is only a nominal decrease in overall device $R_{DS(on)}$.

The datasheets for enhancement-mode GaN transistor devices typically specify this on-resistance with a fully enhanced gate at the maximum rated continuous drain current. In Table 2.5 are data from a 100 V-rated enhancement-mode transistor datasheet that specifies a maximum $R_{DS(on)}$ of 6 mΩ with 5 V from gate to source at 25 °C. In Table 2.6 are data from a datasheet for a 650 V-rated transistor specifying a maximum $R_{DS(on)}$ of 63 mΩ at 25 °C. Also indicated in this datasheet is the typical $R_{DS(on)}$ of 129 mΩ at 150 °C. Both measurements are captured with 6 V from gate to source.

Table 2.5 Data from an Efficient Power Conversion EPC2204 datasheet showing the section relating to $R_{DS(on)}$.

	Static characteristics (T_J = 25 °C, unless otherwise stated)					
	Parameter	Test conditions	Min.	Typ.	Max.	Unit
$R_{DS(on)}$	Drain–source on-resistance	V_{GS} = 5 V, I_D = 16 A	—	4.4	6	mΩ

Source: Adapted from [6].

Table 2.6 Data from a GaN systems GS66508B datasheet showing the section relating to $R_{DS(on)}$.

Parameter	Symbol	Min.	Typ.	Max.	Unit	Conditions
Drain-to-source on-resistance	$R_{DS(on)}$		50	63	mΩ	V_{GS} = 6 V, T_J = 25 °C, I_{DS} = 9 A
Drain-to-source on-resistance	$R_{DS(on)}$		129		mΩ	V_G = 6 V, T_J = 150 °C, I_{DS} = 9 A

Source: Adapted from [7].

2.5 Threshold Voltage

For a power device, the threshold voltage ($V_{GS(th)}$ or V_{th}) is the gate-to-source voltage required to begin conducting drain current in the device. In other words, the threshold voltage defines the voltage below which the device is off. An enhancement-mode or cascode device has a positive threshold voltage, and a depletion-mode device has a negative threshold voltage.

For a GaN power device, the threshold voltage is the gate-to-source voltage at which the 2DEG underneath the gate is fully depleted by the gate electrode [16]. This voltage is dependent upon two components, the voltage created by the piezoelectric strain (defined as $V_{Gstrain}$), plus the built-in voltage due to the specifics of the gate metallurgy. In the case of a Schottky gate device, this built-in voltage is the Schottky barrier height [17] of the gate metal on top of the AlGaN barrier. In the case of a pGaN gate, it is the voltage generated by the built-in field caused by a p-type semiconductor material next to an n-type material.

Because the strain in the AlGaN barrier is relatively constant with temperature, as well as the voltages generated by the internal metallurgy, the threshold voltage in a GaN transistor is relatively constant with temperature, as shown in Figure 2.9. In contrast, the threshold voltage of Si MOSFET will decline much more rapidly with increasing temperature.

Table 2.7 shows data from a 100 V-rated enhancement-mode GaN transistor datasheet. The typical value of 1.1 V is measured at 4 mA, a small amount of current compared with the 16 A continuous rating for this same transistor.

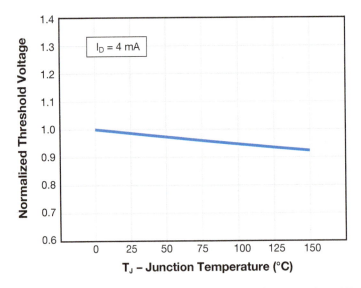

Figure 2.9 EPC2204 normalized threshold voltage versus temperature showing only an 8% change over the normal operating range of this device. *Source:* [6]/Efficient Power Conversion Corporation.

Table 2.7 Data from an Efficient Power Conversion EPC2204 datasheet showing the section relating to V_{th}.

\multicolumn{6}{c}{Static characteristics (T_J = 25 °C, unless otherwise stated)}						
	Parameter	Test conditions	Min.	Typ.	Max.	Unit
$V_{GS(th)}$	Gate threshold voltage	$V_{DS} = V_{GS}$, I_D = 4 mA	0.8	1.1	2.5	V

Source: Adapted from [6].

Table 2.8 Data from a GaN systems GS66508B datasheet showing the section relating to V_{th}.

Parameter	Symbol	Min.	Typ.	Max.	Units	Conditions
Gate-to-source threshold	$V_{GS(th)}$	1.1	1.7	2.6	V	$V_{DS} = V_{GS}$, $I_{DS} = 7$ mA

Source: Adapted from [7].

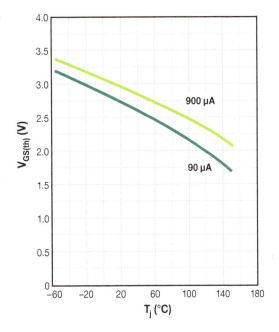

Figure 2.10 Data from an Infineon BSC060N10NS3 G datasheet showing the change in V_{th} versus temperature. *Source:* Adapted from [8].

Table 2.8 presents data from a 650 V rated GaN transistor. In this case, V_{th} is specified at 7 mA, which is a small value compared with the 30 A continuous rating. No information is given on this device's change in threshold voltage as a function of temperature. In Figure 2.10, however, is graphed the threshold voltage versus temperature for a 100 V Infineon BSC060N10NS3 [8] power MOSFET. V_{th} for this silicon device drops about 38% from about 2.75 V to about 2 V, when temperature changes from 25 °C to 125 °C compared with a nominal 8% decrease in the enhancement-mode GaN transistor.

2.6 Capacitance and Charge

A transistor's capacitance is a significant factor in determining the energy that will be lost in the device during a transition from the on- to off-state, or from the off- to the on-state. The capacitance (C) determines the amount of charge (Q) that needs to be supplied to various terminals of the device to change the voltage across those terminals ($Q = C \cdot V$). The faster this charge is supplied, the faster the device will change voltage.

There are three main elements of capacitance related to a FET; (i) gate-to-source capacitance (C_{GS}), (ii) gate-to-drain capacitance (C_{GD}), and (iii) drain-to-source capacitance (C_{DS}).

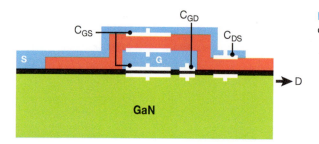

Figure 2.11 Schematic of GaN transistor capacitive sources.

Figure 2.11 illustrates the physical origin of each of these capacitances. Occasionally designers need to look just at the total Thevenin equivalent capacitance seen at either the input terminals ($C_{ISS} = C_{GD} + C_{GS}$), or output terminals ($C_{OSS} = C_{GD} + C_{DS}$).

These capacitances are a function of the voltage applied to various terminals. Figure 2.12 shows how the values change for an enhancement-mode transistor as the voltage from drain-to-source increases. The reason for the drop in capacitance as V_{DS} increases is that the free electrons in the 2DEG of the GaN are depleted. For example, the initial step down in C_{OSS} is caused by the depletion of the 2DEG under the field plate protecting the gate. Higher V_{DS} values extend the depletion region laterally from the field plate edge toward the drain, further depleting the 2DEG and reducing its capacitive component.

The result of integrating the capacitance between two terminals across the range of voltage applied to the same terminals is the amount of charge (Q) that is stored in the capacitor. Since current-over-time equals charge, it is often very convenient to look at the amount of charge necessary to change the voltage across various terminals in the GaN transistor. Figure 2.13 shows the amount of gate charge, Q_G, which must be supplied to increase the voltage from gate-to-source to the desired voltage. Q_G is the integrated value of C_{ISS} as a function of gate-to-source voltage, from the off-state V_{GS} to the on-state V_{GS}. Referring to Figure 2.13, it can be seen that about 5 nC of charge is needed to bring the gate voltage from 0 to 5 V, which

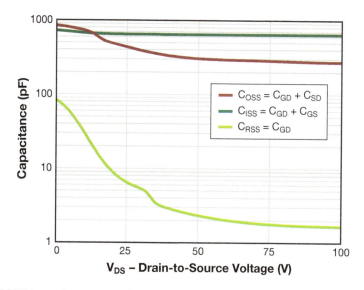

Figure 2.12 EPC2204 capacitance versus drain–source voltage. *Source:* [6]/Efficient Power Conversion Corporation.

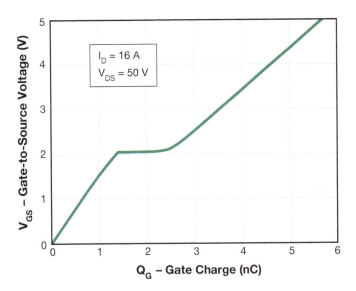

Figure 2.13 EPC2204 gate charge versus gate voltage. *Source:* [6]/Efficient Power Conversion Corporation.

will ensure the device is fully turned on. If the gate drive is capable of supplying 1 A of current, it will take about 5 ns to achieve this voltage.

It is important to note that the gate-to-source capacitance (C_{GS}) and total input capacitance (C_{ISS}) are typically more variable with respect to the gate voltage than the drain voltage, and the capacitance characteristic in Figure 2.12 shows C_{ISS} with $V_{GS} = 0$ V. The C_{ISS}–V_{GS} characteristic is typically not included in device datasheets, but it can easily be calculated as the slope of the Q_G–V_{GS} characteristic.

The Q_{GD} and Q_{GS} ($Q_G = Q_{GD} + Q_{GS}$) are also specified separately because they impact the voltage- and current-switching transition speeds, respectively. Also, the ratio of these two values, Q_{GD}/Q_{GS}, called the Miller Ratio, is often an important metric to determine the point at which a device might turn on due to a voltage transient applied across the drain and source. The Miller Ratio will be discussed in greater detail in Chapter 3.

The capacitances and charges for an enhancement-mode GaN transistor are shown in Table 2.8. The gate-to-drain charge, and its corresponding capacitance (C_{RSS} or C_{GD}), will change with drain-to-source voltage. (There is a detailed discussion about gate charge and its impact on switching characteristics in Chapter 3.) In this example, the values are given at 50 V, which is half the rated BV_{DSS}. This convention is used because, historically, the operating point for transistors in power conversion designs was about half the maximum rated voltage to provide safety margins for overshoot caused during switching transients. In addition, the internal gate resistance (R_G) is specified. This is the resistance that limits the ultimate switching speed of the device, together with any additional resistance in the gate driver circuit. If the device in Table 2.9 is driven by an ideal gate driver circuit that contributes no additional resistance, operating at the specified nominal conditions, the minimum rise time for the gate voltage transition can be calculated as follows:

$$t_{Rise,min} = \frac{Q_G \cdot R_G}{V_{GS}}$$
$$= \frac{(5.9 \text{ nC})(0.6 \text{ }\Omega)}{(5 \text{ V})} \quad (2.6)$$
$$= 708 \text{ picoseconds}$$

Table 2.9 Data from an Efficient Power Conversion EPC2204 datasheet showing the section relating to capacitance, charge, and gate resistance.

	Dynamic characteristics (T_J = 25 °C, unless otherwise stated)						
	Parameter	**Test conditions**	**Min.**	**Typ.**	**Max.**	**Unit**
C_{ISS}	Input capacitance[a]			644	851	
C_{RSS}	Reverse transfer capacitance	V_{DS} = 50 V, V_{GS} = 0 V		2.3		
C_{OSS}	Output capacitance[a]			304	456	pF
$C_{OSS(ER)}$	Effective output capacitance, energy related[b]	V_{DS} = 0–50 V, V_{GS} = 0 V		401		
$C_{OSS(TR)}$	Effective output capacitance, time related[c]			501		
R_G	Gate resistance			0.4		Ω
Q_G	Total gate charge[a]	V_{DS} = 50 V, V_{GS} = 5 V, I_D = 16 A		5.7	7.4	
Q_{GS}	Gate-to-source charge			1.8		
Q_{GD}	Gate-to-drain charge	V_{DS} = 50 V, I_D = 16 A		0.8		nC
$Q_{G(TH)}$	Gate charge at threshold			1.0		
Q_{OSS}	Output charge[a]	V_{DS} = 50 V, V_{GS} = 0 V		25	38	
Q_{RR}	Source–drain recovery charge			0		

All measurements were done with substrate connected to source.
[a] Defined by design. Not subject to production test.
[b] $C_{OSS(ER)}$ is a fixed capacitance that gives the same stored energy as C_{OSS} while V_{DS} is rising from 0 to 50% BV_{DSS}.
[c] $C_{OSS(TR)}$ is a fixed capacitance that gives the same charging time as C_{OSS} while V_{DS} is rising from 0 to 50% BV_{DSS}.
Source: Adapted from [6].

2.7 Reverse Conduction

When current flows into the source electrode of a transistor, this is considered reverse conduction. If the gate–source voltage is high enough to fully enhance the 2DEG, the transistor will conduct this reverse current in its ohmic region of operation, with approximately the same on-resistance as in the forward direction. However, with a gate–source voltage below the threshold, a GaN transistor can also conduct reverse current in another mode of operation.

In a Si MOSFET, there is a p-n junction that forms a diode from the body of the channel to the drain of the transistor. It is, therefore, called a body–drain diode, or simply a body diode. As discussed in Chapter 1, enhancement-mode GaN transistors do not have a p-n diode, but they do conduct in a way similar to a diode in the reverse direction. Figure 2.14 shows how this "body diode" forward voltage drop varies with source–drain current. It should be noted that this reverse-conduction path is formed by turning on the 2DEG in the reverse direction, using the positive gate–drain voltage to enhance the channel. Therefore, if the gate voltage is dropped below 0 V, the reverse-conduction voltage will increase by the same amount. For example, if the gate drive of a circuit turns off the GaN transistor by applying 0 V to the gate,

GaN Transistor Electrical Characteristics | 39

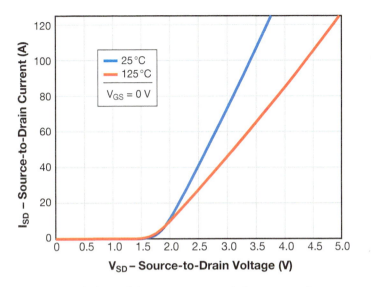

Figure 2.14 EPC2204 body-diode forward drop versus source–drain current and temperature. Source: [6]/Efficient Power Conversion Corporation.

Table 2.10 Excerpt from an efficient power conversion EPC2204 datasheet showing the section relating to the voltage drop when the transistor is conducting in the reverse direction.

	Static characteristics (T_J = 25 °C, unless otherwise stated)					
	Parameter	Test conditions	Min.	Typ.	Max.	Unit
V_{SD}	Source–drain forward voltage	I_S = 0.5 A, V_{GS} = 0 V		1.7		V

Source: Adapted from [6].

the V_{SD} at 0.5 A will be 1.8 V. If the gate drive of a circuit turns off the GaN transistor by applying a negative 1 V to the gate, the V_{SD} at 0.5 A will be 2.8 V.

Because the reverse conduction in a GaN transistor is due to turning on the 2DEG, the forward voltage drop will increase with temperature much the same way as $R_{DS(on)}$ changes with temperature in ohmic operation. In contrast, the body diode voltage drop in a Si MOSFET decreases with temperature. The voltage drop across an enhancement-mode GaN transistor rated at 100 V when conducting in the reverse direction is shown in Table 2.10 [6].

The reverse recovery charge Q_{RR} is a charge related to this reverse-conduction mechanism, representing the amount of charge dissipated when a body diode is turned off. Reverse recovery charge does not directly relate to the device capacitances previously discussed for an enhancement-mode GaN transistor. This charge comes from the minority carriers left over during diode conduction in a p-n junction of a MOSFET. Because there are no minority carriers involved in conduction in an enhancement-mode GaN transistor, there is no reverse recovery charge. Therefore, Q_{RR} is zero, which is a significant advantage compared with power MOSFETs and will be discussed in greater detail in Chapters 3, 7, and 14.

To illustrate this difference, in Table 2.11 are data from an Infineon 100 V OptiMOS™ 3 datasheet [8]. This MOSFET has a maximum $R_{DS(on)}$ of 6 mΩ. In comparison, the EPC2204 device

Table 2.11 Data from an Infineon BSC060N10NS3 G OptiMOS™ 3 datasheet showing the section relating to conduction in the reverse direction.

Parameter	Symbol	Test conditions	Min.	Typ.	Max.	Unit
Dynamic characteristics						
Diode forward voltage	V_{SD}	$V_{GS} = 0\,V, I_F = 50\,A,$ $T_J = 25\,°C$	—	1	1.2	V
Reverse recovery time	t_{rr}	$V_R = 50\,V, I_F = 25\,A,$ $di_F/dt = 100\,A/\mu s$	—	61	—	ns
Reverse recovery charge	Q_{rr}		—	109	—	nC

Source: Adapted from [8].

has a maximum $R_{DS(on)}$ of 6 mΩ. Nevertheless, the MOSFET has a stored charge, Q_{RR}, of 100 nC compared to the EPC2204 Q_{RR} of zero.

2.8 Thermal Characteristics

The maximum temperature a device can withstand during operation, or during prolonged storage is given in Table 2.12. Commercial devices are typically rated for operation between −40 °C and +150 °C. Devices designed to be operated in severe conditions such as cars or satellites in space are typically rated with a range of operating and storage temperatures of −55 to +150 °C [18].

Thermal resistance is a major factor in determining the capabilities of discrete power devices. From a device's thermal characteristics both the maximum power dissipation and maximum current can be derived for user applications. In this section, Figure 2.15 is used as a reference for both chip scale and packaged GaN transistors.

2.8.1 Thermal Resistance – Junction to PCB ($R_{\theta JB}$)

Junction-to-PCB, $R_{\theta JB}$, thermal resistance is perhaps the most important thermal specification since it will be used by the majority of applications. As seen in Figure 2.15, $R_{\theta JB}$ is the thermal resistance from the device junction to the PCB. Therefore, if the end user knows the thermal characteristics and environment of the application, thermal resistances of all the sub-parts can be added algebraically in order to arrive at the total thermal resistance of the total system.

Table 2.12 Excerpt from an Efficient Power Conversion EPC2204 datasheet showing the section relating to the maximum operating and storage temperatures.

Maximum ratings			
T_J	Operating temperature	−40 to 150	°C
T_{STG}	Storage temperature	−40 to 150	

Source: Adapted from [6].

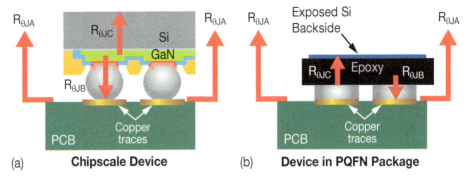

Figure 2.15 Components of thermal resistance in (a) chip-scale device, and (b) packaged device.

2.8.2 Thermal Resistance – Junction to Case ($R_{\theta JC}$)

Junction-to-case, $R_{\theta JC}$, thermal resistance is given for those situations where the end user wishes to add additional heatsinking to the top of the eGaN FET.

2.8.3 Thermal Resistance – Junction to Ambient ($R_{\theta A}$)

Junction-to-ambient, $R_{\theta JA}$, thermal resistance is a way of specifying the thermal resistance between a device and the ambient air with the device mounted onto a PCB. At least three different methods can be found in GaN transistor datasheets for specifying $R_{\theta JA}$. In Table 2.13, $R_{\theta JA}$ is defined in the footnote as being measured when a device is mounted on a PCB made from FR4 material using a one square inch of 2 ounce copper. In more recent datasheets, such as in Table 2.14, two methods are given. The first of these uses the JEDEC 51-2 standard [20] which has a defined layout including thin copper traces. The disadvantage of this method is that it is overly pessimistic stemming from the lack of thick copper traces to remove heat. The third method, also listed in Table 2.14, uses a common PCB layout and is a more realistic assessment of the thermal resistance a user might get in an actual application. A more quantitative and thorough analysis of the many variables that impact the thermal resistance of a device in an actual application is given in Chapter 6.

Table 2.13 Excerpt from an Efficient Power Conversion EPC2204 datasheet showing the section relating to the device thermal resistance.

	Thermal characteristics		
	Parameter	Typ	Unit
$R_{\theta JC}$	Thermal resistance, junction-to-case	1	
$R_{\theta JB}$	Thermal resistance, junction-to-board	2.5	°C/W
$R_{\theta JA}$	Thermal resistance, junction-to-ambient[a]	64	

See https://epc-co.com/epc/documents/product-training/Appnote_Thermal_Performance_of_eGaN_FETs.pdf for detail.
[a] $R_{\theta JA}$ is determined with the device mounted on one square inch of copper pad, single layer 2 oz copper on FR4 board.
Source: Adapted from [6].

Table 2.14 Excerpt from an Efficient Power Conversion EPC2302 datasheet showing the section relating to the device thermal resistance.

	Thermal characteristics		
	Parameter	Typ	Unit
$R_{\theta JC}$	Thermal resistance, junction-to-case (Case TOP)	0.2	
$R_{\theta JB}$	Thermal resistance, junction-to-board (Case BOTTOM)	1.5	
$R_{\theta JA\_JEDEC}$	Thermal resistance, junction-to-ambient (using JEDEC 51-2 PCB)	45	°C/W
$R_{\theta JA\_EVB}$	Thermal resistance, junction-to-ambient (using EPC90142 EVB)	21	

Source: [19]/Efficient Power Conversion Corporation.

2.9 Summary

In this chapter, the basic electrical characteristics of GaN transistors were discussed and related to the physical characteristics of the devices.

The three limitations of a device were explained: maximum drain-to-source voltage, maximum gate voltage, and maximum junction temperature. The characteristics that describe conduction and switching behavior in a circuit are described as on-resistance ($R_{DS(on)}$), threshold voltage ($V_{GS(th)}$), capacitance, and reverse-conduction characteristics. An overview of thermal resistance was also discussed, although a more thorough discussion of this topic can be found in Chapter 6.

The next two chapters are about circuit and layout techniques for GaN transistors. With a step-function improvement in switching speed and power density, designers need to take extra care to properly drive the gate of the GaN device and reduce parasitic elements in the surrounding circuits.

References

1 Lu, B., Piner, E.L., and Palacios, T. (2010). Breakdown mechanism in AlGaN/GaN HEMTs on Si substrate. *Proceedings of the Device Research Conference (DRC)*, 193–194.

2 McPherson, J.W. (1998). Underlying physics of the thermochemical Emodel in describing low-field time-dependent dielectric breakdown in SiO_2 thin films. *J. of Appl. Phys.* 84 (3): 1513–1523.

3 Joh, J. and del Alamo, J.A. (2008). Critical voltage for electrical degradation of GaN high-electron mobility transistors. *IEEE Electron. Device Lett.* 29 (4): 287–289.

4 Arulkumaran, S., Egawa, T., Ishikawa, H., and Jimbo, T. (2003). Temperature dependence of gate-leakage current in AlGaN/GaN high-electron-mobility transistors. *Appl. Phys. Lett.* 82 (18): 3110.

5 Tan, W.S., Houston, P.A., Parbrook, P.J. et al. (2002). Gate leakage effects and breakdown voltage in metal organic vapor phase epitaxy AlGaN/GaN heterostructure field-effect transistors. *Appl. Phys. Lett.* 80 (17): 3207–3209.

6 Efficient Power Conversion Corporation (2018). EPC2204 – Enhancement-mode power transistor, EPC2204 datasheet, December 2022. https://epc-co.com/epc/Portals/0/epc/documents/datasheets/EPC2204_datasheet.pdf.

7 GaN Systems (2018). 650V enhancement mode GaN transistor. GS66508B datasheet, September 2018. https://gansystems.com/wp-content/uploads/2018/07/GS66508B-DS-Rev-180709.pdf

8 Infineon (2009). OptiMOS™ power-transistor. BSC060N10NS3 G datasheet, October 2009 [Revision 2.4]. https://www.infineon.com/dgdl/Infineon-BSC060N10NS3-DS-v02_04-en.pdf?fileId=db3a30431ce5fb52011d1aab7f90133a.

9 JEDEC JEP186 (2021). Guideline to specify a transient off-state withstand voltage robustness indicator in datasheets for lateral GaN power conversion devices, December 2021. https://www.jedec.org/standards-documents/docs/jep186.

10 Zhang, S., Espinoz, A., Gao, H., et al. (2024). Proposing Duty-cycle-based Transient Overvoltage Specifications for GaN HEMTs, ISPSD, Bremen, Germany (2–6 June 2024).

11 Lidow, A. (ed.) (2022). *GaN Power Devices and Applications*. El Segundo, CA: Power Conversion Publications.

12 Sze, S.M. (1981). *Physics of Semiconductor Devices*, 2e, 31. Hoboken, NJ: Wiley.

13 Giancoli, D. (2009) [1984]. Electric currents and resistance. In: Physics for Scientists and Engineers with Modern Physics, 4 (ed. J. Phillips), 658. Upper Saddle River, New Jersey: Prentice Hall.

14 Serway, R.A. (1998). *Principles of Physics*. Saunders College Publications. https://openlibrary.org/books/OL301701M/Principles_of_physics.

15 Cuerdo, R., Pedros, J., Navarro, A., and Braña de Cal, A.F. (2008). High temperature assessment of nitride-based devices. *J. Mater. Sci. Mater. Electron* 19 (2): 189–193.

16 Rashmi, A.K., Haldar, S., and Gupta, R.S. (2002). An accurate charge control model for spontaneous and piezoelectric polarization dependent two-dimensional electron gas sheet charge density of lattice-mismatched. *Solid-State Electron.* 46: 621–630.

17 Liu, Q.Z. and Lau, S.S. (1998). A review of the metal-GaN contact technology. *Solid-State Electron.* 42 (5): 677–691.

18 Efficient Power Conversion Space Division (2023). EPC7003A – 100 V_{DS}, 10 A, 45 mΩ. EPC7003A datasheet, August 2023. https://epc.space/documents/datasheets/EPC7003A-datasheet.pdf.

19 Efficient Power Conversion Corporation (2023). EPC2302 – Enhancement-mode power transistor. EPC2302 datasheet, October 2023. https://epc-co.com/epc/Portals/0/epc/documents/datasheets/EPC2302_datasheet.pdf.

20 JEDEC JESD51-2A (2008). Integrated circuits thermal test method environmental conditions – natural convection (still air). [Revision of JESD51-2, January 2008]. https://www.jedec.org/standards-documents/docs/jesd-51-2A.

3

Driving GaN Transistors

3.1 Introduction

This chapter discusses the basic techniques for using GaN transistors in high-performance power conversion circuits. GaN transistors generally behave like power MOSFETs, but at much higher switching speeds and power densities. A good understanding of these similarities and differences is fundamental to understanding how much existing power conversion systems can be improved by GaN-based device technologies. The next four chapters highlight the benefits of GaN technology, design techniques for maximum performance, and ways to avoid common pitfalls that can result from the new GaN performance capabilities. Techniques to be addressed in these next chapters include:

- How to drive a GaN transistor or GaN IC.
- How to layout a high-efficiency GaN transistor circuit.
- How to predict the lifetime of a GaN device in operation.
- How to design the proper thermal management strategy.

To understand the differences and opportunities offered by these faster-switching devices, the alternative GaN transistor structures need to be considered individually. Two of the structures shown in Chapter 1 will be examined: enhancement-mode transistors using a p-GaN-type gate, and the two-transistor cascode configurations. The gate electrodes of both types of structure have very high input impedance, and control of the device is therefore accomplished by supplying or removing a certain amount of charge from the gate electrode.

To begin the discussion on the switching behavior comparison between GaN transistors and MOSFETs, the classic models for MOSFETs will be used. In Chapters 7 and 8 these models will be specifically adapted for GaN transistors. In Table 3.1 are definitions for examining the various switching properties of the devices. In Figure 3.1a, these definitions are used to segment transistor switching into four regions: (i) the charge required to bring the gate terminal up to the device threshold, (ii) the charge required to complete the current transition (t_{cr}) and reach the plateau voltage (V_{pl}), (iii) the charge required to complete the voltage transition (t_{vf}), and (iv) the charge supplied to drive the gate to the steady-state gate voltage. In Figure 3.1b a gate charge curve for a GaN transistor is shown with the various gate charges.

To better understand why GaN transistors switch so much faster than MOSFETs, these two transistor technologies can be compared quantitatively using figures of merit (FOM). As mentioned in Chapter 1, the theoretical on-resistance versus blocking voltage of a GaN transistor is more than three orders of magnitude lower than that of silicon, and the first-generation

---

GaN Power Devices for Efficient Power Conversion, Fourth Edition. Alex Lidow, Michael de Rooij, John Glaser, Alejandro Pozo, Shengke Zhang, Marco Palma, David Reusch, and Johan Strydom.
© 2025 John Wiley & Sons Ltd. Published 2025 by John Wiley & Sons Ltd.

Table 3.1 Gate charge components and their definitions.

Gate charge components	Definitions
Q_{GS}	Charge required to increase gate voltage from zero to the plateau voltage. ($Q_{GS} = Q_{GS1} + Q_{GS2}$)
Q_{GS1}	Charge required to increase gate voltage from zero to the threshold voltage of the device.
Q_{GS2}	Charge required to commutate the device current.
Q_{GD}	Charge required to commutate the device voltage, at which point the device enters the linear region.

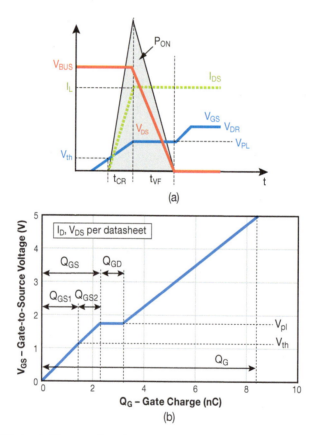

Figure 3.1 (a) Simplified switching waveform for a GaN transistor showing the various stages of turning on. Four regions of transistor switching. (b) Gate charge versus gate voltage showing different gate charge components for an EPC619 GaN transistor. Gate charge curve for a GaN transistor is shown with the various gate charges. *Source:* Adapted from Efficient Power Conversion Corporation [1].

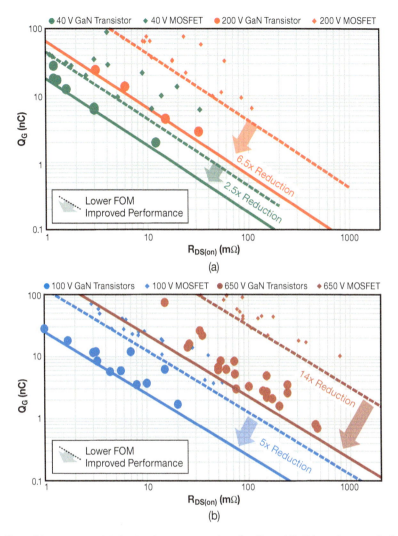

Figure 3.2 On-resistance versus total gate charge comparison for Si- and GaN-based power devices showing (a) 40 V and 200 V, and (b) 100 V and 650 V devices.

production devices were already beyond the silicon limit [2]. In general, these smaller devices have less capacitance when compared to silicon MOSFETs. Although more specialized FOMs will be discussed in the Chapters 7 and 8, the $R_{DS(on)} \cdot Q_G$ product is commonly used for comparing different technologies [3]. Using this FOM to compare GaN to silicon (Figure 3.2) shows, as of the time of this writing, an improvement between almost three times to more than ten times that of silicon, with larger advantages at higher voltages.

3.2 Gate Drive Voltage

All GaN transistors including enhancement-mode, direct-drive [4] depletion-mode, and cascode GaN transistors have maximum positive and minimum negative voltage limits that can be applied to the gate electrode with respect to the source. For a cascode device, such as the

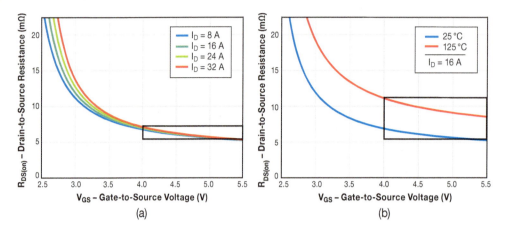

Figure 3.3 Two graphs of the $R_{DS(on)}$ of an enhancement-mode GaN transistor (EPC2045) as a function of gate voltage (a) for various drain currents and (b) at various temperatures, showing the recommended gate drive voltage range (boxed area). *Source:* Ref. [6]/Efficient Power Conversion Corporation.

TP65H015G5WS [5], these voltage limits are ±20 V – similar to that of silicon MOSFETs. For an enhancement-mode device, one manufacturer [6] specifies this maximum voltage as +6 V/−4 V, while other manufacturers [7, 8] specify +7 V/−10 V and +4.2 V/−10 V, respectively. No information is available on direct-drive depletion-mode devices, but they are likely to have similar, if somewhat relaxed, constraints compared to enhancement-mode devices. (Note that different technologies and different manufacturers will rate their devices differently.) Exceeding these limits may damage devices permanently and, therefore, must be avoided. Fortunately, the amount of gate voltage required to drive the device to the on-state is lower than the maximum voltage allowed. With very fast switching speeds, however, care must still be taken to avoid overshoot that might inadvertently take the gates above the maximum voltage limit. For a pGaN enhancement-mode transistor, the device $R_{DS(on)}$ is specified in the datasheet at a recommended voltage, which is typically about 1 V below the absolute maximum rating.

Because of this relatively tight requirement, we will discuss the gate drive requirements for a specific enhancement-mode GaN manufacturer [6], with the understanding that the same basic principles apply to other enhancement and direct-drive manufacturers.

It is possible to drive these enhancement-mode devices with gate voltages as low as 4 V without a significant change in $R_{DS(on)}$, as shown by the rectangular box area in Figure 3.3. Furthermore, it is recommended to keep the gate driver voltage below 5.5 V to allow enough margin between the gate voltage and the absolute maximum gate voltage. These recommended gate voltage limits can be readily achieved by near-critical damping of the gate drive turn-on power loop.

3.3 Gate Drive Resistance

Figure 3.4a shows a typical equivalent circuit for the gate loop during turn-on transient. The gate driver, transistor, gate drive bypass capacitor (C_{VDD}), and inductance of the interconnections between them (L_G) form an LCR-series resonant tank. All the resistive components in the path of the gate current during turn-on, indicated by the red arrow in Figure 3.4a, can be lumped into an equivalent series resistance, $R_{G(eq)}$. This equivalent resistance includes the

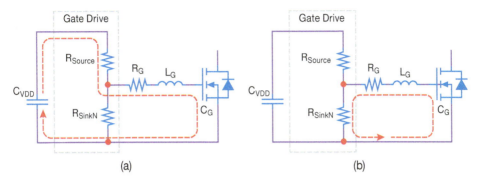

Figure 3.4 Resonant loop formed between the gate driver and GaN transistor during (a) turn-on and (b) turn-off.

transistor gate resistance (R_G), gate drive pull-up resistance (R_{Source}), the high-frequency interconnect resistance between the components, as well as the equivalent series resistance (ESR) of the gate driver supply capacitor. To minimize switching time, it is best to minimize this overall resistance as much as possible, but some resistance is required to avoid voltage overshoot at the gate. So, to critically damp this loop, the overall gate-loop resistance $R_{G(eq)}$ ($R_{G(eq)} = R_G + R_{Source}$) must be equal to, or larger than, the value given in Eq. (3.1). This is achieved by minimizing the gate-loop inductance (L_G) (layout techniques for minimal gate inductance will be discussed in Chapter 4) and adjusting the series gate resistance to limit overshoot. The resultant damped turn-on gate voltage [9] is shown in Figure 3.5.

$$R_{G(eq)} \geq \sqrt{\frac{4 \cdot L_G}{C_{GS}}} \qquad (3.1)$$

For the gate drive voltage falling edge, the negative voltage limit does not present any practical limitation. It is, therefore, possible to drive the GaN transistor faster at turn-off and allow some negative ringing. Furthermore, the gate drive turn-off power loop, shown in Figure 3.4b, will have a smaller inductance as this loop does not include the gate drive bypass capacitor.

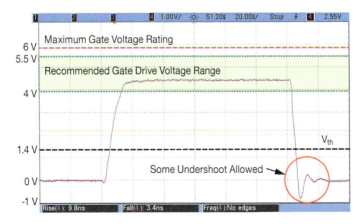

Figure 3.5 Example of an enhancement-mode transistor gate drive voltage showing a critically damped voltage rise and a slightly under-damped voltage fall.

However, care should still be taken to avoid subsequent positive gate voltage ringing beyond the gate threshold of the device, as this will cause the device to turn on again.

Figure 3.5 shows an under-damped turn-off with the subsequent positive ringing of less than half a volt. Since the turn-on and turn-off damping requirements are different, the minimum gate-loop resistance values will also differ. These differences are best addressed by separating the pull-up and pull-down gate driver resistances at the driver output (creating two separate driver outputs), thus allowing the use of two separate gate resistors to independently adjust the turn-on and turn-off gate-loop damping.

3.4 dv/dt Considerations

GaN power devices are exposed to significantly higher voltage and current slew rates that can impact the performance of the transistor. These conditions need to be understood well to fully utilize the technology. In this section, both the impact of the gate driver on generated dv/dt as well as the impact of the switching dv/dt on the gate driver and GaN device will be discussed.

3.4.1 Controlling dv/dt at Turn-On

For GaN transistors, like silicon MOSFETs, the device plateau voltage is a function of the device current, as shown in an example in Figure 3.6. Thus, for a given overall gate-loop resistance, the voltage drop across this resistance is the difference between the gate drive supply voltage and the plateau voltage, and is therefore load-current dependent. Since this voltage drop determines the gate drive current, the length of the plateau interval (and thus dv/dt) is also load-current dependent, with a decreasing gate drive current at higher loads. The implication is that for a voltage source gate drive, the switching loss disproportionately increases with the load current as the gate drive effectively becomes weaker. More detail on these loss calculations will be discussed in Chapter 7.

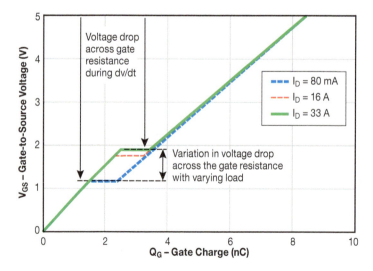

Figure 3.6 Variation of voltage drop across overall gate resistance with changes in drain current for the EPC2619. *Source:* Adapted from Efficient Power Conversion Corporation [1].

To address this effect, the voltage source gate drive can be replaced with a constant current drive. An example of such an approach is used in direct-drive depletion-mode GaN transistors in reference [10]. This results in a power stage that has constant dv/dt, regardless of load current, and by allowing this current to be set externally, the dv/dt of the device can be directly controlled. It should be noted that due to the much higher transconductance of these GaN devices, the variation between plateau voltages is relatively small for different load conditions, but there is still a large voltage variation between the threshold voltage and the plateau voltage at relatively low currents. This is best illustrated in Figure 3.6. Note that the voltage plateau variation between 16 and 33 A is much smaller than between 80 mA and 16 A. Thus, the improvement in dv/dt control that a constant current gate drive provides will be more pronounced at lighter loads.

3.4.2 Complementary Device Turn-On

A high, positive-voltage slew rate (dv/dt) of the drain relative to the source of an off-state GaN device can occur in both hard- and soft-switching applications and is characterized by a quick charging of the device's capacitances, as depicted in Figure 3.7. During this dv/dt event, the drain–source capacitance (C_{DS}) is charged. Concurrently, the gate–drain (C_{GD}) and gate–source (C_{GS}) capacitances in series are also charged. The concern is that, unless addressed, the charging current through the C_{GD} capacitance will flow through and charge C_{GS} beyond Vth and turn the device on. This event, sometimes called a Miller induced turn-on [11], can be very dissipative. Such an unintended turn-on can be avoided by supplying an alternative parallel path across C_{GS} through which the C_{GD} charging current can then flow. With the addition of a gate driver pull-down keeping the device off, some of the current flowing through C_{GD} can be diverted from C_{GS} through the series gate resistor (R_G) to the gate driver pull-down resistor (R_{Sink}). This additional path allows the efficient operation of devices that would otherwise be sensitive to dv/dt turn-on.

However, the effectiveness of such the pull-down resistance comes into question when the impact of the gate-loop inductance is considered [12], as at high enough dv/dt the inductive

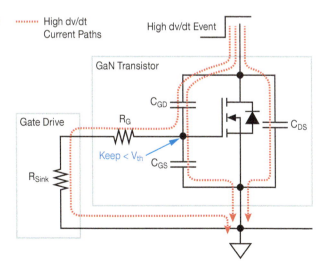

Figure 3.7 Effect of dv/dt on a device in the off-state and requirements for avoiding Miller-induced shoot-through.

impedance could become large enough that the alternative pull-down path no longer becomes functional, and only an improvement in the Miller charge ratio will avoid Miller induced turn-on. This can be achieved through GaN device improvement or through the addition of a negative gate voltage during the off-state, which negatively impacts the device third quadrant operation.

To determine the dv/dt susceptibility of a power device, a Miller charge ratio (Q_{GD}/Q_{GS1}) as a function of drain-to-source voltage needs to be evaluated. A Miller ratio of less than one will guarantee dv/dt immunity [11]. GaN transistors, like MOSFETs, are typically operated up to 80% of rated voltage although datasheets tend to provide data at 50% of rated voltage. At these higher voltages, the Miller charge ratio should remain below one and recent GaN devices have shown significant improvement in this area.

As an example, the Miller charge ratio as a function of drain-to-source voltage for two 100 V-rated parts is plotted in Figure 3.8. The first-generation part, an EPC2016 [13], was dv/dt sensitive above 40% of rated voltage, while a fifth-generation part, EPC2045 [6], is completely dv/dt immune. However, this inherent dv/dt immunity is not available in all GaN devices. As an example, a 650 V device [7], also shown in Figure 3.8, becomes dv/dt sensitive at an even lower voltage and thus still requires at least some pull-down resistor circuit and minimization of gate-loop inductance to keep the device off at higher voltages.

For complementary switching applications (e.g. a half-bridge circuit where one or the other switch is always on), it is possible to improve the dv/dt immunity of the device artificially through adjustment of the dead time between the switching devices. From the gate waveform shown in Figure 3.5, the gate drive voltage briefly becomes negative at turn-off due to the slightly under-damped resonance within the gate loop. By adjusting the turn-on timing of the complementary switch to coincide with this negative voltage dip of the device's gate voltage, the effective charge needed to induce Miller turn-on is increased significantly. Although only applicable when the timing between devices is fixed, this technique allows for an increase in dv/dt switching speed, as well as the use of marginal Miller ratio devices without fear of dv/dt induced turn-on.

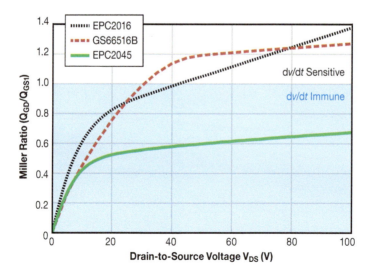

Figure 3.8 Example of Miller charge ratio versus drain-to-source voltage for different devices: First-generation 100 V EPC2016 [13], fifth-generation 100 V EPC2045 [6], and 650 V GS66516B [7]. Results estimated from available datasheet curves.

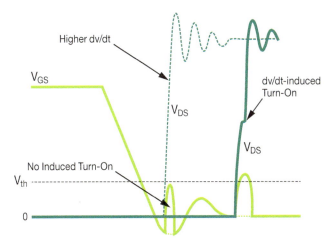

Figure 3.9 Improvement of dv/dt turn-on immunity through controlled gate timing using under-damped gate turn-off.

Figure 3.9 shows two cases where a device with a marginal Miller ratio is turned off and then subjected to a high dv/dt, as induced by the complementary switch turn-on. The solid line in the drain voltage curve shows a dv/dt induced turn-on with a characteristic "knee," where the drain voltage rise time is self-limited due to dv/dt turn-on. In contrast, by turning on the complementary device during the gate voltage dip (dotted line V_{DS}), dv/dt turn-on is avoided and higher dv/dt edge rates are achieved.

3.5 di/dt Considerations

In this section, both the impact of the gate driver on di/dt and the impact of di/dt on the GaN device and gate driver will be discussed.

3.5.1 Device Turn-On and Common-Source Inductance

During device turn-on, the di/dt rate is determined by how quickly the gate drive can increase the gate voltage beyond the device threshold. This rate can be significantly slower than during the dv/dt interval due to the impact of common-source inductance (CSI). The CSI is the inductance on the source side of a device that is common to both the power loop (drain-to-source current) and the gate drive loop (gate-to-source current) and is shown as L_{CS} in Figure 3.10. Once the GaN device starts conducting current, the increasing drain current di/dt will induce a voltage across the CSI that will oppose the gate drive voltage, thereby reducing the gate current used to charge the gate capacitance. This effectively lengthens the current transition period. A more detailed discussion on how to layout a circuit on a PCB to minimize CSI is discussed in Chapter 4. How CSI impacts performance and switching losses is discussed in Chapter 7, Section 7.3, and shows that minimizing the CSI is key to increasing di/dt capability, and thus system efficiency.

For a direct-drive depletion-mode device, the series safety MOSFET can be part of the gate loop as shown in Figure 3.11 and thus the MOSFET, with its related interconnects, also contributes to CSI. For both enhancement-mode and direct-drive devices, it is possible to

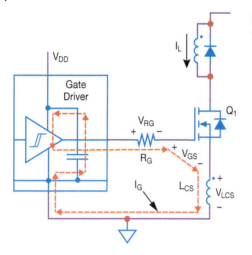

Figure 3.10 Impact of a positive di/dt of an off-state device with common-source inductance.

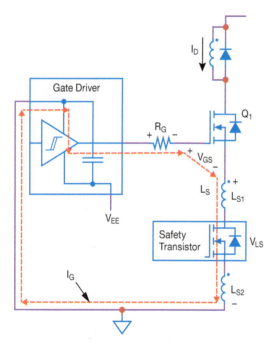

Figure 3.11 Gate circuit including common-source inductance for direct-drive device.

minimize CSI by separating the gate and power loops to as close to the GaN device as possible, and minimize the internal source inductance of the GaN device, which will remain common to both loops. This will also be discussed in more detail in Chapter 4.

3.5.2 Off-State Device di/dt

During the turn-on of a complementary device while the off-state device body diode is conducting, a rising current di/dt through an off-state device with a body diode, as shown in

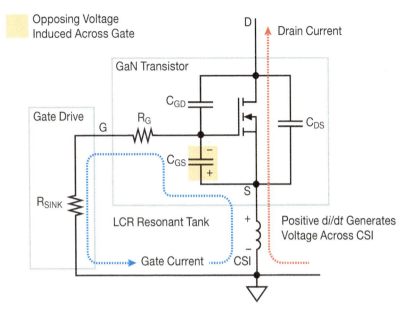

Figure 3.12 Half-bridge circuit with bootstrap supply showing impact of low-side diode conduction.

Figure 3.12, will induce a step voltage across its CSI, like the turn-on case above. This voltage step in turn will generate a gate-loop current that results in a negative voltage across C_{GS}. This negative gate voltage is useful in improving dv/dt immunity during the subsequent voltage step that occurs once the current through the body diode has reduced to zero. However, without sufficient damping of the off-state gate-loop LCR resonant tank, this initial negative voltage step across the gate could induce positive ringing and cause an unintended turn-on and shoot-through, as shown in Figure 3.13.

It is possible to avoid this type of di/dt turn-on by sufficiently damping the gate turn-off loop, but increasing the gate turn-off power loop damping through an increase in gate pull-down resistance would negatively impact the device switching performance. As with turn-on, a better solution is to limit the size of the CSI through improved packaging and device layout.

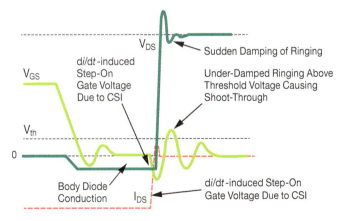

Figure 3.13 di/dt-induced turn-on (shoot-through) of an off-state device with under-damped gate turn-off power loop.

3.6 Bootstrapping and Floating Supplies

In a half-bridge topology, a common solution for generating the floating supply for the high-side transistor is the use of a bootstrap circuit as shown in Figure 3.14. This circuit operates by charging the floating high-side supply capacitor (C_{Boot}) during the on-state of the low-side device from the fixed low-side supply through a high-speed bootstrap diode (D_{Boot}). During the high side on-state, the bootstrap diode blocks the full bus voltage, and the floating capacitor supplies the required gate drive energy at the voltage of the bootstrap capacitor.

In concept, the bootstrap capacitor is charged to the same voltage as the low-side bus (V_{DD}). In practice, the voltage drop across the bootstrap diode, D_{Boot}, and voltage drop across the low side power device, V_{SD}, will affect the resulting bootstrap voltage. For silicon MOSFETs this change in voltage is not important to the power device operation, but for e-mode GaN devices these voltages do matter, as shown in Section 3.2. Prolonged diode conduction of the low-side device will cause the bootstrap supply to charge up to the low-side bus voltage plus the reverse conduction voltage drop (V_{SD}) minus the bootstrap diode voltage, which altogether could be higher than the maximum allowable gate voltage.

As the voltage drop across the power device during conduction is typically much lower than that of the body diode, this overcharging can be improved through reduction of the switching dead time and related low side diode conduction interval as well as through adding an anti-parallel Schottky diode across the power device to reduce this additive voltage drop.

This phenomenon is discussed in more detail in [14] and eight different bootstrap solutions, listed in Table 3.2, are compared. The results from this investigation are shown in Figure 3.15. From this, a number of conclusions can be drawn:

- Adding a series resistor to the bootstrap diode can compensate for some of the increase in bootstrap voltage, although this may be difficult to design in practice as its impact will be load and switching frequency dependent.
- Reducing diode conduction time, adding a Schottky diode, or doing both, are beneficial for limiting the bootstrap voltage as well as reducing system losses. However, in the case of having small GaN devices in a half bridge, it is possible that the capacitive and recovery charge losses of the added Schottky diode outweigh this benefit and results in overall higher losses.

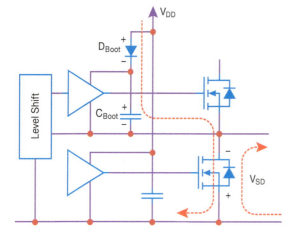

Figure 3.14 Half-bridge circuit with bootstrap supply showing impact of low-side diode conduction.

Table 3.2 Design cases for bootstrap overvoltage management.

Case	Description
Case #1	Large dead time, no capacitor
Case #2	Large dead time, resistor in series with bootstrap
Case #3	Reduced dead time, no capacitor voltage management
Case #4	Reduced dead time, 5.1 V Zener diode clamp
Case #5	Reduced dead time, 4.7 V Zener diode clamp
Case #6	Reduced dead time, Schottky diode in anti-parallel with low-side GaN transistor
Case #7	Reduced dead time, synchronous e-mode GaN transistor bootstrap method
Case #8	Reduced dead time, synchronous e-mode GaN transistor bootstrap method with Schottky diode in anti-parallel with low-side GaN transistor

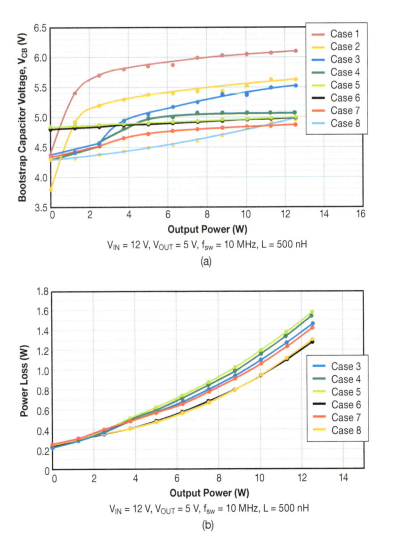

Figure 3.15 Performance comparisons of design cases given in Table 3.2. (a) Bootstrap capacitor DC voltage versus output power (b) total system loss versus output power.

- Using a Zener diode to clamp the bootstrap voltage is effective at limiting the bootstrap supply voltage but can significantly increase system losses at higher switching frequencies.
- Using a synchronous GaN bootstrap device yields the best voltage regulation as well as system efficiency.

Some half-bridge gate driver ICs, such as the one pictured in Figure 3.16, provide an integrated bootstrap voltage clamp to prevent excessive voltage swings [15]. This clamping circuit actively blocks the charging path to the bootstrap capacitor whenever its voltage exceeds a specified maximum level. When the voltage falls by the clamp's hysteresis voltage, the clamping circuit allows charging of the bootstrap capacitor again. In applications where the switching of the half-bridge devices is not complementary (prolonged dead time), the use of a high-side regulator with either a discrete solution using a series voltage regulator or an integrated gate driver IC, as shown in Figure 3.17 [16], should be considered.

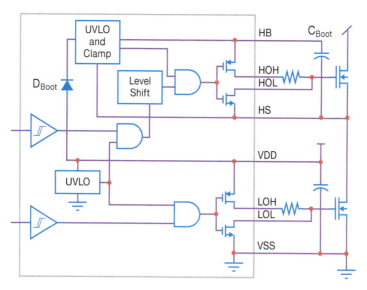

Figure 3.16 Block diagram of a half-bridge GaN transistor driver with integrated high-side supply regulation. *Source:* Adapted from Texas Instruments [15].

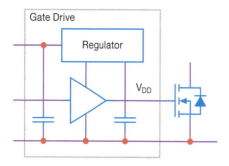

Figure 3.17 Block diagram of a single GaN transistor driver with integrated supply regulation. *Source:* Adapted from Texas Instruments [16].

The bootstrap diode is connected to the switch-node of the half bridge, so it is effectively in parallel with the output capacitance of the low-side transistor. This means that the well capacitance and reverse recovery charge of the bootstrap diode can slow down switching transients and increase switching loss [17, 18]. In a relatively low-voltage half-bridge application (e.g. 100 V), this typically does not cause a problem because Schottky diodes or bootstrapping gate driver ICs are available with minimal impact on the switching speed of the GaN transistors. However, in higher voltage applications, the addition of a diode may no longer be practical, because diodes with the necessary blocking voltage capability may be bulky, lossy, or expensive. In this case, an isolated power supply can be used to supply the floating gate drive of the high-side transistor. The capacitance across the isolation barrier in the floating supply is connected to the switch-node. Like a bootstrap diode, this added switch-node capacitance can potentially impact switching speed. However, many isolated power supplies are available with very low isolation capacitance in the range of a few picofarads.

3.7 Transient Immunity

Ground bounce is a common phenomenon in the world of high-speed logic [19]. The concept is that high-voltage slew rates across capacitors generate large current pulses of short duration. Conceptually, these current pulses generate pairs of dynamic voltage pulses across any layout inductances at the rising and falling edges of the current pulse. These ground bounces can lead to unintended switching, degraded performance, and can potentially damage devices.

An idealized example in Figure 3.18a shows the gate drive in close proximity to the GaN transistor to minimize common-source inductance. By tying the gate drive return directly to the source of the GaN device, the source-side layout inductance is pushed outside the gate drive loop. Any voltage pulses across this source inductance will cause the logic and controller ground to "bounce" relative to the source of the power device (and thus the "ground" of the gate driver). If these pulses are large enough, they can change the logic state of the gate driver input, and thus negatively impact a GaN power device.

The best way to avoid ground bounce is to place the controller on the same ground as the gate driver, as shown in Figure 3.18b, something that may not be practical with multiple low-side switching devices. In those cases, there are two ways to address ground bounce as shown in Figure 3.19. First, the ground bounce noise can be filtered out by placing a small RC low-pass filter (LPF) between the controller and the gate driver. There is a trade-off between too much

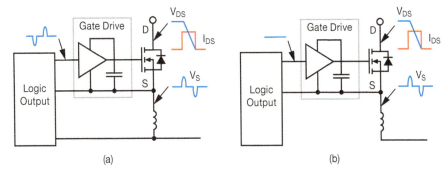

Figure 3.18 Inductance between the gate drive and the power ground causes a "bounce" of the gate driver ground. (a) Tying controller to power ground (b) tying controller to gate driver ground.

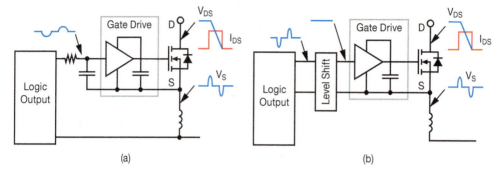

Figure 3.19 Solutions for filtering out ground "bounce" noise from gate drive input. (a) RC filter and (b) level-shifter or isolator.

filtering causing significant delay and pulse width distortion due to variation in the gate driver input thresholds, or not enough filtering maintaining susceptibility to logic glitches.

The second alternative solution is to use a level shifter or isolator between the controller and gate driver. This approach effectively treats the low-side gate driver in the same way as the floating high-side driver. Although a level shifter increases complexity and component count, it does have the added advantage of improving the gate driver propagation delay matching between the high side and low side.

Another mechanism for generating logic glitches is a common-mode current through the level shifter or digital/optical isolator. This can happen to the high-side device in a half-bridge application during high positive or negative switch node dv/dt. For a positive dv/dt event, as shown in Figure 3.20, the high-voltage slew rate across the isolator capacitance causes the generation of a common-mode current flowing in the loops as shown. The common-mode current causes ground bounce within the level shifter and can cause changes in the logic state if the common-mode current is large enough.

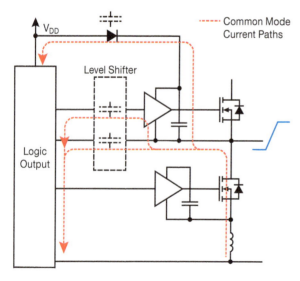

Figure 3.20 High-speed switching causes large common-mode current across level shifter and bootstrap diode capacitance.

With GaN transistors, the switch-node slew rates are likely to be up to hundreds of volts per nanosecond. This issue will need to be addressed to avoid becoming a limiting factor on circuit performance and, since this is a level-shifter issue, it is common to all GaN device applications that have a high-side floating device. Resolution requires minimizing floating high-side-to-ground capacitance as well as increasing dv/dt immunity within the level-shifter. High-side-to-ground capacitance can be minimized by avoiding PCB layout overlap between ground and any of the high side nodes, selecting components with a low inherent capacitance and limiting the size of the high-side copper area.

The capability of the level shifter, isolator, or isolated gate driver to withstand high dv/dt without experiencing common-mode glitches is specified by its common-mode transient immunity (CMTI). In order to guarantee safe operation in a half bridge without common-mode glitches, the level shifter/isolator/driver must have a minimum specified CMTI greater than the expected maximum dv/dt on the switch node.

3.8 Gate Drivers and Controllers for Enhancement-Mode GaN Transistors

The ecosystem of ICs suitable for e-mode GaN transistors has seen exponential growth in recent years. Tables 3.3–3.7 list several commercially available components that are compatible or specifically designed for e-mode GaN transistors. The tables include isolated and non-isolated gate drivers, for low-side or half-bridge topologies, controllers with integrated drivers, and radiation hardened ICs. In some cases, there are footprint-compatible ICs from multiple manufacturers [27, 32, 37, 39].

3.9 Cascode, Direct Drive, and Higher-Voltage Configurations

3.9.1 Cascode Devices

Cascode devices have many unique driving requirements. One of the main characteristics of the cascode device is the fact that it is a hybrid design with two discrete devices, a depletion-mode GaN transistor in-series with a MOSFET, each made with dissimilar technologies that require external connections. Driving the cascode device through the MOSFET gate has many

Table 3.3 Low-side gate drivers for GaN e-mode transistors.

Part number	Description
Analog Devices ADuM4120/1 [20]	Isolated single channel, 2 A output
Infineon 1EDBx275F [21]	Isolated single channel, 5 A (source)/9 A (sink) output
Littelfuse IXD_604 [22]	Dual channel, 4 A output
Texas Instruments UCC27611 [23]	Single channel, 5 A (source)/9 A (sink) output
Texas Instruments LMG1020 [24]	Single channel, 7 A (source)/5 A (sink) output, narrow pulses
Texas Instruments LMG1025-Q1 [25]	Automotive single channel, 7 A (source)/5 A (sink) output, narrow pulses

Table 3.4 Half-bridge gate drivers for GaN e-mode transistors.

Part number	Description
Analog Devices ADuM4221A [26]	Isolated, 1.6 kV DC, half bridge with adjustable dead time, 4 A output
Analog Devices LT8418 [27]	Half bridge, 100 V, 4 A (source)/8 A (sink) output
Infineon 2EDB7259Y [28]	Isolated, 3 kV, independent dual channels, 5 A (source)/9 A (sink) output
Infineon 2EDR7259X [29]	Isolated, 5.7 kV, independent dual channels, 5 A (source)/9 A (sink) output
Infineon 1EDN71x6U [30]	High-side, 200 V, with differential input and up to 2 A output
MPS MPQ1918-AEC1 [31]	Automotive half bridge, 100 V, 1.6 A (source)/5 A (sink) output
MPS MP8699B [32]	Half bridge, 100 V
On Semi NCP51810 [33]	Half bridge, 150 V, 1 A (source)/2 A (sink) output
On Semi NCP51820 [34]	Half bridge, 650 V, 1 A (source)/2 A (sink) output
Skyworks Si827xGB-IM [35]	Automotive, isolated, 2.5 kV, half bridge with adjustable dead time, 1.8 A (source)/4 A (sink) output
ST STDRIVEG600 [36]	Half bridge, 600 V, 1.3/2.4 A (source)/5.5/6 A (sink) output
Texas Instruments LM5113-Q1 [16]	Automotive half bridge, 100 V, 1.2 A (source)/5 A (sink) output
Texas Instruments LMG1205 [37]	Half bridge, 90 V, 1.2 A (source)/5 A (sink) output
Texas Instruments LMG1210 [38]	Half bridge, 200 V, 1.5 A (source)/3 A (sink) output
uPI Semiconductor uP1966E [39]	Half bridge, 80 V

advantages and disadvantages compared to an enhancement-mode device. Some advantages are:

- The cascode device's gate terminal is that of a MOSFET. It has the same MOSFET gate voltage ratings and can be driven, in concept, with traditional MOSFET drivers. It does not necessarily need as much attention to avoid gate overshoot.
- Turning off the GaN transistor with a positive current is self-commutating once the MOSFET turns off. In other words, the load current itself is responsible for generating the necessary depletion-mode gate voltage to turn the device off through charging up the MOSFET output capacitance. The higher the load current, the faster this switching occurs, thus resulting in a turn-off energy that is largely load current independent.

Some disadvantages are:

- There are two loops where common inductance is important. Referring to Figure 3.21, there is the common-source inductance (L_{CSI}), similar to a single enhancement-mode device, and the common cascode inductance (L_{CCI}) in the loop between the MOSFET and the depletion-mode GaN transistor. Both of these loops will negatively impact switching speed.

Driving GaN Transistors | 63

Table 3.5 Controllers with integrated driver for GaN e-mode transistors.

Part number	Topology	Description
Analog Devices LTC7890 [40]	Synchronous buck	Single phase, V_{IN} = 5–100 V, frequency up to 3 MHz
Analog Devices LTC7891 [41]	Synchronous buck	Dual phase, V_{IN} = 5–100 V, frequency up to 3 MHz
Analog Devices LTC7800 [42]	Synchronous buck	Single phase, V_{IN} = 5–60 V, frequency up to 2.25 MHz
Analog Devices LT8390A [43]	Buck-boost	4-switch buck-boost, V_{IN} = 4–60 V, frequency up to 2 MHz
Microchip MIC2132 [44]	Synchronous buck	Dual phase, V_{IN} = 8–75 V, frequency up to 1 MHz
Renesas ISL81806 [45]	Synchronous buck	Dual phase, V_{IN} = 4.5–80 V, frequency up to 2 MHz
Renesas ISL81807 [46]	Synchronous boost	Dual phase, V_{IN} = 4.5–80 V, frequency up to 2 MHz
Renesas ISL8117A [47]	Synchronous buck	Dual phase, V_{IN} = 4.5–60 V, frequency up to 2 MHz
Richtek RT6190 [48]	Buck-boost	Bidirectional 4-switch buck-boost for USB-PD 3.0 SPR and USB-PD 3.1 EPR Modes with I2C. V_{IN} = 4.5–36 V, up to 1 MHz
Texas Instruments LM5141 [49]	Synchronous buck	Single phase, V_{IN} = 3.8–65 V, 440 kHz or 2.2 MHz
Texas Instruments LM5140-Q1 [50]	Synchronous buck	Automotive dual phase, V_{IN} = 3.8–65 V, 440 kHz or 2.2 MHz
Texas Instruments TPS40400 [51]	Synchronous buck	Single phase, V_{IN} = 3–20V, PMBus enabled, up to 2 MHz

Table 3.6 Synchronous rectifier controllers with integrated driver for GaN e-mode transistors.

Part number	Description
NXP TEA1993TS [52]	Single phase with drain sense up to 120 V
NXP TEA1995T [53]	Dual phase with drain sense up to 100 V
NXP TEA1998TS [54]	Single phase drain sense up to 60 V
On Semi NCP4305A [55]	Single phase drain sense up to 200 V and freq. up to 1 MHz
On Semi NCP4306A [56]	Single phase drain sense up to 200 V and freq. up to 1 MHz
On Semi NCP4308A [57]	Single phase drain sense up to 150 V and freq. up to 1 MHz
Texas Instruments UCD7138 [58]	Single phase with drain sense up to 45 V and freq. up to 2 MHz

Table 3.7 ICs for high reliability applications.

Part number	Description
EPC Space FBS-GAM01P-C-PSE [59]	Single-Output eGaN Gate Driver Module
EPC Space FBS-GAM02P-C-PSE [60]	50 V Radiation Hardened High-Speed Multifunction Power eGaN® HEMT Driver
EPC Space FBS-GAM02-P-R50 [61]	50 V/10 A Radiation Hardened Multifunction Power Module
Renesas ISL70040SEH [62]	Radiation Hardened Low-Side GaN FET Driver
Texas Instruments TPS7H6003-SP [63]	Radiation Hardened, QMLV 200-V half-bridge GaN gate driver

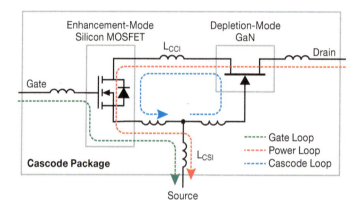

Figure 3.21 Schematic of a cascode GaN device showing parasitic inductances and the different high frequency loops between transistors.

- The use of two discrete devices means the interconnect parasitics are larger than for a single enhancement-mode device. As with most parasitics, this can be addressed through higher levels of package integration and complexity.
- The turn-on speed of the device is limited by the speed of the low-voltage silicon MOSFET and the CSI between the driver and MOSFET. One method to address both of these issues is to integrate the MOSFET and driver into a single module using a low-voltage LDMOS process [64].
- The size of the entire cascode device is at least twice the size of the enhancement-mode GaN transistor.
- The addition of a lower-voltage series MOSFET device impacts the resultant $R_{DS(on)}$ of the cascode device negatively. This impact decreases with increasing device voltage rating, as was shown in Figure 1.12. This makes the cascode structure less suitable for lower-voltage applications.

Another of the concerns with using a cascode structure is the static voltage sharing during the off-state and the dynamic voltage sharing at turn-off and turn-on. For static sharing, the depletion-mode GaN transistor and MOSFET devices must have similar I_{DS} leakage currents. If not well matched, the drain-to-source voltage drop across the MOSFET will either keep increasing or decreasing. If it keeps increasing, the low-voltage MOSFET's maximum voltage will be reached, at which point its I_{DS} will start increasing due to an avalanche breakdown until equilibrium is reached. The voltage across the depletion-mode GaN transistor gate-to-source will then be equal to the MOSFET's rated breakdown voltage. On the other hand, if the MOSFET leakage current is higher than the depletion-mode transistor, the MOSFET drain-to-source voltage will collapse to near zero, at which point the GaN device starts to turn-on, increasing its leakage current and restoring drain leakage equilibrium.

Dynamically, the total output capacitance charge ratio between the MOSFET and the depletion-mode GaN transistor should be similar to the ratio of their rated drain voltages. What complicates matters is the nonlinearity of these capacitances, coupled with the additional parasitic inductances between the devices. These factors can generate significant voltages during current rise and fall intervals. It may even be possible under certain circumstances to dynamically over-voltage the source-to-gate of the depletion-mode GaN transistor.

3.9.2 Direct-Drive Devices

Direct-drive devices are similar to cascode devices in the sense that they are a hybrid design consisting of a depletion-mode (d-mode) GaN transistor in-series with a MOSFET, each made on dissimilar technologies that require external connections. The main difference with direct drive is that the depletion-mode GaN transistor is driven directly in a manner similar to an enhancement-mode device (see Figure 3.22). The added silicon device is only used for "safety" during startup and is fully turned on during normal operation. This means that the gate drive requirements for direct drive are very similar to enhancement-mode devices, with several advantages and disadvantages compared to an enhancement-mode device and cascode devices.

Some advantages are:

- Unlike a cascode device, the direct-drive device does not add the reverse recovery losses of the silicon MOSFET and is therefore considered "zero Q_{RR}," similar to enhancement-mode devices.
- The smallest stand-alone building block includes the gate drive. This may be considered a disadvantage as well, although through integration of the gate drive into a power stage [12] the variability and layout concerns of the gate drive are internally addressed for the customer.

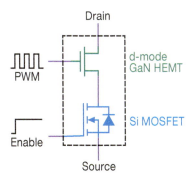

Figure 3.22 Schematic of a direct-drive devices.

- The benefit of the power-stage integration allows the implementation of a constant current drive, over-current and over-temperature protection that would otherwise be difficult to implement with discrete devices.

Some disadvantages are:

- The direct-drive approach requires two supply voltages – a negative supply for the GaN transistor and a positive supply for the safety MOSFET. This negative supply cannot be generated through a bootstrap approach but must either be internally generated [63] or requires an additional isolated DC–DC supply.
- As with the cascode device, the addition of a lower-voltage series MOSFET device negatively impacts the resultant $R_{DS(on)}$ of the direct drive, thus requiring a lower on-resistance GaN device than compared with using an equivalent enhancement-mode GaN device.
- The negative voltage drive requirement for off-state means that the turn-off gate-loop inductance is increased as it has to pass through the supply decoupling capacitor (similar to turn-on for an e-mode device). This reduces the power stage's ability to avoid Miller induced turn-on.

3.9.3 Higher Voltage Configurations

For supporting even higher voltages, it is possible to stack additional depletion-mode devices in series with the two above configurations to generate slightly different super-cascode-based structures [65, 66] shown in Figure 3.23. The basic super-cascode structure is created by adding additional series devices and designing the gate connection impedance network elements (Z in Figure 3.23) to ensure dynamic voltage sharing during switching [67]. For the direct-drive-based super-cascode, the structure is almost identical to the super-cascode, but the

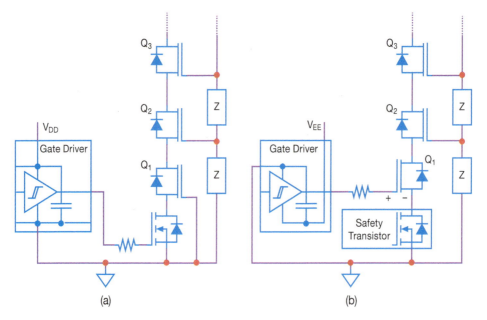

Figure 3.23 Super-cascode structures for higher voltages based on (a) cascode and (b) direct-drive circuit.

bottom GaN device is still directly driven. Regardless of how the bottom device is driven, the series devices are all cascode driven through the gate-tied impedance networks.

3.10 Using GaN Transistors with Drivers or Controllers Designed for Si MOSFETs

The ecosystem of gate drivers and controllers designed for e-mode GaN transistors has grown exponentially in recent years. These ICs enable the easy adoption of the technology by addressing the peculiarities of GaN-based converters discussed in previous sections. Nevertheless, it is also possible to use generic components designed for Si MOSFETs [68], providing that these peculiarities are considered. This section covers the compatibility of generic half-bridge gate drivers or controllers and recommends the necessary circuit changes to combine them with GaN transistors.

Before a component is considered to control GaN transistors, it must meet the following list of requirements:

- Driver-stage voltage supply and under voltage lockout (UVLO) compatibility: As discussed in Section 3.2, the voltage level of the output of the gate driver must remain within the GaN transistor datasheet rating. And the UVLO should provide enough margin to avoid false triggering of the protection while providing protection against too-low drive voltage that can result in high $R_{DS(on)}$. As an example, for a transistor with a V_{GS} rating of +6 V, the UVLO should be approximately 4 and 3.5 V for the low-side and high-side driver stages, respectively.
- Slew Rate Immunity: The IC's dv/dt immunity should exceed the highest value expected in the converter. A value above 50 kV/µs is desired to avoid compromising efficiency in exchange for slower slew rates.
- Bootstrap Power Supply: Most ICs with a half-bridge gate driver use some kind of bootstrap circuit to power the high-side stage, and only those requiring an external bootstrap diode can be adapted for use with GaN transistors. In some cases, the IC uses an LDO to regulate the voltage being supplied to the gate, this feature is preferred but it is not necessary. The requirement for an external diode and the circuit changes recommended in Figure 3.24 are meant to address the issues described in Sections 3.2 and 3.6.
- Dead-time Capability: As discussed in previous sections, GaN transistors are capable of very high switching speeds due to their small parasitic capacitances, and they lack reverse recovery. Both features combined enable switching frequencies beyond 1 MHz, while maintaining high efficiency. At the same time, they have a V_{SD} during reverse conduction, which usually happens during dead times, of 1.5–3 V. Because of this, minimizing dead times to below 10 ns should be a priority to maximize the benefits of the technology. Good propagation delay matching between high-side and low-side is an important feature to enable short dead times without increasing the risk of shoot-through.

Once a MOSFET driver or controller is deemed adequate, the circuit changes pictured in Figure 3.24, and described next, are recommended to realize a harmonious operation between the IC and the GaN transistors:

- Bootstrap Diode: A small Schottky diode with similar voltage rating as the GaN transistors, low capacitance, and small physical size is ideal. A Schottky diode avoids reverse recovery and minimizes the voltage drop between V_{DD} and the voltage across the bootstrap capacitor C_{Boot}. If the high-side driver stage includes an LDO, the resistor in the bootstrap path, R_{Boot},

68 | GaN Power Devices for Efficient Power Conversion

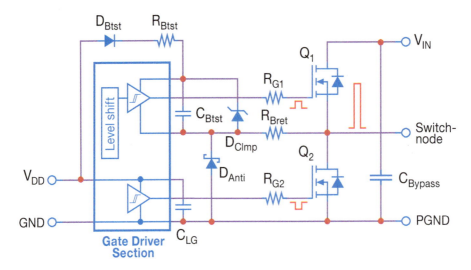

Figure 3.24 Recommended circuit for best compatibility between GaN transistors and Si MOSFET drivers/controllers.

is not required. R_{Boot} is merely a current limiting resistor, so a small value, below 10 Ω, should suffice. Adding resistance in the bootstrap path also increases the charge time of C_{Boot}, so its value should be carefully selected in those applications requiring very short on-times for the low-side transistor.

- **Bootstrap Clamp:** If the driver does not have an integrated LDO, a 5.6 V Zener diode in parallel with C_{Boot} is recommended to clamp the voltage below 6 V, especially during dead times when the low-side transistor experiences reverse conduction.
- **Gate Return Resistor:** A resistor between the switch-node and the IC protects it against large negative switch-node voltages that may be present during dead times. Since this resistor is part of the gate loop it can slow down Q_1's switching speed, unless R_{G1} is reduced to maintain the total gate-loop resistance.
- **Reverse Conduction Clamp:** When the gate return resistor described earlier is not enough to protect the IC against negative switch-node voltages, a Schottky diode in antiparallel with the low-side transistor is recommended. When present, the current during dead times will flow through it instead of the GaN transistor, limiting the voltage drop across it, hence ensuring a lower negative voltage on the switch-node.

Besides the suggestions listed in this section, the layout recommendations covered in Chapter 4 also apply. Because drivers or controllers designed for MOSFET may not allow an optimal layout with GaN transistors, it is important to understand how PCB layout affects parasitic elements. Such understanding shall be the foundation to make the design compromises needed to achieve the best performance.

3.11 Driving GaN ICs

Recently, several monolithic GaN ICs have been released for various applications [69, 70]. As discussed in Chapter 1, these ICs enable integration of logic circuits and power devices at voltages beyond those achievable with a silicon process, typically limited to 30–40 V. As a result, in power conversion applications, lower component count, higher power-density and efficiency, higher reliability, and higher levels of radiation hardness can be achieved. In addition, in light detection and ranging (lidar) applications, the switching speeds of GaN ICs enable much improved space and time resolutions compared to solutions using silicon MOSFETs [71] or even discrete GaN transistors.

In principle, these GaN ICs coexist with other components such as logic gates, microcontrollers and FPGAs, and interact with them in the same manner as their silicon counterparts. This means that GaN ICs are compatible with 5 V or 3.3 V logic levels, they can offer open collector/drain type pins, and provide high impedance for receiving PWM signals coming directly from a controller. In other words, there is no special compatibility requirement between a microcontroller and a GaN IC that did not exist for a Si-based one. In fact, in a GaN power IC with an integrated driver [70] all the concerns and considerations related to the gate driver discussed in this chapter are addressed within the IC, making adoption of GaN technology a simpler task. A detailed example using a GaN power IC [70] with a microcontroller is given in Chapter 10.

In practice, converters using GaN ICs are often exposed to high di/dt and dv/dt that can cause interference issues with other GaN-based or silicon-based components on the board. Most of these issues can be avoided with some layout guidelines. First, sensitive digital signals such as PWM signals, between a controller and a GaN IC, should minimize overlap with large switch-node planes. Failing to do so creates capacitive coupling between the plane and signal traces that can induce currents on the latter and corrupt the signal. Second, a passive first-order filter may be added to reduce sensitivity by placing a small ceramic capacitor close to the pin of the gate driver and a resistor up to $100\,\Omega$ in series with the signal. It should be noted that a filter also introduces delays in the signal that can interfere with the control loop of the converter or change the effective dead time. For this reason, it is recommended to limit the time constant of the filter to less than 1/10 the value of the dead time. Analog signals returning to a controller, especially those used as feedback in a control loop, are also particularly sensitive. To prevent undesired capacitive coupling, it is recommended to shield the signal with a plane connected to its reference voltage.

Integration of multiple components of a power stage into an IC, such as EPC23102 [70], also reduces these potential interference issues. In a GaN IC, the sensitive elements of the power stage like the gate loops or bootstrap circuit are not exposed, hence minimizing capacitive coupling with "noisy" switch-node planes. Loop inductances are also reduced which translates to smaller gate resistance without increasing ringing or a more damped gate waveforms, as described in Eq. (3.1). In addition, the pinout of the IC enables friendlier layouts by providing easier separation between power and signal connections and lower loop inductance that can translate into ground bounce.

3.12 Summary

In this chapter, considerations for operating high-speed GaN transistors and GaN ICs were addressed. These considerations include:

- Gate power loop inductance minimization: The gate driver should be designed to minimize the inductance between the V_{DD} supply capacitor and the actual gate driver power devices (sink and source devices). This will minimize gate driver rise time and maximize driver di/dt.

- Noise immunity: The gate driver design should be made with the assumption that the driver ground and the controller ground can differ significantly, and the input logic pin must be immune to noise-induced changes in logic state.
- High dv/dt immunity for high-side drivers: logic-isolators or level shifters used to transfer the control logic signal to the floating high-side device need to be immune to high dv/dt rise and fall times without changing the logic state.
- Separate control of the turn-on and turn-off: For a general purpose GaN driver, the speed of the driver needs to be matched to the size and speed of the device being driven. This flexibility requires a low-resistance gate driver with the option of additional external resistors. Furthermore, to adjust both the turn-on and turn-off independently for optimum damping, it is preferred to have separate pins for turn-off and turn-on.
- Regulation of gate drive supply voltage: For enhancement-mode transistors in particular, both low-side drivers, and especially high-side drivers, need to regulate the gate drive supply voltage to avoid an over-voltage condition on the transistor gate. A number of alternative solutions are compared in this chapter.
- High-frequency considerations: Pushing the switching time lower will require optimized driver packaging and pin-out, improved driver delays and matching, and decoupling capacitance integration, and even monolithic GaN drivers.
- Alternative gate drive approaches for alternative device structures are presented and evaluated.
- Recommendations for using generic gate drivers and controllers designed for Silicon MOSFETs with GaN transistors are discussed.
- Monolithic GaN ICs with integrated gate drivers eliminate all the concerns discussed in this chapter related to the gate driver. They also offer several performance and layout advantages that will be discussed in Chapters 4 and 10.

The next chapter will focus on layout techniques and ways to minimize the parasitic inductances that have increased importance due to the higher switching speed of GaN transistors.

References

1 Efficient Power Conversion Corporation (2023). EPC2619 – Enhancement-mode power transistor. EPC2619 datasheet, [Revision November 2023]. https://epc-co.com/epc/Portals/0/epc/documents/datasheets/EPC2619_datasheet.pdf.

2 Beach, R. (2010). Master the fundamentals of your gallium-nitride power transistors. *Electronic Design Europe*, (29 April). https://www.electronicdesign.com/technologies/power/article/21791103/master-the-fundamentals-of-your-gallium-nitride-power-transistors.

3 Baliga, B.J. (1989). Power semiconductor device figure-of-merit for high frequency applications. *IEEE Electron Device Lett.* 10: 455–457.

4 Brohlin, P.L., Ramadass, Y., and Kaya, C. Direct-drive configuration for GaN devices. White paper SLPY008A, Texas Instruments, November 2018. http://www.ti.com/lit/wp/slpy008a/slpy008a.pdf.

5 Transphorm (2023). TP65H015G5WS datasheet, March 2023. https://www.transphormusa.com/en/document/datasheet-tp65h015g5ws-650v-gan-fet/.

6 Efficient Power Conversion Corporation (2017). EPC2045 – Enhancement-mode power transistor. EPC2045 datasheet, March 2017 [Revised October 2018]. http://epc-co.com/epc/Products/eGaNFETsandICs/EPC2045.aspx.

7 GaN Systems (2018). GS66516B – Bottom-side cooled 650 V e-mode GaN transistor. GS66516B datasheet, March 2017 [Revised August 2018]. https://gansystems.com/wp-content/uploads/2018/08/GS66516B-DS-Rev-180823.pdf.

8 Panasonic. PGA26E07BA GaN power devices. PGA26E07BA datasheet, December 2016 [Revised March 2019]. https://mediap.industry.panasonic.eu/assets/imported/industrial.panasonic.com/content/data/SC/ds/ds4/PGA26E07BA_E_discon.pdf.

9 Reusch, D., Gilham, D., Su, Y., and Lee, F. (2012). Gallium nitride based 3D integrated non-isolated point of load module. *Applied Power Electronics Conference and Exposition (APEC), Twenty-Seventh Annual IEEE,* Orlando, FL (February 2012), 38–45.

10 Jones, E.A., Yang, Z., Wang, F. et al. (2017). Maximizing the voltage and current capability of GaN FETs in a hard-switching inverter. *Proceedings of IEEE International Conference on Power Electronics and Drive Systems (PEDS),* (December 2017), 740–747.

11 Wu, T. (2001). Cdv/dt induced turn-on in synchronous buck regulators. White paper, International Rectifier Corporation. https://www.google.com/url?sa=t&source=web&rct=j&opi=89978449&url=http://application-notes.digchip.com/014/14-15494.pdf&ved=2ahUKEwjgv76sh52GAxWLI0QIHQFLAmQQFnoECA4QAw&usg=AOvVaw1WJ2c59_Q0OEVaGiK3Qtju

12 Strydom, J.T. (2018). Impact of parasitics on GaN-based power conversion. In: Gallium Nitride-enabled High Frequency and High Efficiency Power Conversion (ed. G. Meneghesso, M. Meneghini, and E. Zanoni), 123–152. Cham: Springer.

13 Efficient Power Conversion Corporation (2013). EPC2016 – Enhancement-mode power transistor. EPC2016 datasheet, [Revised September 2013] https://epc-co.com/epc/Portals/0/epc/documents/datasheets/EPC2016_datasheet.pdf.

14 Reusch, D. and de Rooij, M. (2017). Evaluation of gate drive overvoltage management methods for enhancement mode gallium nitride transistors. *IEEE Applied Power Electronics Conference and Exposition (APEC),* (26–30 March 2017), 2459–2466.

15 Texas Instruments (2017). Automotive 1.2-A/5-A, 100-V half bridge gate driver for GaNFET. LM5113-Q1 datasheet [Revision B]. https://www.ti.com/product/LM5113-Q1.

16 Texas Instruments (2012). UCC27611, 4-A/6-A single-channel gate driver with 4-V UVLO and 5-V regulated output. UCC27611 datasheet [Revised March 2018]. https://www.ti.com/lit/gpn/ucc27611.

17 Strydom, J. and Reusch, D. (2014). Design and evaluation of a 10 MHz gallium nitride based 42 V DC-DC converter. *Applied Power Electronics Conference and Exposition (APEC),* (16–20 March 2014), 1510–1516.

18 Mehta, N. (2014). Design considerations for LMG1205 advanced GaN FET driver during high-frequency operation. Application report SNVA723a, [Revised May 2018]. http://www.ti.com/lit/an/snva723a/snva723a.pdf.

19 Fairchild Semiconductor (1989). Understanding and minimizing ground bounce. Application note [Revised February 2003]. AN-640. https://www.onsemi.com/download/application-notes/pdf/an-640.pdf.

20 Analog Devices (2017). ADuM4120/1 – isolated, precision gate driver with 2 A output. ADuM4120/1 datasheet and product information. https://www.analog.com/media/en/technical-documentation/data-sheets/ADuM4120-4120-1.pdf.

21 Infineon (2022). 1EDBx275F – Single-channel isolated gate driver | EiceDRIVER™ 1EDB family. 1EDBx275F datasheet and product information. https://www.infineon.com/cms/en/product/power/gate-driver-ics/single-channel-isolated-gate-driver-eicedriver-1edb-family/.

22 Littelfuse (2017). IXD_604 – Low-side drivers series. IXD_604 datasheet and product information [Revised March 2018]. https://www.littelfuse.com/products/integrated-circuits/gate-driver-ics/igbt-and-mosfet-driver-ics/low-side-gate-driver-ics/ixd_604.aspx.

23 Texas Instruments (2012). UCC27611 – 7-A/5-A single channel gate driver with 5-V UVLO for nano second input pulses. UCC27611 datasheet and product information [Revised March 2018]. https://www.ti.com/product/UCC27611.

24 Texas Instruments (2018). LMG1020 – 4-A/6-A single-channel gate driver with 4-V UVLO and 5-V regulated output. LMG1020 datasheet and product information [Revised October 2018]. https://www.ti.com/product/LMG1020.

25 Texas Instruments (2019). LMG1025-Q1 – 4 Automotive 7-A/5-A single-channel low-side gate driver with 5-V UVLO for narrow pulse applications. LMG1025-Q1 datasheet and product information [Revised January 2020]. https://www.ti.com/product/LMG1025-Q1.

26 Analog Devices (2020). ADuM4221A – Isolated, half bridge gate driver with adjustable dead time, dual input, 4 A output. ADuM4221A datasheet and product information [Rev. B]. https://www.analog.com/en/products/adum4221.html.

27 Analog Devices (2023). LT8418 – 100V Half-bridge GaN driver with smart integrated bootstrap switch. LT8418 datasheet and product information [Rev. 0]. https://www.analog.com/en/products/lt8418.html.

28 Infineon (2023). 2EDB7259Y – Fast, robust, dual-channel, functional isolated MOSFET gate drivers with accurate and stable timing. 2EDB7259Y datasheet and product information [Rev. 1.3]. https://www.infineon.com/cms/en/product/power/gate-driver-ics/2edb7259y/.

29 Infineon (2023). 2EDR7259X – Fast, robust, dual-channel, functional isolated MOSFET gate drivers with accurate and stable timing. 2EDR7259X datasheet and product information [Rev. 1.4]. https://www.infineon.com/cms/en/product/power/gate-driver-ics/2edr7259x/.

30 Infineon (2023). EiceDRIVER™ 1EDN71x6x – EiceDRIVER™ 200 V high-side TDI gate driver for GaN SG HEMTs and MOSFETs. 1EDN71x6x datasheet and product information. https://www.infineon.com/cms/en/product/power/gate-driver-ics/eicedriver-1edn71x6g/#!documents

31 Monolithic Power Systems (2023). MPQ1918-AEC1 – 100V, High-frequency, half-bridge GaN/MOSFET driver, AEC-Q100 qualified. MPQ1918-AEC1 datasheet and product information [Rev. 1.0]. https://www.monolithicpower.com/en/mpq1918.html.

32 Monolithic Power Systems (2024). MP8699B – Half-bridge GaN/MOSFET driver. MP8699B datasheet and product information [Pre-release]. https://www.monolithicpower.com/en/mp8699b.html.

33 Onsemi (2022). NCP51810 – High performance, 150 V half bridge gate driver for GaN power switches. NCP51810 datasheet and product information [Rev. 3]. https://www.onsemi.com/products/power-management/gate-drivers/NCP51810.

34 Onsemi (2022). NCP51820 – High Performance, 650 V half bridge gate driver for GaN power switches. NCP51820 datasheet and product information [Rev. 6]. https://www.onsemi.com/products/power-management/gate-drivers/NCP51820.

35 Skyworks. Si827x – Isolated gate drivers. Si827x product list [Accessed 20 May 2023]. https://www.skyworksinc.com/en/Products/Isolation/Si827x-Isolated-Gate-Drivers.

36 ST Microelectronics (2021). STDRIVEG600 – High voltage half-bridge gate driver for GaN transistors. STDRIVEG600 datasheet and product information [Rev. 1]. https://www.st.com/en/power-management/stdriveg600.html.

37 Texas Instruments (2017). LMG1205 – 1.2-A, 5-A 90-V, half bridge gate driver with 5-V UVLO for GaNFET and MOSFET. LMG1205 datasheet and product information [Revised April 2023]. https://www.ti.com/product/LMG1205.

38 Texas Instruments (2018). LMG1210 – 1.5-A, 3-A, 200-V half bridge gate driver, 5-V UVLO and programmable dead-time for GaNFET and MOSFET. LMG1210 datasheet and product information [Revised January 2019]. https://www.ti.com/product/LMG1210.

39 uPI SEMI (2021). uP1966E – Dual-channel gate driver for enhancement mode GaN transistors. uP1966E datasheet and product information. https://www.upi-semi.com/en-article-upi-718-2279.

40 Analog Devices (2023). LTC7890 – Low IQ, dual, 2-phase synchronous step-down controller for GaN FETs. LTC7890 datasheet and product information [Rev. 0]. https://www.analog.com/en/products/ltc7890.html.

41 Analog Devices (2022). LTC7891 – 100 V, low IQ, synchronous step-down controller for GaN FETs. LTC7891datasheet and product information [Rev. 0]. https://www.analog.com/en/products/ltc7891.html.

42 Analog Devices (2017). LTC7800 – Low IQ, 60V, high frequency synchronous step-down controller. LTC7800 datasheet and product information. https://www.analog.com/en/products/ltc7800.html.

43 Analog Devices (2022). LT8390A – 60V 2MHz Synchronous 4-switch buck-boost controller with spread spectrum. LT8390A datasheet and product information [Rev. C]. https://www.analog.com/en/products/lt8390a.html.

44 Microchip (2022). MIC2132 – 75V, Dual phase advanced COT controller for multiphase operation. MIC2132 datasheet and product information. https://www.microchip.com/en-us/product/mic2132.

45 Renesas (2022). ISL81806 – 80V Dual synchronous buck controller optimized to drive e-mode GaN FET. ISL81806 datasheet and product information. https://www.renesas.com/us/en/products/power-power-management/dc-dc-converters/step-down-buck/buck-controllers-external-fets/isl81806-80v-dual-synchronous-buck-controller-optimized-drive-e-mode-gan-fet.

46 Renesas (2022). ISL81807 – 80V Dual or 2-phase synchronous boost controller optimized to drive e-mode GaN FET. ISL81807 datasheet and product information. https://www.renesas.com/us/en/products/power-power-management/dc-dc-converters/step-boost/boost-controllers-external-fets/isl81807-80v-dual-or-2-phase-synchronous-boost-controller-optimized-drive-e-mode-gan-fet.

47 Renesas (2024). ISL8117A – Synchronous step-down PWM controllers. ISL8117A datasheet and product information. https://www.renesas.com/us/en/products/power-power-management/dc-dc-converters/step-down-buck/buck-controllers-external-fets/isl8117a-synchronous-step-down-pwm-controllers.

48 Richtek (2022). RT6190 – 36V, 4-Switch bidirectional buck-boost controller with I2C interface. RT6190 datasheet and product information. https://www.richtek.com/Products/Switching%20Regulators/buck-boost-controller/RT6190?sc_lang=en.

49 Texas Instruments (2017). LM5141 – Low Iq, wide input range synchronous buck controller. LM5141 datasheet and product information. https://www.ti.com/product/LM5141.

50 Texas Instruments (2016). LM5140-Q1 – wide Vin dual 2.2MHz low Iq synchronous buck controller. LM5140-Q1 datasheet and product information [Revised December 2016]. https://www.ti.com/product/LM5140-Q1.

51 Texas Instruments (2011). TPS40400 – 3-V to 20-V, 30-A, synchronous buck controller with PMBus™, including Telemetry. TPS40400 datasheet and product information [Revised November 2016]. https://www.ti.com/product/TPS40400.

52 NXP (2016). TEA1993TS – Green Chip synchronous rectifier controller. TEA1993TS datasheet and product information [Rev. 1.1]. https://www.nxp.com/products/power-management/ac-dc-solutions/secondary-side-controllers/greenchip-synchronous-rectifier-controller:TEA1993TS.

53 NXP (2017). TEA1995T – GreenChip dual Synchronous Rectifier controller. TEA1995T datasheet and product information [Rev. 3]. https://www.nxp.com/part/TEA1995T#/.

54 NXP (2017). TEA1998TS – GreenChip synchronous rectifier controller. TEA1998TS datasheet and product information [Rev. 1]. https://www.nxp.com/part/TEA1998TS#/.

55 Onsemi (2016). NCP4305 – Secondary side synchronous rectification driver for high efficiency SMPS topologies. NCP4305 datasheet [Rev. 3]. https://www.onsemi.com/download/data-sheet/pdf/ncp4305-d.pdf.

56 Onsemi (2022). NCP4306 – Secondary Side Synchronous Rectification Driver for High Efficiency SMPS Topologies. NCP4306 datasheet [Rev. 8]. https://www.onsemi.com/download/data-sheet/pdf/ncp4306-d.pdf.

57 Onsemi (2017). NCP4308 – Synchronous Rectifier Controller. NCP4308 datasheet [Rev. 1]. https://www.onsemi.com/download/data-sheet/pdf/ncp4308-d.pdf.

58 Texas Instruments (2015). UCD7138 – Low-side power MOSFET driver with body diode conduction sensing. UCD7138 datasheet and product information [Revised May 2015]. https://www.ti.com/product/UCD7138.

59 EPC Space (2023). FBS-GAM01P-R-PSE – Radiation-hardened single output eGaN® gate driver module. FBS-GAM01P-R-PSE datasheet [Revised February 2023]. https://epc.space/documents/datasheets/FBSGAM01PCPSE-datasheet.pdf.

60 EPC Space (2023). FBS-GAM02P-C-PSE – 50 V_{DC} High-speed multifunction power eGaN® HEMT driver. FBS-GAM02P-C-PSE datasheet [Revised February 2023]. https://epc.space/documents/datasheets/FBSGAM02PCPSE-datasheet.pdf.

61 EPC Space (2023). FBS-GAM02-P-R50 – 50 V_{DC}/10 A radiation-hardened multifunction power module. FBS-GAM02-P-R50 datasheet [Revised February 2023]. https://epc.space/documents/datasheets/FBSGAM02PR50-datasheet.pdf.

62 Renesas (previously Intersil) (2021). ISL70040SEH, ISL73040SEH – Radiation hardened low-side GaN FET driver. ISL70040SEH, ISL73040SEH datasheet [Rev. 10.0]. https://www.renesas.com/us/en/document/dst/isl70040seh-isl73040seh-datasheet

63 Texas Instruments (2023). TPS7H6003-SP – Radiation-hardened, QMLV 200-V half-bridge GaN gate driver. TPS7H6003-SP datasheet and product information [Revised April 2024]. https://www.ti.com/product/TPS7H6003-SP

64 Patterson, G (2015). GaN switching for efficient converters. *Power Electron. Europe* 5: 18–21. www.power-mag.com/pdf/issuearchive/63.pdf.

65 Elpelt, R., Friedrichs, P., Schorner, R. et al. (2004). Serial connection of SiC VJFETs – features of a fast high voltage switch. *REE – Revue de l'Electricite et de l'Electronique* 02: 60–68.

66 Apter, S., Shapiro, D., Verpinsky, V., et al.(2018). Industry's first 1200 V half bridge module based on GaN technology. *Industry Presentations, IS01 at Applied Power Electronics Conference and Exposition (APEC).*

67 Biela, J., Aggeler, D., Bortis, D., and Kolar, J.W. (2008). 5kv/200ns pulsed power switch based on a SiC-JFET super cascode. *IEEE International Power Modulators and High Voltage Conference,* (2008), 358–361.

68 Pozo, A. (2024). Using GaN FETs with controllers and gate drivers designed for silicon MOSFETs. *Bodo's Power Systems*, (February 2024), 32–34.

69 Efficient Power Conversion Corporation (2024). EPC23102 – ePower™ stage IC. EPC23102 datasheet, March 2024. https://epc-co.com/epc/Portals/0/epc/documents/datasheets/EPC23102_datasheet.pdf.

70 Efficient Power Conversion Corporation (2022). EPC23101 – ePower™ chipset. EPC23101 datasheet, December 2022. https://epc-co.com/epc/Portals/0/epc/documents/datasheets/EPC23101_datasheet.pdf.

71 Glaser, J. (2017). How GaN power transistors drive high-performance lidar: generating ultrafast pulsed power with GaN FETs. *IEEE Power Electron. Mag.*, 4, (1): 25–35. https://doi.org/10.1109/MPEL.2016.2643099.

4

Layout Considerations for GaN Transistor Circuits

4.1 Introduction

The capability of GaN devices to switch faster than silicon MOSFETs magnifies the impact of parasitic elements on performance. Moreover, as GaN technology matures and devices improve by increasing their specific on-resistance, parasitic elements will become even more relevant. These parasitic elements consist of resistive, inductive, and capacitive components, and are a result of the circuit layout and printed circuit board (PCB) on which the devices are mounted. Given the fast-switching speeds of GaN transistors, minimizing parasitic inductance is critical to realize the highest benefits of GaN. This chapter will discuss and quantify their importance on the switching transients and will provide layout techniques to reduce them.

For a half-bridge configuration, which is the basic building block of most power converters, there are two main power loops to consider: (i) the high-frequency power loop formed by the two power switching devices along with the high-frequency bus capacitor and (ii) the gate drive loop formed by the gate driver, power device, and high-frequency gate drive capacitor. The common-source inductance (CSI) is defined by the part of the circuit that is common to both the gate loop and power loop. These loops and inductances, numbered by order of importance, are shown in Figure 4.1.

The importance of CSI was briefly discussed in Chapter 3, Section 3.5. In a hard-switching transition, when the current rises during turn-on, a voltage appears across the CSI. This creates a negative voltage feedback in the gate loop that effectively slows down the charging process of the transistor's input capacitance, as shown in Figure 3.10. The result is an increase in the switching time and thus a significant increase in the switching losses, as shown in the graph in Figure 4.2. The impact of the power-loop inductance on performance is not as significant, but it increases the voltage overshoot, which also causes losses, generates EMI, and exposes the transistors and gate driver to harmful voltage spikes.

Gate-loop inductance was also discussed in Chapter 3, Section 3.3. Inductance in the gate loop creates a resonant tank with the device's input capacitance that results in ringing during turn-on and turn-off transients. Equation (3.1) can be used to calculate the external gate resistance required to provide a damped response.

GaN Power Devices for Efficient Power Conversion, Fourth Edition. Alex Lidow, Michael de Rooij, John Glaser, Alejandro Pozo, Shengke Zhang, Marco Palma, David Reusch, and Johan Strydom.
© 2025 John Wiley & Sons Ltd. Published 2025 by John Wiley & Sons Ltd.

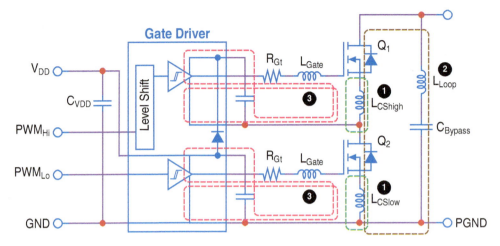

Figure 4.1 Schematic of a half-bridge power stage showing power (2) and gate drive (3) loops with common-source inductance (1).

4.2 Origin of Parasitic Inductance

While minimizing the inductance of the individual elements that make up the loop (i.e. capacitor ESL, device lead inductance, and PCB interconnect inductance) is important, the designer must also focus on minimizing the total loop inductance. As the inductance of the loop is determined by the magnetic energy that is stored within, it is possible to further minimize the overall loop inductance by using the coupling between adjacent conductors to induce magnetic field self-cancellation. For such a coupled structure, consider the theoretical cases shown in Figure 4.3. The inductance from A to B in this loop is given by

$$L_{A-B} = \frac{\mu_R \cdot \mu_0 \cdot (h \cdot l)}{w} \tag{4.1}$$

where μ_0 is the permeability of free air and μ_R is the relative permeability of the PCB material. For the lateral loop case, Eq. (4.1) shows that the inductance is proportional to the area encompassing the path loop ($h \cdot l$ = length × thickness) and is inversely proportional to the thickness (width) of the conductors (w). Likewise, for the vertical loop case, Eq. (4.1) shows that the inductance is proportional to the cross-sectional area of the path loop ($h \cdot l$ = length × thickness) and is inversely proportional to the width of the conductors (w). Basic geometry and Eq. (4.1) favor the vertical loop to achieve the lowest inductance given that the width factor in the lateral loop will always be very small.

To form a low inductance loop, the current must flow in opposite directions in the two adjacent layers of the vertical loop, which causes magnetic field self-cancellation and reduces the inductance. The loop inductance will increase linearly with an increase in conductor spacing (h = thickness). It is therefore recommended to make all high-frequency loops as short and wide as possible, with the two opposing current paths as close as possible.

The following sections will use this concept to propose and quantify different layouts with the goal of reducing CSI and power-loop inductances.

Layout Considerations for GaN Transistor Circuits | 77

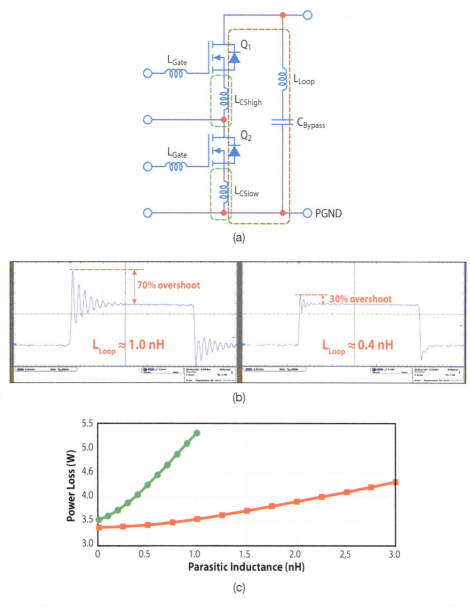

Figure 4.2 Experimental results showing the impact of parasitic inductances on the converter: (a) schematic with location of parasitic inductances, (b) impact of power-loop inductance on voltage overshoot, and (c) power loss dependence on power-loop inductance and CSI.

4.3 Minimizing Common-Source Inductance

As shown in the schematic in Figure 4.1, CSI is the result of a common path for the currents flowing in both the gate and power loops. In a real converter, this common path presents in the form of copper planes, traces, vias, or even elements of the devices' package itself. Because both loops connect to the source of the device, it is impossible to fully eliminate CSI. Nevertheless,

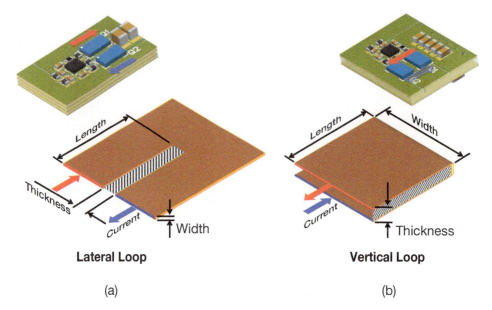

Figure 4.3 PCB parasitic inductance in a lateral loop (a) and a vertical loop (b).

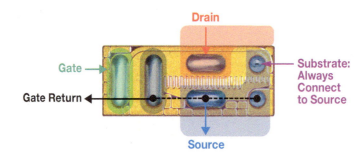

Figure 4.4 Example layout for a GaN transistor in CSP with a dedicated gate-return pin.

some transistors are packaged in such a way that the shared paths for gate and power-loop currents are kept to minimal by using a dedicated gate-return pin [1, 2].

For this strategy to be effective, the gate-return pin should not be connected to the source pads of the device externally, as shown in Figure 4.4. If a dedicated gate-return pin is not available, maintaining orthogonal gate and power-loop paths is a simple solution to minimize CSI for both high-side and low-side FETs in a half-bridge configuration. Figure 4.5 shows how to apply this principle for a packaged PQFN (Power Quad Flat No lead) and a CSP (Chip-Scale Package) GaN transistor.

Many GaN FETs include a substrate connection [2] and when present that substrate connection must always be connected to the source as shown in Figures 4.4 and 4.5.

Layout Considerations for GaN Transistor Circuits | 79

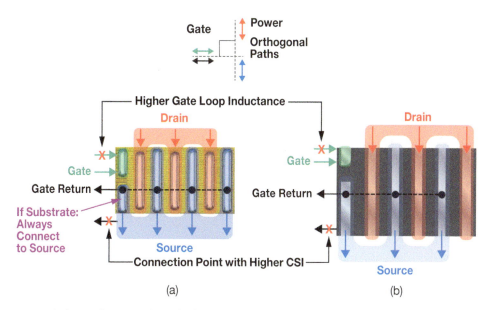

Figure 4.5 Orthogonality principle applied to minimize common-source inductance for a typical GaN transistor in: (a) CSP and (b) PQFN.

4.4 Minimizing Power-Loop Inductance in a Half-Bridge Configuration

The concepts presented in Sections 4.2 and 4.3 can be used to arrange two GaN transistors and their gate driver in a half-bridge configuration. The component placement between top and bottom layers and the current flow within the different layers of the PCB allow different implementations: lateral power loop, external-vertical power loop, and internal-vertical power loop. The advantages and disadvantages of each of these solutions are discussed next.

4.4.1 Lateral Power-Loop Design

The lateral layout places the input capacitors and devices on the same side of the PCB in close proximity to minimize the area of the high-frequency power loop. The high-frequency loop for this design is contained on the same side of the PCB and is considered a lateral power loop, as a result of the power loop flowing laterally on a single PCB layer. An example of the lateral layout using an LGA transistor design is shown in Figure 4.6 with the high-frequency loop highlighted.

Figure 4.6 Conventional lateral power loop for LGA GaN transistor-based converter: (a) top view and (b) side view.

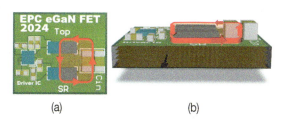

While minimizing the physical size of the loop is important to reduce parasitic inductance, the design of the inner layers is also critical. For the lateral power-loop design, the first inner layer serves as a "shield layer." This layer has a critical role in shielding the internal circuits from the fields generated by the high-frequency power loop. The power loop generates a magnetic field that induces a current in the shield layer that flows in the opposite direction to the power loop. The current in the shield layer generates a magnetic field to counteract the original power-loop's magnetic field. The result is a cancellation of magnetic fields that translates into a reduction in parasitic power-loop inductance.

Having a complete shield plane in close proximity to the power loop yields the lowest power-loop inductance for this approach. For the lateral power-loop design, the high-frequency loop inductance shows little dependence on overall board thickness because the power loop is completely contained on the top layer. The lateral design, however, is strongly dependent on the distance from the power loop to the shield layer contained on the first inner layer [3].

4.4.2 External-Vertical Power-Loop Design

The second conventional layout, shown in Figure 4.7, places the input capacitors and transistors on opposite sides of the PCB, with the capacitors located directly beneath the devices to minimize the physical loop size. This is called a vertical power loop because the loop is connected vertically through the PCB using vias. The LGA transistor design of Figure 4.7 has the vertical power loop highlighted.

For this design, there is no shield layer due to the vertical structure. The vertical power loop uses a magnetic field self-cancellation method (with currents flowing in opposite directions) to reduce inductance, as opposed to the use of a shield plane. For the PCB layout, the overall board thickness is generally much thinner than the horizontal length of the traces on the top and bottom side of the board. As the board thickness decreases, the area of the loop shrinks significantly when compared to the lateral power loop, and the current flowing in opposing

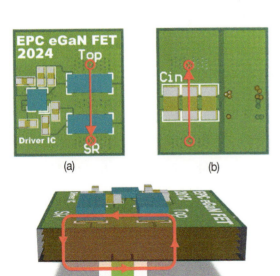

Figure 4.7 Conventional external-vertical power loop for LGA transistor-based converter: (a) top view, (b) bottom view, and (c) side view.

directions on the top and bottom layers begins to provide magnetic field self-cancellation. For a vertical power loop to be most effective, the board thickness must be minimized.

4.4.3 Internal-Vertical Power-Loop Design

An improved layout technique that provides the benefits of reduced loop size, has magnetic field self-cancellation, has inductance that is independent of board thickness, is a single-sided component PCB design, and yields high efficiency for a multilayer structure is shown in Figure 4.8. The design utilizes the first inner layer, shown in Figure 4.8b, as the power-loop return path. This return path is located directly beneath the top layer's power loop, shown in Figure 4.8a, allowing for the smallest physical loop size combined with magnetic field self-cancellation. The side view (Figure 4.8c) illustrates the concept of creating a low-profile magnetic field self-canceling loop in a multilayer PCB structure.

This improved layout places the input capacitors in close proximity to the top device, with the positive input voltage terminals located next to the drain connections of the top transistor. The GaN devices are arranged as in the lateral and vertical power-loop cases. The interleaved inductor node and ground vias are duplicated on the bottom side of the synchronous rectifier transistor.

These interleaved vias provide three advantages:

- The interleaving of the vias with current flowing in an opposing direction reduces magnetic energy storage and helps generate magnetic field cancellation. This results in reduced eddy and proximity effects, reducing AC conduction losses.
- The vias located in between the two transistors provide a shorter high-frequency loop inductance path, leading to lower parasitic inductance.
- The vias located beneath the lower transistor reduce resistance and accompanying conduction losses during the transistor freewheeling period.

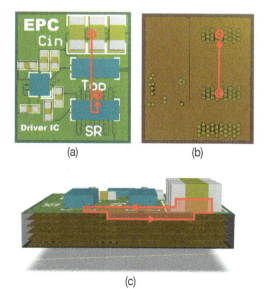

Figure 4.8 Conventional internal-vertical power loop for LGA transistor-based converter: (a) top view, (b) bottom view, and (c) side view.

Table 4.1 Characteristics of conventional and optimal power-loop designs.

	Lateral loop	External-vertical loop	Internal-vertical loop
Single-sided PCB capability	Yes	No	Yes
Magnetic field self-cancellation	No	Yes	Yes
Inductance independent of board thickness	Yes	No	Yes
Shield layer required	Yes	No	No

4.4.4 Comparing Power-Loop Designs

The main characteristics of the three power-loop designs described earlier are compared in Table 4.1.

For a complete comparison of these designs, three boards were designed and tested [3] to quantify the loop inductance. From the results shown in Figure 4.9, it can be concluded that the internal-vertical design presents the best solution to minimize power-loop inductance.

4.4.5 Combining Gate-Loop and Power-Loop in an Internal-Vertical Design

Figure 4.10 shows a layout example combining an internal-vertical power-loop design with the gate loop while minimizing CSI. As described earlier in Section 4.4.3, all components in the power-loop (FETs, gate driver, and bypass capacitors) are placed on the top layer. On this layer, a copper plane connects the capacitors to the drain of the high-side FET (red plane). Similarly, another plane (green plane) connects the source of the high-side FET to the drain of the low-side FET. At the top and bottom extremes of the top layer, ground planes (in gray) connect to the ground terminal of the capacitors and the source of the low-side FET. The loop is then closed using a ground plane in Inner Layer 1 and vias for connections between layers.

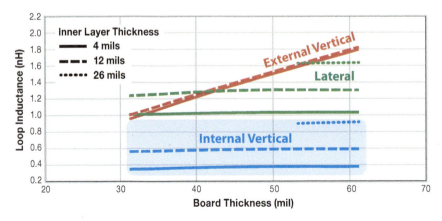

Figure 4.9 Impact of layout technique on power-loop inductance for various board and inner layer thicknesses.

Layout Considerations for GaN Transistor Circuits | 83

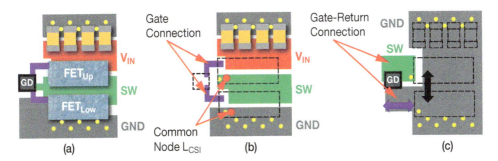

Figure 4.10 Layout example combining power-loop and gate loops, (a) overview, (b) top layer, (c) inner layer 1.

In addition, the top layer is used for the routing between the gate driver and the FETs' gates (purple traces). The gate returns for the high-side and low-side FETs should be kept orthogonal to the power loop to minimize CSI. As shown in Figure 4.10, both layers can be used for this purpose. In fact, the same gate-return planes on Inner Layer 1 can be used for shielding purposes. Shielding for the high-side gate is achieved when a switch-node plane is right below the high-side gate-loop components and traces. Similarly, shielding for the low-side gate is achieved with a ground plane underneath the low-side gate circuit.

4.4.6 Variations of Internal-Vertical Power-Loop Design

There are several variations of the internal-vertical design [4] as shown in Figure 4.11. These variations change the location of the bypass capacitors relative to the FETs, which in turn changes the node used to close the power loop in Inner Layer 1. More precisely, Figure 4.10 shows the capacitors next to the high-side FET, and a ground plane in Inner Layer 1. Instead, the first variation shown in Figure 4.11a places the capacitors next to the low-side FET and uses a V_{IN} plane in the inner layer to close the loop. The second variation, pictured in

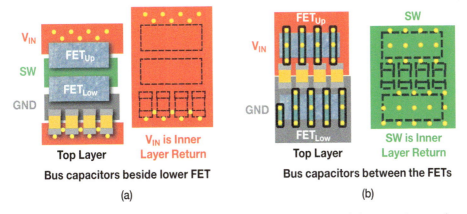

Figure 4.11 Layout variations of an internal-vertical power-loop design (a) with bus capacitors are located next to the low-side transistor, and (b) with the bus capacitors located between the transistors.

Figure 4.11b, uses the capacitors between the FETs, and a SW plane in the inner layer to close the loop.

In these three configurations, the parasitic power-loop inductance is the same because the power-loop components and distances remain essentially unchanged. However, between the three variations, the designer can choose which alternative adapts better to different gate driver pinouts, or multiphase designs.

Thermal performance is another aspect impacted by the location of the capacitors because the capacitors act as a barrier for thermal heat flux paths, so in each variation the FET further away from the capacitors can potentially have lower thermal resistance to heat dissipation areas [5]. Therefore, the case shown in Figure 4.11b offers a symmetrical solution that may be most suitable for bidirectional converters. In addition, because this solution has the switch-node plane buried in the inner layer and shielded by DC voltages (GND and V_{IN}), it enables lower radiated e-field emissions.

4.4.7 Impact of Integration on Parasitics

To further reduce the parasitic inductance of GaN transistor-based designs, various monolithic solutions have been introduced [6–8]. Figure 4.12 shows two examples of CSP GaN devices with different levels of monolithic integration. The first case, pictured in Figure 4.12a, combines high-side and low-side power transistors in a half bridge. The second case, presented in Figure 4.12b, adds the half-bridge gate driver to the power transistors, including the input logic circuits, level shifter for the high-side, bootstrap circuit, and output driver stages.

Monolithic integration offers several advantages. The first is a thermal advantage given that the two power FETs are thermally interconnected with a very low thermal resistance. This serves to reduce the effective thermal resistance of both FETs. In operation, the device dissipating more power will result in a lower temperature rise, and the lower power dissipation device with have a higher temperature. The net effect is that the higher power dissipation device R_{DSon} will reduce given the lower operating temperature and an overall net benefit to the system.

The second benefit of monolithic integration is the significant reduction in the power-loop inductance given that the two power FET simply cannot be placed any closer. As shown in

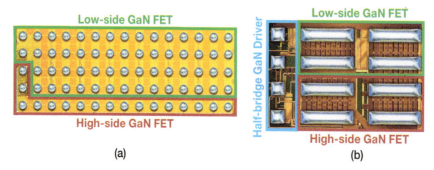

Figure 4.12 Examples of GaN ICs combining multiple devices in a single chip: (a) solder-bump-view of EPC2100, a monolithic half bridge. *Source:* Ref. [6]/Efficient Power Conversion Corporation. (b) Solder-bump-view of EPC2152, a monolithic power stage. *Source:* Ref. [7]/Efficient Power Conversion Corporation.

Figure 4.2 this lower inductance results in reduced voltage overshoot ringing and lower power-loop losses.

The third benefit of monolithic integration stems from including the gate driver [7, 8]. The monolithic integration of the gate driver virtually eliminates CSI, thus essentially eliminating the most detrimental loss mechanism. In addition, the gate driver can be optimally matched to the FETs. In discrete solutions the choice of gate drivers is limited, so very small devices may get paired with a gate driver with too much drive strength requiring higher gate resistances to reduce the switching speed, and likewise for very large FETs where the drive strength may be insufficient to adequately drive with minimum losses. The monolithically integrated gate driver can be well matched to the size of power FETs resulting in optimal converter performance with lowest losses and with minimal design effort.

Figure 4.13 shows a top view of the same GaN monolithic half bridge shown in Figure 4.12a, with its recommended layout. Gate$_1$ is the high-side gate pin. GR$_1$ is the high-side gate-return pin. Gate$_2$ is the low-side gate pin. V_{SW} is the switch-node of the half bridge and consists of 35 solder bumps. V_{IN} is the input voltage supplied to the drain of the upper transistor (Q_1) and consists of eight solder bumps. P_{GND} is the power ground connection at the source terminal of the lower transistor (Q_2) and has 29 solder bumps. The experimentally measured high-frequency loop inductance using the monolithic half-bridge GaN IC, and with filled vias beneath the device on the circuit layout, was measured to be approximately 150 pH, 40% lower than a discrete-based design at 250 pH [9].

A higher level of integration is realized in a monolithic GaN power stage [7, 8], with a half-bridge gate driver designed into the same GaN chip as the high-side and low-side power FETs. As a result, component count is reduced, and layout is greatly simplified while improving power density. Figure 4.14 provides a layout example of a GaN power stage packaged in a PQFN using an internal-vertical power-loop design. The main difference with the example in Figure 4.13 is that the gate signals are not exposed and all the V_{IN}, V_{SW}, and P_{GND} solder bumps are grouped into three large pads, V_{bus}, SWN, and GND, thus simplifying the layout. More detailed examples showing the monolithically integrated device are given in Chapters 10, 14, and 15.

4.5 Paralleling GaN Transistors

The layout considerations presented above assume that a single GaN device is used per switching element. For higher power applications, it may be necessary to place multiple transistors in parallel and have them behave as a single device. Alternatively, multiple devices for a switching element may be required in more complex structures such as a half bridge.

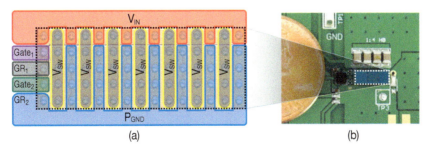

Figure 4.13 Top view pin-out configuration for (a) the asymmetric monolithic half-bridge device and (b) the printed circuit board with device mounted along with capacitors. *Source:* With permission of EPC.

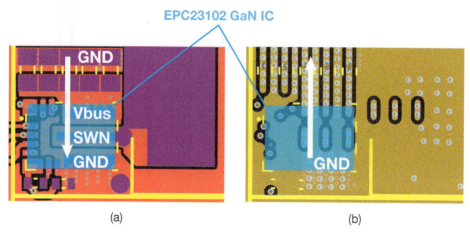

Figure 4.14 Example layout of a monolithic GaN power stage using an internal-vertical power-loop design: (a) top layer and (b) first inner layer.

4.5.1 Paralleling of GaN Transistors – Single Switch Approach

In Figure 4.15, an example is shown with three parallel devices with their interconnections indicated as series resistor and inductor elements. To achieve the best possible overall resistance, matching the drains and sources of each of the devices is connected in a diagonal symmetry such that any additional resistance mismatch in the drain paths is mirrored and compensated for in the source paths.

This configuration is typically used in hot-swap and similar slow-switching applications where DC and low-frequency current sharing is critical. The reason that this implementation cannot be used for high-frequency switching converters becomes clear when considering where to place the gate driver return connection. Choosing a geometrically symmetrical point, for example "E" in Figure 4.15, results in a significant mismatch in CSI in each loop (green ovals and red oval, respectively). At high speeds, CSI is the most important parasitic element, so any high-speed layout must solve the CSI symmetry issue.

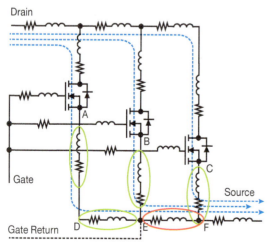

Figure 4.15 Schematic diagram showing three parallel devices with interconnect parasitics. The three drain current paths (blue dashed lines) are designed for matched drain-to-source impedance. Positive common-source inductance is circled in green ovals and negative CSI is circled in red oval.

Layout Considerations for GaN Transistor Circuits | 87

Figure 4.16 Conceptual schematic diagram showing three parallel devices with interconnect parasitics designed for complete symmetry. Common-source inductance is circled in green ovals and drain current paths are shown in blue dashed lines.

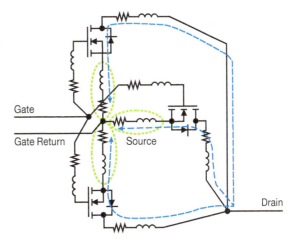

The requirement for the symmetry needed to parallel GaN devices efficiently is illustrated conceptually in Figure 4.16. The symmetry of the power loops, the CSI components, as well as the gate-loop inductances is key to effectively paralleling GaN devices. Even with improved symmetry, when using multiple devices, overall inductances will be higher than for a single device as the gate-return connection point will be pushed further away.

A layout with complete symmetry can be challenging. When considering the priority of these different parasitic components, the need for symmetry in CSI is foremost. Second would be the power loop. Last is the gate drive-loop inductance, as the gate drive speed and power always will be lower than that of the switching device itself. Following the above requirements for complete symmetry and applying the magnetic field self-canceling approach through interleaved vias and opposing currents in neighboring PCB layers, it is possible to generate an adequate solution. One such solution using LGA GaN transistors is shown in Figure 4.17.

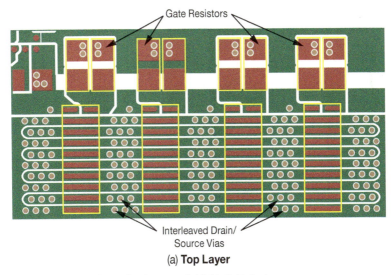

(a) **Top Layer**

Figure 4.17 Layout recommendation for four parallel LGA GaN devices.

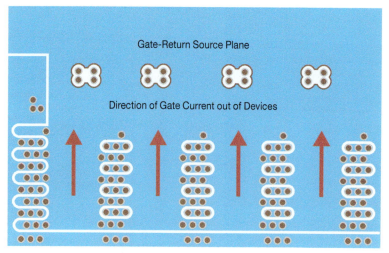

(b) **Second Layer (First Inner Layer)**

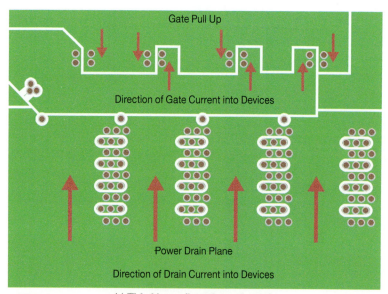

(c) **Third layer (Last Inner Layer)**

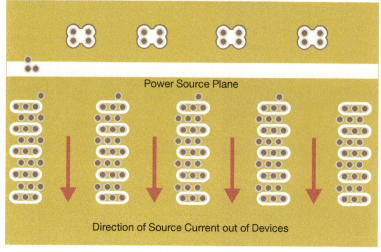

(d) **Bottom Layer**

Figure 4.17 (Continued)

One advantage that GaN transistors have over MOSFETs when placing them in parallel is a positive temperature coefficient in the transfer characteristic over the whole gate voltage range. This means that even during transition, the saturation or triode region of GaN transistors will share current to some extent due to the negative temperature feedback effect. Thus, if one device carries too much current it will heat up relative to the other devices, its on-resistance and plateau voltage will increase, and thus the current in the device reduces accordingly. In a MOSFET during parallel switching at lower currents, the reduction of threshold voltage with increasing temperature can cause the current to increase, and therefore an imbalance between devices can occur. At higher currents, the decrease in transconductance dominates the increase in plateau voltage.

The key elements that should be integrated in a design with two to four GaN transistors connected in parallel to generate a single-switching element can be summarized as follows:

- On the top layer, the devices are placed in a row, with four being the maximum number one would place in parallel before mirroring this design around another axis.
- All the devices have their drain and source terminals extending on both sides of the device to minimize drain and source inductance. These traces then are connected immediately down to all the subsequent PCB layers through a number of parallel and interleaved vias to further reduce inductance as much as possible.
- The gate of each of the devices is connected through a separate pull-up and pull-down resistor, one per device. This allows for independent adjustment of the device switching speed if and as needed.
- On the second layer, the gate-return source connection is made directly to all the source vias and is not externally connected to any other part of the power circuit. This minimizes the CSI and isolates the gate drive from the power loop.
- On the third layer, both pull-up and pull-down gate driver outputs are bused across to each of the devices, and then through the vias up to the gate resistors on the top layer. Thus, the gate-return path on layer two is sandwiched between the gate drive conductors on both adjacent layers.
- The power-loop drain connection is also made on the third layer. The drain current flows up toward the devices before distributing laterally in both directions from the drain vias to the top layer.
- On the bottom layer, the power-loop source connection is made. The source current flows down, away from the devices, after distributing laterally in both directions once through the source vias from the top layer. The third and bottom layers are adjacent, and the drain and source currents in these layers are of equal and opposite magnitude, allowing for magnetic field self-cancellation and minimizing loop inductance.

4.5.2 Paralleling of GaN Transistors – Half-Bridge Approach

For paralleling devices in half-bridge applications, the layout approach shown in Figure 4.17 may work, but does not provide an optimal solution due to some practical limitations. To form a half bridge, consider another switching element with multiple parallel devices placed in mirror image below that of Figure 4.17. This will result in a layout with gate drivers on opposite edges of the power device layout.

This configuration is not suitable for a single half-bridge gate driver, but could be implemented with a separate floating gate driver for each group of parallel devices, as long as a suitable symmetrical layout for bringing out the switch-node can be achieved. Alternatively, the other switching element should be mirrored along the left or right side of the layout, as shown in

Figure 4.17. This would result in a single gate driver along the top edge of the layout, but the high-frequency power loop will run left to right (or right to left), resulting in a mismatch in CSI similar to the equivalent circuit shown in Figure 4.15.

Although both of the above approaches are possible, a superior alternative would be to place in parallel complete half-bridge power loops, rather than individual devices. A conceptual schematic for two such loops is shown in Figure 4.18. At first glance this may seem counter-intuitive, as this solution will result in larger inductances between the DC supply terminals, but these additional inductances do not carry any high-frequency currents. Following the same order of layout requirements as before, a complete half-bridge layout with four parallel power loops is shown in Figure 4.19. An example comparing these paralleling approaches and verifying the performance advantage of the parallel loop approach is given in Chapter 10.

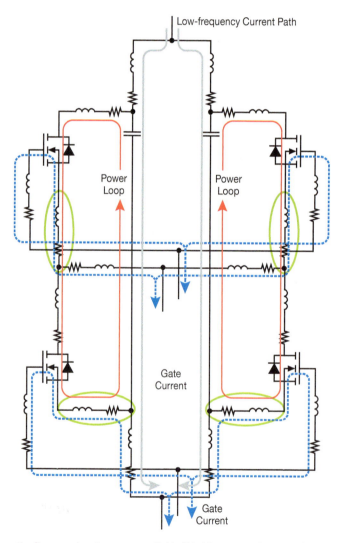

Figure 4.18 Schematic diagram showing two parallel half-bridge power loops with a common gate driver. High-frequency current paths are shown in solid red lines for power loops and blue dotted lines for gate drive loops. Low-frequency current paths are shown in gray solid lines. Common-source inductance is indicated in green ovals.

Layout Considerations for GaN Transistor Circuits | 91

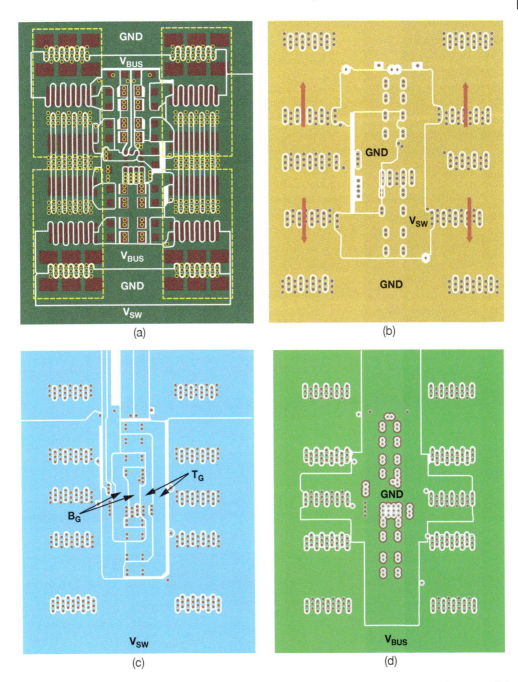

Figure 4.19 Suggested layout for a half-bridge converter with four parallel devices per switch in a parallel power-loop layout. (a) Top layer. (b) Second layer. (c) Third layer. (d) Bottom layer.

As before, the different key elements that have been integrated in the four parallel power-loop design for a half bridge are summarized as follows, with reference to Figure 4.19:

- On the top layer, the internal-vertical layout from Figure 4.8 is mirrored in both the x and y axes with the single gate driver placed in the center of the layout that drives all eight devices.

- The gate of each of the devices is connected through a separate pull-up and pull-down resistor, one per device. This allows for independent adjustment of the device switching speed as and if needed.
- On the second layer, the gate-return source connections are made directly to just one set of the source vias and are not connected to any other part of the power circuit (note the two separate ground planes). This minimizes the CSI, isolates the gate drive from the power loop, and keeps the gate-loop inductances symmetrical.
- Also, on the second layer are the power-loop ground returns that form the basis of the internal-vertical layout. The arrows show the flow of the high-frequency power-loop current from the source of the bottom device toward the ceramic high-frequency bus capacitor.
- On the third layer, both pull-up and pull-down gate driver outputs are bused upward and downward to each of the devices, and then through the vias up to the gate resistors on the top layer. The high-side gate-return path on layer two is sandwiched between the gate drive conductors on both adjacent layers.
- The switch-node connection is also made on the third layer. The low-frequency switch-node current flows down from the devices on either side before combining toward an inductance element (not shown).
- On the bottom layer, the low-side gate-return connection is made in an adjacent layer to the low-side gate driver outputs on layer three, thus preserving the loop magnetic field canceling effects. The low-frequency or DC-current positive bus connections are made on this layer as well.

The actual implementation in Figure 4.20 shows how compact the parallel-loop layout can be.

In DC–DC converters, the half-bridge approach for paralleling GaN transistors results in superior performance over the single switch approach. This is, in part, because DC–DC converters require fast-switching transients and operate at high switching frequencies to reduce converter size and maintain highest efficiency. Since CSI is the most detrimental parasitic inductance, the paralleling approach that minimizes it becomes the deciding choice.

Nevertheless, in other applications such as motor drives, which do not require such high operating frequencies or fast switch-node transients but require higher currents, a combination of the single switch and half-bridge approaches offers a better solution. In motor drives, the switching

Figure 4.20 Half-bridge converter with four parallel devices per switch in a parallel power-loop layout. *Source:* With permission of EPC.

speed of the transistors in the inverter is intentionally reduced to limit eddy currents in the bearings and other elements of the motor. Operating frequencies are also lower due to the computational limitations of the controllers. So, CSI becomes less relevant, and other properties of the combined approach such as lower thermal resistance and lower PCB resistance offer a better overall solution. These benefits will be presented in Chapter 14 through an example.

4.6 Summary

In this chapter, parasitic inductances relevant in GaN-based converters were discussed,; namely, the common-source inductance, the high-frequency power-loop inductance, and the gate-loop inductance. The origin of these parasitic elements and their impact on performance was quantified. Moreover, basic principles and various layout techniques were presented to minimize CSI and power-loop inductance. These concepts were then applied to a wide range of configurations, starting with the most basic single transistor, through a complete half-bridge configuration, and, finally, looking at the placement of multiple devices in parallel to behave as a single device for both a single-switching element and a half-bridge application.

References

1 Infineon (2010). ThinPAK 8X8 new high voltage SMD-Package, Version 1.0, April 2010. https://www.infineon.com/dgdl/Infineon+ThinPAK+8x8.pdf?folderId=db3a304314dca38901 152836c5a412ab&fileId=db3a304327b897500127f6946a286519

2 Efficient Power Conversion Corporation (2023). EPC8010 – Enhancement mode power transistor. EPC8010 datasheet. 2023. https://epc-co.com/epc/Portals/0/epc/documents/ datasheets/EPC8010_datasheet.pdf

3 Reusch, D. and Strydom, J. (2013). Understanding the effect of PCB layout on circuit performance in a high frequency gallium nitride-based point of load converter. *Twenty-Eighth Annual IEEE Applied Power Electronics Conference and Exposition (APEC)*, Long Beach, CA, (16–21 March 2013), 649–655.

4 Glaser, J.S. and Helou, A. (2022). PCB layout for chip-scale package GaN FETs optimizes both electrical and thermal performance. *IEEE Applied Power Electronics Conference and Exposition (APEC)*, Orlando, FL, (20–24 March 2022).

5 Herrera, A.R., Pozo Arribas, A., and de Rooij, M.A. (2024). PCB only thermal management techniques for eGaN® FETs in a half-bridge configuration. *PCIM Europe 2024 International Exhibition and Conference for Power Electronics, Intelligent Motion, Renewable Energy and Energy Management*, Nuremberg, Germany, (11–13 June 2024).

6 Efficient Power Conversion Corporation (2019). EPC2100 – Enhancement mode GaN power transistor half bridge. EPC2100 datasheet, August 2019. https://epc-co.com/epc/Portals/0/epc/ documents/datasheets/EPC2100_datasheet.pdf.

7 Efficient Power Conversion Corporation (2021). EPC2152 – 80V 15V ePower™ stage. EPC2152 datasheet, March 2021. https://epc-co.com/epc/Portals/0/epc/documents/datasheets/ EPC2152_datasheet.pdf.

8 Efficient Power Conversion Corporation (2024). EPC23102 – ePower™ stage IC. EPC23102 datasheet, March 2024. https://epc-co.com/epc/Portals/0/epc/documents/datasheets/ EPC23102_datasheet.pdf.

9 Reusch, D., Strydom, J., and Glaser, J. (2015). Improving high frequency DC–DC converter performance with monolithic half bridge GaN ICs. *Energy Conversion Congress and Exposition (ECCE)*, Montreal, QC, Canada (20–24 September 2015), 381–387.

5

GaN Reliability

5.1 Introduction

The rapid adoption of gallium nitride (GaN) devices in many diverse applications calls for continued accumulation of reliability statistics and research into the fundamental physics of failure in GaN devices, including integrated circuits (ICs). It is also necessary to look for information from real-world experience that either confirms the laboratory-derived data or opens new questions about mission robustness. This chapter documents ongoing analysis using test-to-fail methodology and adds guidelines for improving thermo-mechanical reliability.

5.2 Getting Started with GaN Reliability

GaN high-electron mobility transistors (HEMTs) have revolutionized power conversion technology due to GaN's superior material properties, which led to smaller die size, lower on-resistance, and lower parasitic capacitance than their silicon-based counterparts. In recent decades, GaN has been increasingly deployed in advanced applications, such as light detection and ranging (lidar) for autonomous and commercial vehicles, rooftop solar panels, drones, e-bikes, humanoid robots, and DC–DC converters for servers and data centers, and satellites.

As an emerging technology, the stability, reliability, and robustness of GaN HEMTs attract significant attention. The primary wear-out mechanisms in GaN components include voltage related wear-out, current density driven wear-out, thermo-mechanical wear-out that is predominantly caused by coefficient of thermal expansion (CTE) mismatch, and mechanical stress wear-out that is more assembly and application related. To understand which wear-out mechanisms are of primary concern for a given application, designers who are new to GaN can refer to the general guidelines in Tables 5.1 and 5.2.

5.3 Determining Wear-Out Mechanisms Using Test-to-Fail Methodology

Standard qualification testing for semiconductors typically involves stressing devices at or near the limits specified in their datasheets for a prolonged period, or for a certain number of cycles. The goal of standard qualification testing is to have zero failures out of a relatively large group of parts tested.

GaN Power Devices for Efficient Power Conversion, Fourth Edition. Alex Lidow, Michael de Rooij, John Glaser, Alejandro Pozo, Shengke Zhang, Marco Palma, David Reusch, and Johan Strydom.
© 2025 John Wiley & Sons Ltd. Published 2025 by John Wiley & Sons Ltd.

Table 5.1 GaN primary wear-out mechanism and means of avoidance by application and device type.

Application	CSP or PQFN	Gate related	Drain related	Current density related	Thermo-mechanical related	Mechanical stress related
DC–DC	CSP		C		A	E
	PQFN		C		B	
Lidar	CSP	D		F		E
	PQFN	D		F		
Solar	CSP				A	E
	PQFN				B	
Motor drives	CSP	D	C	F	A	E
	PQFN	D	C	F	B	E

For the meaning of notes, A–F, see Table 5.2.
CSP, chip-scale package and PQFN, power quad-flat no-leads.

Table 5.2 GaN reliability concerns – section references.

Note	General mitigation strategy	For details (see Section)
A	Choose suitable underfill	5.5.4.4
B	Design an assembly process that leads to a part which lays flat and adheres well to the PCB	Reference [1]
C	Remain within datasheet V_{DS} limits, with excursions that remain within drain overvoltage specification	5.5.2
D	Remain within datasheet V_{GS} limits, with excursions that remain within gate overvoltage specification	5.5.1
E	Adhere to recommended mechanical specifications during assembly and handling	Section 3.5 in Reference [2]
F	Adhere to safe operating area (SOA) and electromigration current specifications	5.5.3

This type of qualification testing is inadequate, since it only reports parts that passed a very specific test condition. By testing parts to the point of failure, an understanding of the amount of margin between the datasheet limits can be developed, as well as an understanding of the intrinsic wear-out mechanisms. By knowing the intrinsic wear-out mechanisms, the root cause of failure, and the behavior of this mechanism over time, temperature, electrical, or mechanical stress, the safe operating life of a product can be determined over a more general set of operating conditions (for an excellent description of test-to-fail methodology for testing semiconductor devices, see reference [3]).

As with all power transistors, the key stress conditions involve voltage, current, temperature, and humidity, as well as various mechanical stresses. There are, however, many ways of applying these stress conditions. For example, voltage stress on a GaN transistor can be applied from the gate terminal to the source terminal (V_{GS}), as well as from the drain terminal to the source terminal (V_{DS}). These stresses can be applied continuously as a DC bias, they can be cycled on-and-off, or they can be applied as high-speed pulses. Current stress can be applied as a continuous DC current or as a pulsed current. Thermal stresses can be applied continuously by operating devices at a predetermined temperature extreme for a period, or temperature can be cycled in a variety of ways.

By stressing devices with each of these conditions to the point of generating a significant number of failures, an understanding of the primary intrinsic wear-out mechanisms can be determined. To generate failures in a reasonable amount of time, the stress conditions typically need to significantly exceed the datasheet limits of the product. Care needs to be taken to make certain the excess stress condition does not induce a failure mechanism that would never be encountered during normal operation.

To make certain that excess stress conditions did not cause the failure, the failed parts need to be carefully analyzed to determine the failure root causes. Only by verifying the root cause can a complete understanding of the behavior of a device under a wide range of stress conditions be developed. As the intrinsic wear-out modes in GaN devices are better understood, two facts have become clear: (i) GaN devices can be more robust than Si-based MOSFETs, and (ii) silicon MOSFET intrinsic failure models do not generally apply when predicting GaN device lifetime under extreme or long-term electrical stress conditions.

Table 5.3 lists in the left-hand column all the various stressors to which a transistor can be subjected during assembly or operation. Using the various test methods listed in the third column from the left, and taking devices to the point of failure, the intrinsic wear-out mechanisms can be discovered. The wear-out mechanisms confirmed as of this writing are shown in the column on the right.

Table 5.3 Stress conditions and intrinsic wear-out mechanisms for GaN transistors.

Stressor	Device/package	Test method	Intrinsic wear-out mechanism
Voltage	Device	HTGB	Dielectric failure
			Threshold shift
		HTRB	Threshold shift
			$R_{DS(on)}$ shift
		ESD	Dielectric rupture
Current	Device	DC current (EM)	Electromigration
			Thermomigration
Current + Voltage (Power)	Device	SOA	Thermal runaway
		Short circuit	Thermal runaway
Voltage rising/Falling	Device	Hard-switching reliability	$R_{DS(on)}$ shift
Current Rising/Falling	Device	Pulsed current (lidar reliability)	None found
Temperature	Package	HTS	None found

(Continued)

Table 5.3 (Continued)

Stressor	Device/package	Test method	Intrinsic wear-out mechanism
Humidity	Package	MSL1	None found
		H3TRB	Dendrite formation/Corrosion
		AC	None found
		Solderability	Solder corrosion
		uHAST	None found
Mechanical / Thermomechanical	Package	TC	Solder fatigue
		IOL	Solder fatigue
		Bending force test	Mechanical damage
			Solder strength
		Die shear	Solder fatigue
		Package force	Film cracking
Radiation	Device	Gamma radiation	None found
		Neutron radiation	None found
		Heavy ion bombardment	Crystal displacement damage and ionization damage

5.4 Using Test-to-Fail Results to Predict Device Lifetime in a System

When multiple failure mechanisms or stressors are involved, the total failure rate of a system, commonly known as Failure in Time (FIT), is the sum of the failure rates per failure mechanism [4–6] as shown below,

$$\text{FIT}_{\text{Total}} = \text{FIT}_1 + \text{FIT}_2 + \cdots + \text{FIT}_i \tag{5.1}$$

where FIT is failure in time, which typically represents the number of failures in 10^9 (1 billion) device hours, and the subscript (1, 2, .., i) indicates the different failure mechanisms identified.

FIT is inversely proportional to mean time to failure (MTTF) as described by

$$\text{FIT} = \frac{10^9}{\text{MTTF}} \tag{5.2}$$

Therefore, by plugging Eq. (5.2) into Eq. (5.1), the total MTTF can be described by Eq. (5.3).

$$\frac{1}{\text{MTTF}_{\text{Total}}} = \frac{1}{\text{MTTF}_1} + \frac{1}{\text{MTTF}_2} + \cdots + \frac{1}{\text{MTTF}_i} \tag{5.3}$$

The subscripts are assigned to the reliability stressors that are relevant to the application of interest. Based on Eq. (5.3), it is noted that the smallest denominator (shortest MTTF) dominates the overall lifetime. It is critical to understand which stressor is the limiting factor in reliability because the weakest link warrants the most consideration during design and operations.

In most applications, devices experience various stress conditions over the course of the entire mission lifespan, including a combination of different bias conditions and different temperature profiles, for a given reliability stressor as specified in Eq. (5.3). Each stress condition corresponds to a specific failure rate (failures per unit time) and is specified as FR_a, FR_b, ..., FR_n. The respective duration of each stress condition is denoted as t_a, t_b, ..., t_n. Assuming $t_{\text{total}} = t_a + t_b + \cdots + t_n$ is 10^9 hours, the FIT calculation of total number of failures is then generalized for specific reliability stress conditions as

$$FR \cdot t_{\text{total}} = FR_a \cdot t_a + FR_b \cdot t_b + \cdots + FR_n \cdot t_n \tag{5.4}$$

The time-averaged failure rate FR can be calculated as

$$FR = FR_a \cdot \frac{t_a}{t_{\text{total}}} + FR_b \cdot \frac{t_b}{t_{\text{total}}} + \cdots + FR_n \cdot \frac{t_n}{t_{\text{total}}} \tag{5.5}$$

which can be simplified by introducing fractional operation time,

$$n = \frac{t_n}{t_{\text{total}}} \tag{5.6}$$

noted as a, b, ..., n. The sum of a, b, ..., n is 100%, which is given in Eq. (5.7).

$$a + b + \cdots + n = 100\% \tag{5.7}$$

Now Eq. (5.5) can be simplified to

$$FR = FR_a \cdot a + FR_b \cdot b + \cdots + FR_n \cdot n \tag{5.8}$$

It is known that the failure rate under each sub-stress condition is inversely proportional to the device lifetime LT [5] when the same number of failures is generated. The relation is shown in Eq. (5.9).

$$FR \propto \frac{1}{LT} \tag{5.9}$$

Plugging Eq. (5.9) into Eq. (5.8) yields Eq. (5.10).

$$\frac{1}{LT_{\text{Total}}} = \frac{a}{LT_a} + \frac{b}{LT_b} + \cdots + \frac{n}{LT_n} \tag{5.10}$$

where LT_{Total} is the total projected lifetime and LT_n is the projected lifetime for each sub-stress condition.

Equation (5.10) captures how a mission profile consisting of more than one stress condition results in a system lifetime. The fractional operation time (a, b, ..., n) in the numerators account for the times spent in harsh, moderate, and mild stress conditions.

5.5 Wear-Out Mechanisms

5.5.1 Gate Wear-Out

5.5.1.1 Introduction to Gate Wear-Out Mechanisms

GaN transistors devices require a 5 V_{GS} to properly drive the devices, which leaves a small margin from the recommended gate drive voltage (5–5.5 V) to the datasheet maximum specification ($V_{\text{GS, Max}} = 6$ V) [7–13]. Therefore, it is critical to understand the reliability robustness at different gate bias and temperature conditions. Test-to-fail is implemented to investigate the wear-out mechanisms responsible for gate failures. Through understanding the fundamental wear-out mechanism, a physics-based lifetime model is developed to project lifetime at all given gate–source voltages and temperatures. If the transient gate–source overvoltage rings beyond 6 V_{GS}, a 1% duty cycle-based overvoltage specification is developed that is supported by data, as well as a knowledge of the fundamental failure mechanisms.

5.5.1.2 Gate Reliability Model

To understand gate wear-out mechanisms, four groups of representative GaN HEMTs (EPC2212 [14]) and 32 devices per group were tested using a customized gate reliability testing circuit, shown in Figure 5.1.

Four different accelerated stress conditions were implemented at gate–source voltages of 8, 8.5, 9, and 9.5 V, well exceeding the maximum rated gate voltage ($V_{GS(max)}$) of 6 V. At 9 V and 9.5 , failures occurred very quickly, but it took significantly longer at 8 V and 8.5 V. After the failures were identified, failure analyses were conducted on many failures, and a consistent failure mode was found. Figure 5.2 shows the failure mode observed in all failures analyzed. The location of the gate failures is where the silicon nitride dielectric is sandwiched between gate metal and field plate metal.

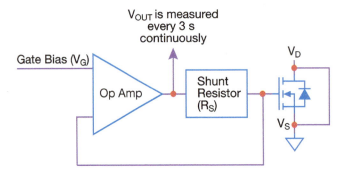

Figure 5.1 A simplified schematic of the customized accelerated gate reliability testing system that can perform *in situ* gate leakage current (I_{GSS}) measurement with three seconds interval, where $I_{GSS} = (V_{out} - V_G)/R_S$.

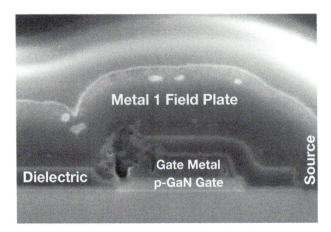

Figure 5.2 Scanning electron microscopy (SEM) image of a gate failure. Dielectric breakdown is observed between the gate metal and the field plate metal. *Source:* With permission of EPC.

There were two main observations that indicated a wear-out mechanism more complicated than a simple dielectric wear-out that would follow a well-known time dependent dielectric failure mechanism (TDDB [15]). These observations were, (i) the failures occurred at voltages well below that which would be normal for the silicon nitride dielectric separating the gate electrode and the field plate and (ii) the failure rate goes down with temperature. This last observation is also in conflict with the established TDDB theory.

To explain all the observations, an impact ionization model was introduced using a two-step process [2, 16–18]. In the first step, electrons within the two-dimensional electron gas (2DEG) enter the p-GaN gate and get accelerated under the influence of positive gate bias within gate. In this process some electrons gain sufficient energy to cause impact ionization and generate holes.

In the second step, the generated holes near the top surface of the gate move away from the gate corner to the field plate metal (source potential) under the positive electric field due to the positive gate bias. As a result, the holes become trapped in the silicon nitride dielectric layer, leading to a growing positive charge density. When the trapped charges accumulate and reach the critical field of the silicon nitride dielectric film, it ruptures catastrophically. Having identified impact ionization as the mechanism responsible for gate wear-out led to the development of the following lifetime model, Eq. (5.11), applicable to p-GaN gates [2, 16–18].

$$\text{MTTF} = \frac{Qc}{G} = \frac{qQc}{\alpha_n J_n} = \frac{A}{(1 - c\Delta T)} \exp\left[\left(\frac{B}{V + V_0}\right)^m\right] \tag{5.11}$$

with parameters listed below:

$m = 1.9$
$V_0 = 1.0 \, \text{V}$
$B = 57.0 \, \text{V}$
$A = 1.7 \times 10^{-6} \, \text{s}$
$c = 6.5 \times 10^{-3} \, \text{K}^{-1}$

The lifetime Eq. (5.11) is plotted against measured acceleration data for EPC2212 in Figure 5.3. To produce this fit, all parameters in Eq. (5.11) were fixed except A and B. The resulting best fit for B (when converted into a field by dividing by the gate thickness) resulted in a value of $7.6 \times 10^6 \, \text{V/cm}$, in very close agreement with literature reported value of 7.2×10^6 V/cm [19].

Figure 5.4 shows the temperature dependence of the lifetime equation at −75 °C, 25 °C, and 125 °C. The temperature dependence (contained in parameter c) is taken directly from Ozbek [20] without fitting to data. Note that at higher temperature, the MTTF is slightly higher than at lower temperatures, which, although counter-intuitive, is consistent with the measured data reported [16, 17].

This lifetime equation is not simply borrowed from the body of standard reliability models developed for MOSFETs. Instead, it represents the first gate lifetime model built up from the root physics of failure, specifically applicable to enhancement-mode GaN transistors.

5.5.1.3 Gate Overvoltage Impact on Lifetime

Gate overvoltage spikes during device turn-on transients are commonly observed in GaN HEMTs under high-frequency, fast-switching conversion applications [21, 22]. As discussed in Chapter 3, the magnitude of the gate overvoltage transients is primarily governed by the gate-loop inductance and the slew rate (V_{GS}/dt), since both are closely related to circuit design

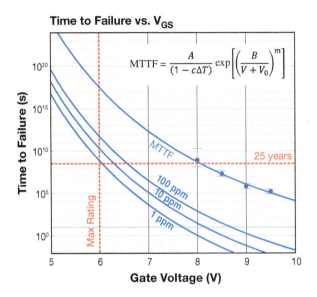

Figure 5.3 EPC2212 *MTTF* vs. V_{GS} at 25 °C (and error bars) are shown for four different voltage legs. The solid line corresponds to the impact ionization lifetime model. Extrapolations of time to failure for 100, 10, and 1 ppm are shown as well.

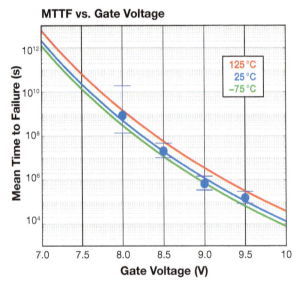

Figure 5.4 MTTF for EPC2212 (25 °C) measured at four different gate biases. Blue line is the lifetime model. Red and green lines are predictions of the lifetime model at 125 °C and −75 °C, respectively.

and PCB layout [23]. Therefore, it is critical to understand the gate overvoltage capability of GaN HEMTs and the associated lifetimes.

The projected lifetime results offered assurance that the failure rate is expected to be less than 1 part per million (ppm) for 25 years if the gate was biased less than the datasheet maximum limit at 6 $V_{GS,\,Max}$. This virtually zero failure rate is also consistent with field experience

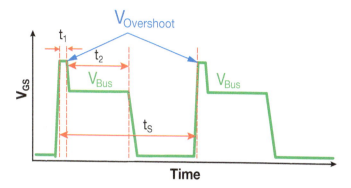

Figure 5.5 An illustration of a normal gate switching waveform in real-world applications. t_1 is the duration of gate overshoot in each period and t_2 is the nominal bus voltage bias duration within each switching period. t_S is the switching period that is dependent on the switching frequency.

[7, 22, 23]. This extremely low projected failure rate helps build confidence in overall gate reliability but does not provide a methodology to estimate the overall lifetime when gate overvoltage spikes are present repetitively during switching.

In real-world applications, the common mission profile can be simplified and illustrated by Figure 5.5. Within each switching period (t_S), it includes two portions which are labeled as overvoltage duration, t_1, and bus voltage duration, t_2.

By following the mathematical approach introduced in Section 5.3 using the durations shown in Figure 5.5, Eq. (5.12) is developed to estimate total lifetime by accounting for gate overshoot period and normal bus voltage period.

$$\frac{1}{T_{Total}} = \frac{a}{T_{VG_{Overshoot}}} + \frac{b}{T_{VG_{Bus}}} \tag{5.12}$$

where a and b are the fractional operation time during a switching period (t_s), as denoted in Eq. (5.7).

Another term, overvoltage duty cycle factor (DC_{Factor}), is introduced in Eq. (5.13), where DC_{Factor} is equal to "a" in Eq. (5.12), which is the ratio between t_1, the gate overvoltage duration and the switching period t_S, which is inversely proportional to the switching frequency.

$$\text{Overvoltage } DC_{Factor} = a = \frac{t_1}{t_s} \tag{5.13}$$

To further demonstrate the model of Eqs. (5.12) and (5.13), two examples are provided based on the data presented in Section 5.5.1.2. Assume the nominal bus voltage for EPC2212 is 5.5 V. The overvoltage is estimated to be 120% of the nominal bus voltage, which is calculated to be 6.6 V ($V_{GS, Max}$ = 6 V). The overshoot duty cycle is expected to be approximately 1% as defined by Eq. (5.13) and Figure 5.5.

To simplify the calculation, it is assumed that the gate is operating 99% of the time at 5.5 V. Therefore, adding the impact of the overshoot to the nominal drive condition, the EPC2212 GaN device lifetime is projected to be 1.69×10^9 seconds, or 53 years, for a 10-ppm failure rate (10 device failures per 1 million tested).

In another example, the GaN devices are used in a poorly designed circuit where the gate terminal sees a 7 V gate voltage spike during turn-on transient repetitively with the same 1% overshoot duty cycle. In this extreme example the bus voltage is still at 5.5 V for 99% of the time. After approximately 25 years, the failure rate is still expected to be only 100 ppm.

5.5.2 Drain Wear-Out

5.5.2.1 Introduction to Drain Wear-Out Mechanisms

The same test-to-fail methodology is adopted to investigate drain-related wear-out mechanism. One of the common concerns among GaN transistor users is dynamic on-resistance ($R_{DS(on)}$). This is a condition whereby the on-resistance of a transistor increases when the device is exposed to high drain–source bias. A comprehensive lifetime model is developed to project dynamic $R_{DS(on)}$ shift with respect to various parameters, including voltage, temperature, frequency, and current. As with the gate overvoltage example, another duty cycle-based overvoltage specification is developed for drain–source overvoltages based on the understanding of the fundamental wear-out mechanisms.

5.5.2.2 Physics-Based Dynamic $R_{DS(on)}$ Lifetime Models

The dominant mechanism causing on-resistance to increase is the trapping of electrons. As the trapped charge accumulates, it depletes electrons from the two-dimensional electron gas (2DEG) in the ON state, leading to an increase in $R_{DS(on)}$.

Figure 5.6 is a magnified image of an early-generation EPC2016C GaN transistor showing thermal emissions in the 1–2 μm optical range. Emissions in this part of the spectrum are consistent with hot electrons, and their location in the device is consistent with the location of the highest electric fields when the device is under high drain–source bias. Knowing that hot electrons in this region of the device are the source of trapped electrons, a better understanding of how to minimize the dynamic on-resistance can be achieved with improved designs and processes.

A resistive load hard switching testing system also known as "fast dR" system [16, 17] was developed, creating orders of magnitude more hot electrons thus accelerating the dynamic $R_{DS(on)}$ shift. A simplified schematic version of the fast dynamic $R_{DS(on)}$ (fast dR) testing system is illustrated in Figure 5.7. The fast dR setup typically switches at 100 kHz with 85% of the time

Figure 5.6 A magnified image of an EPC2016C GaN transistor showing light emission in the 1–2 μm wavelength short-wave infrared light range (SWIR) that is consistent with hot electrons. The SWIR emission (red-orange) has been overlaid on a regular (visible wavelength) microscope image and a semi-transparent image of the design photomask (purple). *Source:* With permission of EPC.

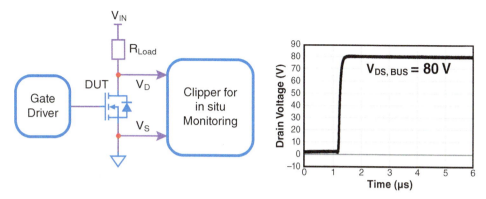

Figure 5.7 Circuit schematic of a resistive-load hard-switching test system with a clipper circuit used for in situ $R_{DS(on)}$ monitoring. In this example a turn-off drain voltage waveform to an 80 V bus voltage is produced by the resistive load hard switching circuit.

where the GaN device under test (DUT) is reverse-biased at a specific drain–source voltage as shown in Figure 5.7.

Figure 5.8 shows how the $R_{DS(on)}$ of a fifth-generation EPC2045 GaN transistor [7], designed with the knowledge that hot electron trapping is accelerated at various voltage stress levels and temperatures. On the left, the devices were tested at 25 °C, at voltages from 60 to 120 V (EPC2045 has a $V_{DS, Max}$ of 100 V). The horizontal axis shows time measured in minutes, with the right side ending at 10 years. The graph on the right shows the evolution of $R_{DS(on)}$ when biased at 120 V at different temperatures. The counter-intuitive result shows that the on-resistance increases faster at lower temperatures. This is consistent with hot-carrier injection because hot electrons travel further between scattering events at lower temperatures and therefore are accelerated to greater kinetic energies by a given electric field. The result is that the electrons scatter with higher energies, reaching layers where they are more likely to become trapped. This suggests that traditional testing methods, where a device is tested at maximum voltage and temperature, may not be adequate to quantify the reliability of a device.

When the applied drain bias is no more than 120% of the $V_{DS, Max}$, and assuming the trapped charge density is significantly smaller than the 2DEG carrier concentration, the model for $R_{DS(on)}$ growth can be simplified as shown in Eq. (5.14).

$$\frac{\Delta R}{R} = a + b \log\left(1 + \exp\left(\frac{V_{DS} - V_{FD}}{\alpha}\right)\right) \sqrt{T} \exp\left(\frac{\hbar \omega_{LO}}{kT}\right) \log(t) \quad (5.14)$$

Independent Variables:

V_{DS} = Drain voltage (V)
T = Device temperature (K)
t = Time (min)

Parameters:

a = 0.00 (unitless)
b = 2.0E-5 ($K^{-1/2}$)
$\hbar \omega_{LO}$ = 92 meV
V_{FD} = 100 V (appropriate for Gen5 100 V products only)
α = 10 (V)

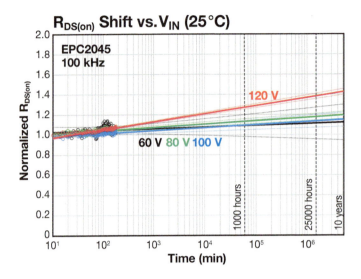

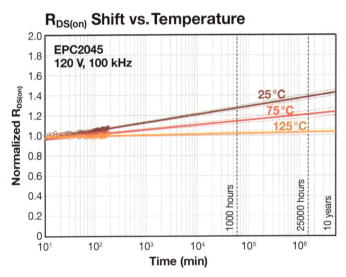

Figure 5.8 The $R_{DS(on)}$ of a fifth-generation EPC2045 GaN transistor over time at various voltage stress levels and temperatures. On the left, the devices were tested at 25 °C with voltages from 60 to 120 V. The graph on the right shows the evolution of $R_{DS(on)}$ at 120 V at various temperatures.

5.5.2.3 Effect of Switching Frequency and Switching Current

In the analysis so far, the effects of switching frequency (f) and switch current (I) on the $R_{DS(on)}$ growth characteristics have been ignored. The current directly impacts the number of electrons injected into the high-field region during the hard-switching transition and therefore has a linear effect on the hot carrier density. Likewise, the switching frequency determines the number of hot carrier pulses seen at the drain in a given time interval and therefore also has a linear effect on the surface trapping rate.

By assuming that the surface trapping rate is linearly proportional to both frequency (f) and current (I), the effects of f and I are included in Eq. (5.15), where a simple scaling term is derived to relate the $R_{DS(on)}$ growth in one switching condition (f_1, I_1) to another (f_2, I_2).

$$R(t;f_2,I_2) = R(t;f_1,I_1) + b\left(\log\left(\frac{f_2}{f_1}\right) + \log\left(\frac{I_2}{I_1}\right)\right) \quad (5.15)$$

Based on Eq. (5.15), the effect of changing the switching frequency or current is to simply offset the $R_{DS(on)}$ growth curve vertically by a small amount. Therefore, if the device is operated under conditions with low $R_{DS(on)}$ rise (low slope b), the effect of changing frequency or current will be negligible.

Figure 5.9 compares the modeled $R_{DS(on)}$ vs. time for an EPC2045 at three different switching frequencies, from 10 kHz to 1 MHz. Note that the curves are simply offset from each other vertically. The same would be true had we compared different switch currents. Because the offset changes as the logarithm of *frequency* or current, even a 10× increase in switching frequency (or current) would be difficult to observe experimentally owing to ±10% noise in the measurement and projection [16–18].

5.5.2.4 Drain–Source Overvoltage Impact on Lifetime

In the case where the amount of trapped charge approaches the number of electrons available in the 2DEG (the surface trapped charges (Q_S) approaches the built-in 2DEG piezoelectric charge (Q_P)), the simplifying assumption used to develop Eq. (5.14) is no longer valid. This situation could occur when devices are taken to voltages well above their design limits. Figure 5.10 shows results for EPC2045 devices tested up to 150 V at 75 °C and 125 °C. Note how the straight-line extrapolation that would occur with a simple log(time) dependence is no longer applicable. By removing the simplified assumption that only a small fraction of Q_P is trapped and transform into Q_S, the result shown in Eq. (5.16) is obtained. Calculating Eq. (5.16) using the expanded list of parameters yields the solid lines in Figure 5.10, providing further evidence of the validity and applicability of this physics-based model.

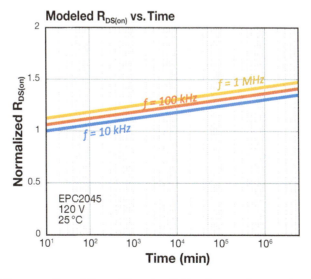

Figure 5.9 Modeled $R_{DS(on)}$ vs. *Time* at three different switching frequencies, covering two orders of magnitude. Note that the effect of frequency change is a small vertical offset in the growth characteristic. The same offset would occur at different switch currents.

108 | GaN Power Devices for Efficient Power Conversion

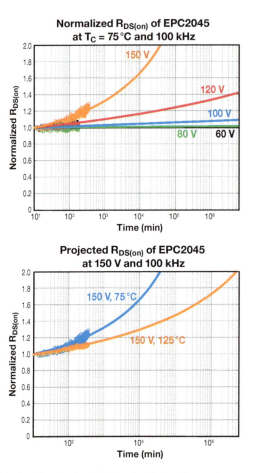

Figure 5.10 100 V EPC2045 devices in a hard-switching circuit at various voltages up to 150% of design rating (top), and at two different temperatures, also at 150% of design rating (bottom). The solid lines are the model predictions, and the dots represent measurement points.

$$\frac{\Delta R}{R} = a_1 \left[\frac{a_2 \, \Psi \log(1 + a_3 t/\Psi)}{1 - a_2 \, \Psi \log(1 + a_3 t/\Psi)} \right]$$

where

$$a_1 \equiv \frac{C}{Q_P} \qquad a_2 \equiv \frac{1}{Q_P} \qquad a_3 \equiv B \qquad (5.16)$$

With the following expanded list of parameters:

$a_1 = 0.6$ (unitless)
$a_2 = b/a_1$ (where $b = .0\text{E-}5 \; \text{K}^{-1/2}$ from [6, 8])
$a_3 = 1000 \; (\text{K}^{1/2} \, \text{min}^{-1})$
$b = 2.0\text{E-}5 \; (\text{K}^{-1/2})$
$\hbar\omega_{LO} = 92 \; \text{meV}$
$V_{FD} = 100 \; \text{V}$ (appropriate for Gen5 100 V products only)
$\alpha = 10 \; (\text{V})$

T = Device temperature (K)
t = Time (min)

5.5.2.5 200 V Transistor Model

A similar analysis was developed for 200 V GaN transistors. The resultant variables are as follows:

a_1 = 0.6 (unitless)
a_2 = 2.8 b/a_1 (where b = 2.0E-5 $K^{-1/2}$)
a_3 = 1000 ($K^{1/2}$ min^{-1})
b = 2.0E-5 ($K^{-1/2}$)
$\hbar\omega_{L0}$ = 92 meV
V_{FD} = 210 V (appropriate for Gen5 200 V products only)
α = 25 (V) (appropriate for Gen5 200 V products only)
T = Device temperature (K)
t = Time (min)

Figure 5.11 compares this model to measurements of 200 V devices. On the left is the normalized $R_{DS(on)}$ for the 200 V rated EPC2215 [8] at three voltages. The highest voltage, 280 V, is 40% above the maximum rating. On the right are measurements compared with the model at two different temperatures and the maximum rated voltage.

5.5.2.6 Drain Overvoltage Specification

Transient drain voltage overshoot is commonly observed in GaN-based converters due to high slew rate and fast switching applications. A survey of transient overvoltage specification from a suite of GaN suppliers was conducted by JEDEC JC-70 committee and presented in JEP186 [9]. Most of the transient overvoltage specifications describe it as a device robustness indicator. In addition, many of them consider drain voltage overshoot as a single rare event or an atypical occurrence. Therefore, an application-driven and user-friendly repetitive transient off-state drain overvoltage specification on datasheets is desired.

The resistive load hard switching system in Figure 5.7 was deployed to study dynamic $R_{DS(on)}$ shift under 120 V_{DS} overvoltage stress. 20% of $R_{DS(on)}$ shift compared to the initial $R_{DS(on)}$ value is used as the failure criteria to determine time-of-failure. Equation 5.14 is used to extrapolate the time-of-failure when the in situ monitored $R_{DS(on)}$ shifts more than 20%. EPC2045 was first subjected to testing under such accelerated conditions and the testing result is shown in Figure 5.12, projecting a lifetime of 2×10^5 minutes by adding 10% confidence level. By multiplying 85% during which the 120 V is applied, it yields 1.7×10^5 minutes, representing the total lifetime when the DUT is biased continuously under 120 V. When compared with 25 years of expected lifetime (1.3×10^7 minutes), it translates to 1.3% of total lifespan in mission. To add more margin, the number is rounded to 1%, which becomes the proposed drain overvoltage specification that was first discussed in Chapter 2.

To further validate this total time-based specification, the same testing conditions were applied to 100 V rated GaN products including EPC2204 [10], EPC2218 [11], EPC2071 [12], and EPC2302 [24]. Figure 5.13 summarizes the testing results of the listed products, where they are all projected to outperform the 1.3×10^5 minutes of lifetime.

This total time-based specification can be scaled to a shorter duration that occurs repetitively, cycle-by-cycle occurring in each switching period. Equation (5.17) defines the proposed 1% transient overvoltage duty cycle factor (DC_{Factor}),

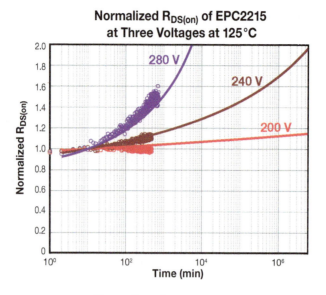

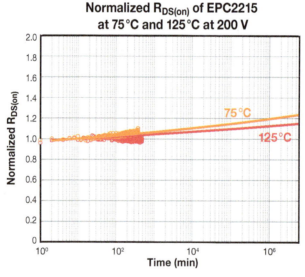

Figure 5.11 (Top) 200 V EPC2215 normalized $R_{DS(on)}$ at three voltages. Note that 280 V is 40% above the maximum rated voltage. (Bottom) EPC2215 at 75 °C and 125 °C and 200 V. The solid lines are the model results using variables for 200 V devices, and the dots are actual measurements.

$$\text{Overvoltage DC}_{\text{Factor}} = \frac{120\% \text{ Overvoltage Duration }(T_O)}{\text{Switching Period }(T_S)} \leq 1\% \quad (5.17)$$

where T_O is the overvoltage duration within each switching period and T_S is the switching period. This mathematical relation is demonstrated in Eq. (5.17) and further illustrated in Figure 5.14.

For instance, if a converter operates at 100 kHz, equivalent of 10 μs per switching period, it suggests that the GaN devices should withstand a repetitive 120 V overvoltage spike with a 100 ns duration in each switching cycle over 25 years of lifetime. To verify this proposed

GaN Reliability

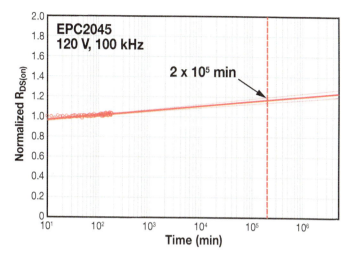

Figure 5.12 Evolution of $R_{DS(on)}$ of a representative EPC2045 device, a fifth-generation 100 V rated GaN transistor, tested at 120 V and 75 °C. It is projected to exceed 20% $R_{DS(on)}$ shift at 2×10^5 minutes.

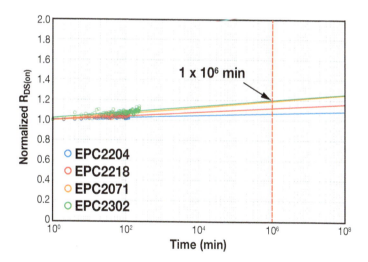

Figure 5.13 Evolution of $R_{DS(on)}$ of representative EPC2204, EPC2218, EPC2071, and EPC2302 GaN transistors, rated at 100 V and tested at 120 V and 75 °C. They are projected to have less than 20% $R_{DS(on)}$ shift at a minimum of 1×10^6 minutes, significantly exceeding the 2×10^5 minutes lifetime based on EPC2045.

overvoltage specification, an unclamped inducive switching (UIS) circuit was developed [24, 25]. Figure 5.15 shows the resulting overvoltage pulse that is generated by UIS.

A number of 100 V-rated GaN transistors from different wafer lots are stressed by a 120 $V_{DS, Peak}$ overvoltage spike at 100 kHz operation frequency, and 75 °C junction temperature. Figure 5.16 shows that representative EPC2218 devices from three different wafer lots were tested to over billions of switching cycles showing very small dynamic $R_{DS(on)}$ shift [24, 25].

Figure 5.17 shows the 120 V overvoltage testing results of another representative 100 V rated GaN transistor EPC2302 in a power quad flat no-lead (PQFN) package. The device was tested to approximately 10 billion cycles at ambient temperature (25 °C), where very little $R_{DS(on)}$ shift

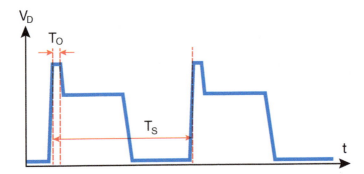

Figure 5.14 Illustration of the 1% overshoot duty cycle overvoltage specification. 1% is the ratio between T_O (overvoltage duration) and T_S (one switching period).

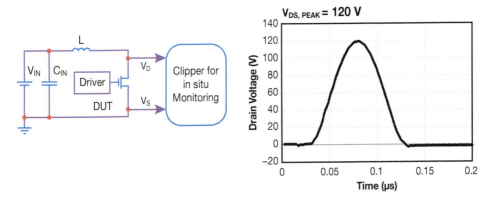

Figure 5.15 Simplified schematic of the unclamped inductive switching circuit and the resulting overvoltage pulse with $V_{DS,\ Peak}$ of 120 V under 100 kHz operating frequency.

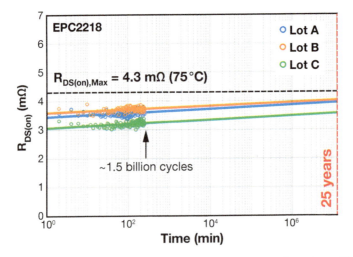

Figure 5.16 Evolution of dynamic $R_{DS(on)}$ of a representative EPC2218 DUTs from three different wafer lots under 120 $V_{DS,\ Peak}$ and 75 °C UIS testing for more than 1.5 billion cycles.

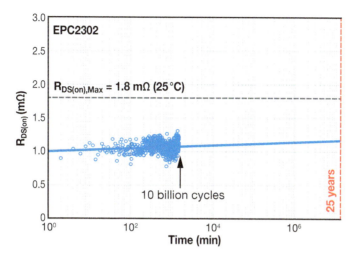

Figure 5.17 Evolution of dynamic $R_{DS(on)}$ of a representative EPC2302 DUT under 120 $V_{DS, Peak}$ UIS testing for approximately 10 billion switching cycles.

was seen. A good agreement between 10 billion data points and the lifetime model (blue fit line) was also observed, proving the validity and versatility of the lifetime model. The projection demonstrates the excellent robustness of GaN devices under 120% overvoltage stress over long-term continuous operation. The voltage waveform shown in Figure 5.15 is more representative of real-time circuit applications, in which transient overvoltage ringing might be expected.

To summarize, a repetitive drain overvoltage specification is proposed. This duty cycle-based specification offers a more quantitative and easy-to-implement guideline for design engineers. This work also demonstrates the extreme overvoltage robustness of GaN HEMTs.

5.5.3 Current Density Wear-Out

5.5.3.1 Introduction to Current Density Wear-Out Mechanisms

Thermal limits can be of concern for GaN devices when high current and high drain–source voltage occur simultaneously. For certain applications, the capability of withstanding short-circuit fault conditions is a must. Therefore, short-circuit testing was performed, where GaN demonstrated excellent robustness under such extreme stress conditions. When the devices are exposed to continuous high current at elevated temperatures, electromigration (EM) robustness becomes a frequently asked question. Accelerated EM testing was conducted on PQFN devices, demonstrating excellent robustness against EM.

5.5.3.2 Safe Operating Area

Safe operating area (SOA) testing exposes the GaN transistors to simultaneous high current (I_D) and high voltage (V_{DS}) for a specified pulse duration. The primary purpose is to verify the transistor can be operated without failure at every point (I_D, V_{DS}) within the datasheet SOA graph. It is also used to probe the safety margins by testing to fail outside the safe zone. During SOA tests, the high-power dissipation within the die leads to a rapid rise in junction temperature and the formation of strong thermal gradients. For sufficiently high power or pulse duration, the device simply overheats and fails catastrophically. This is known as thermal overload failure.

In Si MOSFETs, another failure mechanism known as secondary breakdown (or Spirito effect [26]) has been observed in SOA testing. This failure mode, which occurs at high V_D and low I_D, is caused by unstable feedback between junction temperature and threshold V_{TH}. As the junction temperature rises during a pulse, V_{TH} drops, which can cause local current to rise. The rising current, in turn, causes temperature to rise faster, thereby completing a positive feedback loop that leads to thermal runaway and ultimate failure. The goal of this study is to determine if the Spirito effect exists in GaN transistors.

For DC, or long-duration pulses, the SOA capability of the transistor is highly dependent on the heatsinking of the device. This can present a huge technical challenge to assess the true SOA capability, often requiring specialty water-cooled heatsinks. However, for short pulses (<1 ms), the heatsinking does not impact SOA performance. This is because on short timescales the heat generated in the junction does not have sufficient time to diffuse to any external heatsink. Instead, all the electrical power is converted to raising the temperature (thermal capacitance) of the GaN film and nearby silicon substrate. As a result of these considerations, SOA tests were conducted at two pulse durations: 1 ms and 100 μs.

Figure 5.18 shows the SOA data of 200 V EPC2034C [27]. In this plot, individual pulse tests are represented by points in (I_D, V_{DS}) space. These points are overlaid on the datasheet SOA graph. Data for both 100 μs and 1 ms pulses data are shown together. Green dots correspond to 100 μs pulses where a part passed, whereas red dots indicate where a part failed. A broad area of the SOA was interrogated without any failures (all green dots), ranging from low V_{DS} all the way to $V_{DS, max}$ (200 V). All failures (red dots) occurred outside the SOA, indicated by the green line in the datasheet graph. The same applies to 1 ms pulse data (purple and red triangles); all failures occurred outside of the datasheet SOA.

Figure 5.19 provides SOA data for three more parts, AEC EPC2212 (fourth-generation automotive 100 V), EPC2045 (fifth-generation 100 V), and EPC2014C (fourth-generation 40 V) [28]. In all cases, the datasheet safe operating area has been interrogated without failures, and all failures occur outside of SOA limits, often well outside the limits.

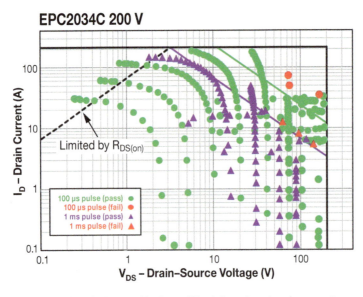

Figure 5.18 EPC2034C SOA plot. The "Limited by $R_{DS(on)}$" line is based on datasheet maximum specification for $R_{DS(on)}$ at 150 °C.

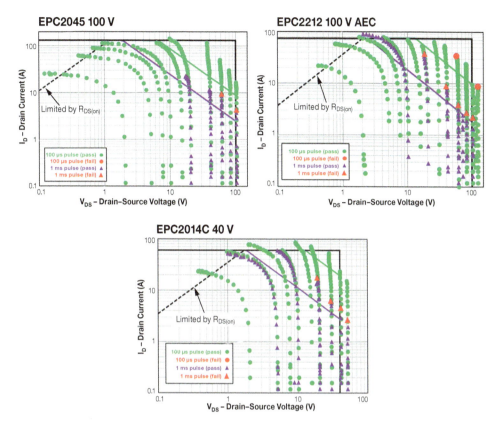

Figure 5.19 SOA results for EPC2045, EPC2212, EPC2014C. Measurements for 1 ms (purple triangles) and 100 µs (green dots) pulses are shown together. Failures are denoted by red triangles.

The datasheet SOA graph is generated with finite element analysis (FEA), using a thermal model of the device including all relevant layers along with their heat conductivity and heat capacity. Based on transient simulations, the SOA limits are determined by a simple criterion – for a given pulse duration, the power dissipation must be such that the junction temperature does not exceed 150 °C before the end of the pulse. This criterion results in limits based on constant power, denoted by the 45° green (100 µs) and purple (1 ms) lines in the SOA graph. This approach leads to a datasheet graph that defines a conservative safe operating zone, as evidenced by the extensive test data in this study. In power MOSFETs, the same constant power approach leads to an overestimate of capability in the high-voltage regime, where failure occurs prematurely due to thermal instability (Spirito effect).

While the exact physics of failure is yet to be determined, the main outcome of this study is clear – GaN transistors will not fail when operated within their datasheet SOA.

5.5.3.3 Short-Circuit Robustness

Short-circuit robustness refers to the ability of a transistor to withstand unintentional fault conditions that may occur in an application while in the ON (conducting) state. In such an event, the device experiences the full bus voltage combined with a current that is limited only by the inherent saturation current of the transistor and the circuit parasitic resistance, which varies with the application and location of the fault. If the short-circuit state is not quenched by protection circuitry, the extreme power dissipation will ultimately lead to thermal failure of the transistor. The goal of short-circuit testing is to quantify the "withstand time" the part can survive under these conditions.

Typical protection circuits (e.g. de-saturation protection for IGBT gate drivers) can detect and react to over-current conditions in 2–3 µs. It is therefore desirable if the GaN transistor can withstand unclamped short-circuit conditions for about 5 µs or longer. The two main test circuits used for short-circuit robustness evaluation are described in [29]. They are

- Hard-switched fault (HSF): gate is switched ON (and OFF) with drain voltage applied.
- Fault under load (FUL): drain voltage is switched ON while gate is ON.

For this study, devices were tested in both fault modes and no significant differences in the withstand time were found. Therefore, the focus will be on FUL results for the remainder of this discussion. However, it is important to note that from HSF testing, GaN transistors did not exhibit any latching or loss of gate control that can occur in silicon-based IGBTs [30]. This result was expected given the lack of parasitic bipolar structures with the GaN devices. Until the transistors fail catastrophically, the short circuit can be fully quenched by switching the gate low, an advantageous feature for protection circuitry design.

Two representative GaN transistors were tested:

1) EPC2203 (80 V): a fourth-generation automotive grade (AEC) device [31].
2) EPC2051 (100 V): a fifth-generation device [32].

These devices were chosen because they are the smallest in their product families. This simplified the testing owing to the high currents required for short-circuit evaluation. However, based on simple thermal scaling arguments, the withstand time is expected to be identical for other in-family devices.

Figure 5.20 shows fault-under-load (FUL) data on EPC2203 for a series of increasing drain voltages. With V_{GS} at 6 V (the datasheet maximum), and a 10 µs drain pulse, the device did not fail up to V_{DS} of 60 V. Under these conditions, over 1.5 kW is dissipated in a 0.9 mm × 0.9 mm die. At the higher V_{DS}, the current is seen to decay over time during the pulse. This is a result of rising junction temperature within the device and does not signify any permanent degradation.

Using a longer pulse duration (25 µs), the parts eventually fail from thermal overload. Representative waveforms are shown in Figure 5.21. The time of failure is marked by the abrupt sharp rise in drain current. After this event, the devices are permanently damaged. The withstand time is measured from the beginning of the pulse to the time of failure.

To gather statistics on the withstand time, cohorts of eight parts were tested to failure using this approach. Table 5.4 summarizes the results. EPC2203 was tested at both 5 V (recommended gate drive) and 6 V ($V_{GS(max)}$), with mean withstand time of 20 µs and 13 µs, respectively. Note that the device survives less time at 6 V because of the higher saturation current. EPC2051 exhibited a slightly lower time-to-fail (9.3 µs) compared with the EPC2203 at 6 V. This is expected because of the more aggressive scaling and current density of fifth-generation products. However, in all cases, the withstand time is comfortably long enough for most short-circuit protection circuits to respond and prevent device failure. Furthermore, the withstand time showed small part-to-part variability.

The lower rows in Table 5.4 provide pulse power and energy relative to die size. To gain insight into the relationship between these quantities and the time to failure, time-dependent heat transfer was simulated to determine the rise in junction temperature ΔT_J during the short-circuit pulse. The results are shown in Figure 5.22. Note that EPC2203 fails catastrophically at a ΔT_J of around 475 °C, whereas EPC2051 fails around 575 °C. The simulated ΔT_J is

GaN Reliability

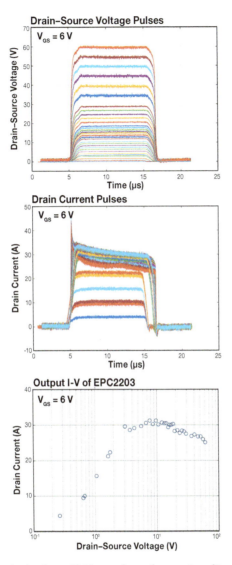

Figure 5.20 EPC2203 fault under load test (FUL) waveforms for a series of increasing drain voltages. Drain pulse is 10 μs and V_{GS} = 6 V. The device did not fail for this pulse width. In the V_{DS} vs. time plot (top), V_{DS} is Kelvin-sensed directly at the device terminals. In the I_{DS} vs. time plot (center), it is noted that I_{DS} decreases over time due to self-heating. Resulting output curve for this test sequence (bottom). Drain current is reported as the average current during the pulse. Drain current rolls over in the saturation region owing to device heating at higher V_{DS}.

well fit by a simple square root dependence on time (heat diffusion), as shown in the equation. P denotes the average power per unit area, and $k = 6.73 \times 10^{-5}$ Km²/Ws$^{1/2}$.

The intense power density during the pulse leads to rapid heating in the GaN layer and underlying silicon substrate. Because the pulse is short and heat transfer is relatively slow, only a small thickness of semiconductor substrate (<~100 μm in depth) can help to absorb the energy. The temperature grows as the square root of time (characteristic of heat diffusion), and linearly with the pulse power. As can be seen in Figure 5.22 for EPC2203, both the 5 V

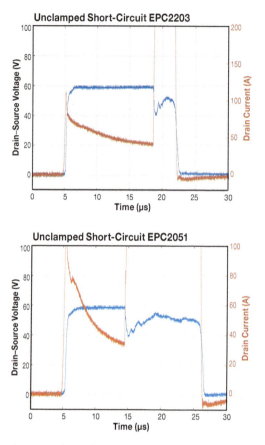

Figure 5.21 Fault-under-load test waveforms for a typical EPC2203 (top) and EPC2051 (bottom) at V_{DS} = 60 V, V_{GS} = 6 V, and a 25 μs drain pulse. The abrupt rise in drain current marks the time of catastrophic thermal failure.

Table 5.4 Short-circuit withstand time statistics for EPC2203 and EPC2051.

Short-circuit pulse V_{DS} = 60 V	EPC2203 (Gen 4) V_{GS} = 6 V	EPC2203 (Gen 4) V_{GS} = 5 V	EPC2051 (Gen 5) V_{GS} = 6 V	EPC2051 (Gen 5) V_{GS} = 5 V
Mean TTF (μs)	13.1	20.0	9.33	21.87
Standard deviation (μs)	0.78	0.37	0.21	2.95
Minimum TTF (μs)	12.1	19.6	9.08	18.53
Average pulse power (kW)	1.764	1.4	3.03	2.03
Energy (mJ)	23.83	27.6	27.71	42.49
Die area (mm^2)	0.9025		1.105	
Avg power/area (kW/mm^2)	1.95	1.55	2.74	1.84
Energy/area (mJ/mm^2)	26.4	30.59	25.08	38.46

Note: Statistics derived from eight devices in each condition. Withstand times are tightly distributed around mean value. Average pulse power and energy correspond to a typical part within the population.

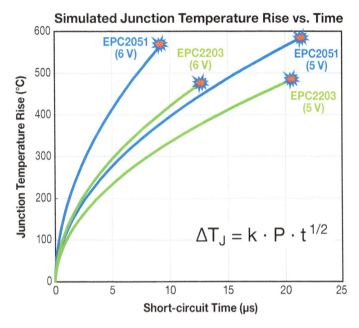

Figure 5.22 Simulated junction temperature rise versus time during the short-circuit pulses for both EPC2051 and EPC2203 at both 5 V and 6 V V_{GS}. Measured failure times are indicated by red markers.

and 6 V conditions fail at the same junction temperature rise of ~475 °C. The same is true for EPC2051, where both conditions fail at the same ΔT_J of ~575 °C.

Three key conclusions stem from these results:

1) For a given device, the time to failure is inversely proportional to the power dissipation squared (P^{-2}). This applies for short-circuit and SOA pulses of duration <1 ms.
2) The intrinsic failure mode resulting from high power pulses is directly linked to the junction temperature exceeding a certain critical value.
3) Wide bandgap GaN devices can survive, and maintain control, at junction temperatures greater than 400 °C. This capability is inaccessible to silicon devices owing to free-carrier thermal runaway.

5.5.3.4 Electromigration (EM) for Copper Interconnect

With electronic devices seeing an increase in power density but with a reduction in size, copper pillars have emerged as one of the more popular new interconnect solutions due to their excellent electrical characteristics and heat dissipation [33]. Figure 5.23 illustrates how the copper pillars connect to a lead frame-based package and a chip.

The pillar consists of two main structures including a solder cap and a cylindrical or elliptical shaped copper. The solder cap serves as the interconnect between the device and the package, mainly composed of tin (Sn) with trace amounts of silver (Ag), gold (Au), and copper (Cu) [33–38].

EM has been identified as one of the main wear-out mechanisms in the interconnects of PQFN package devices. EM is defined as the movement of atoms in metal structure, leading to void formation [39, 40]. The primary cause of EM is electron "wind" generated from the transfer of momentum between moving electrons and metal ions in the crystal. When the

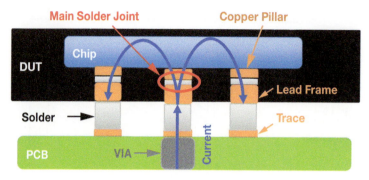

Figure 5.23 Illustration of the EM copper pillar test structure.

momentum surpasses the diffusion threshold that is governed by an activation energy [39, 40], metal atoms can move and create voids.

High current density and high temperature are the two primary stressors responsible for EM void formation. Current density is expressed as the current divided by the area of contact between the two-metal interconnect. The initial void formation leads to decreasing cross-sectional area, which further increases current density, accelerating the void formation. This process forms a positive feedback loop, which is characterized by the current density term in Black's model [39] as shown in Eq. (5.18). High temperature is the second main stressor. Joule heating causes an increase in junction temperature, which further accelerates the movement of atoms, resulting in more void formation. This process forms a second positive feedback loop, which can be modeled by the exponential term in Eq. (5.18).

$$\text{MTTF} = Aj^{-n}e^{\frac{Q}{kT}} \tag{5.18}$$

where A is a constant, j is current density, n is an exponent, Q is the activation energy, k is Boltzmann's constant at 8.62×10^{-5} eV/K, and T is the temperature in Kelvin unit.

Both feedback loops could lead to an open circuit from void formation or electrical shorts from the melting metal interconnect. Due to EM being a slow mechanism that can take years to develop under use conditions, testing under accelerated stress conditions is necessary to generate EM related failures within a reasonable amount of time.

A custom test chip was designed by following JEDEC standard, JEP154 [41]. The test setup is placed in a temperature chamber with the DUT card placed in the center. Two thermocouples are deployed. One is mounted at the center of the oven to monitor the ambient temperature. The other one is placed directly on the backside of the DUT where the Si substrate is exposed. The test chip is covered with thermal putty and sandwiched between two copper heat sinks to maintain a constant temperature of the test chip. The temperature difference between the copper pillar interconnect and the backside of the device, where the second thermocouple is placed, is calculated to be 0.64 °C by using the $R_{\text{th, JC}}$ of 0.2 °C/W and a total of 3.2 W of power dissipated at 125 °C. The copper pillar interconnect of interest has an elliptical shape with a cross-section area of 5271 µm².

Test conditions of 27 kA/cm² at 125 °C and 55 kA/cm² at 150 °C were selected, based on previous research studies focusing on copper pillar interconnects [33–38]. A failure criterion of 10% resistance increase was adopted according to the recommendations in JEP154 [41]. Both test conditions yielded zero failures as shown in Figures 5.24 and 5.25. Figure 5.24 shows that after 480 hours of testing, no devices exceeded a 2% resistance increase. Similarly, Figure 5.25 shows that after approximately 800 hours of testing under such extreme stress

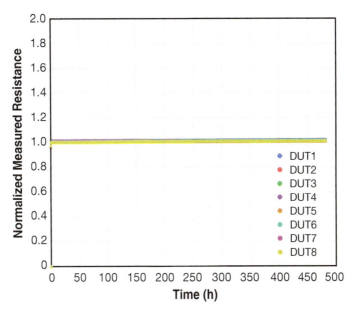

Figure 5.24 Normalized measured resistance of copper interconnect for 24 kA/cm² at 125 °C.

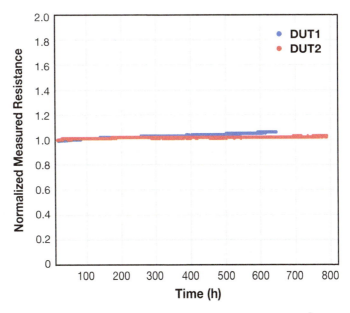

Figure 5.25 Normalized measured resistance of copper interconnect for 55 kA/cm² at 150 °C.

condition (55 kA/cm² at 150 °C), no parts exceeded the failure criterion of 10% resistance increase. These results are consistent with various studies that focus on EM copper interconnects [33–38].

A current density power exponent of 2 has been frequently reported for copper pillar interconnects by various studies [33–38]. An activation energy of 1 eV is commonly accepted for SnAg solder cap through previous works [33–38]. By using the values of $n = 2$ and $Q = 1$ eV and assuming the time to failure of 645 hours at 0.1% failure rate, the constant A from Black's equation is calculated to be 2.39. By having the constant A, the lifetime at 0.1% failure rate at any given temperature and current density can be projected. Many GaN-based PQFN devices are based on an EM current density limit of 10 kA/cm^2 [33–38]. By plugging in a current density of 10 kA/cm^2 and a junction temperature of 125 °C, greater than 11 years of lifetime at 0.1% failure rate is projected.

5.5.4 Thermo-Mechanical Wear-Out

5.5.4.1 Introduction to Thermo-Mechanical Wear-Out Mechanisms

Solder joint cracking, which occurs due to a mismatch in the CTE between materials, has emerged as a common concern in applications that demand frequent and large temperature swings. A general temperature cycling (TC) lifetime model is developed in this section based on testing parts to failure. The lifetime model encompasses device dimensions, bump shape, stand-off height, and various PCB properties including modulus, Poisson's ratio, and PCB thickness. When the expected lifetime of chip-scale packaged (CSP) devices is less than the customers' specifications, underfill with the right materials properties is recommended to improve TC lifetime.

5.5.4.2 Impact of Die Size and Bump Shape on Temperature Cycling (TC) Reliability

The main failure mode under TC stress is identified as solder joint cracking [2, 16–18]. CTE mismatch between the materials, namely the device, solder and PCB, is attributed as the fundamental wear-out mechanism. The CTE value of a typical FR4 PCB is CTE = 18 ppm/°C [42], but a chip-scale packaged (CSP) device typically has a CTE of 4 ppm/°C [43]. They are connected by SAC305 solder (96.5%Sn-3%Ag-0.5%Cu) that usually has a CTE of 23 ppm/°C [44].

A suite of CSP GaN transistors with various dimensions are evaluated for temperature cycling performance, including EPC2204, EPC2218, EPC2069 [45], EPC2071, and EPC2206 [46], shown in Figure 5.26.

The TC experiment was constructed to ensure that the only variables are the device dimensions and bump shape. These devices were mounted on identical test PCB boards using identical solder (SAC305). The PCB boards consist of two-layer Cu, 1.6-mm thick, FR4 board. The standoff height (i.e. the solder height after assembly) of ∼130 μm was maintained during the assembly process. This was verified by performing a physical cross-section of the assembled boards. The TC range was from −40 °C to 125 °C, with a ramp rate of 15 °C/min and soak time of 10 minutes at the end points following industry standard JESD22-A104F [47]. After every

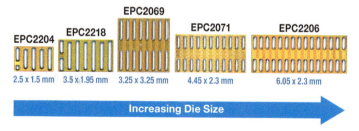

Figure 5.26 Devices tested in this study: EPC2204, EPC2218, EPC2069, EPC2071, and EPC2206. Source: With permission of EPC.

temperature cycling interval (~200 cycles), an electrical screening was performed, where an increase in $R_{DS(on)}$ exceeding datasheet limits was identified as the main electrical failure mode.

The failure distribution was analyzed using a two-parameter Weibull distribution for each device using maximum likelihood estimation (MLE) [48]. The resulting Weibull fits are indicated by solid lines in the graph of Figure 5.27, and the Weibull characteristics are included in Table 5.5. Solder joint cracking at the gate connection was found to be the consistent wear-out mode throughout all devices analyzed by physical cross-sectioning and SEM inspection, establishing that wear-out of the smallest solder bump at the farthest extreme is the limiting factor for TC lifetime.

As an initial attempt to model the failure rate, the Mean-Time-To-Fail (MTTF) data from the Weibull distribution, measured in number of cycles, was compared to die area to check for die size correlation with TC lifetime. The data is fit to Eq. (5.19),

$$\text{MTTF} = A(\text{Die Area})^{-n} \tag{5.19}$$

where A is a constant, Die Area is the area of die, and n is the exponent. The resultant fit is judged by the goodness-of-fit (R^2). Fitting MTTF from Table 5.2 with Eq. (5.11) yielded an R^2 of 0.67, indicating a poor fit. It suggests that die area alone is unable to provide a good

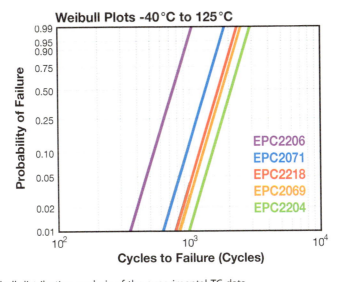

Figure 5.27 Weibull distribution analysis of the experimental TC data.

Table 5.5 Weibull statistics for tested devices.

Device	Weibull slope parameter	Mean time to fail (cycles)
EPC2206	5.6	737
EPC2071	5.6	1309
EPC2218	5.6	1630
EPC2069	5.6	1737
EPC2204	5.6	2208

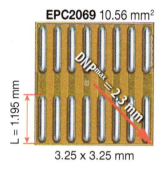
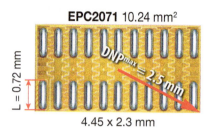

Figure 5.28 Example of gate length and DNP^{max} for EPC2069 and EPC2071. *Source:* With permission of EPC.

correlation with TC lifetime by following the commonly accepted lifetime models in literature [49–51].

The concept of maximum distance from neutral point (DNP^{max}) is introduced as shown in Figure 5.28. During TC stress, the center point of the device experiences the least stress compared to extremities of the device. This center point is defined as the neutral point, while the distance from the neutral point to the farthest extremity of solder bump is defined as DNP^{max}.

Equation (5.20) is developed by combining Norris–Landzberg modified Coffin–Manson TC lifetime model [52] and the concept of DNP [53].

$$\text{MTTF} = A(\text{DNP}^{max})^{-n} \quad (5.20)$$

The best fit to Eq. (5.20) yielded an R^2 value of 0.79, slightly improved compared with simply using the device area. However, it is still not considered a good fit.

Because the corner gate solder joint cracking is responsible for TC failures, the length of corner gate bump (L) must be factored into DNP^{max} for a more accurate TC lifetime model development. Therefore, effective DNP^{max} (DNP^{eff}) is developed and shown in (Eq. 5.21).

$$\text{DNP}^{eff} = \text{DNP}^{max} + a \cdot L \quad (5.21)$$

The resulting fit is shown in Figure 5.29 and results in an R^2 value of 0.99 using gate length factor $a = -0.65$, and power exponent $n = 1.4$.

The fitted power exponent of 1.4 shown in Figure 5.29 is consistent with other literature results [49–51], where exponents between 1 and 2 are frequently reported in SAC305 solder joint cracking failures under TC stress with similar test conditions.

In summary, a TC lifetime model is proposed considering the device size and corner gate bump shape in Eq. (5.22).

$$\text{MTTF} = A(\text{DNP}^{max} - 0.65 \cdot L)^{-n} \quad (5.22)$$

5.5.4.3 Effect of PCB Properties on Temperature Cycling (TC) Lifetime

High-density power modules frequently utilize high-layer count and thick PCBs. This raises concerns about solder joint reliability under TC due to the increased stiffness of these complex PCBs. Three types of PCBs and a single type of PQFN packaged GaN transistors are evaluated for TC reliability testing under the test conditions of −40 °C to 125 °C with 10 minutes dwell time at hot and cold temperature extremes. Both PCB type-1 and type-2 use 2-layer Cu with a total PCB thickness of 1.6 mm but with 1 oz Cu per layer and 2 oz Cu per layer, respectively, where 1 oz Cu is equivalent of approximately 35 μm in thickness. PCB type-3 utilizes 16-layer Cu with

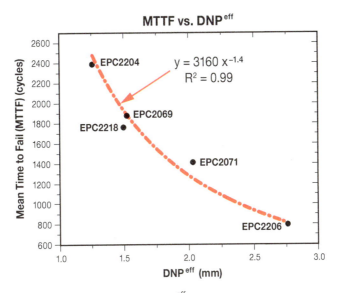

Figure 5.29 Comparing MTTF with effective DNP (DNP^{eff}).

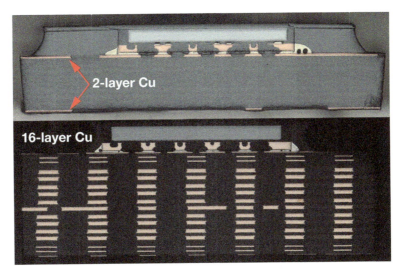

Figure 5.30 Cross-sectional view of the two-layer PCB (2 oz Cu per layer) and 16-layer PCB. *Source:* With permission of EPC.

varying Cu thickness and the total PCB thickness is 3.2 mm. Figure 5.30 shows the cross-sectional result of the 2-layer PCB and 16-layer PCB.

Product datasheet limits are used as the failure criteria, and $R_{DS(on)}$ exceeding the datasheet maximum limit is found to be the primary electrical failure mode. At each TC testing interval, parts were randomly selected for failure analysis. The gate corner solder joint cracking is the consistent failure mode as shown in Figure 5.31.

The failure distribution was analyzed using a two-parameter Weibull distribution plot with solid lines, as shown in Figure 5.32. The Weibull shape (or slope) parameter was constrained to

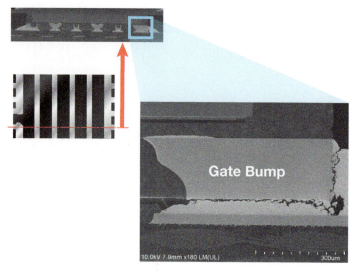

Figure 5.31 SEM images of a TC failure, where the gate solder joint at the PQFN device corner is responsible for the TC failure. *Source:* With permission of EPC.

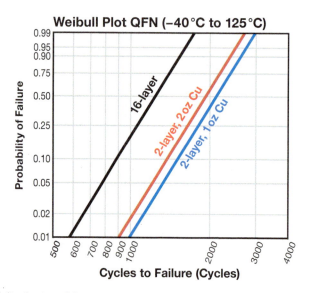

Figure 5.32 Weibull distribution of three testing legs. Leg 1 (blue) uses 2 layers of Cu with 1 oz Cu per layer. Leg 2 (red) still uses two layers of Cu but with 2 oz of Cu per layer. Leg 3 (black) uses 16-layer PCB with varying Cu thickness.

be the same for all TC legs because a single failure mode was found through failure analysis. Table 5.6 summarizes the details of the Weibull analysis.

Table 5.6 shows that the two-layer PCB with 1 oz Cu layer outperformed the other group with 2 oz Cu per layer by approximately 10% in MTTF. When comparing the MTTF of the 2-layer PCB and 16-layer PCB, an approximately 40% lifetime reduction is measured.

In the following discussions, a first-principles model is adopted to investigate the fundamental mechanisms responsible for the TC lifetime differences that were found in this study.

Table 5.6 Weibull characteristics of TC experiment of three different types of PCB.

PCB types	Weibull slope parameter	Measured MTTF (Cycles)
2-layer Cu PCB 1 oz Cu per layer	5.6	2086
2-layer Cu PCB 2 oz Cu per layer	5.6	1903
16-layer PCB	5.6	1224

In literature, it has been widely reported that thicker boards generally lead to worse TC performance and shorter lifetime caused by solder joint wear-out failure mode [54–57]. The influence of PCB's properties on solder joint lifetime under TC stress can be modeled by Clech's "board thickness" model [55]. Clech's model is developed by modeling the mechanical coupling between component and PCB. Although it is commonly referred to as the "board thickness" model, it is a comprehensive model that accounts for all critical parameters involving the component, board, and assembly.

The overall lifetime, N_{Total}, consists of three parts of life which associate three different mechanical coupling mechanisms. The first part, N_1, is the lifetime that is characterized by the in-plane tensile shear force, acting on the device. Figure 5.33 illustrates the evolution of the dimensional changes of a device and a PCB when the ambient temperature increases from a low temperature, where the stress on the solder joints is neutral, to the hot temperature extreme where the device expands significantly less than the PCB due to the CTE mismatch. As a result, the solder joints are stretched laterally as shown in Figure 5.33.

N_1 represents the in-plane tensile stiffness of the mounted device as shown by the green arrow in Figure 5.33. Equation (5.23) specifies the lifetime caused by such in-plane stencil shear force,

$$N_1 = \frac{F}{\Delta\alpha^2} \times \frac{1-\gamma_{\text{QFN}}}{E_{\text{QFN}}h_{\text{QFN}}} = \frac{F}{\Delta\alpha^2} \times C_1 \qquad (5.23)$$

where F is a constant for a specific device-PCB system and under a given TC stress condition, $\Delta\alpha$ is the CTE mismatch between the device and PCB, γ_{QFN} is Poisson's ratio of the device, E_{QFN} is its Young's modulus, and h_{QFN} is the height of the device. C_1 is denoted as the axial compliance of the device, $C_1 = \frac{1-\gamma_{\text{QFN}}}{E_{\text{QFN}}h_{\text{QFN}}}$.

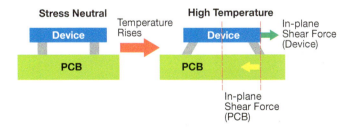

Figure 5.33 Illustration of the in-plane tensile shear forces acting on the device and PCB.

The second term, N_2, is controlled by the in-plane tensile shear force that acts on the PCB as highlighted by the yellow arrow in Figure 5.33. Equation (5.24) characterizes the corresponding lifetime that is related to such tensile stiffness of the PCB.

$$N_2 = \frac{F}{\Delta \alpha^2} \times \frac{1-\gamma_{PCB}^2}{2E_{PCB}h_{PCB}} = \frac{F}{\Delta \alpha^2} \times C_2 \quad (5.24)$$

where F and $\Delta \alpha$ are the same as in Eq. (5.23), γ_{PCB} is Poisson's ratio of the PCB, E_{PCB} is its Young's modulus, and h_{PCB} is the PCB thickness. Equation (5.25) defines the axial compliance of the PCB, C_2.

$$C_2 = \frac{1-\gamma_{PCB}^2}{2E_{PCB}h_{PCB}} \quad (5.25)$$

Lastly, N_3 represents the bending moments of the bimetallic strip of the device and PCB. Figure 5.34 shows the FEA simulation result of such bending motion. This part of lifetime, N_3, is dominated by the flexural modulus of the QFN device and the PCB.

$$N_3 = \frac{F}{\Delta \alpha^2} \times \frac{H^2}{\frac{E_{QFN}^f h_{QFN}^3}{12(1-\gamma_{QFN})} + \frac{E_{PCB}^f h_{PCB}^3}{6(1-\gamma_{PCB}^2)}} = \frac{F}{\Delta \alpha^2} \times C_3 \quad (5.26)$$

where E_{QFN}^f and E_{PCB}^f are the flexural Young's moduli of the QFN device, respectively. C_3 is the bending compliance of the bimetallic strip assembly of the device and PCB, defined by Eq. (5.27).

$$C_3 = \frac{H^2}{\frac{E_{QFN}^f h_{QFN}^3}{12(1-\gamma_{QFN})} + \frac{E_{PCB}^f h_{PCB}^3}{6(1-\gamma_{PCB}^2)}} \quad (5.27)$$

H is further defined by Eq. (5.28),

$$H = \frac{h_{QFN}}{2} + h_{standoff} + \frac{h_{PCB}}{2} \quad (5.28)$$

where $h_{Standoff}$ is the standoff height of the solder joint, post-assembly.

Therefore, the total lifetime N_{Total} is determined by the sum of all three parts, as shown in Eq. (5.27).

$$N_{Total} = N_1 + N_2 + N_3 = \frac{F}{\Delta \alpha^2} \times (C_1 + C_2 + C_3) \quad (5.29)$$

Table 5.7 summarizes all the key parameters that are used for the TC lifetime modeling of the QFN device and a two-layer Cu PCB system. The modulus of three different types of PCB is measured by dynamic mechanical analyzer (DMA) [59]. The standoff height ($H_{standoff}$), PCB thickness (h_{PCB}), and QFN device thickness (h_{QFN}) are measured through physical

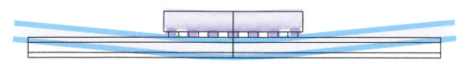

Figure 5.34 COMSOL [58] finite element analysis (FEA) simulation illustrates the flexural bending between the QFN device and PCB. *Source:* Adapted from COMSOL [58].

Table 5.7 Key parameters that are used for the TC lifetime projection.

	2-layer PCB1 Oz Cu per layer	2-layer PCB2 Oz Cu per layer	16-layer PCB	Unit
E_{QFN}	112	112	112	GPa
h_{QFN}	0.65	0.65	0.65	mm
v_{QFN}	0.22	0.22	0.22	
E_{PCB}	14.4	15.5	9.6	GPa
h_{PCB}	1.57	1.6	3.2	mm
v_{PCB}	0.16	0.16	0.16	
$h_{standoff}$	0.087	0.087	0.087	mm
H	1.197	1.212	2.017	mm
C_1	0.011	0.011	0.011	
C_2	0.022	0.020	0.016	
C_3	0.112	0.104	0.071	
$C_1+C_2+C_3$	0.144	0.134	0.097	

cross-section. Poisson's ratios and modulus of the QFN device are estimated based on the respective weighted percentage of the composing materials.

Based on the parameters listed in Table 5.7, the common constant F in Eqs. (5.23)–(5.26) can be estimated. Therefore, every part of the lifetime (N_1, N_2, N_3) and the overall lifetime, N_{Total}, can be modeled, as summarized in Table 5.8. Table 5.8 shows that the modeled lifetime agrees well with the measured MTTF. N_3 lifetime, which represents the bending flexural rigidity contribution, accounts for approximately 90% of the overall lifetime. The axial shear stress acting on the PCB contributes to approximately 10% of the overall TC lifetime, and shear stress from the QFN device is minimal, less than 1%.

The analyses revealed that 10% MTTF difference between two-layer PCB with 2 oz Cu per layer and 1 oz Cu per layer is primarily (80%) caused by the bending life of N_3. The modulus difference between two types of PCB dominates as summarized in Table 5.8. The more significant TC lifetime difference from 2-layer PCB to 16-layer PCB is mainly attributed to the PCB thickness. Such trend is consistent with previously reported works [54–57].

Table 5.8 Measure MTTF vs. modeled lifetimes of three different PCB assemblies.

PCB type	Measured MTTF (Cycles)	Modeled life (Cycles)	Modeled N_1 (Cycles)	Modeled N_2 (Cycles)	Modeled N_3 (Cycles)
2-layer 1 oz Cu PCB (−40 °C to 125 °C)	2086	2088	9	209	1870
2-layer 2 oz Cu PCB (−40 °C to 125 °C)	1903	1901	9	173	1719
16-layer PCB(−40 °C to 125 °C)	1224	1225	9	103	1113

5.5.4.4 Criteria for Choosing a Suitable Underfill

The selection of underfill material should consider a few key properties of the material as well as the die and solder interconnections. Firstly, the glass transition temperature of the underfill material should be higher than the maximum operating temperature in the application. Also, the CTE of the underfill needs to be as close as possible to that of the solder since both will need to expand/contract at the same rate to avoid additional tensile/compressive stress in the solder joints. As a reference, typical lead-free SAC305 and Sn63/Pb37 have CTEs of approximately 23 ppm/°C. Note that when operating above the glass transition temperature (Tg), the CTE increases significantly. Besides Tg, and CTE, the Young (or storage) modulus is also important. Three underfill are recommended in Table 5.9 because they are proven to improve TC reliability.

The main guidelines for choosing an underfill for use with GaN transistors are listed below:

- Underfill CTE should be in the range of 16–32 ppm/°C, centered around the CTE of the solder joint (23 ppm/°C). Lower values within this range are preferred because they provide better matching to the die and PCB.
- Glass transition temperature (Tg) should be comfortably above the maximum operating temperature. When operated above Tg, the underfill loses its stiffness and ceases to protect the solder joint.
- Young's (or storage) modulus in the range of 6–13 GPa. If the modulus is too low, the underfill is compliant and does not relieve stress from the solder joints. If it is too high, the high stresses begin to concentrate at the die edges.

To better understand the key factors influencing thermo-mechanical reliability when using underfills, COMSOL FEA simulations of EPC2206 under temperature cycling stress were conducted. Figure 5.35 shows the simulation deck used for this analysis. The die is placed on a 1.6 mm FR4 PCB, and the temperature change is $\Delta T = +100\,°C$ above the neutral (stress free) state. Two key underfill parameters were varied: Young's modulus and CTE. As shown in Figure 5.35, stress is analyzed along the cut line shown, providing visibility into the stress within the solder bars, die, and underfill.

Figure 5.36 shows the von Mises [60] peak shear stress in the edge-most solder bar along the cutline. For clarity, only stress in the solder bar is shown. In addition, mechanical deformations are exaggerated by 20 times to illustrate the shear displacement in the joint. Four distinct underfill conditions are simulated by changing Young's modulus (E) or the CTE of the underfill. As can be seen, the solder bar in the no underfill case has by far the most extreme shear stress and deformation. The addition of underfill significantly alleviates stress from the joint.

Table 5.9 Material properties of three recommended underfill materials.

| Manufacturer | Part number | Tg (TMA) (C) | CTE (ppm/C) | | Storage modulus (DMA) at 25 °C (N/mm²) | Viscosity at 25 °C | Poisson's ratio |
			Below Tg	Above Tg			
HENKELS LOCTITE	ECCOBOND-UF 1173	160	26	103	6000	7.5 Pa*S	
NAMICS	U8437-2	137	32	100	8500	40 Pa*S	0.33
NAMCIS	XS8410-406	138	19	70	13000	30 Pa*S	

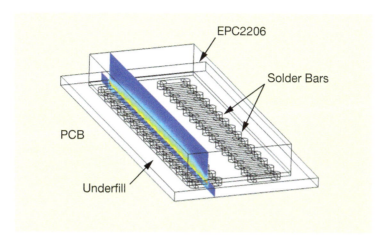

Figure 5.35 Simulation deck for finite element analysis of stresses inside EPC2206 under temp cycling stress. Die with underfill sitting on 1.6 mm FR4 PCB. Stress is analyzed along the colored cut line shown.

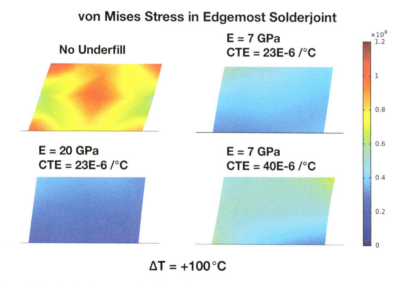

Figure 5.36 von Mises (peak shear stress) in the edge-most solder bar under a temperature cycle change of $\Delta T = +100\,°C$. Four different underfill conditions are simulated, with changing Youngs modulus (E) of the underfill and different CTE as well. Note that mechanical deformation has been exaggerated by 20× in all cases. *Source:* With permission of EPC.

Higher Young's modulus reduces this stress further. For underfills with poor CTE matching to the solder joint, stress can also build up in the joint. FEA analysis shows that there is an optimal Young's modulus in the range of ~6–13 GPa, providing a good compromise between protecting the solder joint and protecting the die edge. Regarding CTE, the analysis shows that high underfill CTE (>32) should be avoided.

The effect of underfill on temperature-cycling reliability was studied using EPC2218A [61]. Three different combinations of temperature cycling stress conditions, with and without underfill material were studied. Two temperature cycling ranges were tested: temperature cycle 1 (TC1): −40 °C to 125 °C and temperature cycle 2 (TC2): −40 °C to 105 °C. Over the temperature range of TC1, two cases were compared: one with and one without underfill material. The underfill material selected was from Henkels Loctite (part number: Eccobond-UF 1173) which showed good performance in Table 5.9. For all cases, the parts were mounted on DUT cards consisting of a two-layer, 1.6 mm thick, FR4 board using SAC305 solder paste, and water-soluble flux. All underfilled devices were subjected to a plasma clean process prior to the underfill application. Industry standard (JESD22-A108F [47]) as well as other customers' specifications were followed for this study.

A group of 88 EPC2218A devices were tested for each test leg, and all three legs used similar ramp rate and dwell time at the two temperature extremes. After every temperature cycling interval, electrical screening was performed. Exceeding datasheet limits was used as the criterion for failure. Physical cross-sectioning and SEM inspection were followed to further examine the electrical test failures. Solder joint cracking was found to be the consistent failure mode throughout all failures analyzed. The experimental results from the test-to-fail approach are summarized in Weibull plots in Figure 5.37 and Table 5.10.

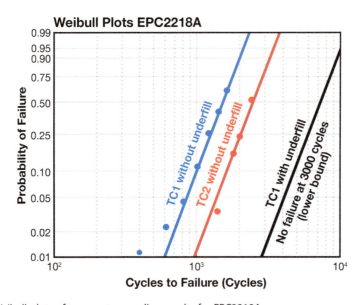

Figure 5.37 Weibull plots of temperature cycling results for EPC2218A.

Table 5.10 Temperature cycling profile and MTTF determined by Weibull plots in Figure 5.37.

TC condition	ΔT (°C)	Cycle duration (min)	Frequency (cycles per day)	MTTF (cycles)
TC1 without underfill	165	40	36	1505
TC2 without underfill	145	30	48	2430
TC1 with underfill	165	40	36	7230 (lower-bound confidence level)

The TC1 (−40 °C to 125 °C) tests without underfill material reached more than 50% cumulative failures at 1600 cycles, where physical failure analysis found that solder joint cracking was the single failure mode for all failures at various read points. The TC2 (−40 °C to 105 °C) tests without underfill achieved 50% failure rate after 2400 cycles.

The data in Figure 5.37 shows that a larger temperature range accelerates the time of failure in TC stress. After 3000 cycles of TC1 (−40 °C to 125 °C) with Henkel underfill, no outlier devices were found in the absolute $R_{DS(on)}$ value, nor in $R_{DS(on)}$ shift post electrical testing. All parameters examined showed very tight distributions throughout all temperature cycling intervals.

Physical cross-sectioning was conducted randomly on the 3000-cycle passing devices, where no solder joint cracking was observed. This shows that applying proper underfill material can significantly improve the TC capability of the chip-scale package devices. Therefore, the Weibull fit line for the TC1 with the underfill leg is merely the lower bound confidence level based on the current test results.

To understand the main failure mechanisms involved in board-level temperature cycling, the Norris–Landzberg lifetime model was adopted and shown in Eq. (5.30) [52].

$$N = A \cdot f^{-\alpha} \cdot \Delta T^{-\beta} \cdot \exp\left(\frac{E_a}{kT_{\text{Max}}}\right) \tag{5.30}$$

where N is the number of cycles to fail, f is the cycling frequency, and α is the cycling frequency exponent. This frequency term f is to describe the frequency of usage.

In this study, the cycling frequency is determined by counting the total number of cycles per day. The cycling frequency exponent α is −1/3 [62–66]. ΔT is the range of temperature change in one cycle and β is the temperature range exponent. The temperature range exponent β is typically around 2. Since SAC305 solder is used in this study, a value of 2.3 for β is assumed [62–66].

The last variable is an Arrhenius term that focuses on the creep failure mechanism at the maximum temperature, T_{Max} in each cycle, where E_a is the activation energy, k is the Boltzmann constant, and T_{Max} is the maximum temperature of the high-temperature dwell stage in Kelvin (K). The activation energy (E_a) was calculated to be 0.18 eV.

This study forms the basis for the temperature-cycling reliability analysis of solar and DC–DC converters presented in Sections 5.6.1.6 and 5.6.2.5 respectively.

5.6 Mission-Specific Reliability Predictions

Section 5.4 introduced a theoretical framework for analyzing device lifetimes in applications that include stress conditions of different intensities and durations. In this section that framework is applied to three example applications: solar systems, DC–DC conversion, and lidar.

5.6.1 Solar-Application-Specific Reliability

5.6.1.1 Introduction

Microinverters and power optimizers are widely utilized in modern solar panels to maximize energy efficiency and conversion. Such topologies and implementations usually require a minimum of 25 years of lifetime, which is becoming a critical challenge for market adoption. Low-voltage GaN power devices (V_{DS} rating < 200 V) are a promising solution and are being used extensively by an increasing number of solar manufacturers.

In this section, a test-to-fail approach is adopted and applied to investigate the intrinsic underlying wear-out mechanisms of GaN transistors. The study enables the development of physics-based lifetime models that can accurately project the lifetimes under the unique demands of various mission profiles in solar applications.

5.6.1.2 Trends in Photovoltaic Power Conversion

The ever-increasing demand for renewable energy sources has led to a rapid growth in rooftop solar installations across residential and commercial sectors. Traditionally, string inverters have been widely employed in solar installations, where multiple solar panels are connected in series. The inverter is responsible for converting direct current (dc) output from solar panels to alternating current (ac) electricity that can be used to power homes.

String inverters have served as a reliable choice for years. However, they also face many challenges, including reduced performance due to shading, panel mismatch issues, and a lack of module-level monitoring. Most importantly, due to the series configuration of the string inverters, the lowest performing panel dominates the energy conversion rate of the entire system, which could significantly lower the system efficiency.

The Department of Energy released the $1/watt photovoltaic (PV) system initiative in 2010, where developing higher efficiency and more reliable module-level integrated inverters was highlighted as the key area of improvement to meet the target [67]. The SunShot 2030 PV program envisions a similar cost target by 2030 [68]. To meet the goals and maximize energy production, emerging technologies such as microinverters and power optimizers have gained significant attention.

Microinverters are small, individual inverters that are attached to each solar panel, allowing for dc to ac power conversion at the panel level. This enables each solar panel to function at its peak performance by using independent maximum power point tracking (MPPT). Even if a tree branch shades certain panels, all the neighboring panels can still convert at their full capacity. The drop in efficiency only affects the panels in the shade.

Independent tracking also allows solar users to monitor the health of each panel easily. If a panel requires repair, it won't bring down the whole system. In addition, microinverters make it easy to add panels to increase power output. Microinverters can be more expensive than string inverters but can pay off over time by getting more power from the system. Therefore, microinverters in the market need to match panel guarantees with 25-year warranties [69, 70].

Power optimizers are dc–dc converters integrated into the solar panel wiring, enabling MPPT of each individual solar panel by continually regulating the dc characteristics to maximize energy output. A power optimizer is a good solution for situations where shading is an issue, or the panels must be placed on multiple roof surfaces with different orientations. Therefore, power optimizers generally are a more energy efficient solution than string inverters. The power optimizer also requires 25 years of warranty [71, 72].

5.6.1.3 Applying Test-to-Fail for Solar

After reviewing the benefits that are driving the switch from string inverters to microinverters and power optimizers in photovoltaic systems, the test-to-fail methodology is introduced and the three device "stressors" most likely responsible for device failure are identified – gate bias, drain bias, and temperature cycling. The total MTTF can be described by Eq. (5.31).

$$\frac{1}{\mathrm{MTTF_{Total}}} = \frac{1}{\mathrm{MTTF_{Gate}}} + \frac{1}{\mathrm{MTTF_{Drain}}} + \frac{1}{\mathrm{MTTF_{TC}}} \tag{5.31}$$

Therefore, it is critical to understand which stressor is the limiting factor in reliability and warrants more consideration during design and operation. Each stressor is studied independently by using this test-to-fail approach, where the individual intrinsic wear-out mechanism is successfully identified, and the corresponding lifetime is determined.

5.6.1.4 Gate Bias

GaN HEMTs are used in dc–ac (microinverters) or dc–dc (power optimizers) topologies in their solar applications. The gate terminal must be biased periodically during switching. Hence, gate reliability over time is the first stressor to examine. As shown in Figure 5.3 (Section 5.5.1.2), GaN HEMTs have an approximately 1 ppm failure rate projected after 25 years of continuous dc bias at $V_{GS,\,max} = 6\,V$.

5.6.1.5 Drain Bias

The low on-resistance ($R_{DS(on)}$) and small die size of GaN HEMTs significantly increase the power conversion efficiency and reduce the power losses in microinverter and dc–dc converter applications. However, one common concern for GaN is dynamic on-resistance.

The flyback is one of the more popular topologies for microinverters in solar applications. When selecting the appropriate GaN transistors for the primary side, three main contributing factors to the drain voltage are considered. These are (i) the bus voltage, (ii) the flyback voltage, and (iii) the voltage overshoot due to ringing caused by the parasitic inductance in the design. The typical bus voltage for a microinverter is 60 V in a solar application. The flyback voltage is determined by the product of the system's output voltage and the turns ratio of the transformer. By adding some margin for the voltage overshoot and derating, a 170 V maximum V_{DS} rating is frequently desired by the solar customers using such topology.

The EPC2059 [73] is a 170 V maximum V_{DS} rated product that meets the general requirements for microinverters in solar applications. Figure 5.38 shows the in situ $R_{DS(on)}$ test results of a representative EPC2059 device that was operated under continuous hard switching at 136 V (80% of the max rated drain bias of 170 V) while the case temperature was modulated at 80 °C. This temperature is used because it is considered the nominal operating temperature for solar panels. As shown in Figure 5.38, the lifetime model is plotted against the measured data.

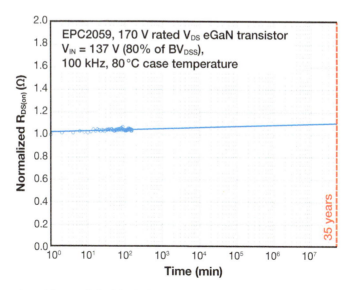

Figure 5.38 The projected $R_{DS(ON)}$ shift of the EPC2059, a 170-V rated device, in 25 years of 100-kHz continuous hard-switching operation at 136 V is approximately 10%. The blue circles represent measured data.

The model predicts the $R_{DS(on)}$ increase due to continuous hard switching in 25 years to be approximately 10%.

Another popular option for solar systems is to use a dc–dc converter in a power optimizer. This has been adopted by many solar providers due to its superior efficiency. EPC's GaN devices such as the 100-V rated EPC2218 and EPC2302 among others are good fits for this application.

Figure 5.39 plots the results obtained with the lifetime model alongside the in-situ measured data for two representative devices – EPC2218 and EPC2302. A shift of less than 10% in 25 years of continuous hard switching at 80% of the max rated drain bias and 100 kHz is expected. This result suggests that dynamic $R_{DS(on)}$ failure is not the dominant factor determining the lifetime for EPC's GaN devices in solar applications.

5.6.1.6 Temperature Cycling

Temperature cycling is a third critical area of particular interest for solar applications. Solar panels are placed outside and experience significant ambient temperature changes each day. Therefore, devices mounted on the PCBs in the solar panels must be capable of surviving 25 years of continuous ambient temperature change.

In real-world applications, solar panels experience varying ambient temperatures, and the amount of temperature change varies significantly depending on the location and season. As a result, a more general lifetime model for thermo-mechanical stress is warranted to account for all mission profiles over the 25 years of lifetime. A TC lifetime model is developed below to account for different ΔT at different seasons of the year, as shown in Eq. (5.32).

$$\frac{1}{N_{Total}} = \frac{a}{N_{\Delta T_a}} + \frac{b}{N_{\Delta T_b}} + \cdots + \frac{i}{N_{\Delta T_i}} \tag{5.32}$$

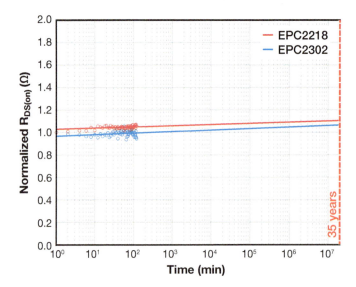

Figure 5.39 The projected $R_{DS(on)}$ shifts of the EPC2218 and the EPC2302, which both are 100-V rated devices, under continuous hard-switching operation at 80 V, 100 kHz are plotted here. The blue and red circles represent measured data.

where N_{Total} is the total calculated lifetime number of cycles, $N_{\Delta T_a}$ corresponds to cycles-to-failure for the condition of ΔT_a and a is the fraction of time the device was operational under the condition of ΔT_a, $N_{\Delta T_b}$ corresponds to cycles-to-failure for the condition of ΔT_b and b is the fraction of time the device was operational under ΔT_b, and $N_{\Delta T_i}$ corresponds to cycles-to-failure for the condition of ΔT_i and i is the fraction of time the device was operational under ΔT_i.

There are three main factors that need to be considered to model the lifetime of the solder joints:

1) The duration of each mission profile needs to be separated. This effect is accounted for by the fractional coefficient in the numerator of each term in Eq. (5.32), such as a, b, ..., and i.
2) The temperature change (ΔT) in each mission profile needs to be estimated. This term is addressed by the Norris–Landzberg model plotted in Figure 5.40. The solder joints experience the most stress during the period when the devices are subjected to the largest ΔT, which translates to the shortest cycles-to-failure. The overall lifetime of the device essentially will be dominated by the most stressful period. This effect is addressed by putting the cycles-to-failure terms ($N_{\Delta T}$) in the denominator and then summing them up collectively.
3) The hottest temperature extreme of each cycle or the baseline temperature needs to be estimated. For instance, the solder joints may experience different stress levels given an identical ΔT in the winter or in the summer.

Each of these factors is included in the analysis that follows, which is based on the board-level thermomechanical reliability study presented in Section 5.5.4.4, assuming a 0.1% failure rate for the EPC2218A with underfill.

The projected lifetime curves using the Norris–Landzberg model are plotted in Figure 5.40 assuming T_{Max} is 125 °C, which is the maximum recommended operating temperature for

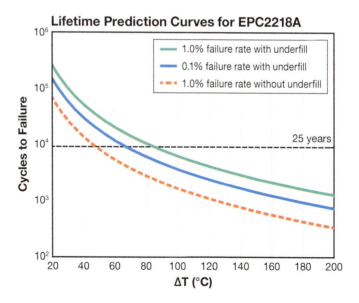

Figure 5.40 Lifetime prediction curves for EPC2218A with respect to ΔT using the Norris–Landzberg model.

most of the power modules. The horizontal, black-dashed line at 9125 cycles represents a duration of 25 years of continuous operation assuming one thermal cycle per day.

Figure 5.40 shows that after 25 years of continuous operation under a constant temperature swing of 72 °C from hot to cold, or vice versa, only 0.1% of the EPC2218A devices with underfill material would fail the datasheet limit due to an increase in $R_{DS(ON)}$ value. At a 1% failure rate, 99% of the devices should be capable of surviving 25 years of continuous operation when subjected to a constant ΔT of 95°C. Even without underfill material, 99% of the parts should survive a fixed ΔT of approximately 51 °C over 25 years of lifetime.

Now let us examine a real-world example of the lifetime model from Eq. (5.32) (Section 5.5.4.4). Assume that system is installed outdoors near solar panels in Phoenix, Arizona, where the climate is well suited for solar, but also experiences extreme temperature changes over time. Figure 5.41 shows the historical weather report of Phoenix, Arizona, over a full year [74].

In addition to the ambient temperature, 30 °C of device self-heating is added for the total lifetime calculations. For the 0.01% failure rate, or 100 ppm, which means 100 devices failed in 1 million parts used, the EPC2218A with underfill is projected to withstand 18,218 cycles to failure, equivalent to 49.9 years of lifetime operation considering one cycle per day.

If we extrapolate to a 0.001% failure rate, or 10 ppm, suggesting only 10 failures out of 1 million devices tested, now the total lifetime is calculated to be 10,971 cycles. This is equivalent to approximately 30 years of continuous operation with one cycle per day.

The results imply that temperature cycling is the most critical stressor that could be limiting the overall lifetime for GaN used in solar applications. However, by using proper underfill materials TC reliability can be significantly improved to exceed the required 25 years of continuous operation with a low failure rate under nominal solar mission profiles.

5.6.1.7 Conclusions

The test-to-fail results and physics-based lifetime projections show that neither gate bias nor drain bias are major reliability concerns for microinverters or power optimizers in solar applications. Using appropriate underfill materials can vastly reduce thermal cycling reliability risk, resulting in lifetimes exceeding 25 years.

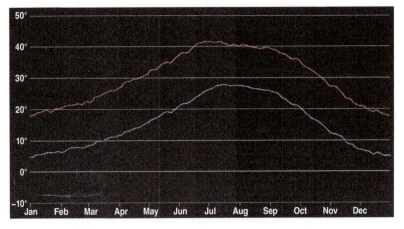

Figure 5.41 Yearly historical temperature profile of Phoenix, Arizona. The red curve is the daily high temperature, and the blue curve is the daily low temperature from reference. *Source:* Adapted from Ref. [74].

5.6.2 DC–DC-Application-Specific Reliability

5.6.2.1 Introduction

DC–DC converters exist in virtually every application of modern power electronics. Due to small die size, low on-resistance, and low parasitic capacitance, GaN power devices have demonstrated better switching performance and power density with figures of merit (FOM) 3–10 times superior to comparable silicon devices. This trend will only accelerate as GaN FETs continue to improve while Si MOSFET are already very close to their theoretical limits.

GaN devices have enabled easy-to-use topologies like the synchronous buck converter to reach new levels of efficiency and power densities. Taking advantage of reduced switching losses and no reverse recovery, designers can increase switching frequencies while also reducing power losses. This increase in switching frequency allows for smaller, more efficient inductors that in turn can increase efficiencies by further lowering resistive losses while reducing overall volume. The amount of capacitance can also be cost reduced and with better transient response. Overall, this leads to designs with higher power density, higher efficiency, and lower system cost, hence the broad adoption trends seen throughout various end markets.

GaN HEMTs are particularly valuable where power density is the goal. For example, designers have taken advantage of EPC chip-scale packaging (CSP) to significantly increase the power density of intermediate bus converters (IBC) for server applications migrating to a 48 V distribution rail. Many designers have chosen an LLC topology operated as a DC transformer (DCX) with GaN in both primary and secondary sides. On the primary side the small size of GaN allows the devices to reduce conduction and gate drive losses in the same footprint as a power MOSFET, while the small C_{OSS} allows the LLC to operate with a higher power delivery cycle and better transformer utilization. On the secondary side GaN enables the lowest conduction losses in a given area while minimizing gate drive losses thanks to the very small Q_G. This combination of best-in-class power devices and advanced packaging technologies has allowed for record power densities [75].

In this chapter, the test-to-fail methodology is adopted to investigate the intrinsic wear-out mechanisms that would be experienced in common DC–DC converters. Devices are stressed under gate bias, drain bias, and temperature cycling individually. The lifetime of each stressor is therefore projected based on the physics-based model developed from test-to-fail results and an understanding of the unique stress conditions in DC–DC converters.

5.6.2.2 Test-to-Fail Methodology

In DC–DC applications, three key stressors are identified: gate bias, drain bias, and temperature cycling (TC). The total $MTTF$ can be described by Eq. (5.33).

$$\frac{1}{\text{MTTF}_{\text{Total}}} = \frac{1}{\text{MTTF}_{\text{Gate}}} + \frac{1}{\text{MTTF}_{\text{Drain}}} + \frac{1}{\text{MTTF}_{\text{TC}}} \tag{5.33}$$

5.6.2.3 Gate Bias

In DC–DC converters, the gate terminal of GaN HEMTs must be biased periodically during switching. GaN HEMTs have approximately 1 ppm failure rate projected after 25 years of continuous DC bias at $V_{GS(\text{max})} = 6$ V. This shows that gate bias stress is not the dominant stressor limiting the overall lifetime.

5.6.2.4 Drain Bias

A frequently discussed reliability concern for GaN under drain bias is dynamic on-resistance. This is a wear-out mechanism where the $R_{DS(on)}$ of GaN HEMTs rises when subjected to high drain–source voltage (V_{DS}). One of the dominant mechanisms responsible for the increase in $R_{DS(on)}$ is hot electron-induced trapping effects [2, 16–18]. As the trapped charges accumulate, electrons from the 2DEG are depleted, leading to an increase in $R_{DS(on)}$. The detailed lifetime model derivation is discussed in Section 5.5.2.

The next sections address the following knowledge gaps:

1) How can a representative drain voltage waveform of a common DC–DC converter be correlated with various reliability testing topologies (stressors)?
2) What are the projected lifetimes of each individual reliability testing topology (stressor) based on the lifetime model developed from the electron trapping effect?
3) How does individual reliability lifetime prediction determine the overall lifetime of GaN devices?

A SPICE simulation [76] was first conducted for a buck converter using an EPC9078 demonstration board featuring 100 V EPC2045 GaN transistors [77]. To include the corner conditions for a real-world application, an intentionally poorly designed buck converter was simulated, where abnormally high parasitic inductances were added to emulate a worst-case scenario. Figure 5.42a shows the simulated turn-off voltage waveform, where the drain voltage immediately rings to a peak voltage of approximately 120 V and then the amplitude of ringing drops off quickly to stabilize at a bus voltage of 80 V. The simulated voltage waveform in Figure 5.42a can be deconvoluted by two separate voltage waveforms as shown in Figure 5.42b,c. Figure 5.42b illustrates that the overvoltage ringing can be fitted with a set of half-sinusoidal voltage waveforms. After the ringing subdued and reaches the bus voltage, the equilibrium part of the waveform can be modeled by a voltage waveform as shown in

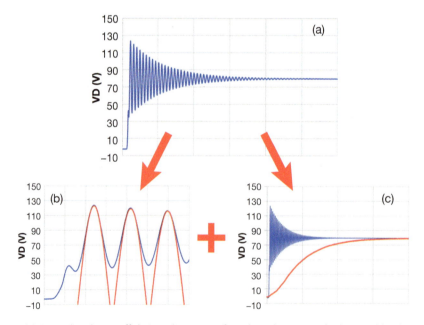

Figure 5.42 (a) A simulated turn-off drain voltage waveform based on a poorly designed buck converter, where a 120 V ringing and 80 V bus voltage are shown. (b) Ringing can be fitted with a set of half-sinusoidal waveforms. (c) The equilibrium portion of the waveform can be fitted by a different voltage waveform shown in red.

Figure 5.42c. Waveforms in Figure 5.42b,c can be realized by two different reliability testing circuits, which will be discussed separately in the following discussions.

Transient overvoltage ringing is commonly observed in GaN HEMTs under high dv/dt switching conditions. Because GaN HEMTs lack avalanche mechanisms, the reliability impact under such transient overvoltage stress needs to be understood. To properly address this issue, the unclamped inductive switching (UIS) test circuit in Figure 5.15 was deployed again.

Figure 5.43 shows normalized $R_{DS(on)}$ of a representative EPC2218 device under 120 V peak overvoltage testing to 1.5 billion pulses. The case temperature of the DUT was maintained at 75 °C throughout the experiment by an active temperature control system. Results presented in Figure 5.43 show excellent overvoltage robustness of GaN HEMTs with 120% of $V_{DS,\,Max}$ applied. In a typical turn-off voltage waveform, there are usually multiple overvoltage oscillations before it stabilizes at the bus voltage. However, the first spike typically has the highest voltage. First-principles modeling estimates that the very first overvoltage pulse causes the most trapped charges, which dominates the dynamic $R_{DS(on)}$ shift in every switching period [8–10].

Figure 5.42c shows how the equilibrium portion of the waveform can be fitted. The resistive hard switching topology circuit in Figure 5.7 was used again in this study. Figure 5.43 shows the test result of another representative EPC2218 device under 80 V, 100 kHz testing. The model predicts less than 10% $R_{DS(on)}$ increase over 100 years of continuous switching at 100 kHz and 80 $V_{DS,\,Bus}$, revealing good robustness of GaN HEMTs under hard-switched stress conditions with 80 V_{DS}.

Because two different testing topologies address different spectrums of a common turn-off voltage waveform from a buck converter, the reliability impact of each individual stressor can be combined as shown in Eq. (5.32).

$$\frac{1}{\mathrm{MTTF}_{\mathrm{Total\_Drain}}} = \frac{1}{\mathrm{MTTF}_{\mathrm{Overvoltage}}} + \frac{1}{\mathrm{MTTF}_{\mathrm{Bus\ Voltage}}} \qquad (5.34)$$

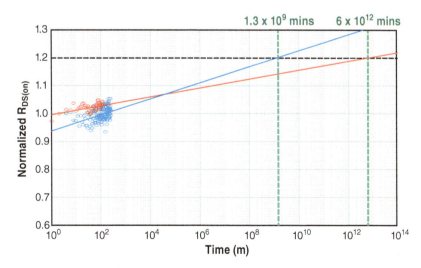

Figure 5.43 Normalized $R_{DS(on)}$ of two EPC2218 devices were projected to a time where $R_{DS(ON)}$ shifts 20% as compared to first read point. The blue data points and lifetime project represent the 120 V overvoltage testing by UIS. The red one was tested by the resistive load hard switching circuit at 80 V.

Previously, 25 years of continuous operation was used as targeted lifetime for general DC–DC converter applications. However, the projected $R_{DS(on)}$ values at the end of 25 years are still notably less than the datasheet maximum limit in both cases. Therefore, 20% *in situ* $R_{DS(on)}$ drift failure criterion is implemented to project time-of-failure. Figure 5.43 shows the projected time-to-failure for EPC2218 under UIS (120 $V_{DS,\ Peak}$) and resistive load hard switching (80 $V_{DS,\ Bus}$) is 1.3×10^9 minutes and 6×10^{12} minutes, respectively.

By plugging the time-of-failure results into Eq. (5.32), the total lifetime is dominated by the overvoltage contribution because it is orders of magnitude less than the resistive load switching testing result. The total lifetime is calculated to be approximately 2470 years, based on 100 kHz testing data. Designers can scale the projected lifetime with respect to the operating frequency, as discussed in Figure 5.9. If 1 MHz operating frequency is used, the projected lifetime would be 247 years.

The projected total lifetime results show that even under an extreme drain bias condition with severe overshoot in a common buck converter, GaN HEMTs demonstrated excellent robustness. In summary, dynamic on-resistance wear-out mechanism should not be a critical concern for GaN FETs used in common DC–DC converters.

5.6.2.5 Temperature Cycling

Temperature cycling is another critical area of interest for DC–DC converter applications. This analysis is based on the board-level thermomechanical reliability study presented in Section 5.5.4.4, which showed that proper underfill material improves the temperature cycling lifetime of chip-scale packaged GaN devices by a factor of at least 4.8 times. In the following discussions, only TC1 condition with underfill data is used.

For an upper limit in this analysis, T_{Max} is assumed to be 125 °C, which is the typical maximum design temperature for power modules. The number of cycles to failure (N) at 100 ppm, or 0.01%, failure rate for EPC2218A with underfill can be plotted as a function of ΔT using Eq. (5.28) (Section 5.5.4.4), while the Arrhenius term is a constant coefficient. The result is shown by the black line in Figure 5.44. The horizontal axis (ΔT) only includes a range of

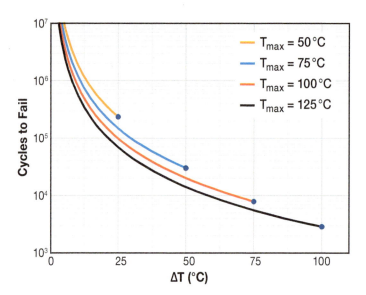

Figure 5.44 Number of cycles to fail at 100 ppm or 0.01% failure rate vs. ΔT at T_{Max} of 50 °C (yellow), 75 °C (blue), 100 °C (red), and 125 °C (black).

0–100 °C because power modules in real-world applications are typically kept at 25 °C ambient temperature when not in operation, which yields a maximum ΔT of 100 °C.

In some of the DC–DC converters that are designed for a lower T_{Max} of 100 °C during normal operation, the Arrhenius term should now be slightly larger due to a smaller denominator (T_{Max}) in the exponential equation. The red line in Figure 5.44 shows the number of cycles to fail at 100 ppm extracted from the Weibull distribution as a function of ΔT, where the red curve is slightly above the black curve ($T_{\text{Max}} = 125$ °C). Because T_{Max} is lowered by 25 °C, the red curve is now plotted from 0 °C to 75 °C on the horizontal ΔT axis.

For some applications that are designed for a T_{Max} of 75 °C, the model is plotted in blue, where a longer lifetime is expected because of the larger Arrhenius term. A T_{Max} of 50 °C is also included in Figure 5.44 as shown in the yellow line.

How can designers use Figure 5.44 to determine the TC lifetime for their DC–DC converter design? By way of example, take a converter that will be operating in the desert climate of Phoenix, Arizona. Referring to Figure 5.41 the ambient outside temperature in the summer can be as high as 50 °C (122 °F). This hypothetical converter generates another 75 °C of heat during operation, which gives a T_{Max} of 125 °C. By following the black curve in Figure 5.44 and finding the vertical intercept where ΔT of the horizontal axis is 75 °C, the estimated number of cycles to 100 ppm failure rate is a little more than 5000 cycles, hopefully representing decades of operation when also considering the more moderate temperature seasons. This approach provides a practical method to correlate lab generated TC reliability results to real-world applications.

5.6.2.6 Conclusions

Three main stressors are identified for the DC–DC converter example above, gate bias, drain bias, and temperature cycling.

For gates, a 1 ppm failure rate was projected after 25 years of continuous DC gate bias at the maximum rated voltage ($V_{\text{GS}} = 6$ V).

For drain–source stress a physics-based model based on hot electron-trapping mechanism was used to explain the dynamic $R_{\text{DS(on)}}$ wear-out mechanism under drain bias. The measured data and the lifetime model predict that the $R_{\text{DS(on)}}$ shift is expected to be less than 20% over the lifetime of the part. This is unlikely to limit the product's lifetime.

The wear-out mechanism responsible for temperature cycling (TC) failure is solder joint cracking. A third lifetime calculation that includes TC range, temperature extreme, and cycling speed was introduced also demonstrating good robustness under extreme conditions.

Combining the wear-out rates of all three stressors shows that neither gate bias nor drain bias is of significant reliability concern in DC–DC converter applications. Thermo-mechanical stress due to TC is deemed to have the highest risk that warrants careful considerations. Using appropriate underfill materials can vastly reduce TC reliability risk while giving excellent lifetimes.

5.6.3 Lidar Application Reliability

5.6.3.1 Introduction to Lidar Reliability

Compared to other applications, GaN transistors used for light detection and ranging (lidar) are often subject to long durations of reverse bias and short pulses of relatively high current. This section evaluates the reliability of devices used in lidar applications, both discrete transistors and GaN lidar ICs which include low-voltage driver circuits.

5.6.3.2 Long-Term Stability Under High Current Pulses

The concept of this test method is to stress parts in an actual lidar circuit for a total number of pulses well beyond their ultimate mission profile. The mission profiles for automotive lidar vary from customer to customer. A typical automotive profile would call for a 15-year life, with two hours of operation per day, at 100 kHz pulse repetition frequency (PRF). This corresponds to approximately four trillion total lidar pulses. Some worst-case scenarios might call for as many as 10–12 trillion pulses in service life.

By testing a population of devices well beyond the end of their full mission profile while verifying the stability of the system performance and the device characteristics, this test method directly establishes the suitability of GaN devices for lidar applications. To achieve the large number of pulses, parts are stressed continuously, rather than in bursts as used in typical lidar circuits.

For this section, two popular AEC grade parts were put under test: EPC2202 (80 V) and EPC2212 (100 V). Four parts of each type were tested simultaneously. During the stress, two key parameters were continuously monitored on every device: (i) peak pulse current and (ii) pulse width. These parameters are critical to both the range and resolution of a lidar system.

Figures 5.45 and 5.46 show the results of this test over the first 13 trillion pulses. The cumulative number of pulses well exceeds a typical automotive lifetime and covers worst-case use conditions. Note that there is no observed degradation or drift in either the pulse width or height. While this is an indirect monitor of the health of the GaN devices, it indicates that no degradation mechanisms have occurred that would adversely impact lidar performance. These results demonstrate the excellent stability of GaN in lidar applications.

5.6.3.3 Monolithic GaN-on-Si Laser Driver ICs

Lidar systems often use discrete GaN transistors separate from a gate driver chip due to the benefits of GaN's small footprint and superior switching performance. However, GaN laser drive ICs that integrate a high-speed GaN driver with a discrete GaN transistor are also available [78]. These integrated monolithic lidar solutions offer even higher performance, smaller form factor, and lower cost than the existing discrete solutions. As a result, these ICs enable a wider range of lidar applications including robotics, surveillance systems, drones, autonomous cars, vacuum cleaners, and many more.

5.6.3.4 Key Stressors of eToF Laser Driver IC for Lidar Application

The integration of the gate driver and power transistor onto a chip-scale package greatly reduces the parasitic inductances and further improves the speed, minimum pulse width, and power dissipation. It also introduces challenges in isolating the key electrical stressors because many of the IC's voltages and currents cannot be accessed directly. The first step of the analysis is to identify the key stressors that affect the IC in lidar applications. The lidar IC's operating conditions are best emulated through high-temperature operating life (HTOL) testing. EPC21601 [78] is selected as the test vehicle for this test-to-fail study.

Three key stressors are identified:

- Logic supply voltage V_{DD} that supplies the drive voltage to the low-voltage (LV) GaN FETs in laser driver circuit as well as the gate of the high-voltage (HV) GaN output FET.
- Laser drive voltage V_D that is predominantly applied to the drain terminal of the HV output FET.
- Operating frequency which stresses both the LV laser driver circuits and the HV output FET.

GaN Reliability | 145

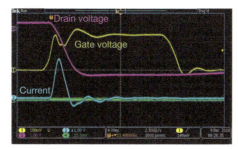

AEC-Q101 series of discrete FETs
- 8 samples (>7000 h)
- 0 failures and perfect pulse stability

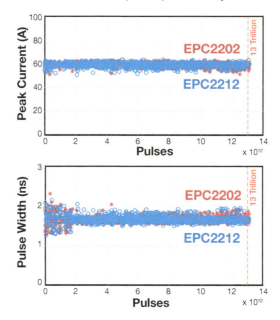

Figure 5.45 Long-term stability of pulse width (bottom) and pulse height (center) over 13-trillion lidar pulses. Data for four EPC2202 (red) devices and four EPC2212 (blue) devices are overlaid in the plots.

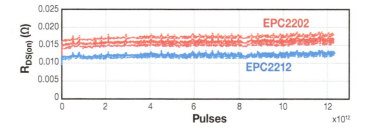

Figure 5.46 Long-term stability of $R_{DS(on)}$ during lidar reliability testing. These parameters are measured at six-hour intervals on every part by briefly interrupting the lidar stress. Data for four EPC2202 (red) devices and four EPC2212 (blue) devices are overlaid in the plots.

5.6.3.5 Effect of V_{DD}, Logic Supply Voltage

When the EPC21601 generates a burst of short pulses, the logic supply voltage (V_{DD}) is applied to the gate terminals of the low-voltage GaN FETs in the laser driver circuits and the gate of the HV GaN power transistor. It is equivalent of performing a dynamic gate test for all GaN FETs with a burst frequency of 1 kHz, very low duty cycle (~0.02%), and high operating frequency (30 MHz). When not pulsed, the part is in the OFF state and the gate bias is nearly zero (see Figures 5.47 and 5.48).

In the qualification HTOL test, V_{DD} was biased at the absolute maximum rating of 5.5 V, and no issue was found after 1000 hours of testing at 125 °C junction temperature. To test the device's robustness, the V_{DD} voltage was first increased to 7 V, which is more than 125% of the absolute maximum rating. This stress condition addresses the worst possible overvoltage ringing on the V_{DD} pin during normal operation by customers.

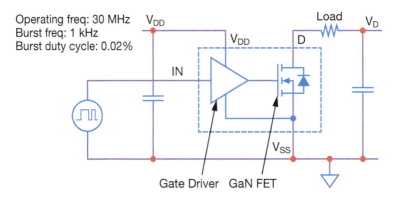

Figure 5.47 Block diagram of EPC21601 laser drive integrated circuits.

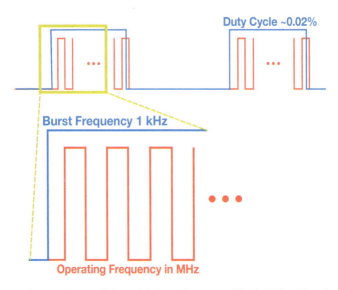

Figure 5.48 Diagram of operating conditions with burst frequency (Blue) 1 kHz with a duty cycle of ~0.02% and operating frequency in MHz.

Sixteen devices were tested for more than 1000 hours at 7 V_{DD} and 125 °C junction temperature, and no failures occurred, indicating a significant margin designed in the laser driver IC products. However, the test result cannot provide a quantifiable design margin or predict the lifetime at a given mission profile. Therefore, more stringent stress conditions must be applied to test the devices to failure where the goal is to fail the parts quickly and conduct failure analysis to understand the underlying failure modes and mechanisms. To determine the voltage acceleration of the V_{DD} stress a matrix of tests was conducted from 8.5 V to 9.5 V at 25 °C.

Figure 5.49 shows the Weibull analysis results of two different V_{DD} voltages at room temperature. The data was analyzed using a two-parameter Weibull distribution for each voltage leg using maximum likelihood estimation (MLE). The fits are indicated by solid lines in the graphs. The Weibull shape (or slope) parameter was constrained to be the same for all voltage legs because a single failure mode was found through failure analysis.

Failure analysis determined that all failures were "soft" parameter failures in which quiescent current exceeded the 20 mA maximum datasheet limit, with V_{DD} = 5 V and the measurement conducted during the OFF state [78]. Under closer examination, the quiescent current only exceeded datasheet limits when V_D = 20 V was provided. When the quiescent current soft failures were subjected to lidar operation, the integrity of their pulses was uncompromised.

Figure 5.50 shows the waveforms of the input signal (blue) of V_{IN} (the logic input to EC21601) and the corresponding output signals from V_D of the quiescent current failures (green and yellow), where no pulse distortion or missing pulses were observed. This suggests even when a device was damaged by extremely high V_{DD} stress, it was still functional, and the repeatability of current pulses was not adversely impacted.

Since all failures at different voltages and temperatures showed identical "soft" electrical failures, physical failure analysis was conducted to determine the underlying root cause. Gate rupture of the low-voltage GaN FETs in the driver circuit was found to be the failure mechanism regardless of stress voltages and temperatures. This result is expected based on the circuit

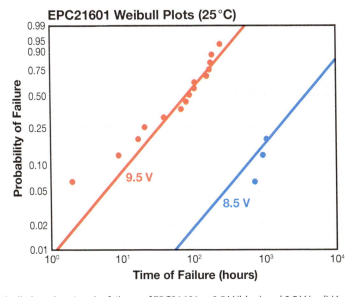

Figure 5.49 Weibull plots showing the failures of EPC21601 at 8.5 V (blue) and 9.5 V (red) V_{DD}, respectively, and T_J = 25 °C.

148 | GaN Power Devices for Efficient Power Conversion

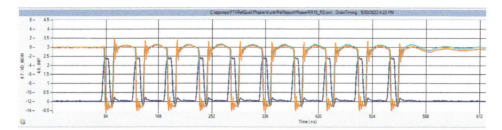

Figure 5.50 The input (blue) waveform and the corresponding output waveforms of the quiescent current "soft" failures (green and yellow).

analysis because the V_{DD} voltage is applied to the gates of the LV and HV GaN FETs. The gate lifetime model in Eq. (5.11) fits the measured data for V_{DD}.

Figure 5.51 shows the lifetime projection against the measured acceleration data for EPC21601 at 25 °C. The fit projected greater than 25 years of lifetime with less than 1 ppm failure rate at the 5.5 V maximum V_{DD} voltage rating. This result agrees with the extrapolated lifetime for gate at 5.5 V under static DC gate bias, demonstrating the extraordinarily robustness of the laser driver IC under accelerated V_{DD} stress.

5.6.3.6 Effect of V_D, Laser Drive Voltage

By examining the circuits that connect to the V_D pin in detail, the accelerated V_D HTOL can cause two potential failure modes in EPC21601.

1) V_D primarily goes to the drain terminal of the HV GaN FET. Due to the nature of lidar operation, the high-voltage output FET is under reverse drain bias most of the time. When the laser pulses are generated, the HV FET turns on and conducts current. Accelerated V_D

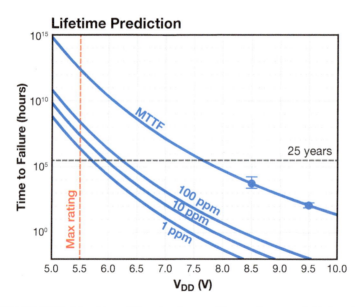

Figure 5.51 EPC21601 MTTF data at two different voltages with error bars are plotted against V_{DD} at 25 °C. The solid line corresponds to the impact ionization lifetime model. Extrapolations of time to failure for 100, 10, and 1 ppm are shown as well.

HTOL testing of the IC therefore resembles a dynamic HTRB test of the output FET with a high duty cycle. Therefore, the intrinsic failure modes due to accelerated drain bias test for a discrete GaN transistor apply.

2) Besides connecting to the drain node of the HV FET, the V_D pin also connects to a single laser driver circuit, which affects the number of pulses generated by the device. If that path was compromised by the accelerated V_D stress, it could lead to missing pulses, which is another crucial failure mode for lidar application.

The HTOL qualification test was conducted at 30 V_D, the maximum recommended voltage [78]. To accelerate failures under laser drive voltage stress, HTOL with 60 V_D and 125 °C T_J was carried out. No failures were found after more than 1000 hours of testing, and all parts met the datasheet limits.

Figure 5.52 recorded the input and output pulse waveforms of the parts that completed 1000 hours of V_D = 60 V and T_J = 125 °C HTOL testing, where no degradation in pulse waveforms was observed. It is also important to note that the HV output transistor experienced more than 25 V overshoot at the end of each pulse during HTOL resulting from the short pulses. This suggests that the device saw repetitive >85 V transient overvoltage stress on V_D. This also demonstrates excellent overvoltage robustness of the device under V_D stress.

At this point, the most rigorous testing corner is covered by the testing matrix. Further increasing the drain bias might introduce a different intrinsic wear-out mechanism for the HV GaN transistor that is not applicable to the lidar application or the reliability of the laser drive IC. In short, no failure mode was found to be associated with the laser supply voltage (V_D).

5.6.3.7 Effect of Operating Frequency

Output waveform distortion or missing pulses is a valid concern when the lidar ICs are operating under extremely high frequencies. HTOL testing at 48 MHz and 96 MHz operating frequencies was conducted for more than 1400 hours with 16 devices from each test condition. No failures were reported. Please note that 48 MHz and 96 MHz are 160% and 320% of the 30 MHz, maximum recommended operating frequency per datasheet.

In addition, input and output waveforms of devices after 1400 hours of testing are measured and no waveform distortion or missing pulses were observed, demonstrating the robustness of the laser driver IC products.

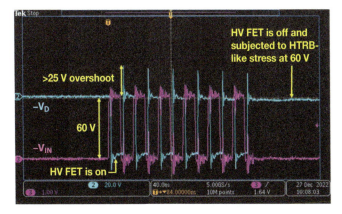

Figure 5.52 Output waveforms (blue) of a representative passing part after it was subjected to 1005 hours of HTOL testing at 60 V V_D and 125 °C. The purple waveform is the corresponding input signal from V_{IN}.

In conclusion, of the three stressors unique to lidar ICs considered in this section – logic supply voltage V_{DD}, laser drive voltage V_D, and operating frequency – only the logic supply voltage was observed to generate device failures under extremely accelerated conditions. Lidar ICs operated within datasheet limits perform reliably.

5.7 Summary

As GaN device production continues to increase and applications diversify, separate reliability concerns arise which may depend on the use case. By understanding the wear-out mechanisms that affect a system in each phase of its mission profile, GaN device lifetimes can be calculated analytically for each specific application. The failure rate of each wear-out mechanism, which is confirmed by testing to failure, can be minimized by following the guidelines provided in this chapter.

References

1 Adriano G., Li D., and Zhang, S. (2024). Solder stencil design guidelines for reliable assembly of PQFN GaN devices. Appl. note AN029. EPC El Segundo, CA (2024). https://epc-co.com/epc/DesktopModules/DnnSharp/SearchBoost/FileDownload.ashx?file=13927&sb-bhvr=1&portalid=0.

2 Gajare, S., Li, D., Garcia, R. et al. (2024). GaN reliability and lifetime projections: phase 16. EPC El Segundo, CA (2024). https://epc-co.com/epc/design-support/gan-fet-reliability.

3 ZVEI: Die Elektroindustrie ZVEI Robustness Validation Working Group (2015). *Handbook for Robustness Validation of Semiconductor Devices in Automotive Applications*, 3e. Germany: ZVEI: Die Elektroindustrie (German Electrical and Electronic Manufacturers' Association).

4 Volosencu, C. (2017). *System Reliability*. German Electrical and Electronic Manufacturers' Association https://doi.org/10.5772/66993.

5 Method for calculating failure rates in units of FITS, JEDEC Standard JESD85A, January 2014.

6 Bernstein, J.B., Bensoussan, A., and Bender, E. (2024). *Reliability Prediction for Microelectronics*. Hoboken, NJ: Wiley.

7 Efficient Power Conversion Corporation (2022). EPC2045 – Enhancement-mode power transistor. EPC2045 datasheet. https://epc-co.com/epc/Portals/0/epc/documents/datasheets/EPC2045_datasheet.pdf.

8 Efficient Power Conversion Corporation (2023). EPC2215 – Enhancement-mode power transistor. EPC2215 datasheet. https://epc-co.com/epc/Portals/0/epc/documents/datasheets/EPC2215_datasheet.pdf.

9 Guideline to specify a transient off-state withstand voltage robustness indicator in datasheets for lateral GaN power conversion devices, JEDEC Standard JEP186. December 2021. https://www.jedec.org/standards-documents/docs/jep186.

10 Efficient Power Conversion Corporation (2024). EPC2204 – Enhancement-mode power transistor. EPC2204 datasheet. https://epc-co.com/epc/Portals/0/epc/documents/datasheets/EPC2204_datasheet.pdf.

11 Efficient Power Conversion Corporation (2024). EPC2218 – Enhancement-mode power transistor. EPC2218 datasheet. https://epc-co.com/epc/Portals/0/epc/documents/datasheets/EPC2218_datasheet.pdf.

12 Efficient Power Conversion Corporation (2024). EPC2071 – Enhancement-mode power transistor. EPC2071 datasheet. https://epc-co.com/epc/Portals/0/epc/documents/datasheets/EPC2071_datasheet.pdf.

13 Efficient Power Conversion Corporation (2024). EPC2071 – Enhancement-mode power transistor. EPC2071 datasheet. https://epc-co.com/epc/Portals/0/epc/documents/datasheets/EPC2302_datasheet.pdf.

14 Efficient Power Conversion Corporation (2023). EPC2212 – Enhancement-mode power transistor. EPC2012 datasheet, 2023 [Revised February 2023]. https://epc-co.com/epc/products/gan-fets-and-ics/epc2212.

15 Scarpulla, J., Eng, D.C., Olson, S.R., and Wu, C.S. (1999). A TDDB model of Si3N4-based capacitors in GaAs MMICs. *IEEE 37th Annual International Reliability Physics Symposium*, San Diego, CA, 1999.

16 Lidow, A. (ed.) (2022). *GaN Power Devices and Applications*. El Segundo, CA: Power Conversion Publications.

17 Pozo, A., Zhang, S., Stecklein, G., et al. (2022). GaN reliability and lifetime projections: phase 14. EPC El Segundo, CA (2022). https://epc-co.com/epc/design-support/gan-fet-reliability.

18 Garcia, R., Gajare, S., Espinoza, A. et al. (2023), GaN reliability and lifetime projections: phase 15. EPC El Segundo, CA (2023). https://epc-co.com/epc/design-support/gan-fet-reliability.

19 Ooi, T.L.W., Chean, P.L., You, A.H. et al. (2020). Mean multiplication gain and excess noise factor of GaN and Al0.45Ga0.55N avalanche photodiodes. *Eur. Phys. J. Appl. Phys.* 92: 10301.

20 Ozbek, A.M. (2011). Measurement of impact ionization coefficients in GaN. Ph.D. thesis. North Carolina State University, Raleigh, NC.

21 Wang, B., Zhang, R., Wang, H. et al. (2023). Gate lifetime of p-gate GaN HEMT in inductive power switching. *35th International Symposium on Power Semiconductor Devices and ICs (ISPSD)*, Hong Kong, China (28 May–01 June 2023), 20–23.

22 Kozak, J.P., Zhang, R., Porter, M. et al. (2023). Stability, reliability, and robustness of GaN power devices: a review IEEE transactions on power electronics. *IEEE Trans. On Power Electron.* 38 (7): 1–31.

23 Dymond, H.C.P., Wang, J., Liu, D. et al. (2017). A 6.7-GHz active gate driver for GaN FETs to combat overshoot, ringing, and EMI. *IEEE Trans. on Power Electron.* 33 (1): 581–594.

24 Zhang, R., Garcia, R., Strittmatter, R. et al. (2023). In-situ $R_{DS(on)}$ characterization and lifetime projection of GaN HEMTs under repetitive overvoltage switching. *IEEE Trans. Power Electron.* 3 (9): 10589–10594.

25 Zhang, S., Gahare, S., Garcia, R. et al. (2023). Projecting GaN HEMTs lifetimes under typical stresses commonly observed in DC-DC converters. *Power Electron. Devices Compon.* 6: 100051.

26 Spirito, P., Breglio G., d'Alessandro et al. Analytical model for thermal instability of low voltage power MOS and S.O.A. in pulse operation. *14th International Symposium on Power Semiconductor Devices & ICS*, Santa Fe, NM (4–7 June 2002), 269–272.

27 Efficient Power Conversion Corporation (2023). EPC2034C – Enhancement-mode power transistor. EPC2034C datasheet. https://epc-co.com/epc/Portals/0/epc/documents/datasheets/EPC2034C_datasheet.pdf.

28 Efficient Power Conversion Corporation (2021). EPC2014C – Enhancement-mode power transistor. EPC2014C datasheet. https://epc-co.com/epc/Portals/0/epc/documents/datasheets/EPC2014C_datasheet.pdf.

29 Mishra, S. (2008). Fault current limiting and protection circuit for power electronics used in a modular converter. M.S. thesis. University of Tennessee, Knoxville, TN.

30 Glaser, J. (2017). An introduction to lidar: a look at future developments. *IEEE Power Electron. Mag.* 4 (1): 25–35.

31 Efficient Power Conversion Corporation (2022). EPC2203 – Enhancement-mode power transistor. EPC2203 datasheet. https://epc-co.com/epc/Portals/0/epc/documents/datasheets/EPC2203_datasheet.pdf.

32 Efficient Power Conversion Corporation (2021). EPC2051 – Enhancement-mode power transistor. EPC2051 datasheet. https://epc-co.com/epc/Portals/0/epc/documents/datasheets/EPC2051_datasheet.pdf.

33 Kim, B.J., Lim, G.T., Kim, J. et al. (2009). Microstructure evolution in Cu pillar/eutectic SnPb solder system during isothermal annealing. *Met. Mater. Int.* 15: 815–818.

34 Ma, H.C., Guo, J.D., Chen, J.Q. et al. (2015). Reliability and failure mechanism of copper pillar joints under current stressing. *J. Mater. Mater. Sci.: Mater. In Electron.* 26 (10): 7690–7697.

35 Ding, M., Wang G., Chao, B. et al. A study of electromigration failure in Pb-Free solder joints. *Proceedings of 43rd IEEE International Reliability Physics Symposium*, San Jose, CA, (April 2005), 518–523.

36 Nah, Jae-Woong, Suh, J.O., Tu, K.N. et al. Electromigration in Pb-free solder bumps with Cu column as flip chip joints, *56th Electronic Components and Technology Conference 2006*, San Diego, CA, USA, 2006, 6.

37 Madanipour, H., Kim, H., Kim, Y. et al. (2021). Study of electromigration in Sn-Ag-Cu micro solder joint with Ni interfacial layer. *J. Alloys Compd* 862: 158043.

38 Islam, N., Kim, G., Kim, K. et al. Electromigration for advanced Cu interconnect and the challenges with reduced pitch bumps, *2014 IEEE 64th Electronic Components and Technology Conference (ECTC)*, Orlando, FL, USA, 2014, pp. 50–55.

39 Black, J.R., Metallization failures in integrated circuits. Rome Air Development Center. Air Force Systems Command, Griffiss Air Force Base, 1968, https://doi.org/10.1002/nav.3800080206.

40 Lienig, J. and Thiele, M. (2018). *Fundamentals of Electromigration – Aware Integrated Circuit Design*. New York, NY: Springer.

41 Guideline for characterizing solder bump electromigration under constant current and temperature stress, JEDEC Standard JEP154, Version 1.0, 2011.

42 Coonrod, J. (2011). Understanding when to use FR-4 or high frequency laminates. *OnBoard Technology*, September 2011: pp. 26–30.

43 Michaelides, S. and Sitaraman, S.K. (1999). Die cracking and reliable die design for flip-chip assemblies. *IEEE Trans. Adv. Packag.* 22 (4): 602–613.

44 Wu, B., Yang, Y.H., Han, B. et al. (2018). Measurement of anisotropic coefficients of thermal expansion of SAC305 solder using surface strains of single grain with arbitrary orientation. *Acta Mater.* 156: 196–204.

45 Efficient Power Conversion Corporation (2022). EPC2069 – Enhancement-mode power transistor. EPC2069 datasheet. https://epc-co.com/epc/Portals/0/epc/documents/datasheets/EPC2069_datasheet.pdf.

46 Efficient Power Conversion Corporation (2022). EPC2206 – Enhancement-mode power transistor. EPC2206 datasheet. https://epc-co.com/epc/Portals/0/epc/documents/datasheets/EPC2206_datasheet.pdf.

47 Temperature cycling, JEDEC Standard Test Method JESD22-A104F, November 2020.

48 Cramér, H. (1999). *Mathematical Methods of Statistics*, 43. Princeton, NJ: Princeton University Press.

49 Clech, J-P. (2015). Board, package and die thickness effects under thermal cycling conditions. *Proceedings of SMTA International 2015*, pp. 40–50.

50 Tee, T.Y., Ng, H.S., Yap, D. et al. (2003). Comprehensive board-level solder joint reliability modeling and testing of QFN and PowerQFN packages. *Microelectron. Reliab.* 43 (8): 1329–1338.

51 Farooq, M., Goldmann, L., and Bergeron, C. (2003). Thermo-mechanical fatigue reliability of Pb-free ceramic ball grid arrays: Experimental data and lifetime prediction modeling. *53rd Electronic Components and Technology Conference*, New Orleans, LA (27–30 May 2003). pp. 827–833.

52 Norris, K.C. and Landzberg, A.H. (1969). Reliability of controlled collapse interconnections. *IBM Journal of Research and Development* 13 (3): 266–271.

53 Han, B. (1996). Determination of an effective coefficient of thermal expansion of electronic packaging components: a whole-field approach. *IEEE Trans. Compon. Packag. Manuf. Technol.: Part A* 19 (2): 240–247.

54 Darveaux, R. (2005). Effect of assembly stiffness and solder properties on thermal cycle acceleration factors. THERMINIC 2005, September 2005, Belgirate, Lago Maggiore, Italy. pp. 192–203

55 Clech, J.P. (2016). The combined effect of assembly pitch and distance to neutral point on solder joint thermal cycling life. *Proceedings of SMTA International Conference, Rosemont, IL*, pp. 25–29.

56 de Vreis, J., Jansen, M., and van Driel, W. (2009). Solder-joint reliability of HVQFN-packages subjected to thermal cycling. *Microelectronics Reliability* 49 (3): 331–339.

57 Rahangdale, U., Srinivas, R., Krishnamurthy, S. et al. (2017). Effect of PCB thickness on solder joint reliability of Quad Flat no-lead assembly under Power Cycling and Thermal Cycling. *33rd Thermal Measurement, Modeling & Management Symposium (SEMI-THERM)*, San Jose, CA, USA, pp. 70–76.

58 COMSOL, Inc. COMSOL multiphysics reference manual, (6.1). https://www.comsol.com

59 Groenewoud, W.M. (2001). *Characterization of Polymers by Thermal Analysis*. Amsterdam, Netherlands: Elsevier Science.

60 Mises, R.V. (1986). The mechanics of solids in the plastically-deformable state. *No. NAS* 1 (15): 88448.

61 Efficient Power Conversion Corporation (2023). EPC2218A – Enhancement-mode power transistor. EPC2218A datasheet. https://epc-co.com/epc/Portals/0/epc/documents/datasheets/EPC2218A_datasheet.pdf.

62 Vasudevan, V. and Fan, X. (2008). An acceleration model for lead-free (SAC) solder joint reliability under thermal cycling, *58th Electronic Components and Technology Conference (ECTC)*, Lake Buena Vista, FL, (27–30 May 2008), 139–145.

63 Sun, F., Liu, J., Cao, Z. et al. (2016). Modified Norris–Landzberg model and optimum design of temperature cycling alt. *Strength Mater.* 48: 135–145.

64 Lall, P., Shirgaokar, A., Arunachalam, D. et al. (2012). Norris–Landzberg acceleration factors and Goldmann constants for SAC305 lead-free electronics. *J. Electron. Packag.* 134 (3): 031008.

65 Deshpande, A., Jiang, Q., Dasgupta, A. et al. (2019). Fatigue life of joint-scale SAC305 solder specimens in tensile and tensile and shear mode. *18th IEEE Intersociety Conference on Thermal and Thermomechanical Phenomena in Electronic Systems (ITherm)*, Las Vegas, NV (28–31 May 2019). pp. 1026–1029.

66 Cui, H., Accelerated temperature cycle test and Coffin-Manson model for electronic packaging. *Annual Reliability and Maintainability Symposium* (RAMS) (2005), Alexandria, VA (24–27 January 2005). pp. 556–560.

67 U.S Department of Energy (2010). $1/W Photovoltaic Systems: white paper to explore a grand challenge for electricity from solar, (August 2010). https://www1.eere.energy.gov/solar/sunshot/pdfs/dpw_white_paper.pdf.

68 Cole, W., Frew, B., Gagnon, P. et al. (2017). SunShot 2030 for Photovoltaics (PV): Envisioning a Low-cost PV Future, National Renewable Energy Laboratory, Golden, CO (2017).

69 Emphase Energy Systems (2023). Enphase IQ 7-based M-series microinverters. Enphase Energy datasheet, February 2023. https://enphase.com/download/m215-m250-iq7-series-platform-data-sheet

70 Emphase Energy Systems (2023). IQ8M and IQ8A microinverters. Enphase Energy datasheet, April 2023. https://enphase.com/download/iq8m-iq8a-microinverter-data-sheet.

71 SolarEdge Technologies (2023). SolarEdge power optimizer module embedded solution. OPJ300-LV, SolarEdge datasheet, January 2023. https://knowledge-center.solaredge.com/sites/kc/files/se-pb-csi-datasheet.pdf.

72 SolarEdge Technologies (2023). Ppower optimizer frame-mounted P370/P401/P404/P500. SolarEdge datasheet, May 2023. https://knowledge-center.solaredge.com/sites/kc/files/se-p-series-add-on-frame-mounted-power-optimizer-datasheet.pdf.

73 Efficient Power Conversion Corporation (2020). EPC2059 – Enhancement-mode power transistor. EPC2059 datasheet. https://epc-co.com/epc/Portals/0/epc/documents/datasheets/EPC2059_datasheet.pdf.

74 MSN weather data for Phoenix, Arizona. https://www.msn.com/en-us/weather/forecast/in-Phoenix,AZ?loc=eyJsIjoiUGhvZW5peCIsInIiOiJBWiIsInIyIjoiTWFyaWNvcGEgQ291bnR5IiwiYyI6IlVuaXRlZCBTdGF0ZXMiLCJpIjoiVVMiLCJ0IjoxMDIsImciOiJlbi11cyIsIngiOiItMTEyLjA3MyIsInkiOiIzMy40NDgyIn0%3D&weadegreetype=F.

75 Efficient Power Conversion Corporation (2021). How to design synchronous buck converter using GaN FET compatible analog controllers with integrated gate drivers. Application Note How2AppNote025. https://epc-co.com/epc/Portals/0/epc/documents/application-notes/How2AppNote025%20How%20to%20Design%20Synchronous%20Buck%20Converter%20Using%20GaN%20FET.pdf.

76 Analog Devices, LTspice®. https://www.analog.com/en/resources/design-tools-and-calculators/ltspice-simulator.html.

77 Efficient Power Conversion Corporation (2017). EPC9078 – development board. EPC9078 quick start guide. https://epc-co.com/epc/products/evaluation-boards/EPC9078.

78 Efficient Power Conversion Corporation (2022). EPC21601 – eToF laser driver IC. EPC21601 datasheet. https://epc-co.com/epc/desktopmodules/dnnsharp/searchboost/filedownload.ashx?file=12923&sb-bhvr=1&portalid=0.

6

Thermal Management of GaN Devices

6.1 Introduction

The power consumed in a device during operation is dissipated in the form of heat. It is therefore important to understand the ability of a device to transfer the heat to its surrounding environment. Thermal design becomes increasingly critical with the latest generation of GaN transistors, where smaller die sizes and chip-scale packages (CSPs) are often harnessed to increase power density and improve electrical performance. Heat extraction from a small GaN transistor, or integrated circuit, can be as effective as from a bulky Si MOSFET with a good thermal design.

This chapter will review the thermal models for GaN devices in conventional bottom-cooled packages as well as top-cooled and CSPs. Thermal management for an entire GaN-based power conversion system will also be discussed, including equivalent circuit models, temperature measurements, and experimental characterization techniques. Then thermal management techniques for cooling devices without using a heatsink will be discussed. Finally, how to select and attach a heatsink for these different thermal design schemes, as well as the expected cooling results will be presented.

6.2 Thermal Equivalent Circuits

In general, the thermal modeling of GaN transistors is like that of MOSFETs. However, it is useful to separately consider the thermal models for GaN transistors in conventional bottom-cooled packages versus chip-scale GaN transistors with multi-sided cooling capability.

6.2.1 Thermal Resistances in a Lead-frame Package

For high-voltage parts, traditional lead-frame packages are often used, either through-hole (e.g. TO-220) or surface-mount (e.g. TO-252). For lower voltage parts, smaller dual flats no leads (DFN) packages are popular. These plastic over-molded packages are thermally insulated with plastic encapsulant, with the main direction of heat removal through the case to the PCB. For such devices, heat flow is mainly one-dimensional. Recent package developments include a high thermal conductivity back-side surface used for cooling, such as the TOLT package [1] and ACEPACK SMIT [2]. GaN devices have also been introduced in partially encapsulated power quad flat no-lead (PQFN) packages [3] with exposed substrate. An experimental thermal example of a GaN device in a PQFN package is presented in Section 6.6.3.

GaN Power Devices for Efficient Power Conversion, Fourth Edition. Alex Lidow, Michael de Rooij, John Glaser, Alejandro Pozo, Shengke Zhang, Marco Palma, David Reusch, and Johan Strydom.
© 2025 John Wiley & Sons Ltd. Published 2025 by John Wiley & Sons Ltd.

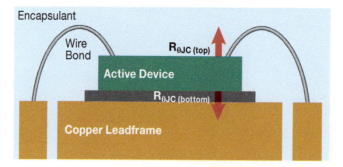

Figure 6.1 Cross section of a transistor mounted on a copper lead frame.

Figure 6.1 illustrates the thermal paths for a transistor mounted onto a copper lead-frame. The device may be soldered directly to the lead-frame, or a ceramic substrate may be inserted for electrical isolation, and the top side is over-molded with an encapsulant. While the encapsulant may be somewhat thermally conductive, the thermal path through the bottom is much more efficient than through the top. The junction-to-case thermal impedance ($R_{\theta JC}$) for this configuration shall be considered as the thermal resistance from the top of the die, through the bottom soldered surface of the die, through the copper lead-frame, to the mating surface between the lead-frame and the PCB or heatsink, which is defined as the "case" in this package. In this case, this is the only significant path for heat flow. A larger transistor area lowers this resistance. A thicker substrate or case has a longer thermal path to the heatsink, and consequently a higher $R_{\theta JC}$.

With this type of package, most of the heat can only be transferred through the lead frame. The copper lead frame is typically directly soldered to a PCB for surface-mount packages. Optionally, a heatsink can be attached to the PCB for additional bottom-side cooling using a thermal interface material (TIM) between the two mated surfaces. The lead frame of a through-hole package, such as the TO-220, can be attached to a heatsink independently from the through-hole leads that provide electrical connection to a PCB.

Figure 6.2 gives a simple schematic model describing the thermal resistance in steady state for the structure shown in Figure 6.1. The figure shows the thermal resistance broken down by location in the structure, starting from the junction to the bottom of the lead frame ($R_{\theta JC}$), then through the thermal interface material ($R_{\theta CS}$), and finally from the mating surface of the heatsink to the ambient air ($R_{\theta SA}$). The sum of these three is the total thermal resistance from the junction to the ambient ($R_{\theta JA}$).

The bottom-side junction-to-case thermal resistance is sometimes specified in datasheets as $R_{\theta JC(bottom)}$ to distinguish it from the much higher top-side thermal resistance through the

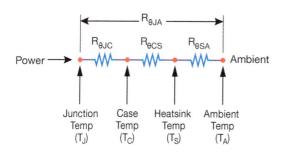

Figure 6.2 Steady-state thermal resistance schematic model for the physical device structure given in Figure 6.1.

encapsulant, which is similarly specified as $R_{\theta JC(top)}$. For example, Infineon BSC035N10NS5 lists an $R_{\theta JC(bottom)}$ of 0.8 °C/W and an $R_{\theta JC(top)}$ of 20 °C/W [4].

6.2.2 Thermal Resistances in a Chip-Scale Package

As discussed in Chapter 2, GaN devices are available in both a CSP and partially encapsulated PQFN. Figure 6.3 shows a cross-section of a GaN transistor mounted with the junction facing the PCB and separated by the solder bumps or bars that form the device terminals. For this package configuration, there are two main thermal heat dissipation paths: (i) down through the BGA/LGA pads or lead-frame, solder joints, and into the PCB; and (ii) up through the top side of the die. This configuration is thermally like other "flipped" device packages, such as DirectFET MOSFETs [5].

The thermal resistance of the path through the solder bumps to the copper traces on the PCB is defined as the thermal resistance from junction to board ($R_{\theta JB}$). The thermal resistance through the top of the die is defined as the thermal resistance from junction to case ($R_{\theta JC}$). $R_{\theta JB}$ and $R_{\theta JC}$ are intrinsic to the GaN device itself and cannot be changed with different applications or thermal techniques.

Figure 6.4 shows the thermal circuit model schematic of the heat flux paths for the cross-section presented in Figure 6.3. The thermal resistances $R_{\theta JB}$ and $R_{\theta JC}$ for a GaN transistor package is analogous to the previously mentioned thermal resistances $R_{\theta JC(bottom)}$ and $R_{\theta JC(top)}$ for a lead-frame package. However, in contrast to a lead-frame package, both the top and bottom sides of a chip-scale GaN transistor are effective cooling paths.

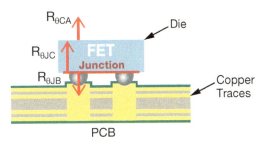

Figure 6.3 Cross section of a GaN transistor, highlighting the dominant heat flow paths from the junction to the top and bottom of the die.

Figure 6.5 shows thermal resistance comparisons between Si MOSFETs and CSP GaN transistors. The bottom-cooling thermal resistance, $R_{\theta JB}$, in Figure 6.5a follows an inverse trend with the device area for both GaN and Si devices. However, the top-cooling thermal resistance, $R_{\theta JC}$, in figure 6.5b is significantly lower for GaN transistors because of the low thermal resistance from the junction to the backside of the substrate (case).

6.2.3 Junction-to-Ambient Thermal Resistance ($R_{\theta JA}$)

The thermal resistance between the junction and ambient ($R_{\theta JA}$) is typically given in a device datasheet, standardized to a one square inch single layer PCB of 2 oz (70 μm) thick copper and without a heatsink or forced air flow. Recently, additional standards used to specify $R_{\theta JA}$, such as the JEDEC 51-2 [18], have been adopted. These configurations are shown in Figure 6.6a,b respectively.

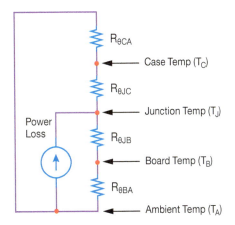

Figure 6.4 Steady-state thermal resistance schematic models for the physical device structures given in Figure 6.3.

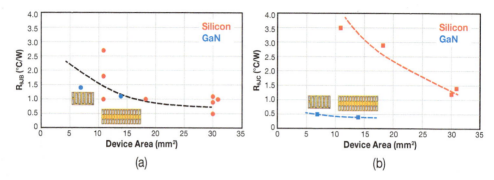

Figure 6.5 Thermal resistances for some typical LGA GaN transistors and packaged Si MOSFETs with (a) junction-to-board showing the trend in black dashed trace and (b) junction-to-case showing the trend in red for MOSFETs and blue for GaN transistors [4, 6–17].

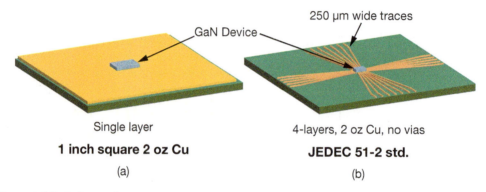

Figure 6.6 Typical setup for reporting $R_{\theta JA}$ in MOSFET and GaN FET datasheets (a) using the 1 inch square standard, and (b) the recent JEDEC 51-2 alternative.

For the JEDEC standard, the four-layer board design comprises a double signal, double buried power plane (2S2P), and uses 250 μm wide traces. All copper is 2 oz thick, and the inner layers are solid planes. There are also no vias connecting the planes. It should be noted that the $R_{\theta A}$ reported by the one inch square PCB method will under estimate the thermal resistance because it assumes a solid single structure connection to the FET. The JEDEC standard, however, will over-estimate thermal resistance due to the thin traces for conducting heat away from the part.

Figure 6.7 plots the datasheet $R_{\theta JA}$, of the one inch square standard, as function of device area for various GaN FETs and MOSFETs [6]. Notable in Figure 6.7 is the strong dependency between device area and thermal resistance for the GaN FETs. This dependency will be covered in detail in Section 6.3.1. For MOSFETs this dependency is lower mainly due to the larger package size.

It should also be noted that $R_{\theta JA}$ includes the PCB-to-ambient thermal resistance ($R_{\theta BA}$) as well as a minor case-to-ambient thermal resistance ($R_{\theta CA}$) contribution.

In a practical setting, the thermal resistance between junction and ambient $R_{\theta JA}$ will depend on the heat flux of surrounding components, PCB size and construction, layout, air flow, and whether a heatsink is used or not. If the practical board is of similar size and construction to the two standard cases (one inch square and JEDEC), then the thermal resistance between junction and ambient $R_{\theta JA}$ will lie between the two values given in the datasheets.

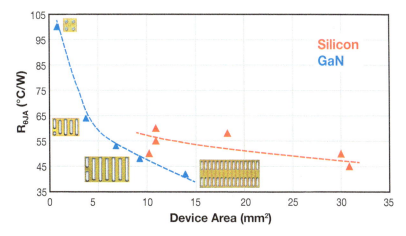

Figure 6.7 Junction-to-ambient thermal resistances for various GaN transistors and packaged Si MOSFETs based on the one inch square standard [4, 6–17].

6.2.4 Transient Thermal Impedance

GaN transistors are rarely operated with a continuous DC current, and most switching converter operations involve on and off times, such as pulses, of various durations and repetition rates. The measure of the thermal impact due to these pulses or repetition of pulses is given in typical datasheets as a transient thermal impedance graph. Figure 6.8 shows an example transient thermal impedance graph for the EPC2619 [17] of the junction-to-board $Z_{\theta JB}$. The transient thermal impedance $Z_{\theta JB}$ is normalized to the thermal resistance to the board $R_{\theta JB}$.

As an example, a GaN transistor is mounted on a PCB without a heatsink attached. The circuit repeatedly pulses the transistor with 10 W instantaneous power dissipation (P_{DM}) at a 10% duty cycle with 100 μs duration. The board is assumed to absorb the transient heat flux with little temperature impact, and thereby maintain a constant steady-state temperature. The effective thermal resistance is

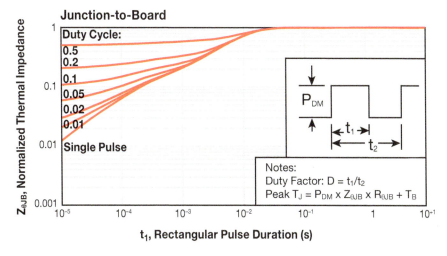

Figure 6.8 Transient thermal response curve for the EPC2619 GaN transistor. *Source:* Ref. [17]/Enhancement Mode Power Transistor.

$$R_{\theta\text{JB(effective)}} = Z_{\theta\text{JB}} \cdot R_{\theta\text{JB}} = (0.1) \cdot (2.0°C/W) = 0.2°C/W \tag{6.1}$$

and the junction temperature is

$$T_\text{J} - T_\text{B} = R_{\theta\text{JB(effective)}} \cdot P_\text{DM} = (0.2°C/W) \cdot (10\,W) = 2.0°C \tag{6.2}$$

At very low switching frequencies (<10 Hz), the normalized thermal impedance, as shown in Figure 6.8, converges to 1 because the circuit reaches thermal steady-state during each pulse. The normalized transient thermal impedance converges to the duty cycle when the switching frequency exceeds 100 kHz. This is equivalent to multiplying the steady-state thermal resistance by the average power of the pulse. Therefore, the transient thermal impedance can be neglected in some designs, such as a DC–DC converter with minimal fluctuation in output power.

Similarly, datasheets may also provide the transient thermal impedance for junction to case $Z_{\theta\text{JC}}$.

If the thermal resistances or steady-state temperatures for the board and heatsink or TIM are known, then the circuit in Figure 6.4 can be used to solve for the transient junction temperatures as well. The steady-state thermal resistances $R_{\theta\text{JB}}$ and $R_{\theta\text{JC}}$ are simply multiplied by the normalized transient thermal impedances presented earlier in this chapter to obtain the effective thermal resistances.

6.3 Cooling Methods

At its essence, a thermal design is about heat flux exchange. Heat is generated in one location and exchanged at another. The most common methods to exchange heat are (i) thermal radiation [19], (ii) natural convection, (iii) forced air convection, and (iv) liquid-assisted forced convection.

Convection-based heat exchange is predominantly driven by area, the larger the area, and the more spread out the heat source is, the more effective the exchange of heat. The cooling system should thus be designed to maximize the area and heat-spreading into that area.

6.3.1 Natural Convection Cooling

Natural convection relies on the temperature difference between the cooling surface and ambient to exchange heat flux, usually in still air. The large temperature difference generates local air movement which then pulls in cooler air from the surroundings. The larger the temperature difference, the more effective natural convection becomes. There are obvious limits to how high the temperature can become, which is the main limiting factor of natural convection cooling.

6.3.2 Forced-Air Convection Cooling

Forced-air convection is like natural convection except that it includes some means, such as a fan, to force the movement of air flow across the cooling surface. The rate of air movement is given in linear feet per minute (LFM), which is most popular, or other equivalent terms such as cubic feet per minute (CFM) or in SI units, cubic meters per second (m^3/s). Most power electronic systems specify a flow rate requirement of 400 LFM and anything above 1000 LFM is considered high, with little to be gained at these higher rates.

Natural convection cooling systems by their nature still moves air, but is minor in comparison to force air-cooled systems. This small movement of air overlaps with forced air cooling in the range between 50 LFM and 100 LFM.

6.3.3 Heatsink Heat Exchangers

The primary purpose of a heatsink is to increase the area for heat exchange. This is accomplished by using high thermally conductive materials, such as aluminum or copper, and by designing protrusions called fins or pins that conduct heat away from the main contact surface. Heatsinks are designed to meet a specific thermal resistance from the heat accepting surface to ambient. Tools are available to simplify a heatsink design [20, 21].

Key considerations in fin geometry include height, width, spacing, angle, and the method of attachment to the baseplate [22–28]. Some common heatsink configurations include the following:

- Extruded fin: linear fins and baseplate are fabricated as one solid piece, which may limit the options for fin width and spacing [23–25]
- Bonded fin: linear fins are attached to the base plate with epoxy, which may allow more options for fin geometry [25]
- Swaged fin: linear fins are attached to the base plate through swaging, eliminating the need for epoxy [26]
- Pin-fin or cross-cut: fins are either cylindrical (pin fin) or squares projected from the base plate (cross-cut), which improved surface area and natural convection air flow [27]

The selection of heatsink geometry is partly determined by available air flow. Heatsink thermal resistance to ambient is inversely related to both surface area and air flow. Wider fin spacing may reduce overall surface area, but it may also improve ducted air flow in a forced-air configuration [24]. Section 6.4.2 will delve into the details of GaN device cooling surface area and how that impacts heat flux flow into the heatsink.

Heatsinks suitable for natural convection cooling tend to have thicker and taller fins because the thermal resistance to ambient is strongly related to the air flow generated by the thermal gradient across the length of each fin. The naturally produced air flow across the fins enables heat exchange through convection, which is also supplemented by radiation. For a natural convection thermal design, a vertically oriented pin-fin or cross-cut heatsink may provide the best thermal resistance for a given heatsink volume [27].

Forced-air-cooled heatsinks, on the other hand, tend to have thinner and shorter fins which allows heat flux to move more quickly and do not require the temperature gradient needed for natural convection cooling.

Examples of natural convection cooled and forced-air-cooled heatsinks are as shown in Figure 6.9.

More advanced heatsink options include embedded heat pipes or liquid-cooled plates, which use internal fluid flow to improve the performance of the heatsink [26]. A detailed example of an experimental design using a heatsink will be given in the experimental section of this chapter.

6.3.4 Liquid Cooling Heat Exchangers

A liquid can be used to transport heat flux away from a small power circuit to a location with more room to accommodate a larger cooling system. The configuration of a liquid heat exchanger is like an air-cooled heatsink in that there is a contact surface that accepts the heat flux from the power devices. The main difference is that heat flux is now exchanged into a liquid and transported to another location where it is then dissipated to the ambient by means of a radiator. Such systems are more complex and add cost because they require plumbing and a pump to move the liquid through an exchanger to the ambient (see Figure 6.10). Liquid-cooled systems are thus only used in special cases [29].

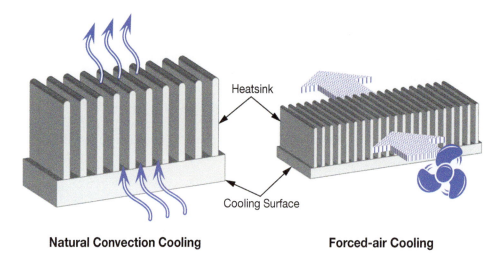

Figure 6.9 Differences in heatsink design between natural and force-air cooling.

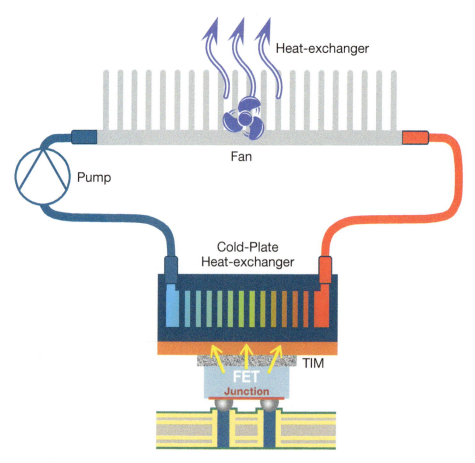

Figure 6.10 Closed-loop liquid-based cooling system.

6.4 System-Level Thermal Overview: Single FET

The previous section presented the two main heat dissipation paths from a GaN FET: (i) through the solder joints ($R_{\theta JB}$), and (ii) into the PCB or through the back of the die ($R_{\theta JC}$). These thermal resistances are fixed and intrinsic to the device and its construction. The heat transfer however can be improved by design using the PCB itself or aided by a heatsink, yielding several benefits such as higher heat flux capability or reduced temperature rise.

This section introduces the basic techniques, applied to a single device, to reduce thermal resistance to ambient, $R_{\theta JA}$. The first part focuses on the $R_{\theta JB}$ path dominated by the PCB. The second part focuses on the path dominated by back-side cooling, $R_{\theta JC}$, aided by a heatsink.

6.4.1 Single Device PCB-Only Cooling

GaN devices are almost exclusively soldered onto a printed circuit board (PCB) constructed of interleaved layers of copper and fiberglass weave. Copper is an excellent electrical and thermal conductor, and this intrinsically links the electrical and thermal designs. Electrical layout techniques for the PCB were presented in Chapter 4, and this section expands on those layout techniques to include the thermal impact of these layout choices.

Layout choices covered include PCB area, copper trace design, and how to use both sides of the PCB for heat-spreading. Figure 6.11 shows a cross-section of a GaN device mounted to a PCB with heat flux paths, temperature locations, and thermal resistances, that will be discussed in this section.

Without a heatsink mounted to the backside of the device, the main heat path is through the solder joints into the PCB, which will function as the heat-spreader and heatsink. Figure 6.12 shows a cross-sectional and top view image of how heat propagates into and through the PCB.

6.4.1.1 PCB Area for Improved Cooling
Conducting heat flux into the ambient uses two intrinsically linked fundamental characteristics – board area and temperature-difference. Each of these characteristics is limited by maximum operating temperatures or design specification temperature limits.

Figure 6.13 shows the relationship between board area and temperature rise for a fixed heat flux of 1 W per EPC2218 [7] device. Increasing the surface area of the board increases the convective area for heat transfer to the ambient environment, thus allowing for more heat flux to flow off the board and into the ambient. Figure 6.13 also shows that the relationship between

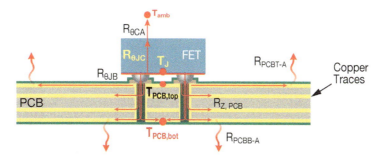

Figure 6.11 PCB-only cooling cross-section showing heat flux paths, temperature locations, and thermal resistances.

164 | GaN Power Devices for Efficient Power Conversion

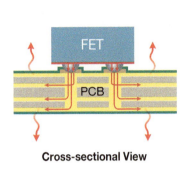

Figure 6.12 Heat-spreading of a GaN device mounted to a PCB. *Source:* With permission of EPC.

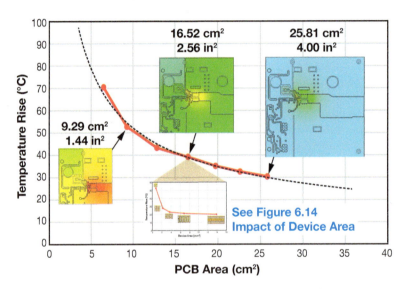

Figure 6.13 Relationship between PCB area and board temperature rise for a fixed 1 W dissipation per EPC2218 [7] device in a half-bridge configuration. *Source:* Adapted from Efficient Power Conversion Corporation [7].

PCB area and temperature rise is non-linear and as the PCB area increases, the reduction in temperature rise diminishes.

Similarly, the size of the GaN device also impacts the heat flow. Smaller devices are less effective at spreading their heat into the surrounding copper than larger devices as they have fewer bumps connecting to the copper plane. The effect of device size on temperature rise is shown in Figure 6.14 for a fixed board area of 2.56 in^2 (16.52 cm^2) and a ½ W heat dissipation in each device.

The design process thus needs to balance thermal performance against cost due to board size. Analytical tools [30] become valuable in predicting the thermal performance of a PCB design.

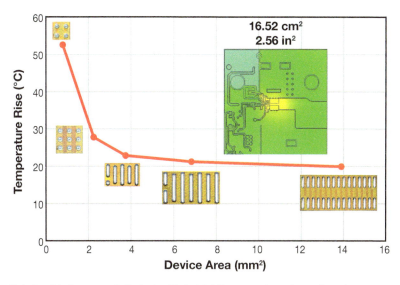

Figure 6.14 Relationship between GaN device [7, 8, 16, 17] area and maximum board temperature rise for a fixed board area of 2.56 in^2 (16.52 cm^2) and a ½ W heat dissipation in each device.

6.4.1.2 Impact of Copper Traces and Planes on Cooling

The copper traces are the main paths for the heat to spread on the PCB and dissipate into the ambient surroundings. Three design properties of the copper layer that directly impact thermal performance are thickness, trace width, and area.

The layer on which the GaN device is soldered is the primary heat conduction and spreading layer and is also the most effective at heat-spreading due to its direct thermal connection to the device. It is thus imperative to make sure that the thermal resistance into this layer is designed to be as low as possible.

The most important factors to consider are trace width in conjunction with path length and area beyond the device [31]. Figure 6.15 illustrates two layout designs for a PQFN packaged GaN device. In Figure 6.15a narrow traces are used to connect to the bars of the device, and the path to a connecting plane area is made at some distance from the device. In addition, the connecting plane for those traces is small. The combined effect of this design is high thermal resistance to the plane area and a high thermal resistance of the plane to the environment, thus limiting the system power dissipation. In addition, there are non-connected planes in proximity to the connecting plane but do not contribute to heat exchange to ambient temperature.

Figure 6.15b, on the other hand, uses wider traces that connect "as soon as possible" to the large planes. The non-contributing area of Figure 6.15a is reassigned and used to combine and expand the area of the heat-exchange zone. The combined effect of this design is a significant reduction in thermal resistance to ambient conditions over the design in (a).

As with PCB area, copper plane area also exhibits a diminishing thermal improvement as area increases, so the designer needs to find a balance between performance and cost due to the increase board area. Should additional cooling be required, back-side cooling methods using a heatsink, covered in Section 6.4.2, should be considered.

The thickness of the copper layers is another choice when designing a PCB. Table 6.1 shows simulation results for different copper thicknesses in a PCB with a fixed area of 25 cm^2 and two EPC2218 [7] devices, each dissipating 1 W of power.

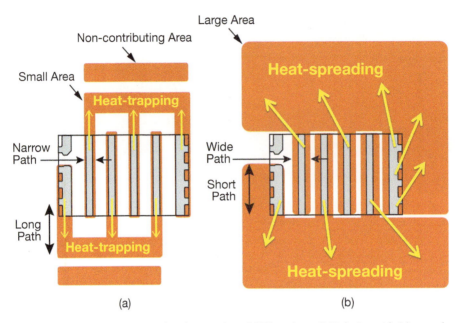

Figure 6.15 Top view of copper equivalent layouts for a PQFN package GaN device with (a) poor thermal performance due to small trace width, long path length and planes and (b) improved layout that uses wide trace widths, short paths, and large planes.

Table 6.1 Temperature rise (ΔT) for various copper thicknesses on a fixed area PCB of 25 cm².

Copper thickness	ΔT still air	ΔT 400 LFM
1 oz/35 μm	64 °C (baseline)	42 °C (baseline)
1.5 oz/53 μm	57 °C (−13%)	39.5 °C (−6%)
2 oz/70 μm	53 °C (−21%)	37.5 °C (−12%)
3 oz/105 μm	51 °C (−25%)	35.5 °C (−18%)

With natural convection cooling and increasing the copper thickness of the layers in the PCB from 35 to 105 μm, the temperature rise of the devices can be decreased by as much as 25%. Many GaN devices however have fine pitch bumps/bars that pose PCB manufacturability challenges and limit these designs to 70 μm copper thickness [32].

6.4.1.3 Using the Bottom Side of the PCB for Additional Heat-Spreading

The dielectric layers of a PCB have low thermal conductivity and thus make poor thermal contact to the inner and bottom copper layers. Furthermore, the inner layers are essentially buried "inside" the PCB where they can only spread heat but cannot exchange it to the ambient temperature. Despite these challenges, the inner and bottom copper layers can still be utilized in the thermal system by designing thermal interconnects or "heat pipes" between the layers that can improve the conduction of heat flux between the layers and allow the bottom side of the PCB to be used for heat exchange to the surrounding area.

Thermal interconnects can be designed in the form of vias, which are essentially copper tubes connecting one copper layer to another. Design considerations for vias include hole diameter, non-conductive or conductive fill, and wall plating thickness, all of which are covered in the IPC4761 standard [33].

Placing vias close to the heat-source ensures their effectiveness at conducting heat. This is not always possible due to electrical or cost constraints. There are two recommended practices for via placement: (i) under the GaN device bumps or bars and (ii) adjacent to the GaN device.

Placing vias under the GaN device bumps will minimize the thermal resistance path to the via. There are several design parameters to be considered for under-bump vias: (i) hole and annular ring diameters, (ii) tenting, and (iii) assembly-related constraints.

A via has three main dimensions: (i) hole diameter, (ii) annular ring diameter, and (iii) wall plating thickness. The copper annular ring diameter cannot exceed the width of the copper trace for the GaN device bump/bar, or it could violate design clearance rules. In addition, there are limits on the smallest hole diameter that can be drilled into the board, which is approximately 7.8 mils (200 μm). The minimum viable annular ring for 7.8 mil holes is 13.8 mils (350 μm) when using 2 oz thick copper and is also the minimum copper trace width that can be used for the bump/bar under the GaN device. When using 2 oz thick copper, this minimum spacing is 7.8 mils (200 μm). This combination yields a minimum of 21.6 mils (550 μm) bump-to-bump center spacing. The via plating wall thickness is based on manufacturing class, defined in the IPC4761 standard [33], and for class 2 is typically 0.787 mils (20 μm).

Vias must be tented, i.e. covered with solder-mask. This is required because un-tented vias alter the solder-mask design of the GaN device land-pattern which negatively affects the bump shape and seating height of the part during assembly, and thus impacts the reliability of the solder-joint.

Finally, via holes cannot be left open because open holes can drain solder during the reflow process. To prevent solder draining into the via requires a special construction of via defined in the IPC4761 standard [33] as a type VII via. Such a via is shown in Figure 6.16 and shows a non-conductive fill added to plug the via. The plug is then plated to seal it. This seal prevents

Via-In-Pad-Plated-Over (VIPPO)

- Wall thickness = 0.78 mil per IPC standard class 2
- Hole diameter (typical) = 7.8 mil
- Annular ring = 13.8 mil diameter min.
- Plated over
- Non-conductive filled
- Tented on both sides of the board
- Used for under bump and close to component pads
- Usable up to 2 oz (2.8 mil / 70 μm) copper thickness

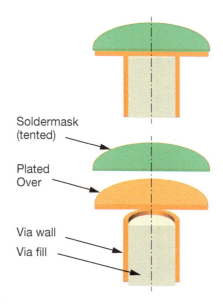

Figure 6.16 IPC 4761 Type VII via construction. *Source:* Adapted from IPC International [33].

outgassing during reflow and thus eliminates voids from forming in the solder joint. The nonconductive filled via also helps improve the reliability when subjected to temperature cycling, as the thermal coefficient of expansion closely matches that of the dielectric in the board, normally FR4 [34].

For devices that have finer pitch, or designs that preclude the cost of using type VII vias, the vias need to be placed adjacent to the device. The key is to place the vias as physically close as possible to the device to ensure that the thermal resistance path from the device bump to the via is minimized. For designs with adjacent vias, it is still imperative to tent as solder paste can migrate toward the hole where it can drain solder during reflow.

Using multiple vias will significantly reduce the overall thermal resistance into the PCB and down to the bottom layer. Larger-diameter vias can be used for adjacent vias designs, but there is a performance penalty as less vias can be accommodated in the same area. The inner copper layers can also be utilized to improve heat-spreading and thus reduce the overall thermal resistance of the path down to the bottom layer by increasing the area where temperature differences are the greatest.

The final design criterion for vias is layout staggering. Placing vias in a single line or "square" matrix tends to weaken the fiberglass weave of the PCB by creating "zip" lines [35]. For this reason, it is good practice to stagger the vias in a zig-zag pattern to break these straight lines.

Figure 6.17 shows the performance benefits of using vias and utilizing the bottom copper layer as additional heat exchange to ambient using two EPC2619s [17] in a half-bridge configuration. The thermal simulations shown in Figure 6.17 use the internal vertical layout

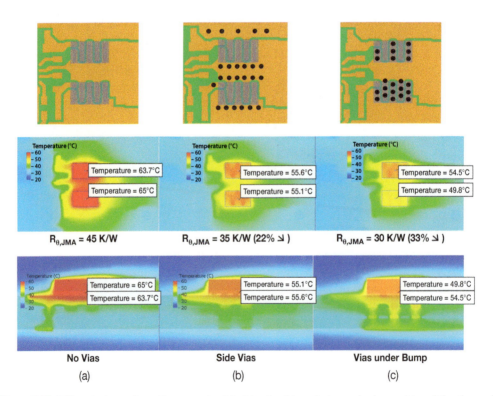

Figure 6.17 Different via configurations: no vias (a), side vias (b), and vias under bump (c) and the thermal performance for each of those configurations. *Source:* With permission of EPC.

technique presented in Chapter 4, and the results report the thermal resistance from junction to ambient in moving air ($R_{\theta JMA}$). Figure 6.17a shows a no-vias case, which is the baseline design, Figure 6.17b shows an adjacent via design, and Figure 6.17c shows a via under the bump design. Compared to the baseline design, the adjacent via design has a thermal resistance reduction that exceeds 20%, while using the via under the bump design, lowered the thermal resistance by over 30%. The cross-sectional images of Figure 6.17 show how adding vias reduce the device temperature and help transport more heat to the bottom of the PCB compared to the baseline design.

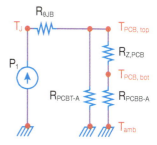

Figure 6.18 Steady-state thermal resistance schematic model for a double PCB-sided cooled single GaN FET mounted to a PCB with no heatsink attached.

6.4.1.4 Single FET PCB-Only Cooling Model

Figure 6.18 shows a simplified thermal resistance network for a single FET in a PCB-only cooling configuration as shown in Figure 6.11. The device area is small and thus the thermal resistance from the case to ambient $R_{\theta CA}$ is high enough to be ignored. This means all the heat flux flows through $R_{\theta JB}$ and onto the top layer of the PCB with corresponding temperature $T_{PCB, top}$. There are two paths for the heat to exchange into the ambient environment: (i) transversely from the top copper layer, with thermal resistance R_{PCBT-A} and (ii) down through the PCB, and vias, with thermal resistance $R_{Z, PCB}$, and into the bottom copper layer with transverse thermal resistance R_{PCBB-A}.

6.4.1.5 Joule Heating Effect on Device Heat Flux

Up to this point, the discussion has focused on heat generated in the GaN device only. The superior performance of GaN devices has shrunk their size and led to a concentration of connections to the device. This concentration in turn leads to increased current density and higher electrical resistance. The higher resistance in the connection leads to losses, and those losses further increase the heat flux associated with the GaN device, particularly in high current designs. This phenomenon is called joule heating [36].

Designers can reduce the effects of joule heating in their designs by keeping the concentrated paths as short as possible, creating parallel connections to the device, and balancing the layers in the board that enable those parallel connections. Tools are now available that can quickly show designers where current concentrates [30, 37] so designs can be improved before being manufactured.

Figure 6.19 shows an image of the EPC9186 [38] demo board that highlights joule heating in the top layer carrying 37.5 A_{RMS} through the traces. The highest current densities can be seen around the shunts on the right side of the board, indicated by the red circle, which is due to current necking down into the shunt connection. There are also elevated current densities around the devices themselves, indicated by the green circle, yet those are not showing any potential issues to this design.

6.4.2 Single GaN Device Heatsink-Assisted Cooling

As with most converter designs, the thermal performance of GaN transistors can be improved by attaching a heatsink. This section presents the selection of key thermal components and heatsink attachment schemes for both top-side and bottom PCB-side.

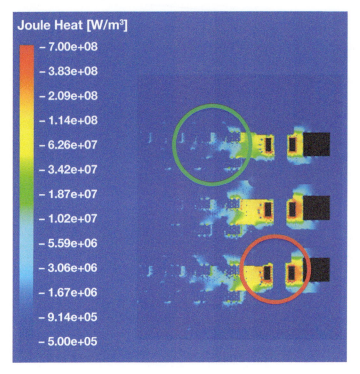

Figure 6.19 Shows Joule heating due to power dissipation in the copper traces in the top layer of the EPC9186 demo board [38]. The red circle shows the joule heating around the current shunts, while the green circle shows the heating surrounding the GaN transistors. *Source:* With permission of EPC.

6.4.2.1 Top-Side Heatsink Cooling

The discussion up to now has focused on the thermal resistance into the PCB ($R_{\theta JB}$). The exposed substrate of GaN devices has a much lower thermal resistance to the top side of the device ($R_{\theta JC}$) than into the PCB, as shown in Figure 6.5. This thermal path can be utilized by adding a heatsink to the top side of the GaN device.

A heatsink cannot simply be attached to a GaN device and therefore requires an interface. This interface is a material that is a good thermal conductor that mates the thermal surfaces of the GaN device and heatsink and is called a TIM. A simple method to attach a heatsink to the top side of the GaN device using a TIM, spacers, and screws is shown in Figure 6.20.

The spacers are used to seat the heatsink thermal surface at a specific distance from the surface of the PCB that is higher than the upper height of the seated GaN devices. This leaves a small gap for the TIM between the GaN device and thermal surface of the heatsink. The spacers are threaded, so screws can be used to secure the heatsink to the PCB. All components under the heatsink must be lower in height than the distance between the PCB and the thermal surface of the heatsink.

Figure 6.21 shows a cross section of the thermal approach given in Figure 6.20 with a bus capacitor placed under the heatsink. The top-side-cooled GaN device's thermal performance is dominated by the sum of $R_{\theta JC}$, R_{TIM}, and $R_{\theta SA}$, where the latter is the highest thermal resistance. The key to a good thermal design is to spread the heat flux as quickly as possible. This can be achieved by reducing the contribution of the TIM R_{TIM}, which will be covered in the next section, and using an excellent heat-spreading material for the heatsink.

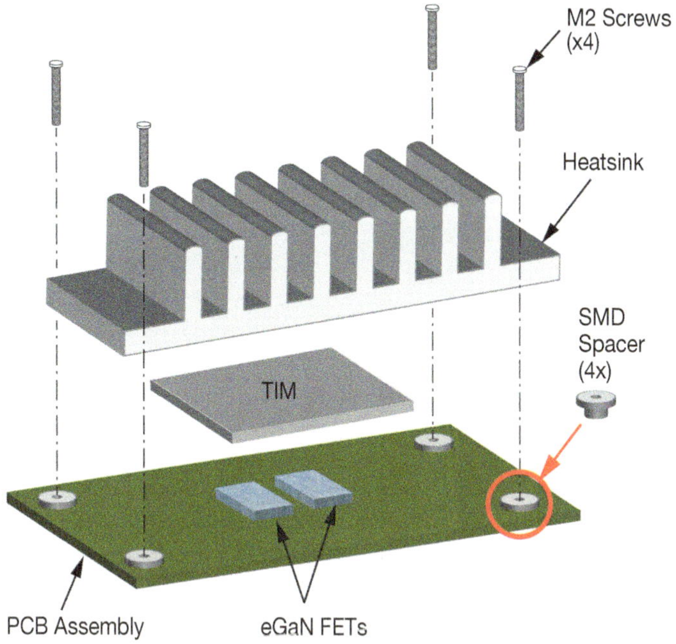

Thermal Management of GaN Devices | **171**

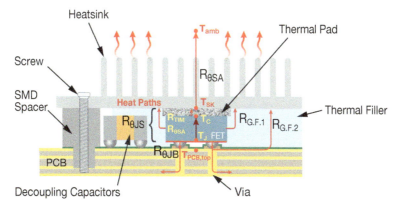

Figure 6.20 Exploded image of a heatsink attachment to a PCB for top-side cooling a GaN device.

Figure 6.21 Cross-sectional diagram of a GaN device mounted on a PCB with a heatsink attached for top-side cooling.

6.4.2.2 Thermal Interface Material Overview

A TIM is an important interface between the GaN device and heatsink as it performs several functions such as: (i) providing a low thermal resistance past between the GaN device and heatsink, (ii) electrical isolation in most cases, (iii) limiting the mechanical force on the GaN device, and (iv) providing a low thermal interface impedance between the GaN device and heatsink.

The small area of the top side of the GaN device requires a high-performance TIM material with high thermal conductivity [39] to ensure the lowest thermal resistance. This is because thermal resistance is intrinsically related to area [39] where the smaller the area, the higher the thermal resistance. Thermal conductivity values higher than 3.5 W/make are preferred and materials exceeding 15 W/m·K are considered excellent. However, there is a cost aspect linked to the performance of the TIM, where higher performance materials cost more, but this is offset by the small quantity needed for GaN devices given their small size.

The TIM occupies a finite space between the GaN device and heatsink surface. Given dimensional tolerances and thermally induced mechanical expansion and contraction, the TIM needs to be mechanically compliant under compression. TIM compliance is the characteristic that allows it to adhere to both surfaces well called surface wetting, which is important to reduce the thermal impedance at the interface. Mechanical compliance also means that the TIM will compress, introducing a force on the GaN device. The force is proportional to the distance between the GaN device and heatsink surface, the starting thickness of the TIM and the hardness of the TIM. The maximum permissible force on the chip-scale GaN device to prevent cracking and solder bump deformation is 50 psi (345 kPa or kN/m^2) [40]. This means an appropriate thickness of TIM needs to be selected based on the gap it is filling and its hardness, typically given in shore durometer [41]. Suitable TIMs will lie in the shore harness range of 00 [41] and come in various thicknesses ranging from 0.5 to 1 mm. In most cases a thermal design will compress the TIM by around 40–60%.

A TIM may also need to be electrically insulating, and in some cases have a high dielectric breakdown voltage. In a discrete FET half-bridge converter, this means the low-side device substrate is connected to ground and the high-side device substrate is connected to the switch-node. Figure 6.22 shows the TIM compression and location of the electrical isolation for a top-side heatsink thermal solution for a GaN device.

Some TIMs come in two-part liquid form, also termed gap filler, when applied and solidifies when cured. Liquid TIMs exert virtually zero force on the GaN device and are useful for wetting the side walls of CSP GaN devices for additional cooling as shown in Figure 6.23.

As can be seen in Figure 6.23, small CSP devices benefit the most from side wall cooling where the side wall area is of the same order of magnitude or larger than the top-side area. A liquid TIM is a convenient option for achieving multi-sided cooling. A combination of solid and liquid TIMs may be used, such as a pad for the top side and a liquid TIM for the side walls.

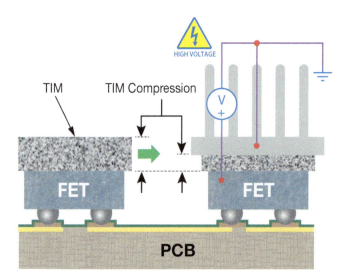

Figure 6.22 Illustration of a heatsink attachment with a TIM pad for GaN FETs and ICs including the electrical isolation location.

Thermal Management of GaN Devices | 173

Perimeter of die adds additional surface area

Part number	Die area (mm²)	Perimeter area (mm²)
EPC2619	3.75	4.00
EPC2218A EPC2065	6.83	5.45
EPC2215	7.36	6.20
EPC2054	1.69	2.60
EPC2036	0.81	2.25

Figure 6.23 Diagram of a small CSP GaN transistor showing the surface area of the top and sides of the die. The table on the right calculates the area of the device perimeter area compared with the area of the back surface for various sizes of GaN FETs.

Table 6.2 Examples of commercially available thermal interface materials.

Part number	Form factor	Thermal conductivity	Electrical volume resistivity	Dielectric strength	Hardness
Bergquist LIQUIFORM TLF 6000HG [42]	Pre-cured Gel	6 W/m-K	4.37×10^{11} Ω-m	10.5 kV/mm	Not applicable
Parker Chromerics THERM-A-GAP Gel 75 [43]	Silicone Gel	7.5 W/m-K	10^{12} Ω-m	8.7 kV/mm	Not applicable
Lipoly SH-PUTTY3-100 [44]	Silicone Grease	8.0 W/m-K	10^{13} Ω-m	12 kV/mm	Not applicable
t-Global TG-PP-10 [45]	Thermal Putty	10.0 W/m-K	Not specified	Not specified	50 Shore OO
t-Global TG-A-1780 [46]	Gap Filler Pad	17.8 W/m-K	6×10^{12} Ω-m	8 kV/mm	70 Shore OO
Laird TPCM-580 [47]	Phase-change material	3.8 W/m-K	3×10^{10} Ω-m	Not specified	Not applicable

Table 6.2 lists suitable TIMs that can be used with GaN devices. Most datasheets specify the material's thermal conductivity, which can be used to calculate the thermal resistance for a given volume of the material. The datasheet for a TIM pad may also specify the thermal resistance as a function of area and applied pressure. If specified, the thermal resistance also considers the thermal bond on both surfaces of the pad ("surface wetting"), as well as the impact of

mechanical compression. Table 6.2 lists the specified thermal conductivity for each example, as well as the electrical volume resistivity and dielectric breakdown voltage.

Given the importance of the TIM for GaN devices, the following summary can be used as a guide to achieving the lowest thermal resistance:

1) Keep the distance between the top side of the GaN device to the heatsink surface to a minimum,
2) select the highest thermally conductive TIM,
3) add a liquid TIM to the sidewalls of small CSP GaN devices for small size devices.

6.4.2.3 Heat-Spreader-Assisted Thermal Solutions

Some thermal designs require the heatsink to be connected to Earth. Given that there is electrical capacitance between the switch-node and the heatsink, this poses an EMI risk and a potential safety issue. To overcome this risk, a dual heatsink and heat-spreader can be used. In such a solution, a flat copper heat-spreader is placed over the GaN devices with a high-performance TIM between them, and a heatsink is installed over the heat-spreader using a TIM between them.

In this case the performance of the TIM between the heatsink and heat-spreader is enhanced by the larger area of the heat-spreader. This can help reduce the overall cost of the TIM material. The heat-spreader is thus both a thermal interface and serves to isolate the high dv/dt of the switch-node from the Earthed heatsink by shunting high-frequency current back to ground at the GaN device as shown in Figure 6.24.

6.4.2.4 Top-Side Heatsink Cooling Model

Figure 6.25 shows a simplified thermal resistance network for a single FET in a top-side-cooled configuration, including gap filler (see Figure 6.21). In this configuration, it is assumed that the thermal resistance to heatsink is the dominant path, and thermal resistance to the bottom side of the PCB, $R_{PCB, bot}$, is very high and can be ignored. For a more accurate analysis, the analysis given in Section 6.4.1 can be added to the model as shown in gray in Figure 6.25.

The heat flux flows from both GaN device paths $R_{\theta JC}$ and $R_{\theta JB}$. There are three main paths for the heat flux to exchange into the ambient: (i) up from the top side of the GaN device, through the TIM and into the heatsink, $R_{\theta JS1}$, (ii) from the side walls of the GaN device through the gap filler and into the heatsink, $R_{G.F.1}$, and (iii) transversely from the bottom of the GaN device to the top copper layer of the board and through the gap filler and into the heatsink, $R_{\theta JS2}$.

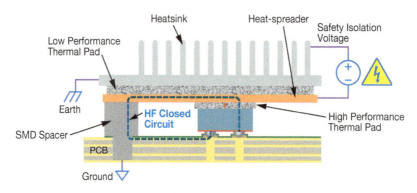

Figure 6.24 Cross-sectional view of a dual heatsink and heat-spreader thermal system with two types of TIMs and circuit for ground current.

The path from junction through the top side of the GaN device and into the heatsink, $R_{\theta JS1}$, includes the junction to case thermal resistance, $R_{\theta JC}$, and thermal resistance of the TIM, R_{TIM}.

The path from junction through the side walls of the GaN device and into the heatsink, $R_{G.F1}$, includes the $R_{\theta JC}$ and thermal resistance of the gap filler (liquid TIM). Since the heat flux path is both lateral and perpendicular to the junction, the junction to case side wall thermal resistance becomes a function of position along the vertical path of the side wall, $R_{\theta JC} \cdot d$, where d is the vertical distance from the junction. Adding this complexity to the calculation is unnecessary given that the contribution of other thermal resistances in the path is more significant, and it can safely be assumed the variation of thermal resistance as function of d will be very small because the GaN device substrate is a good thermal conductor.

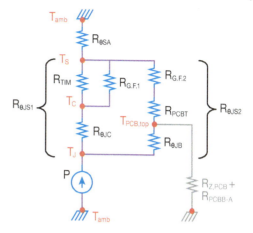

Figure 6.25 Steady-state thermal resistance model for a single GaN FET with heatsink attached to the top side including side wall cooling using a gap filler. The grayed circuit indicates where the PCB-only cooling thermal circuit model given in Figure 6.18 would be located.

The final heat flux path is from junction into the top layer of the board, $R_{\theta JB}$, then transversely to the gap filler, R_{PCBT}, and up through the gap filler and into the heatsink, $R_{G.F2}$. The thermal resistance from the heatsink to ambient air temperature is given by $R_{\theta SA}$ and can be calculated using the techniques given in Section 6.3.3.

Figure 6.25 also shows the heat flux path from the top side of the PCB down to the bottom side of the PCB in gray, which is the same as the corresponding path shown in Figure 6.18.

6.4.2.5 PCB Bottom-Side Heatsink Cooling

Some converter designs may require enhanced cooling but may not be able to use the top side of the GaN device for various reasons. One option is to cool the GaN device though the bottom side of the PCB, where a heatsink is mounted directly onto the PCB using a TIM. This typically requires multiple thermal vias, or a copper inlay in the PCB that allows heat flux to flow through the PCB [48] like that presented in Section 6.4.1.3 for PCB-only bottom-side cooling. The configuration for bottom-side heatsink cooling is shown in Figure 6.26.

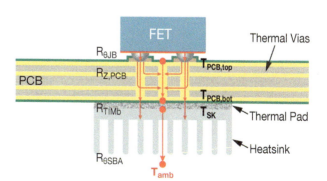

Figure 6.26 Cross-sectional diagram of a GaN device mounted on a PCB with a heatsink and thermal vias for bottom-side cooling.

Bottom-side cooling has several design challenges. First, the ability of the vias to conduct the heat flux to the bottom side of the board is limited, as shown in Section 6.3.3, requiring many vias which can disrupt the electrical circuit. Second, using a copper inlay adds cost to the PCB and also blocks the return path for the internal vertical layout power loop discussed in Chapter 4. Third, no components can be placed on the bottom side of the board, thus making layout on the top side more complex and potentially reducing performance.

Should components be required on the bottom side of the board, it would require milling pockets into the heatsink to contour around those components and which reduces the surface area in contact with the bottom-side copper and thus reducing thermal performance. Alternatively, using a thicker TIM would also degrade thermal performance.

6.5 System-Level Thermal Analysis: Multiple FETs

Up to this point, the discussion has focused on a single GaN device as a heat source. While there are power circuits that use a single device only, most configurations use two or more devices, such as the half-bridge configuration and when paralleling of devices.

The half-bridge configuration is a common building block in many power electronics circuits, and consists of two transistors Q1 and Q2, as well as the supplementary components such as gate drivers, regulators, and passive components. Paralleling adds additional devices near each other to increase the current capability of the converter.

Multiple heat sources lead to an exchange of heat flux between the sources. One device will invariably be at a higher temperature than the other, and heat flux will flow from the higher-temperature device to the lower-temperature device. This is called co-heating. In addition, there is a higher total heat flux due to the combination of the multiple heat sources.

All the previously discussed cooling techniques still apply for a half-bridge configuration, but now include the interactive element of heat flux exchange. The temperature rise of multiple heat sources can be approximated by adding the temperature rise at locations of all the heat sources with only one active heat source at a time [31]. This section leverages the thermal techniques already presented and adds the thermal interaction of multiple devices for both PCB-only and heatsink-cooled configurations.

6.5.1 Half-Bridge PCB-Only Cooling

Figure 6.27 shows a cross-section of a half-bridge configuration for PCB-only cooling that expands on Figure 6.11. It uses the internal vertical layout presented in Chapter 4, Section 4.4.3.

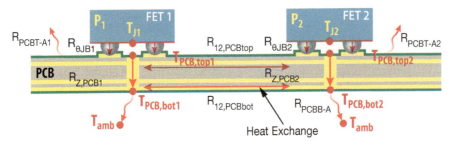

Figure 6.27 Cross-section diagram of a half-bridge configuration of GaN devices mounted on a PCB with a heatsink attached for top-side cooling.

Thermal Management of GaN Devices | 177

The internal vertical layout technique can be implemented in either of two basic configurations: (i) adjacent bus capacitor [49], shown in Figure 6.27, and (ii) center bus capacitors [50]. With multiple heat sources, the closer to each they are, the higher their interaction will become. This concentration of heat flux makes it more difficult to conduct the heat from the source and leads to higher net temperatures for the devices. Physically spreading out the heat sources can reduce co-heating, and in turn decrease the thermal resistance of the system.

An example of the difference in performance between two GaN devices near each other compared to those with a large separation is shown in Figure 6.28 using two EPC2218s [7]. In Figure 6.28a the bus capacitor is placed adjacent to one of the GaN devices, and in Figure 6.28b, the bus capacitors are placed between them. By adding more distance between the parts, $R_{\theta JA}$ can be reduced by around 5% for both FETs with no additional cost or electrical performance penalties.

Figure 6.29 shows a simplified equivalent thermal resistance circuit model for two FETs in a half-bridge configuration for PCB-only cooling. The model features two of the same circuits from Figure 6.18 to describe each power source, with added resistance elements for modeling heat exchange on the top $R_{12,PCBtop}$ and bottom $R_{12,PCBbot}$ layers of the PCB.

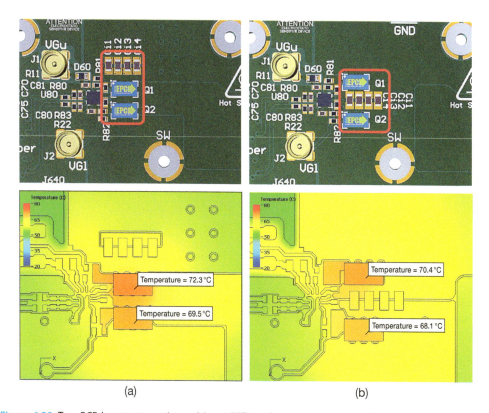

Figure 6.28 Two PCB layout comparisons: (a) two FETs in close proximity with adjacent bus capacitors, and (b) FETs spread out with bus capacitors between them. Lower image shows the corresponding simulated thermal performance. *Source:* With permission of EPC.

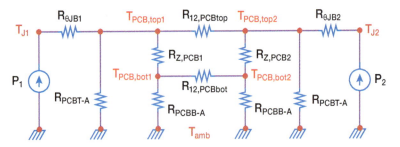

Figure 6.29 Steady-state thermal resistance schematic model for a half-bridge configuration of GaN devices for PCB-only cooling.

6.5.2 Half-Bridge Top-Side Heatsink Cooling

Section 6.4.2 presented a method to attach a heatsink to the top side of a single GaN device and Section 6.5.1 augmented the discussion by showing the exchange of heat flux between the GaN device via the copper layers in the PCB. In this section the methodology is expanded by adding a heatsink to the top side of the devices that not only enhances thermal performance, but also adds an additional heat exchange path between the heat sources.

Given that the heatsink and TIM are excellent thermal conductors, the heat exchange between the heat sources is now dominated by the thermal resistance of the heatsink in the lateral direction. In most cases, the GaN devices will be located close enough to each other that the thermal resistance between them from the heatsink approaches zero and the temperature of the heatsink in the vicinity of the devices can be assumed constant. Figure 6.30 shows the various heat paths in a half bridge with back-side cooling.

Like the single GaN device case, heatsink cooling two or more GaN devices with their top-side exposed substrates connected to their respective sources means that the TIM must electrically isolate the heatsink from the top side of the GaN devices as discussed in Section 6.4.2.2.

Given that a heatsink and TIM are more thermally conductive than the PCB, the effect of heat exchange is less pronounced compared to a PCB-only case. Figure 6.31 shows two examples of heat exchange between multiple devices using a large heatsink in an ideal scenario where over 90% of the heat dissipated from the devices conducts into the heatsink. Both

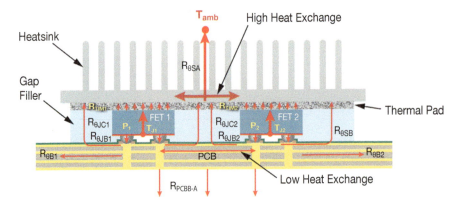

Figure 6.30 Cross-section of a half-bridge configuration of GaN devices mounted on a PCB with a heatsink attached for top-side cooling.

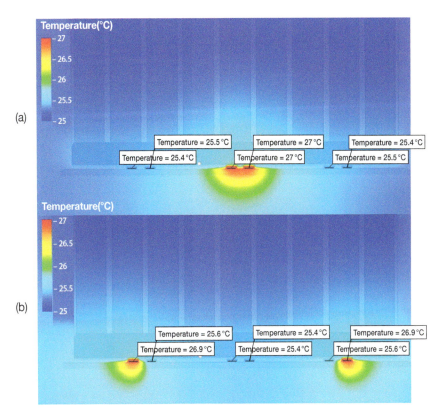

Figure 6.31 (a) Co-heating between two closely located devices with a large heatsink attached. (b) Co-heating between two parts with a large separation. *Source:* With permission of EPC.

scenarios have three groups of two paralleled FETs under one heatsink, but with different devices dissipating heat.

In this example the FETs dissipate 1 W each to show the effect of heat exchange between them. In both situations, the effect of heat exchange increases the temperature of the passive device, i.e. the device not dissipating heat, by about 0.5 °C regardless of location. The example shown in Figure 6.31a with both dissipating FETs in close proximity to one another, also shows only an additional 0.1 K/W thermal resistance compared to the case where both dissipating FETs are separated by a large distance as shown in Figure 6.31b.

Figure 6.32 shows a simplified equivalent thermal resistance circuit model for a half-bridge configuration with heatsink attached for top-side cooling. The model features two of the same circuits from Figure 6.25, one for each heat source. This model assumes the heatsink has a uniform temperature (T_S) across its surface.

6.5.3 Monolithically Integrated GaN Integrated Circuit (IC) Thermal Model

The advent of GaN monolithic integration, where two or more devices are fabricated on the same substrate, intrinsically couples their thermal resistances because they share the same substrate and several solder bumps or bars. In the example of Figure 6.33, an early monolithic integrated circuit combined both the high-side and low-side FETs onto the same chip in a half-bridge configuration.

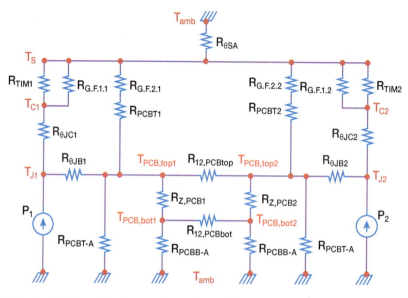

Figure 6.32 Steady-state thermal resistance schematic model for a single half-bridge configuration of GaN FET devices with heatsink based on the cross-section in Figure 6.30.

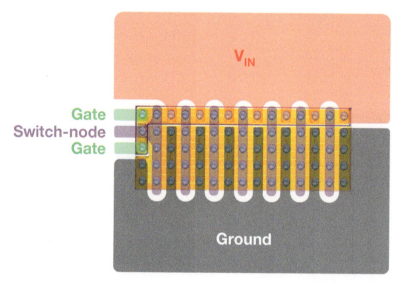

Figure 6.33 Image of the monolithic half-bridge EPC2105 [51] showing the three power connections that share the same thermally coupled substrate. *Source:* Adapted from Efficient Power Conversion Corporation [51].

Figure 6.34 shows a system model revised to show the thermal coupling between the junctions of Q1 and Q2 for a monolithic half-bridge IC. The power losses of both transistors are summed and dissipated through the same thermal paths, with very little deviation in junction temperature between them. This thermal coupling was experimentally verified, based on infrared observations of the EPC2105 asymmetric half-bridge IC during operation [6].

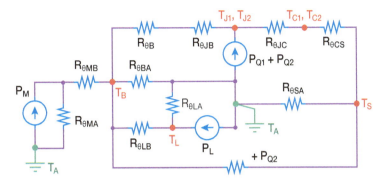

Figure 6.34 Thermal equivalent circuit for a half-bridge IC power stage utilizing top-side cooling, with power losses dissipated by one half-bridge GaN IC, a filter inductor, and the surrounding system.

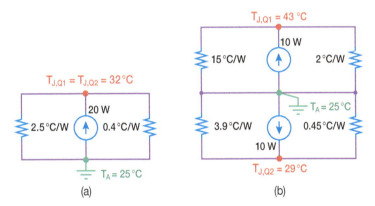

Figure 6.35 Example comparing the thermal coupling in an asymmetric half bridge composed of (a) two discrete GaN transistors and (b) one GaN IC with similar overall die area and thermal resistance to the two discrete transistors.

Figure 6.35 demonstrates the potential thermal benefit of using a half-bridge IC over a comparable pair of discrete transistors in a synchronous buck converter. In this example, the junction temperatures of Q1 and Q2 are theoretically calculated with an ambient temperature of 25 °C, and an assumption that the thermal resistances of the PCB, heatsink, and TIM are low enough that they can be neglected.

In Figure 6.35a, the EPC2105 is operated with power losses of 10 W each in Q1 and Q2, for a total power loss of 20 W [51]. For comparison, Figure 6.35b shows the same operating condition for a discrete half bridge composed of EPC2052 (Q1) [52] and EPC2029 (Q2) [53] transistors. The combined die area and electrical parameters of the discrete transistors are similar to the EPC2105 half-bridge IC.

In the discrete example, the Q1 and Q2 junction temperatures are 43 °C and 29 °C, respectively. In the IC example, both transistor junctions are 32 °C. Although the synchronous switch Q2 operates 3 °C hotter, the hard-switching transistor Q1 operates 11 °C cooler, which is a significant improvement to the maximum die temperature in the system.

In a practical system, the thermal resistances of the PCB, heatsink, and TIM are not negligible, which means that the two discrete transistor junction temperatures will likely deviate even more due to the separate thermal paths in the PCB ($R_{\theta B1}$ and $R_{\theta B2}$) and through the TIM ($R_{\theta CS1}$ and $R_{\theta CS2}$). Furthermore, if the switching frequency is relatively high, the hard-switching transistor Q1 power loss may be higher than the synchronous switch Q2, and the thermal benefit of a monolithic GaN IC over discrete transistors will be even more pronounced.

6.6 Experimental Thermal Examples

The discussion up to this point has focused on the background of thermal systems and techniques to minimize the thermal resistance between the GaN devices and ambient temperature for both PCB-only and top-side heatsink configurations. This section delves into the experimental aspects and will cover measurement techniques and present some examples using the techniques presented.

6.6.1 Temperature Measurement

In-circuit temperature measurement is an important element of thermal modeling validation, performance evaluation and design, and the junction temperature is often the most challenging to measure. Other temperature measurements include the ambient temperature of the surrounding air, the case temperature of each transistor, the board temperature, and the heatsink temperature. Figure 6.36 shows some potential temperature measurement locations in a half-bridge converter with a top-mounted heatsink.

Temperature measurement techniques of electronics systems include infrared optical measurement using a thermal camera and physical contact measurement using thermal couples and other similar sensors [54].

6.6.1.1 Optical Thermal Measurement

An infrared (IR) thermal camera is a very popular temperature measurement tool because of its ease of use. The primary limitation of IR measurement is accessibility to specific locations that may be optically blocked by elements of the circuit. If the PCB or transistor is buried under a TIM or covered by a heatsink, it is no longer directly accessible for measurement using a thermal camera.

The accuracy of IR measurement is also strongly dependent on the emissivity of the surface being measured. Shiny metallic surfaces typically have an emissivity of 0.1 or less, and IR measurement is most accurate with a target emissivity of 0.7 or higher. In addition, reflective

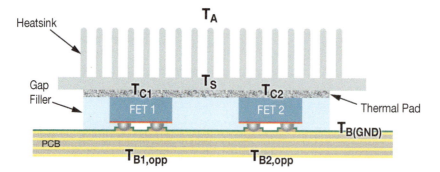

Figure 6.36 Cross-section of potential temperature sensor locations in a half-bridge power stage with a heatsink attached.

surfaces tend to act as IR mirrors, thus reading temperatures emanating from irrelevant objects near the board under test. Various high-emissivity materials can be added to the target to mitigate this problem, such as matte-black electrical tape, polyimide film tape, white correction fluid, and non-metallic paint. Emissivity error was demonstrated in [55], where a metal-packaged ASIC was observed with a thermal camera. The reading of the bare package was 27.5 °C, and the same ASIC was measured at 43.9 °C with polyimide tape on top.

6.6.1.2 Physical Contact Temperature Measurement

A number of contacting temperature sensor technologies are available for temperature measurement of electronic circuits. Thermocouples and resistive temperature detectors (RTDs) can be used for ambient temperature measurements. For embedded PCB temperature measurements, thermistors or semiconductor-based sensors are more common. Another option for accurate temperature measurements are fiber optic thermal sensors, such as the OpSense® OTG-F sensor [56].

There are a number of errors that can be introduced in physical contact measurement. The most common error in physical contact temperature measurement is to assume that the sensor is reading the target. Figure 6.37 illustrates the problem using a thermal sensor placed next to the GaN device that contracts the side wall. Figure 6.37a shows that the sensor is poorly contacting the side wall of the GaN device and makes excellent contact with the gap filler. This situation introduces a high thermal resistance to the GaN device wall and a low thermal resistance to the gap filler. Furthermore, the sensor is also part of the main heat flux path. This setup renders poor accuracy for the temperature measurement.

To rectify the temperature measurement issue, the thermal sensor can be thermally isolated from the gap filler. It is also good practice to improve the contact to the side wall of the GaN device where possible, and the gap filler can be used for that purpose. Thermal isolation can be achieved using low thermal conductivity materials such as polystyrene foam. This issue is not specific to the GaN device measurement and is relevant to all thermal measurements in the system.

Thermal couple sensors rely on the bi-metal contact voltage at the measurement point that is temperature dependent. The voltage across a thermal couple is very small and thus prone to noise injection, particularly when placed against the high voltage, high dv/dt of an upper FET in a half-bridge configuration. To overcome this issue, it is best to avoid using thermal couples to measure the upper device, and instead, measure the lower device whose exposed substrate is

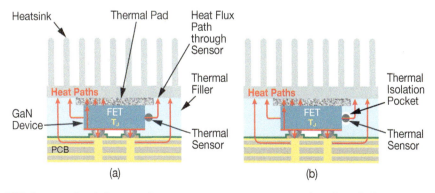

Figure 6.37 Cross-sectional diagram of a temperature sensor contacting the side wall of a GaN device where (a) the sensor makes good contact with the gap filler and (b) the sensor is inside a thermal isolation zone.

connected to ground and electrically quiet. To measure the upper device, the fiber optic sensor is a superior choice as it is immune to induced electrical voltage.

6.6.2 PCB-Only Cooling Experimental Example

This example experimentally validates the impact of the various PCB-only cooling techniques, such as copper thickness and via placement presented in Section 6.4.1.

A half-bridge configuration using EPC2619 [17] is mounted on a 25 cm^2 board made of four copper layers. The techniques presented in Sections 6.4.1 and 6.5.1, and specifically copper thickness, device separation and via placement, will be experimentally tested and compared with the simulation results [30] in three distinct designs.

The first experimental board is the baseline and uses ½ oz (18 μm) thick copper layers, minimal vias to conduct heat flux to other layers and is designed using the internal vertical layout technique with the bus capacitors adjacent to the upper FET. The second experimental board improves on the baseline design by using 2 oz (70 μm) thick copper and vias adjacent to the devices. The third experimental board further improves on the second board design by using vias under the device bumps and places the bus capacitor between the FET to separate them thermally without impacting the electrical circuit.

In the experiment, the GaN FETs are reverse biased using a current of 0.75 A and the voltage measured across the drain and source to determine the power dissipation in each case, which is 1.15 W. A thermal IR camera was used to measure the temperature of each device five minutes after commencement of testing to allow for thermal settling. An OPSense® OTG-F fiber optic thermal sensor [56] was fed through a specifically designed hole in the bottom of the board allowing near direct contact with each FET as shown in Figure 6.38. The thermal resistance was then calculated from the differences of the various measured temperatures and the heat flux determined from the electrical measurement. The experimental validation was conducted using natural convection cooling.

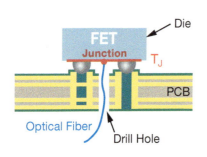

Figure 6.38 Schematic of fiber optic sensor setup for accurate thermal readings.

Figure 6.39 shows the measured thermal images for each of the three test cases. Compared to the baseline, shown in Figure 6.39a, adding adjacent vias and using thicker copper decreased the

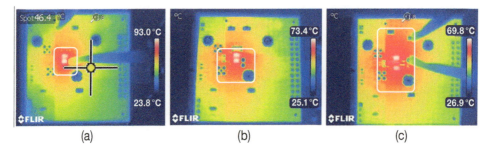

Figure 6.39 Measured thermal images for a 25 cm^2 board fitted with EPC2619 [17] in a half-bridge configuration (a) with ½ oz (18 μm) Cu, minimal vias, and bus capacitor adjacent to the upper FET (b) with 2 oz (70 μm) Cu, vias adjacent to the FETs, and bus capacitor adjacent to the upper FET (c) with 2 oz (70 μm) Cu, vias under the FET bumps, and bus capacitors between the FETs. *Source:* With permission of EPC.

GaN device temperature rise from 70.1 °C down to 48.3 °C as shown in Figure 6.39b. This represents a 31% decrease in temperature rise over the baseline. Compared to the baseline shown in Figure 6.39a, using vias under the FET bumps, thicker copper and moving the FETs apart, decreased the GaN device temperature rise from 70.1 °C down to 42.9 °C as shown in Figure 6.39c. This represents a 39% decrease in temperature rise over the baseline. Also shown in Figure 6.39 are rectangles showing the heat-spreading area around the FETs and how that area increases based on the techniques employed.

Finite element thermal simulations of the same board designs were conducted, and the thermal resistance from junction to ambient $R_{\theta JA}$ results compared with the experimental results. The baseline design simulation yielded 66.3 °C/W and the experimental 61 °C/W. For the design using 2 oz (70 µm) copper thickness and adjacent vias, the simulation yielded 46.8 °C/W and the experimental 43 °C/W. Finally, the design using 2 oz (70 µm) copper, vias under the FET bumps and separating the FETs, the simulation yielded 39 °C/W and the experimental 37.3 °C/W.

While the simulation results correlate with the experimental, in each case the experimental results are lower than predicted by the simulation. The main contribution to this is the junction temperature measurement that is very difficult to measure accurately. The actual junction temperature in all cases is most likely higher than measured, which when used would increase the thermal resistance calculation. In addition, the higher the temperature of the junction becomes, the greater the error, as the data shows.

Additional simulations were conducted using finer changes in variations of the layout design and summarized in Table 6.3. The finer variations include: (i) only changing the copper thickness while keeping the layout and via design fixed, (ii) only changing the use of vias while keeping the layout and copper thickness fixed, and (iii) only changing the layout configuration while keeping copper thickness and via use fixed. The results in Table 6.3 show that via under the bump yield the highest improvement in thermal performance followed by copper thickness. Spacing the FETs out has a smaller impact on performance because of the influence of the other techniques that make heat-spreading more effective.

6.6.3 Top-Side Cooling Experimental Example using a GaN IC

This second experimental example demonstrates the effectiveness of attaching a heatsink and heat-spreader to a GaN device. This example uses a 100 V rated monolithic GaN IC, EPC23102 [57], mounted on an EPC9177 [58] demonstration board.

Table 6.3 Temperature rise and thermal resistances for each PCB configuration.

Configuration	Design change	ΔT	$R_{\theta JA}$
Change in copper thickness	½ oz Cu	53 °C Baseline	46 °C/W Baseline
	2 oz Cu	44.6 °C (−19%)	37.3 °C/W (−23%)
Change in via design	Min vias	60.4 °C Baseline	48.9 °C/W Baseline
	Adjacent vias	49.6 °C (−18%)	40.4 °C/W (−21%)
	Vias under bump	43.8 °C (−27%)	40.4 °C/W (−32%)
Change in layout configuration	Adjacent FETs	46.6 °C Baseline	40.6 °C/W Baseline
	Spaced out FETs	44.6 °C (−5%)	37.3 °C/W (−9%)

186 | *GaN Power Devices for Efficient Power Conversion*

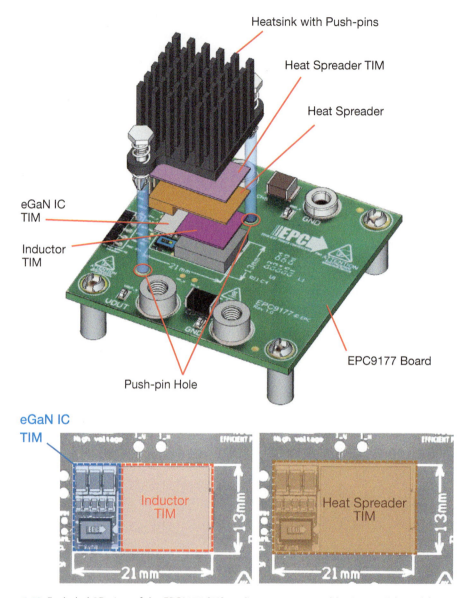

Figure 6.40 Exploded 3D view of the EPC9177 [58] cooling system assembly. *Source:* Adapted from [58].

Figure 6.40 shows an exploded 3D image of the EPC9177 [58] cooling solution. First, a high thermal conductivity TIM, the TG-A1780 [46] with a thickness of 0.5 mm and a thermal conductivity of 17.8 W/m · K, is placed over the GaN IC. A second, lower-cost TIM, the TG-A6200 [59] with a thickness of 0.3 mm and a thermal conductivity of 6.2 W/m · K is placed over the inductor. A contoured copper heat-spreader is then placed over both TIMs. A third low-cost TIM, which is the same material as used on the inductor, is then installed between the heat-spreader and heatsink that completes the assembly.

All the TIMs in this example are electrically insulating. However, because of the unique construction of the GaN IC, its exposed substrate is connected to the GND terminal of the IC allowing an even higher thermal conductivity TIM to be used that may be electrically conducting. Using an electrically conducting TIM in this case allows the heat-spreader to be connected to the GND of the circuit. The other two TIMs must still be electrically insulating.

The heatsink is a 21 mm × 21 mm × 18 mm tall cross-cut fin style with spring loaded pop-in pins making it easy to install without the need for screws.

The EPC9177 [58] board was operated in a standard configuration as a buck converter with 48 V input and powering a 12 V load. A thermocouple was placed in contact with the top surface of EPC23102. The load current was ramped in 1 A increments, at 20 second intervals, up to the maximum rating of 20 A.

Four different thermal conditions were tested: (i) no heatsink or heat-spreader installed and with natural convection cooling, (ii) with a heatsink and heat-spreader installed and with natural convection cooling, (iii) no heatsink or heat-spreader installed and with 400 LFM forced air cooling, and (iv) with a heatsink and heat-spreader installed and with 400 LFM forced air cooling. The maximum temperature rise for this example was limited to 80 °C, based on the restrictions given for the target application [58].

Figure 6.41 shows the experimental results for the different thermal tests performed on the EPC9177. In the first group of results with natural convection only cooling, the addition of a heatsink increased the current capability of the board from 15 to 21 A, a change of 40%. In the second group, using 400 LFM of forced air cooling, the temperature rise of the GaN IC decreased from 80 °C down to 35 °C, a change of 56%. Figure 6.41 also shows an unusual result where the natural convection cooling with a heatsink installed is virtually identical in thermal performance to the 400 LFM force air cooling case with no heatsink attached.

6.6.4 Multiple Paralleled GaN Devices Cooling Experimental Example for a High-Current Motor Drive

The third, and final, experimental example demonstrates the thermal performance of multiple distributed GaN devices across a board with and without a heatsink, and with natural convection and forced air cooling. This example uses four paralleled 100 V rated EPC2302 [3] per switch position mounted on an EPC9186 [38] three-phase brushless DC (BLDC) motor drive demonstration board capable of continuous operation while delivering 150 A_{RMS} current to each motor phase. It is operated at 100 kHz switching frequency. An image of the EPC9186

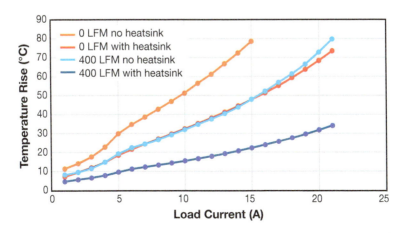

Figure 6.41 Temperature rise of EPC23102 under four test condition variations as a function of load current.

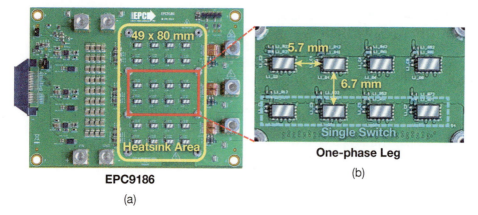

Figure 6.42 Photo of the EPC9186 [38] three-phase motor drive board showing the distribution and separation of the EPC2302 [3] GaN FETs. *Source:* (a) Ref. [38]/Efficient Power Conversion Corporation and (b) With permission of EPC.

[38] board is shown in Figure 6.42 along with the location and separation distances of the FETs in a phase leg.

The heatsink used for testing was manufactured by Alpha Novatech Inc., part number S08FTS02, and is shown in Figure 6.43. A high thermal conductivity TIM, the TG-A1780 [46], with a thickness of 0.5 mm and a thermal conductivity of 17.8 W/m·K, is placed over

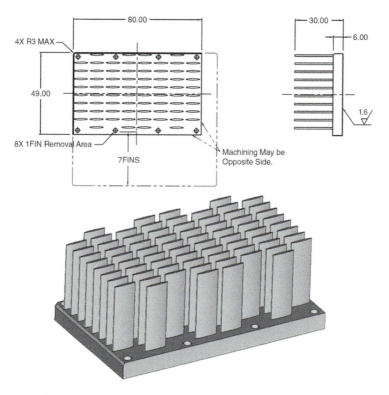

Figure 6.43 Details of the heatsink used on the EPC9186 [38] motor drive during testing.

the GaN devices. The heatsink is affixed with screws into 1 mm tall SMD resting posts or spacers soldered on the PCB. These spacers ensure the correct stand-off height to compress the TIM material by 40%. The TIM also provides electrical isolation between the exposed GaN FET substrates and heatsink. With this setup, each FET is able to dissipate up to 3 W of power and remain below the maximum temperature rise of 75 °C.

During experimental testing, the motor phase current was ramped while allowing the board to reach thermal stability for each measurement until a maximum observable temperature of 100 °C or temperature rise of 75 °C was reached.

Four different thermal conditions were tested: (i) no heatsink installed and with natural convection cooling, (ii) with a heatsink installed and with natural convection cooling, (iii) no heatsink installed and with approximately 400 LFM forced air cooling, and (iv) with a heatsink installed and with approximately 400 LFM forced air cooling.

Figure 6.44 shows experimental thermal performance results taken with EPC9186 [38] in various cooling configurations. For the natural convection cooling case, adding a heatsink shows an improvement in thermal performance by decreasing the temperature rise by 25 °C. When using 400 LFM of forced air to cool the board, the increase in motor phase current almost triples over that of the natural convection cases.

The EPC9186 [38] board was also simulated under DC current conditions for two thermal setup cases: (i) no heatsink installed and using natural convection cooling, and (ii) with a heatsink installed and with 400 LFM forced air cooling. The same setup was experimentally tested with the EPC9186 [38] switching at 20 kHz to significantly reduce the switching loss component of the GaN FETs such that it can be compared to the equivalent DC loss analysis case.

The thermal heat map for the simulation results is shown in Figure 6.45. For the natural convection with no-heatsink case, the temperature rise was around 26–30 °C with 35 A motor phase current. For the 400 LFM forced air with heatsink installed case, the temperature rise was approximately 45–52 °C with 130 A motor phase current. Figure 6.45 also shows the good heat-spreading in the copper layers of the board, particularly for the no-heatsink case. Note that the gradient from left to right is due to the increasing current density the copper conducts.

Figure 6.46 shows the experimental results over a wide motor phase current range and adds the corresponding simulation points that show excellent agreement.

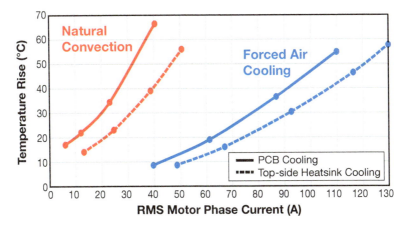

Figure 6.44 Thermal performance results as function of motor phase current of four paralleled EPC2302 GaN devices per switch position operating in a three-phase BLDC motor drive, EPC9186 [38], switching at 100 kHz and powered from a 48 V DC source. *Source:* Adapted from [38].

190 | *GaN Power Devices for Efficient Power Conversion*

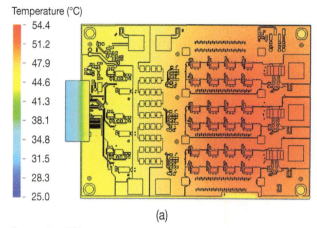

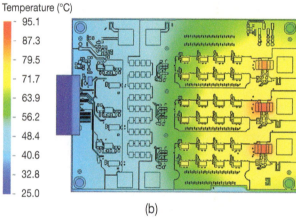

Figure 6.45 Simulated heat map plot of the EPC9186 [38] with (a) no heatsink installed and with natural convection cooling, and (b) with heatsink installed and 400 LFM forced air cooling. *Source:* Adapted from [38].

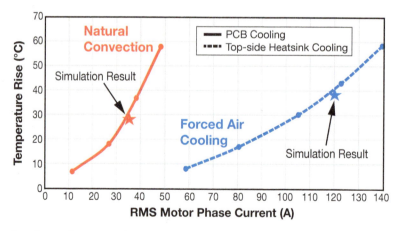

Figure 6.46 Experimentally measured temperature rise as function of motor phase current for no heatsink installed and with natural convection cooling (red) and, with heatsink installed and 400 LFM forced air cooling (blue dashed), with the simulation points for each case.

6.7 Summary

This chapter reviewed thermal management techniques for GaN-based converter designs. The thermal characteristics typically included in device datasheets, such as thermal resistances and transient thermal impedances, were explained. Thermal design schemes for PCB-only cooling and multi-sided cooling were presented, including selection of the heatsink and TIM, as well as key design considerations for attaching a heatsink. The typical options for temperature measurement were reviewed.

Finally, the chapter gave three application examples: (i) the effectiveness of PCB-only cooling design techniques for CSP GaN devices, (ii) a monolithic GaN IC with heatsink and heatspreader attached, and (iii) paralleled PQFN GaN devices in motor drive applications with various thermal conditions. The thermal techniques presented give designers the ability to choose the appropriate techniques that maximize the thermal performance while balancing the cost tradeoffs.

References

1 Infineon. TOLT– top-side cooling package, TOLT – the TO-leaded top-side cooling package for superior thermal performance. Product information. https://www.infineon.com/cms/en/product/power/mosfet/n-channel/optimos-and-strongirfet-latest-packages/tolt/.

2 ST Micro Electronics, Automotive-grade surface-mount (SMD) packages with top-side cooling. Product information. https://www.st.com/content/st_com/en/campaigns/automotive-grade-surface-mount-smd-packages-with-top-side-cooling.html

3 Efficient Power Conversion Corporation (2024). EPC2302 – Enhancement mode power transistor. EPC2302 datasheet [Revised March 2024]. https://epc-co.com/epc/Portals/0/epc/documents/datasheets/EPC2302_datasheet.pdf.

4 Infineon (2022). BSC035N10NS5 – OptiMOS^{TM} 5 power-transistor, 100 V. BSC035N10NS5 datasheet [Rev. 2.5 September 2022]. https://www.infineon.com/dgdl/Infineon-BSC035N10NS5-DS-v02_01-EN.pdf?fileId=5546d4624ad04ef9014ae8b5f3bc1b6f.

5 International Rectifier (2010). DirectFET© technology thermal model and rating calculator. Application note [September 2010]. AN-1059. http://application-notes.digchip.com/014/14-15379.pdf.

6 Reusch, D., Strydom, J., and Lidow, A. (2016). Thermal evaluation of chip-scale packaged gallium nitride transistors. *IEEE J. Emerg. Sel. Top. Power Electron.* 4 (3): 738–746.

7 Efficient Power Conversion Corporation (2024). EPC2218 – Enhancement mode power transistor. EPC2218 datasheet [Revised March 2024]. https://epc-co.com/epc/Portals/0/epc/documents/datasheets/EPC2218_datasheet.pdf.

8 Efficient Power Conversion Corporation (2023). EPC2066 – Enhancement mode power transistor. EPC2066 datasheet [Revised April 2023]. https://epc-co.com/epc/Portals/0/epc/documents/datasheets/EPC2066_datasheet.pdf

9 Infineon (2014). BSF134N10NJ3 G – n-Channel power MOSFET. BSF134N10NJ3 G datasheet [January 2014]. https://www.infineon.com/dgdl/Infineon-BSF134N10NJ3_G-DS-v02_06-en.pdf?fileId=db3a30432e779412012e7afa4a6c3834.

10 Infineon (2009). BSB012N03LX3 G – OptiMOS^{TM} power-MOSFET. BSB012N03LX3 G datasheet [November 2009]. https://media.digikey.com/pdf/Data%20Sheets/Infineon%20PDFs/BSB012N03LX3G.pdf.

11 Alpha & Omega Semiconductor (2012). AON7280 – 80V N-Channel MOSFET. eAON7280 datasheet [December 2012]. http://www.aosmd.com/pdfs/datasheet/AON7280.pdf.

12 Infineon (2020). BSZ075N08NS5 – MOSFET OptiMOS^{TM} 5 power-transistor, 80 V. BSZ075N08NS5 datasheet [Rev. 2.3 October 2020]. https://www.infineon.com/dgdl/Infineon-BSZ075N08NS5-DS-v02_01-en.pdf?fileId=5546d461454603990145ccd5743e61f8.

13 Texas Instruments (2010). CSD16323Q3C – N-channel NexFET^{TM} power MOSFETs. CSD16323Q3C datasheet [August 2010]. https://mm.digikey.com/Volume0/opasdata/d220001/medias/docus/635/CSD16323Q3C.pdf.

14 Texas Instruments (2009). CSD16321Q5C datasheet – Dual-Cool^{TM} N-channel NexFET^{TM} power MOSFET. CSD16321Q5C datasheet Revised May 2010]. https://mm.digikey.com/Volume0/opasdata/d220001/medias/docus/648/CSD16321Q5C.pdf.

15 Infineon (2020). BSC010N04LS – OptiMOS^{TM} power-MOSFET, 80 V. BSC010N04LS datasheet [Rev. 2.4 March 2020]. https://www.infineon.com/dgdl/Infineon-BSC010N04LS-DS-v02_02-EN.pdf?fileId=db3a3043353fdc16013552c1c63647c4.

16 Efficient Power Conversion Corporation (2022). EPC2036 – Enhancement mode power transistor. EPC2036 datasheet [Revised February 2022]. https://epc-co.com/epc/Portals/0/epc/documents/datasheets/EPC2036_datasheet.pdf.

17 Efficient Power Conversion Corporation (2023). EPC2019 – Enhancement mode power transistor. EPC2019 datasheet [Revised November 2023]. https://epc-co.com/epc/Portals/0/epc/documents/datasheets/EPC2619_datasheet.pdf.

18 JEDEC JESD51-2A (2008). Integrated circuits thermal test method environmental conditions – natural convection (still air). [Revision of JESD51-2, January 2008]. https://www.jedec.org/standards-documents/docs/jesd-51-2A.

19 Thermal radiation. Wikipedia. [Accessed 21 May 2024]. https://en.wikipedia.org/wiki/Thermal_radiation.

20 Efficient Power Conversion (2022). GaN FET thermal calculator. Design tool information. https://epc-co.com/epc/design-support/gan-power-bench/gan-fet-thermal-calculator

21 MyHeatSinks, Inc. (2024). Thermal resistance calculator – plate fin heat sink. https://myheatsinks.com/calculate/thermal-resistance-plate-fin/.

22 Texas Instruments (2011). Understanding thermal dissipation and design of a heatsink. Application report [May 2011]. http://www.ti.com/lit/an/slva462/slva462.pdf.

23 Lee, S., (1995). How to select a heat sink. [June 1995]. https://www.electronics-cooling.com/1995/06/how-to-select-a-heat-sink/#.

24 Edmunds, L. (2018). Heatsink characteristics. International Rectifier, Application note AN-1057. [Accessed December 2018] https://www.infineon.com/dgdl/an-1057.pdf?fileId=5546d462533600a401535591d3170fbd.

25 onsemi (2015). Heat sink selection guide for thermally enhanced S08-FL. Application note AND9016/D. [Rev. 1 Feb. 2015]. https://www.onsemi.com/pub/Collateral/AND9016-D.PDF.

26 Mersen (2018). Cooling of power electronics. [June 2018]. https://www.mersen.com/sites/default/files/publications-media/4-spm-cooling-of-power-electronics-mersen.pdf.

27 Advanced Thermal Solutions Inc. (2010). Heat sink selection methodology in electronics cooling. [December 2010]. https://www.google.com/url?sa=t&source=web&rct=j&opi=89978449&url=https://www.qats.com/cms/wp-content/uploads/2013/09/Qpedia_Dec10_HS-Selection-Application.pdf&ved=2ahUKEwis2uKspqCGAxVmJ0QIHaBMC_YQFnoECA8QAw&usg=AOvVaw1obGHf6awPOUYhkdPT8mmc

28 Ning, P., Li G., Wang F., and Ngo, K. (2008). Selection of heatsink and fan for high-temperature power modules underweight constraint. *Proceedings of the IEEE Applied Power Electronics Conference and Exposition (APEC)*, Austin, TX, (2008), 192–198.

29 Picotest. P2124A hHigh sspeed lline mmodulator (PSRR/PSMR/PSNR wwater-ccooled GaN pprobe. https://www.picotest.com/product/p2124a-high-speed-line-modulator-psrr-psmr-psnr-water-cooled-gan-probe/.

30 Cadence 2023. Celsius EC solver – Electronics cooling software for optimizing the thermal efficiency of complete systems. https://www.cadence.com/en_US/home/tools/system-analysis/thermal-solutions/celsius-ec-solver.html.

31 Texas Instruments (2010). Thermal design by insight, not hindsight. Application Report AN-2020. [Revised April 2013]. https://www.ti.com/lit/an/snva419c/snva419c.pdf.

32 IPC International (2020). IPC-6012E – Qualification and performance specification for rigid printed boards. [March 2020]. https://www.google.com/url?sa=t&source=web&rct=j&opi=89978449&url=https://www.ipc.org/TOC/IPC-6012E-toc.pdf&ved=2ahUKEwjWtdLOqaCGAxVAJkQIHaEHB8sQFnoECAYQAQ&usg=AOvVaw0Hk8B-FWdnHMzpnoEWDem7.

33 IPC International (2006). IPC-4761 – design guide for protection of printed board via structures. [July 2006]. https://www.google.com/url?sa=t&source=web&rct=j&opi=89978449&url=https://www.ipc.org/TOC/IPC-4761.pdf&ved=2ahUKEwjX-f6tq6CGAxWmIUQIHU8ADA0QFnoECAYQAQ&usg=AOvVaw266fqnbXlG_iQNAD67CntO

34 McCoy, B.S. and Zimmermann, M.A. (2004). Performance evaluation and reliability of thermal vias. *IEEE Applied Power Electronics Conference and Exposition (APEC)*, Anaheim, CA, (2): 1250–1256.

35 Shen, Y., Wang, H., Blaabjerg, F. et al. (2020). Thermal modeling and design optimization of PCB vias and pads. *IEEE Trans. on Power Electron.* 35 (1): 882–900.

36 Joule heating. Wikipedia. [Accessed 21 May 2024]. https://en.wikipedia.org/wiki/Joule_heating.

37 Keysight Technologies (2022). Power aanalyzer by kKeysight. Altium Extension application. [Updated 27 July 2023]. https://www.altium.com/documentation/altium-designer/power-analyzer-keysight.

38 Efficient Power Conversion Corporation (2023). EPC9186 – 150 A_{RMS}, wide input voltage 3-phase BLDC motor drive inverter. EPC9186 Reference design board information. [Revision 1.1]. https://epc-co.com/epc/products/evaluation-boards/epc9186.

39 Thermal conductivity and resistivity. Wikipedia. [Accessed 21 May 2024]. https://en.wikipedia.org/wiki/Thermal_conductivity_and_resistivity.

40 Garcia, R., Gajare, S., Espinoza, A. et al. (2023). GaN reliability and lifetime projections: phase 15. EPC El Segundo, CA (2023). EPC Reliability Report, 2023. https://epc-co.com/epc/documents/product-training/Reliability%20Report%20Phase%2015.pdf.

41 Shore durometer. Wikipedia. [Accessed 21 May 2024]. https://en.wikipedia.org/wiki/Shore_durometer

42 Bergquist (2020). 6000HG – Liqui Form TLF. 6000HG datasheet, March 2020. https://datasheets.tdx.henkel.com/BERGQUIST-LIQUI-FORM-TLF-6000HG-en_GL.pdf.

43 Parker Chomerics (2022). THERM-A-GAP GEL 75 high performance fully cured dispensable gel. THERM-A-GAP GEL 75 datasheet, [August 2023]. https://www.parker.com/Literature/Chomerics/Catalogs/Parker%20Chomerics%20THERM-A-GAP%20GEL%2075.pdf.

44 LiPoly (2021). SH-putty3 thermal conductive putty. SH-putty3 datasheet, March 2021. https://lipoly.com/pdf/SH-putty3.pdf.

45 T-Global Technology (2021). TG-PP-10 silicone thermal putty. TG-PP-10 datasheet, February 2021. https://www.tglobaltechnology.com/wp-content/uploads/2023/01/TG-PP-10.pdf.

46 T-Global Technology (2023). TG-A1780 Ultra soft thermal pad. TG-A1780 datasheet, July 2023. https://www.tglobaltechnology.com/wp-content/uploads/2023/08/TG-A1780_UK.pdf.

47 Laird Technology (2012). Tpcm™ 580 series phase change material. Tpcm™ 580 datasheet, November 2012. https://www.laird.com/sites/default/files/2022-11/THR-DS-Tpcm%20580%20Data%20Sheet%20.pdf.

48 Zhang, L., Liu P., Guo S., and Huang, A.Q. (2016). Comparative study of temperature sensitive electrical parameters (TSEP) of Si, SiC, and GaN power devices. *Proceedings of the IEEE Workshop on Wide Bandgap Power Devices and Applications*, 302–307.

49 Reusch, D. and Strydom, J. (2013). Understanding the effect of PCB layout on circuit performance in a high frequency gallium nitride-based point of load converter. *Twenty-Eighth Annual IEEE Applied Power Electronics Conference and Exposition (APEC)*, Long Beach, CA, (16–21 March 2013), 649–655.

50 Glaser, J.S. and Helou, A. (2022). PCB layout for chip-scale package GaN FETs optimizes both electrical and thermal performance. *IEEE Applied Power Electronics Conference and Exposition (APEC)*, Orlando, FL, (20–24 March 2022).

51 Efficient Power Conversion Corporation (2020). EPC2105 – Enhancement mode GaN power transistor half bridge. EPC2105 datasheet, June 2020. https://epc-co.com/epc/documents/datasheets/EPC2105_datasheet.pdf.

52 Efficient Power Conversion Corporation (2022). EPC2052 – Enhancement mode GaN power transistor. EPC2052 datasheet, [Revised September 2022]. https://epc-co.com/epc/documents/datasheets/EPC2052_datasheet.pdf.

53 Efficient Power Conversion Corporation (2021). EPC2029 – Enhancement mode GaN power transistor. EPC2029 datasheet, April 2021. https://epc-co.com/epc/documents/datasheets/EPC2029_datasheet.pdf.

54 Blackburn, D.L. (2004). Temperature measurements of semiconductor devices – a review. *Proceedings of the IEEE Semiconductor Thermal Measurement and Management Symposium*, (11 March 2004), 70–80.

55 Flir (2015). Use low-cost materials to increase target emissivity. [November 2015]. https://www.flir.com/discover/rd-science/use-low-cost-materials-to-increase-target-emissivity/.

56 OPSense Solutions, OTG-F fiber optic temperature sensor. Product information. https://opsens-solutions.com/products/fiber-optic-temperature-sensors/otg-f.

57 Efficient Power Conversion Corporation (2024). EPC23102 – ePower™ Stage IC. PC23102 datasheet, [Revised April 2024]. https://epc-co.com/epc/Portals/0/epc/documents/datasheets/EPC23102_datasheet.pdf.

58 Efficient Power Conversion Corporation (2023). EPC9177: Small area, low-profile, synchronous buck converter. Demonstration board product information. https://epc-co.com/epc/products/evaluation-boards/epc9177

59 T-Global Technology (2023). TG-A6200 ultra soft thermal pad. Product information. [October 2018]. https://www.tglobaltechnology.com/product/tg-a6200-ultra-soft-thermal-pad

7

Hard-Switching Topologies

7.1 Introduction

All efficient electrical power converters operate under the same basic principle: an energy storage element is connected to a power source, from which it obtains and stores energy, and subsequently it is connected to a power sink, often called a load, to which it then transfers energy. This cycle is repeated periodically so that on average, a controlled amount of power can be transferred from the source to the load. A key benefit is that the source and the load can have much different characteristics, so much so that they could not normally connected to each other without something breaking or catching fire. An everyday example of this is the much-maligned "wall-wart" that powers and charges portable electronics.

The connections are made with switches that open and close, and the energy is usually stored in a magnetic or electric field. There are many possible variations on this, but in modern power converters, the switches are implemented with field-effect transistors (FETs) and the energy storage with inductors and capacitors. A key point with switching power converters is that for a given power throughput, for shorter and more frequent switching cycles, less energy per cycle needs to be stored, and thus energy storage components become smaller.

The simplest type of these converters is the pulse-width modulated (PWM) controlled hard-switched converter [1]. These have many positive features:

- Simple operating behavior,
- simple transfer functions that can be analytically derived,
- small number of operating modes,
- are bi-directional capable,
- rigorous design methods,
- are well understood,
- wide operating range,
- continuous power flow regulation from full design power to zero power.

Because of these features, hard-switched PWM has the most versatility and is the most common type of switching power converter. Recalling that increasing the switching frequency can reduce the size of the energy storage components, one can understand the drive to increase switching frequency as much as possible, but before long one runs into some limits, the main limit being energy lost on each switching transition due to parasitic components and non-ideal switching. Switching loss varies inversely with conduction loss when the switches are on, so it is important to be able to estimate these losses to make informed design tradeoffs.

GaN Power Devices for Efficient Power Conversion, Fourth Edition. Alex Lidow, Michael de Rooij, John Glaser, Alejandro Pozo, Shengke Zhang, Marco Palma, David Reusch, and Johan Strydom.
© 2025 John Wiley & Sons Ltd. Published 2025 by John Wiley & Sons Ltd.

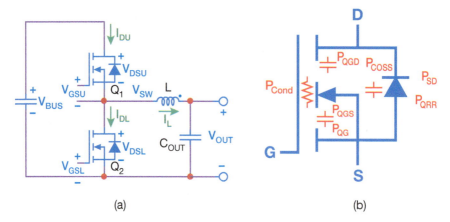

Figure 7.1 Half-bridge building block configuration using GaN devices (a). Parasitic device elements that impact switching design and performance (b).

The half-bridge configuration of transistors is a widely used building block of switching power converters. It is comprised of two series connected transistors powered by a source. The mid-point of the series-connected transistors is linked to a load via a filter. This configuration is shown in Figure 7.1a for MOSFETs or GaN devices. The transistors are switched alternately, where the ratio between their conducting states determines the change on output voltage. This popular configuration is favored by designers because it is easy to control by simply changing the duty ratio, also known as pulse-width modulation (PWM).

However, the non-ideal characteristics of the switching devices, shown in Figure 7.1b, lead to losses for the half-bridge configuration when operated in PWM mode. These losses stem from the non-ideal characteristics of the device such as capacitance, resistance, and finite switching transition time and are defined as hard-switching losses. Understanding these loss mechanisms enables designers to optimize performance to specific requirements such as maximum power density or highest efficiency. In this chapter, losses stemming from hard-switching of a half-bridge configuration are presented for buck converter operation. The conclusion is that the superior properties of GaN transistors yield significant performance improvements over MOSFET-based converters.

7.2 Hard-Switching Loss Analysis

The main metrics of any switching converter performance are: (i) efficiency, where higher is better, (ii) size, where smaller is better, and (iii) cost, where lower is better. Efficiency can be increased through improvements in the switching (dynamic) and conduction (static) characteristics of the devices. The improvement in the device's dynamic characteristics, such as that for GaN devices, allows for a higher operating frequency that leads to a reduction in the size of the transformers, inductor, and capacitors.

In hard-switching converters, transistors are turned on and off rapidly while there is simultaneously voltage and current between the drain and source of the device. These switching transitions can lead to significant power losses during the switching event and are defined as switching transition losses.

The transitions of current and voltage within the device are not the only switching loss contributors. Other factors are output capacitance losses (P_{OSS}), gate-charge losses (P_G), reverse conduction losses (P_{SD}), and reverse recovery losses (P_{RR}).

The output capacitance losses (P_{OSS}) are associated with the output capacitance of the device. The charging or discharging of this capacitor requires energy of which some significant fraction can be dissipated during operation of the converter.

Gate-charge losses (P_G) are like the output capacitance losses in that the energy required to charge the gate-to-source capacitance is dissipated in the resistance of the current path of the capacitance for both turn-on and turn-off. The resistance path includes the gate driver, gate drive resistor, device series gate resistance (R_G), and other parasitic resistances in the circuit.

In most converters, an anti-parallel diode is present across the drain-to-source terminals of the transistor. In some cases, it is inherent to the device structure, such as in MOSFETs. In cascode GaN transistors, as described in Chapter 1, there is a body diode in the Si MOSFET, which conducts in series with the GaN transistor's channel during reverse conduction. In the case of enhancement-mode GaN transistors, there is a mechanism to conduct reverse current when the device is off (commonly known as dead time), allowing operation like the PN junction body diode of a MOSFET. Since diode conduction occurs in each switching cycle during the dead time when both transistors are turned off simultaneously, the dynamic loss calculation includes these reverse conduction losses (P_{SD}). In Si MOSFETs, these losses are commonly referred to as diode losses (P_D).

Reverse recovery losses (P_{RR}) are due to the recombination time of excess minority carrier charge present in the depletion region of PN junctions and are only present in MOSFET and cascode GaN transistors. They are absent in enhancement-mode GaN transistors because dead-time current utilizes majority carriers and therefore have no reverse recovery effects. Since this chapter deals specifically with enhancement-mode GaN, reverse recovery will not be discussed in detail. However, it is important to note that it is a major contributor to switching loss with Si transistors, frequently dominating all other dynamic loss components under certain common operating conditions [2].

7.2.1 Switching Transition Losses

To drive frequency higher in a hard-switching converter, power devices must have very low dynamic (switching) losses. The dominant component of these losses is the hard-switching "event" where in a turn-on switching transition current flows through the device before the voltage across that device commutates to zero as shown in Figure 7.2a [3]. The reverse sequence occurs when the device turns off as shown in Figure 7.2b.

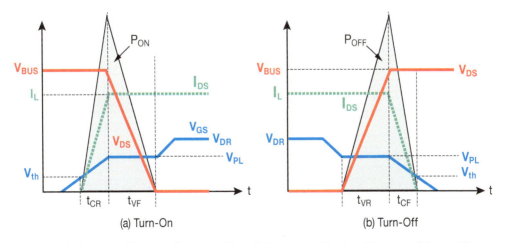

Figure 7.2 Idealized switching waveforms used for calculating switching loss (a) turn-on (b) turn-off.

The switching transition power loss can be determined graphically from Figure 7.2. The power dissipation is simply the area under the power curve, the latter being the $V \cdot I$ product. Summing the voltage transition power losses (P_{Vt}) and the current transition power losses (P_{Ct}) yields the following equation for transition power loss:

$$P_{sw} = P_{Vt} + P_{Ct} = \tfrac{1}{2} \cdot V_{BUS} \cdot I_L \cdot (t_{CR} + t_{VF} + t_{VR} + t_{CF}) \cdot f_{sw} \tag{7.1}$$

where t_{CR}, t_{VF}, t_{VR}, and t_{CF} are the switching commutation times as seen in Figure 7.2, V_{BUS} is the voltage across the device during the off-state, and I_{DS} is the on-state drain-to-source current, and where one parameter (either voltage or current) is always in transition while the other is fixed. This leads to the factor of ½ in Eq. (7.1).

The switching times are not given in the circuit and need to be determined from the gate-charge (Q_G) characteristics of device based on the circuit operating conditions. GaN transistors are driven in a similar manner to MOSFETs. The gate electrodes have very high input impedance, and control of the device is accomplished by supplying or removing charge to/from the gate electrode. Transistor switching can be segmented into four regions, as shown in Figure 7.3, aided with the numbered items as follows: (i) the charge required to bring the gate electrode up to device threshold (Q_{GS1}), (ii) the charge required to transition the current from zero to the load current (Q_{GS2}), (iii) the charge required to transition the drain-to-source voltage (Q_{GD}), and (iv) the incremental additional charge to fully enhance the gate ($Q_G - (Q_{GS1} + Q_{GS2} + Q_{GD})$). The following sections analyze each of these regions so that the transition times can be determined and the corresponding losses calculated. The figures and numerical values are from the datasheet of the EPC2619 [4], but are qualitatively representative of typical FET behavior. Note that Q_{GS1} does not affect transition losses, so it is only used as part of the calculations (Section 7.2.1.2).

7.2.1.1 Miller Charge (Q_{GD}): The Voltage Transition Period

The voltage commutation period of a traditional hard-switching commutation is based on the Miller charge. The Miller charge is characterized by the plateau voltage on the gate waveform and can be used to determine the switching losses during the voltage transition period. At turn-

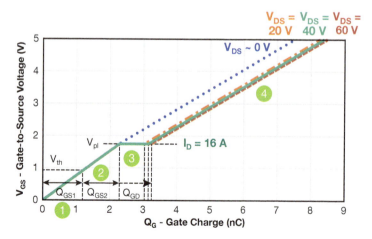

Figure 7.3 Impact of drain–source voltage and drain current on gate charge and gate voltage for EPC2619. *Source:* Adapted from Efficient Power Conversion Corporation [4].

on, the period begins after the current has fully transitioned and is complete when the drain-to-source voltage has reached zero. The reverse process occurs for turn-off. The larger the voltage swing, the longer it will take for the transition, and the higher the losses. Figure 7.3 is a graph of the measured gate voltage as a function of gate charge, highlighting the impact of various drain-to-source voltages.

In general, the time (t) it takes to charge a capacitor to a specific charge (Q) is given by

$$t = \frac{Q}{I} \tag{7.2}$$

where I is the current used to charge the capacitor (assumed constant during the period t).

Q_{GD} for any given drain-to-source voltage can be calculated using Eq. (7.3) [3], if the function of $C_{RSS}(v_{DS})$ is known.

$$Q_{GD} = \int_0^{V_{BUS}} C_{RSS}(v_{DS}) \cdot dv_{DS} \tag{7.3}$$

Alternatively, the $C_{RSS}(v_{DS})$ function can be captured from the graph provided in device datasheets. It can be seen from Figure 7.4 that all the devices' capacitances have a non-linear relationship with drain-to-source voltage.

Substituting Eq. (7.2) into Eq. (7.1) for the voltage transition only and using a linear approximation for the gate voltage and current transitions, the power losses during the voltage transition can then be approximated with Eq. (7.4).

$$\begin{aligned} P_{Vt} &\cong \frac{V_{Bus} \cdot I_L}{2} \cdot t_{Vx} \cdot f_{sw} \\ &= \frac{V_{Bus} \cdot I_L}{2} \cdot \frac{Q_{GD}}{I_G} \cdot f_{sw} \end{aligned} \tag{7.4}$$

As can be seen in Eq. (7.4), the gate current appears in the power loss equation. The current into the gate thus affects the transition time, and increasing the gate current will reduce this time. However, gate driver impedance and gate circuit inductance may limit this gate current and therefore the transition period.

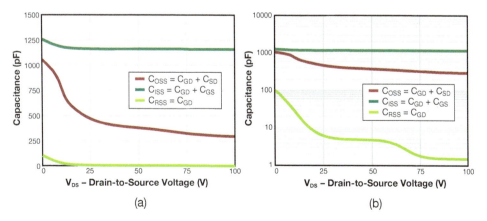

Figure 7.4 Device capacitances as a function of drain-to-source voltage for the EPC2619 [4]: (a) on a linear scale; (b) on a semi-log scale. *Source:* Adapted from Efficient Power Conversion Corporation [4].

The gate current (I_{Gvon}) during the turn-on voltage transition period (t_{VF}) can be estimated as

$$I_{Gvon} = \frac{V_{DR} - V_{PL}}{R_{Gon}} \tag{7.5a}$$

where V_{DR} is the gate driver on-state output voltage.

The gate current (I_{Gvoff}) during the turn-off voltage transition period (t_{VR}) can be estimated as

$$I_{Gvoff} = \frac{V_{PL}}{R_{Goff}} \tag{7.5b}$$

The gate currents I_{Gvon} and I_{Gvoff} can be equated to I_G in Eq. (7.4) to determine the voltage transition power loss.

7.2.1.2 Gate Charge (Q_{GS2}): The Current Transition Period

The charge that determines the current transition time is Q_{GS2}. It can be used to calculate the switching losses during this period. For turn-on, the period begins after the gate voltage reaches the threshold voltage and current begins to flow. It is complete when the drain-to-source voltage begins to transition. For turn-off the sequence occurs in reverse. The larger the current swing, the longer it will take for the transition, and power losses will increase. Figure 7.5 shows a graph of the gate voltage as a function of gate charge for various current levels. The higher the drain current, the longer the period lasts.

The relationship between the gate voltage and drain current is highly non-linear and therefore requires a graphical technique to estimate Q_{GS2}.

The typical datasheet provides information for only one switching condition, but it can be used to determine the Q_{GS2} needed for the loss calculations. The plateau voltage in Figure 7.5 needs to be noted, as well as Q_{GS} for the same conditions. In this example, I_{DS} is 29 A and the plateau voltage is about 1.9 V. The threshold voltage for these devices is about 0.93 V, giving an estimated value of Q_{GS1} of 1.19 nC, and Q_{GS2} of about 1.3 nC at this drain current. In Figure 7.6

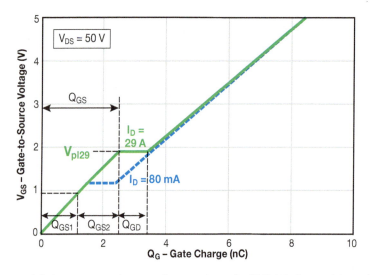

Figure 7.5 Impact of drain current on the gate plateau voltage for EPC2619. *Source:* Adapted from Efficient Power Conversion Corporation [4].

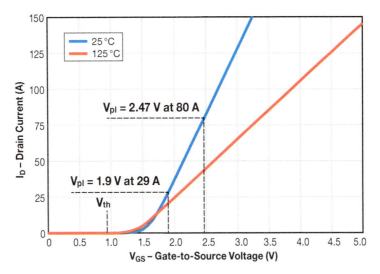

Figure 7.6 Transfer characteristic for EPC2619. *Source:* Adapted from Efficient Power Conversion Corporation [4].

we have highlighted this same operating condition. If we do this same calculation at 80 A, the plateau voltage changes to 2.47 V, and Q_{GS1} is unchanged, but Q_{GS2} increases to 2.2 nC.

In general, Q_{GS1} can be calculated using Eq. (7.6).

$$Q_{GS1} = \left(\frac{Q_{GS}}{V_{pl}}\right) \cdot V_{th} \tag{7.6}$$

Since V_{DS} is constant until $v_{GS} = V_{pl}$, the input capacitance is approximately linear and the ratio of Q_{GS} and V_{pl} is virtually constant. This fact allows the computation of gate charges for other operating values of V_{DS} and I_D.

The $Q_{GS(op)}$, which is the Q_{GS} at the operating value of I_{DS}, can also be calculated for the operating conditions by reading off the plateau voltage $V_{pl(op)}$ from the transfer characteristic graph, as shown in Figure 7.6, and using Eq. (7.7):

$$Q_{GS(op)} = \left(\frac{Q_{GS}}{V_{pl}}\right) \cdot V_{pl(op)} \tag{7.7}$$

with $Q_{GS(op)}$ and Q_{GS1} determined, Q_{GS2} can be calculated using Eq. (7.8).

$$Q_{GS2} = Q_{GS(op)} - Q_{GS1} \tag{7.8}$$

Substituting Eq. (7.2) into Eq. (7.1) for the current transition only, the losses can be approximated using Eq. (7.9):

$$\begin{aligned} P_{ct} &\cong \frac{V_{Bus} \cdot I_{DS}}{2} \cdot t_{Cx} \cdot f_{sw} \\ &= \frac{V_{Bus} \cdot I_{DS}}{2} \cdot \frac{Q_{GS2}}{I_G} \cdot f_{sw} \end{aligned} \tag{7.9}$$

The gate current again appears in the power loss equation. Thus, the current into the gate affects the transition time, and increasing the gate current will reduce this time.

The gate current during the current turn-on and turn-off transition periods can be estimated by using the average of v_{pl} and v_{th}:

$$I_{Gcon} = \frac{V_{DR} - \left(\dfrac{V_{pl} + V_{th}}{2}\right)}{R_{Gon}} \tag{7.10a}$$

$$I_{Gcoff} = \frac{\left(\dfrac{V_{pl} + V_{th}}{2}\right)}{R_{Goff}} \tag{7.10b}$$

The gate currents I_{Gcon} and I_{Gcoff} can be equated to I_G in Eq. (7.9) to determine the current transition power loss.

Using Eqs. (7.4), (7.5a), (7.9), and (7.10a), the total device turn-on power loss (P_{on}) then can be determined using Eq. (7.11):

$$P_{on} = P_{Vt} + P_{Ct}$$

$$= \frac{V_{Bus} \cdot I_{DS} \cdot f_{sw} \cdot R_{Gon}}{2} \cdot \left[\frac{Q_{GD}}{V_{DR} - V_{pl}} + \frac{Q_{GS2}}{V_{DR} - \left(\dfrac{V_{pl} + V_{th}}{2}\right)}\right] \tag{7.11}$$

Similarly using Eqs. (7.4), (7.5b), (7.9), and (7.10b), the total turn-off power loss (P_{off}) can be determined using Eq. (7.12):

$$P_{off} = P_{Vt} + P_{Ct}$$

$$= \frac{V_{Bus} \cdot I_{DS} \cdot f_{sw} \cdot R_{Goff}}{2} \cdot \left[\frac{Q_{GD}}{V_{pl}} + \frac{Q_{GS2}}{\left(\dfrac{V_{pl} + V_{th}}{2}\right)}\right] \tag{7.12}$$

The total transition switching losses can now be summarized as the sum of P_{on} and P_{off}:

$$P_{sw} = P_{on} + P_{off} \tag{7.13}$$

7.2.2 Output Capacitance Losses

All transistors have a highly nonlinear voltage-dependent capacitance that appears across the drain-to-source terminals, as shown in Figure 7.4. This is termed output capacitance (C_{OSS}). The power loss due to the charging and discharging of this output capacitance is termed the output capacitance loss (P_{OSS}).

All switching converters will have some form of switch and rectifier, so this discussion will explain the output capacitance-related losses based on a half-bridge topology. Operating conditions for other converter topologies can then be inferred from similar operating conditions.

Power loss due to the output capacitance can be calculated for the specific working voltage using Eq. (7.14):

$$P_{OSS} = f_{sw} \cdot E_{OSS\_total} \tag{7.14}$$

where E_{OSS_total} is the total dissipated energy related to output capacitance losses for one switching cycle.

The topology and operating conditions will determine whether P_{OSS} losses are present. As an example, self-commutation transitions have zero P_{OSS} losses, but only if the load current is sufficient to completely charge C_{OSS} to V_{BUS} or discharge to zero, and if the time to complete the self-commutation transition is much longer than the current transition time of the device itself.

Since capacitance can store energy, the stored energy can be either dissipated or recycled to the supply or output, requiring an assessment of how output capacitance losses are being calculated by looking into the relationship between the output energy (E_{OSS}) and output charge (Q_{OSS}) under various conditions and at the operating bus voltage. Figure 7.7 illustrates a basic half-bridge topology showing the output capacitances that will be used to explain the relationship between output charge and energy to correctly determine the output capacitance-related losses when operating as a buck converter. Figure 7.7a shows the state of the half-bridge capacitance prior to the switching transition with the exchange and dissipation during the transitions, and Figure 7.7b shows the state of the capacitors after the transition.

To begin, some simplifying assumptions need to be made which are:

- Q2 is naturally commutated.
- The output capacitances are lossless (C_{OSS_Q1} and C_{OSS_Q2}).
- The bus voltage (V_{BUS}) source is zero impedance.
- Turn-off and turn-on are referenced to the control switch Q1.
- Only hard-switching transitions are considered.

Many published analyses make assumptions that are only valid for linear capacitances, but C_{OSS} is highly nonlinear. Hence, the roles of energy and charge must be accounted for individually. For this discussion, the output capacitance charge (Q_{OSS}) and energy (E_{OSS}) will need to be determined using Eqs. (7.15) and (7.16), respectively:

$$Q_{OSS} = \int_0^{V_{BUS}} C_{OSS}(v_{DS}) \cdot dv_{DS} = C_{OSS,tr} \cdot V_{BUS} \tag{7.15}$$

$$E_{OSS} = \int_0^{V_{BUS}} v_{DS} \cdot C_{OSS}(v_{DS}) \cdot dv_{DS} = \frac{1}{2} \cdot C_{OSS,er} \cdot V_{BUS}^2 \tag{7.16}$$

and the incremental output capacitance $C_{OSS}(v_{DS})$ function can be captured from the graph typically provided in the datasheet for the device being analyzed. In the case of asymmetrical configurations, the $C_{OSS}(v_{DS})$ must be captured from the respective data sheets for each

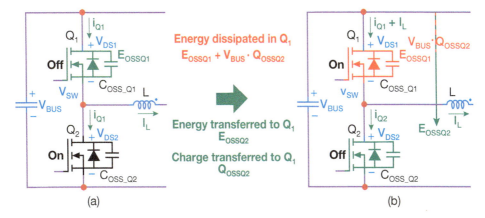

Figure 7.7 Half-bridge topology with output capacitance shown, with the state of the capacitors prior to the switching transition (a) and on completion of the switching event (b).

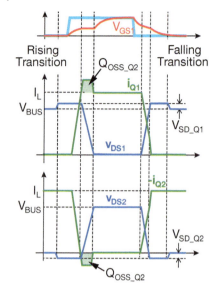

Figure 7.8 Idealized buck converter switching waveforms including effects of the synchronous rectifier (Q2) C_{OSS_Q2}.

transistor, Q1 and Q2. Figure 7.4 gives an example capacitance characteristic from the datasheet of the EPC2619, an enhancement-mode GaN transistor rated at 100 V.

The output capacitance charge (Q_{OSS}) can also be derived from the time-related equivalent capacitance ($C_{OSS,tr}$) at the specific bus voltage (V_{BUS}). Similarly, the output capacitance energy (E_{OSS}) can be derived from the energy-related equivalent capacitance ($C_{OSS,er}$) at the specific bus voltage (V_{BUS}). These parameters are typically specified in the device datasheet, but only at one bus voltage value. If a different bus voltage is used, they cannot be linearly scaled to higher or lower voltages due to the nonlinearity of C_{OSS} as a function of drain-to-source voltage. In these instances, the integral forms of Eqs. (7.15) and (7.16) must be used. In some device datasheets, Q_{OSS} and E_{OSS} characteristics are provided, and can directly be used without the need for further integration.

The conservation of energy and charge can be applied to determine the effect of C_{OSS} on switching loss.

The simpler case of the buck converter switch-node falling transition will be considered first. The inductor current I_L charges C_{OSS_Q1} and discharges C_{OSS_Q2} during this period as shown in Figure 7.8. This process is lossless, hence requiring no further analysis.

$$E_{OSS\_Q1\_charge} = E_{OSS\_Q2\_discharge} = E_{OSS\_fall} = 0 \qquad (7.17)$$

Next, the more complex case of the buck converter switch-node rising transition will be considered where both C_{OSS_Q1} and C_{OSS_Q2} play a role. As the switch-node voltage (v_{SW}) transitions from near zero to V_{BUS}, the energy E_{OSS_Q1} stored in C_{OSS_Q1} is dissipated, but simultaneously C_{OSS_Q2} must be charged. Since I_L is positive, this requires that $i_{Q1} > I_L$ because C_{OSS_Q2} is being charged, as can be seen in Figure 7.8. The current required to charge C_{OSS_Q2} is thus supplied by the bus and results in additional power dissipation in Q1. The energy from the bus can then calculated using Eq. (7.18) as follows:

$$E_{BUS\_rise} = V_{BUS} \cdot Q_{OSS\_Q2} \qquad (7.18)$$

Not all of $E_{BUS,rise}$ will be dissipated since charging C_{OSS_Q2} stores energy as E_{OSS_Q2} and does not contribute to losses. It must therefore be subtracted from the energy that flows from the bus. Q1 however is turning on and as such the stored energy E_{OSS_Q1} will be dissipated, so it must be included in the losses. Putting this altogether, the total energy dissipated due to output capacitance during turn-on and turn-off can then be compiled into Eq. (7.19) as

$$E_{OSS\_total} = E_{BUS\_rise} - E_{OSS\_Q2} + E_{OSS\_Q1} + E_{OSS\_fall} \qquad (7.19)$$

which can be simplified as

$$E_{OSS\_total} = V_{BUS} \cdot Q_{OSS\_Q2} - E_{OSS\_Q2} + E_{OSS\_Q1} \qquad (7.20)$$

when Q1 and Q2 are identical ($E_{OSS\_Q1} = E_{OSS\_Q2}$), Eq. (7.20) reduces to:

$$E_{OSS\_total} = V_{BUS} \cdot Q_{OSS\_Q2} \qquad (7.21)$$

From Eq. (7.21), we see that only using E_{OSS} will underestimate the losses due to output capacitance in a half-bridge configuration. Additional information on output capacitance losses is available in [2].

7.2.3 Gate-Charge (Q_G) Losses

The power loss associated with the gate charge is calculated as follows:

$$P_G = Q_G \cdot V_{GSon} \cdot f_{sw} \tag{7.22}$$

where V_{GSon} is the voltage across the gate when fully turned on by the gate driver.

The gate power losses become an important consideration at higher frequencies and at lower output power levels. It is important to note that all the gate energy is supplied during the charging phase, half of which is dissipated resistively at that time, and the remaining half of the energy is dissipated during the discharge phase.

It should also be noted that the gate charge Q_G must be calculated separately for the control switch, Q1, and the synchronous rectifier, Q2, because of the differences that can be present due to drain–source voltage exposure and differences in devices, such as in asymmetrical half-bridge configurations, that lead to differences in Q_{GD}.

7.2.4 Reverse Conduction Losses (P_{SD})

Reverse or diode conduction occurs in a transistor when its power terminals, such as drain and source, are subject to a negative bias induced by a reverse current and the device control terminal, such as the gate, is held off. The reverse characteristics for GaN FETs were given in Section 2.7 and the discussion in this section will be limited to the scenario of zero volts across the gate. The voltage drop, V_{SD}, can be extracted from the device datasheet I_{SD}–V_{SD} characteristic, at the operating current and temperature with example given in Figure 7.9.

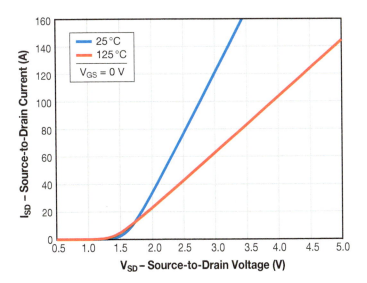

Figure 7.9 Reverse conduction characteristic for the EPC2619. *Source:* Adapted from Efficient Power Conversion Corporation [4].

In a half-bridge configuration reverse conduction typically occurs in the device acting as a rectifier. There are two conditions that can result in reverse conduction: (i) reverse conduction prior to the control device turn-on, defined as synchronous rectifier turn-off, and (ii) reverse conduction following a self-commutated switch-node transition, defined as synchronous rectifier turn-on.

It is possible to have two synchronous rectifier turn-on occurrences in a switching cycle, which can occur in a zero-voltage switched (ZVS) buck converter with long dead times, or one synchronous rectifier turn-on and one synchronous rectifier turn-off, which occurs in a hard-switching buck or boost converter. However, it is not possible for two synchronous rectifier turn-off events to occur in a single switching period. because the synchronous rectifier does not actively change the state of the switch-node.

The power loss due to reverse conduction can be calculated using the following basic equation:

$$P_{SD} = V_{SD} \cdot I_{DS} \cdot (t_{rc\_off} + t_{rc\_on}) \cdot f_{sw} \qquad (7.23)$$

The total reverse conduction time $t_{rc} = t_{rc\_off} + t_{rc\_on}$ needs to be determined from the operating conditions as it is dependent on load current, supply voltage, temperature, and device parameters. The determination of t_{rc} is different for rectifier turn-on and turn-off, but both rely on dead time in some form. Voltage-commutated converters require dead time between turn-off of one device and turn-on of the complementary device. The time between when the rectifier turns off and the active device turns on is defined as charging dead time (t_{dt_chrg}), and the time between when the active device turns off and the rectifier turns on is defined as the discharging dead time (t_{dt_dis}) and corresponds to how energy is stored in the inductor element of the circuit.

The discussion in this section for a buck converter is applicable to boost mode operation where the upper device becomes the rectifier device.

7.2.4.1 Reverse Conduction Time for Synchronous Rectifier Turn-Off (t_{rc_off}).

Figure 7.10 shows both device gates voltage and switch-node voltage waveforms for a synchronous rectifier turn-off configured for a buck converter in a half-bridge topology as shown in Figure 7.7 and where Q2 is the rectifier device. Figure 7.10 shows several timing events related to the synchronous rectifier turn-off:

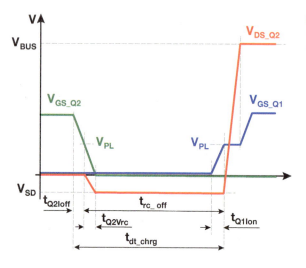

Figure 7.10 Gate and switch-node voltage transitions of a half-bridge buck converter operation for synchronous rectifier turn-off.

a) Current turn-off time for the synchronous rectifier (t_{Q2Ioff}).
b) Voltage transition time for the synchronous rectifier (t_{Q2Vrc}).
c) Current turn-on time for the control device (t_{Q1Ion}).
d) Turn-off reverse conduction time (t_{rc_off}).
e) Dead time to transition into inductor current charging (t_{dt_chrg}). This time is set by the controller.

Synchronous rectifier turn-off assumes that the synchronous rectifier is on and conducting a negative current prior to changing state and occurs within the dead time between the gating signals of the control and rectifier devices. The first change at the beginning of the dead-time event occurs with turning the synchronous rectifier Q2 off, triggering the current turn-off followed by the voltage transition into reverse conduction state. During the voltage transition phase, the voltage magnitude increases (becomes negative) from the fully on state level of R_{DSon} to the reverse voltage (v_{SD}), and can be linearly approximated with little error. The reverse conduction state ends when the dead-time period ends plus the current turn-on time of the control device.

The reverse voltage (v_{SD}) can be determined by using the value of the inductor current at the instant of turn-off of the synchronous rectifier and the reverse conduction characteristic, similar to that shown in Figure 7.9. The reverse voltage also corresponds to the plateau voltage (V_{PL}).

To determine the reverse conduction loss, the reverse conduction time (t_{rc_off}) needs to be found:

$$t_{rc\_off} = \left(t_{dt\_chrg} + t_{Q1Ion}\right) - \left(t_{Q2Ioff} + t_{Q2Vrc}\right) \tag{7.24}$$

where

$$t_{Q2Ioff} = R_{Goff} \cdot \frac{Q_{GS2\_Q2}}{\dfrac{V_{pl} + V_{th}}{2}} \tag{7.25}$$

and

$$t_{Q2\,Vrc} = \frac{Q_{OSSQ1\,rc} + Q_{OSSQ2\,rc}}{2 \cdot I_{turn\_on}} \tag{7.26}$$

$$Q_{OSSQ1rc} = \int_{V_{BUS}}^{V_{BUS} + V_{PL}} C_{OSSQ1}(v_{DS}) \cdot dv_{DS} \tag{7.27a}$$

$$Q_{OSSQ2rc} = \int_{0}^{V_{PL}} C_{OSSQ2}(v_{DS}) \cdot dv_{DS} \tag{7.27b}$$

which assumes that the device output capacitance is fully symmetrical, and the time period is divided in half to account for the linear transition of the voltage. Finally:

$$t_{Q1on} = R_{Gon} \cdot \frac{Q_{GS2\_Q1}}{V_{DR} - \left(\dfrac{V_{pl} + V_{th}}{2}\right)} \tag{7.28}$$

At low switching frequency operation, one can ignore t_{Q2off}, t_{Q2Vrc}, and t_{Q1on} calculations as those times are very short relative to the switching period and overall errors in loss calculations become negligible. The time calculation simplifies to just dead time (t_{dt_chrg}).

7.2.4.2 Reverse Conduction Time for Synchronous Rectifier Turn-On (t_{rc_on})

Figure 7.11 shows both device gates voltage and switch-node voltage waveforms for a synchronous rectifier turn-on configured for a buck converter in a half-bridge topology as shown in Figure 7.7, and where Q2 is the rectifier device. Figure 7.11 shows several timing events related to the synchronous rectifier turn-off:

a) Current turn-off time for the control device (t_{Q1Ioff}).
b) Self-commutation zero-voltage transition time (t_{ZVS}).
c) Voltage transition time for the synchronous rectifier (t_{Q2Vrcz}).
d) Current turn-on time for the synchronous rectifier (t_{Q2Ion}).
e) Turn-on reverse conduction time (t_{rc_on}).
f) Dead time to transition into inductor current discharging (t_{dt_dis}). This time is set by the controller.

The synchronous rectifier turn-on is the more complex case as it must meet specific requirements to occur and is driven by the self-commutation current of the inductor. Figure 7.12 shows three possible outcomes resulting from a self-commutating event which are: (i) Partial self-commutating voltage switching (PVS), (ii) zero-voltage switching (ZVS) and, (iii) reverse conduction following a ZVS transition.

Turn-on reverse conduction can only occur following a self-commutation transition if the timing between the transitions exceeds that needed to establish zero-voltage switching (ZVS), as shown in Figure 7.12, and satisfying ($t_{dt\_dis} + t_{Q2Ion}$) > t_{ZVS} defined in Figure 7.11. Should the requirements not be met, then no reverse conduction will occur and hard-switching or partial zero-voltage switching techniques will need to be used to determine losses for this transition event. Hard-switching techniques have already been presented and partial zero-voltage switching will be covered at the end of this section.

Synchronous rectifier turn-on assumes that the control device is on and conducting a positive current prior to changing state and occurs within the dead time between the gating signals of the control and rectifier devices.

The first change, at the beginning of the dead-time event, occurs with turning the control device Q1 off, triggering the current turn-off followed by the self-commutating switch-node voltage transition period and after the switch-node crosses zero volts. The voltage then transitions into reverse conduction state. During the voltage transition phase the voltage increases

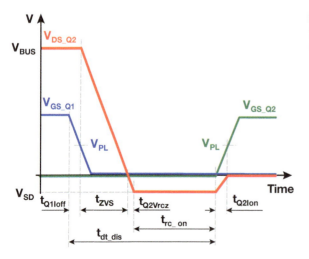

Figure 7.11 Gate and switch-node voltage transitions of a half-bridge buck converter operation for synchronous rectifier turn-on.

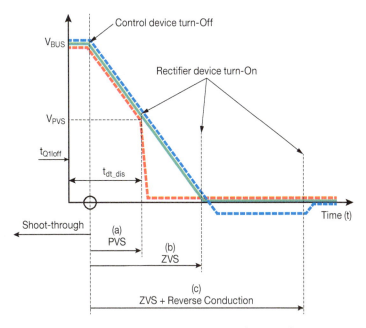

Figure 7.12 Switch-node voltage commutation with the same load current for various dead times; (a) PVS, (b) ZVS, and (c) ZVS plus diode conduction.

from the fully on state level of R_{DSon} times current to the reverse voltage (v_{SD}) and can be linearly approximated with little error. Similarly for the current where the inductor current transitions from the output capacitances of the devices to the channel of the synchronous rectifier. The reverse conduction state ends when the dead-time period ends plus the current turn-on time of the synchronous rectifier.

The reverse voltage (v_{sd}) can be determined by using the value of the inductor current at the instant of turn-off of the control device and the reverse conduction characteristic, like that shown in Figure 7.9. The reverse voltage also corresponds to the plateau voltage (V_{PL}).

To determine the reverse conduction loss immediately prior to the control FET turning on, the reverse conduction time (t_{rc_on}) needs to be found:

$$t_{rc\_on} = (t_{dt_{dis}} + t_{Q2Ion}) - (t_{ZVS} + t_{Q1loff} + t_{Q2Vrcz}) \tag{7.29}$$

where

$$t_{Q1loff} = R_{Goff} \cdot \frac{Q_{GS\,2\_Q1}}{\frac{V_{pl} + V_{th}}{2}} \tag{7.30}$$

with ZVS transition

$$t_{ZVS} = \frac{Q_{OSSQ1} + Q_{OSSQ2}}{I_{turn\_off}} \tag{7.31}$$

$$Q_{OSSQ1} = \int_0^{V_{BUS}} C_{OSSQ1}(v_{DS}) \cdot dv_{DS} \tag{7.32a}$$

$$Q_{OSSQ2} = \int_{V_{Bus}}^0 C_{OSSQ2}(v_{DS}) \cdot dv_{DS} \tag{7.32b}$$

and

$$t_{Q2Vrcz} = \frac{Q_{OSSQ1\,rc} + Q_{OSSQ2\,rc}}{4 \cdot I_{turn\_off}} \tag{7.33}$$

$$Q_{OSSQ1rc} = \int_{V_{BUS}}^{V_{BUS} + V_{PL}} C_{OSSQ1}(v_{DS}) \cdot dv_{DS} \tag{7.34a}$$

$$Q_{OSSQ2rc} = V_{PL} \cdot C_{OSSQ2}(0\,V) \tag{7.34b}$$

Equation (7.34b) assumes that the device output capacitance between 0 V and $-V_{PL}$ remains constant. This assumption is reasonable because the datasheets do not provide the output capacitance in the negative bias region, the change in voltage is small, and hence the error will be small. Finally:

$$t_{Q2\,Ion} = R_{Gon} \cdot \frac{Q_{GS\,2\_Q2}}{V_{DR} - \left(\dfrac{V_{pl} + V_{th}}{2}\right)} \tag{7.35}$$

At low switching frequency operation, one can ignore t_{Q1off}, t_{Q2on}, and t_{Q2off} calculations as those times are very short relative to the switching period, and overall errors in loss calculations become negligible. The time calculation simplifies to just dead time (t_{dt_dis}) and ZVS transition time (t_{ZVS}).

7.2.4.3 Partial Self-Commutation Voltage Switching (PVS)

With a self-commutating switching event, and if the requirement of $(t_{dt\_dis} + t_{Q2Ion}) > t_{ZVS}$ defined in Figure 7.11 cannot be met, then the switching transition becomes a partial self-commutating voltage switching (PVS) transition, an incomplete ZVS transition. In this case, a hard-switching event is established with associated losses but occurs at lower voltage than the bus voltage. There is no reverse recovery loss, regardless of whether the device is a MOSFET or a GaN transistor, as no diode is conducting.

The basics of PVS are shown in Figure 7.12a with the dashed red trace. Before calculating any PVS losses, one must first determine the incomplete self-commutating time. Figure 7.13 shows the drain–source and gate–source voltages for both Q1 and Q2 for three PVS scenarios. Focusing on any one scenario, the PVS time is dependent on three factors: (i) the time for Q1 to turn-off and stop conducting current (t_{Q1Ioff}), (ii) the inductor current discharge dead time (t_{dt_dis}) defined earlier, and (iii) the time for Q2 to turn-on and conduct the inductor current (t_{Q2Ion}). Only the dead time will impact PVS time as the current transition times are inductor current dependent.

Key differences in calculating PVS losses compared to standard hard-switching loss calculations are:

a) The voltage across the drain–source of Q1 transitions from 0 V (on-state) to $V_{BUS} - V_{PVS}$.
b) The voltage across the drain–source of Q2 transitions from V_{BUS} to V_{PVS}.

These differences result in different voltages across the devices at the PVS instant requiring unique calculations for each device for both Q_{OSS} loss and hard-switching loss.

The first case in Figure 7.13 shows a high PVS voltage (V_{PVS1}), the second shows an equal case with the PVS voltage at half the bus voltage (V_{PVS2}), and the third with low PVS voltage (V_{PVS3}). Only the second case results in a symmetrical switching event. The other two cases both result in asymmetrical voltages across the devices and require the power loss calculations be adjusted accordingly. Both hard-switching (P_{HS}) and output capacitance loss (P_{OSS}) will need to be adjusted.

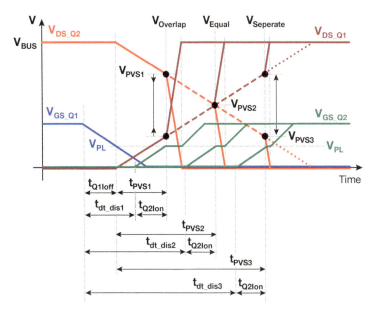

Figure 7.13 Zoomed-in drain–source and gate–source voltages for various partial self-commutation switching voltage transition scenarios in a half-bridge buck converter configuration.

Hard-switching loss calculations are straightforward in that in Eqs. (7.11) and (7.12) are used, and the bus voltage (V_{BUS}) are replaced with V_{PZVS} and $V_{BUS} - V_{PZVS}$ for Q2 and Q1, respectively.

The situation for output capacitance loss is more complicated and requires separate calculations for each device.

The PVS switch-node voltage (V_{PVS}) when Q2 turns on can be determined by calculating the amount of charge transferred during the self-commutating time relative to the total charge in the circuit using Eq. (7.36):

$$V_{PVS} = \frac{I_{turn\_off} \cdot t_{PVS} \cdot V_{Bus}}{Q_{OSSQ1} + Q_{OSSQ2}} \qquad (7.36)$$

Equation (7.36) is based on Eq. (7.15), and is the ratio in time of the total charge for full ZVS and the PVS time (t_{PVS}) which can be determined as follows:

$$t_{PVS} = t_{dt\_dis} + t_{Q2Ion} - t_{Q1off} \qquad (7.37)$$

Figure 7.14 shows the output charge for the EPC2619 [4] device used in both the Q1 and Q2 positions and assumes a 60 V bus voltage for case scenarios shown in Figure 7.13. For case one with a high PVS voltage, Q1 only discharges a small amount while Q2 has a high voltage at the instant of switching. Equations (7.38a) and (7.38b) are then used to calculate the respective charge for each device:

$$Q_{OSSPVSQ1} = \int_{0}^{V_{Bus} - V_{PVS}} C_{OSSQ1}(v_{DS}) \cdot dv_{DS} \qquad (7.38a)$$

$$Q_{OSSPVSQ2} = \int_{V_{Bus}}^{V_{PVS}} C_{OSSQ2}(v_{DS}) \cdot dv_{DS} \qquad (7.38b)$$

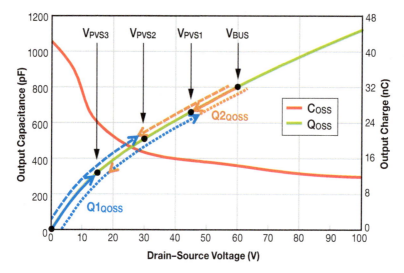

Figure 7.14 Example of EPC2619 [4] symmetrical FET configuration buck converter change in output charge under three PVS switching scenarios with 60 V bus. *Source:* Adapted from Efficient Power Conversion Corporation [4].

Similarly, for the equal case where the integrals use the same voltage for V_{PVS} and $V_{BUS} - V_{PVS}$. However, the charge integrals do not necessarily result in the same values and depends on the non-linearity of the output charge in the operating region. The same process is used for the low PVS case.

Once the PVS point has been determined it can be used to determine the output capacitance losses for the PVS event. Figure 7.15 shows a revision of Figure 7.7 to include the changes for the PVS transition. Notably different for this event and for a buck converter is that both Q1 and Q2 already have charge stored prior to the switching transition and affects the calculations of energy transfer and dissipation. The second difference is the active switch turning on now is Q2.

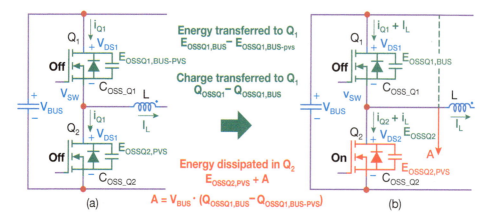

Figure 7.15 Half-bridge topology with output capacitance shown, the partial voltage switching state of the capacitors prior to the switching transition (a) and on completion of the switching event (b).

The PVS voltage and charge differences nearly always result in different conditions for each of the devices, so the asymmetrical E_{OSS} loss calculation methodology given in Section 7.2.2 and based on Eq. (7.20) must be modified for the partial voltage switching event. Figure 7.16 shows the ideal waveforms for the partial switching voltage event of a buck converter falling switch-node transition. As discussed in Section 7.2.2, there are three changes that occur: (i) the stored energy is dissipated in the device turning on, in this case Q2, (ii) energy is transferred to the device turning off, in this case Q1, and (iii) charge is transferred to the device turning off, in this case Q1.

Equation (7.14) is still valid for P_{OSS} loss but the details of E_{OSS} need to be revised. Based on the three changes that occur due to output capacitance during a partial switching event, the total $E_{OSS\_total\_PVS}$ loss can be expressed as follows:

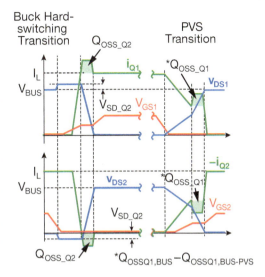

Figure 7.16 Idealized buck converter switching waveforms showing the transitions impacted by C_{OSS} for both the buck converter switch-node rising transition and for the specific case of partial voltage switching for the falling edge of the switch-node.

$$E_{OSS\_total\_PVS} = E_{OSSchrgXfrQ1} - E_{OSSstoredSQ1} + E_{OSSQ2,PVS} \quad (7.39a)$$

where

$$E_{OSSchrgXfrQ1} = V_{BUS} \cdot \left(Q_{OSSQ1,BUS} - Q_{OSSQ1,BUS-PVS}\right) \quad (7.39b)$$

and

$$E_{OSSstoredSQ1} = E_{OSSQ1,BUS} - E_{OSSQ1,BUS-PVS} \quad (7.39c)$$

With $Q_{OSSQ1,BUS}$ can be computed using Eq. (7.15) for C_{OSS} for Q1 and changing the integral parameters of Eq. (7.15) for $Q_{OSSQ1,BUS-PVS}$ as follows:

$$Q_{OSSQ1,BUS-PVS} = \int_{V_{Bus}-V_{PZVS}}^{V_{Bus}} v_{DS} \cdot C_{OSS}(v_{DS}) \cdot dv_{DS} \quad (7.40a)$$

Similarly for the stored energy. The final energy stored in Q1, $E_{OSSQ1,BUS}$, will be at the BUS voltage, so Eq. (7.16) can be used for that calculation using C_{OSS} for Q1, and changing the integral parameters of Eq. (7.16) for $E_{OSSQ1,BUS-PVS}$ as follows:

$$E_{OSSSQ1,BUS-PVS} = \int_{V_{Bus}-V_{PZVS}}^{V_{BUS}} v_{DS} \cdot C_{OSS}(v_{DS}) \cdot dv_{DS} \quad (7.40b)$$

And finally, the $E_{OSSQ2,PVS}$ can be determined as follows:

$$E_{OSSQ2,PVS} = \int_{0}^{V_{PZVS}} v_{DS} \cdot C_{OSS}(v_{DS}) \cdot dv_{DS} \quad (7.40c)$$

7.2.4.4 Adding an Anti-Parallel Schottky Diode

Figure 7.17 shows the forward voltage drop for both a MOSFET and a typical enhancement-mode GaN transistor when operating in reverse conduction. There is about 1.5 V difference between the two devices and, as the temperature increases, it can increase to as high as 2 V. This graph, however, does not account for dynamic behavior, where in the case of the enhancement-mode GaN transistor, there is no reverse recovery (Q_{RR} = 0), in addition to lower output capacitance.

Sections 7.2.4.1 and 7.2.4.2 showed that optimal timing between the transistors can yield very low losses under specific conditions. Those conditions are dynamic and depend on operating conditions such as load current and bus voltage. For most circuits, it is challenging to actively control the dead time to the precision needed to minimize losses. However, a simple anti-parallel Schottky diode can be connected across the GaN transistor to reduce reverse conduction loss P_{SD} allowing reduced reliance on precise dead-time control management.

One of the most critical requirements for the addition of the anti-parallel Schottky diode is the minimization of the connection inductance between the two devices. This comes down to three factors: (i) the parasitic inductance between the drain and source of the GaN transistor, (ii) the parasitic inductance of the Schottky diode, and (iii) the layout inductance connecting the GaN transistor to the Schottky diode. The low parasitic inductance of the GaN transistors with LGA or BGA packages makes the addition of an external Schottky diode simple and efficient.

Using the definition of reverse conduction time (t_{rc}) previously discussed, the power losses associated with the diode conduction as a function of reverse conduction time are shown in Figure 7.18 for a typical GaN transistor and equivalent MOSFET. Due to the lower forward voltage drop of the MOSFET body diode, the losses as a function of the reverse conduction time increase at a lower rate than those for a GaN transistor. It is worth noting that this comparison does not include the reverse recovery loss of the Si MOSFET, which varies significantly with dead time as well [2].

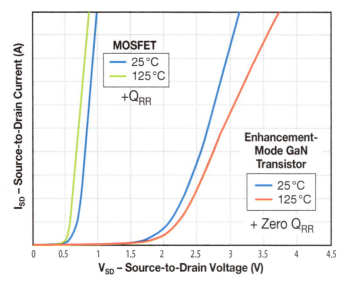

Figure 7.17 Reverse transfer characteristics of 100 V MOSFET (typical) and enhancement-mode GaN transistor (typical).

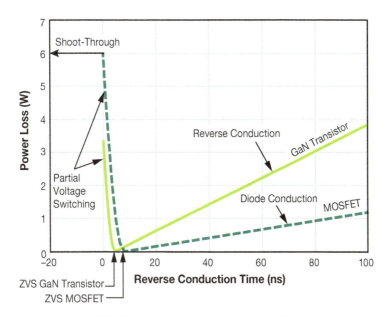

Figure 7.18 GaN transistor and MOSFET comparison of the impact of effective dead time on power loss of the synchronous rectifier device in a buck converter operating with $V_{BUS} = 48$ V, $V_{Out} = 12$ V, $I_{OUT} = 16$ A, $f_{sw} = 1$ MHz. The MOSFET losses due to diode reverse recovery are not included.

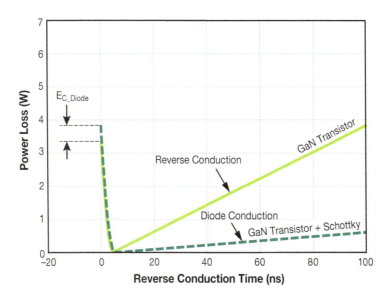

Figure 7.19 Comparison of the impact of effective dead time on the power loss of an enhancement-mode GaN transistor-based buck converter operating with $V_{BUS} = 48$ V, $V_{Out} = 12$ V, $I_{OUT} = 16$ A, $f_{sw} = 1$ MHz, with and without an anti-parallel Schottky diode.

The addition of an anti-parallel Schottky diode across the GaN transistor synchronous rectifier will reduce the conduction voltage during the diode conduction period as shown in Figure 7.19. A buck converter was tested with various reverse conduction times and the effect of adding an anti-parallel Schottky diode was measured. The results shown in Figure 7.20 were

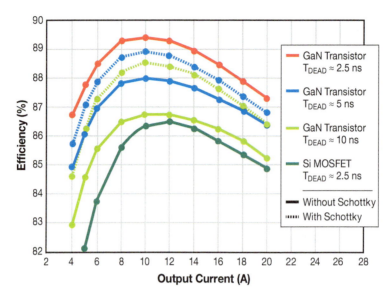

Figure 7.20 Experimental verification of the impact of adding an anti-parallel Schottky diode to an enhancement-mode transistor in a buck converter with V_{BUS} = 12 V, V_{OUT} = 1.2 V, f_{sw} = 1 MHz.

obtained with a buck converter operating with a V_{BUS} of 12 V, output voltage of 1.2 V, and operating at a switching frequency of 1 MHz.

However, the addition of an anti-parallel Schottky diode does add some output capacitance, with an associated increase in output capacitance loss.

This difference is most clearly seen in Figure 7.19 at 0 ns dead time, where the diode adds an energy loss equivalent to its own E_{OSS}. However, the diode also increases the C_{OSS}-related loss in Q_1 given in Eq. (7.20), and the overall increase in switch-node capacitance increases the output capacitance loss during turn-on voltage fall time. Therefore, the net benefit of adding an antiparallel Schottky diode may diminish at higher bus voltages, where the additional capacitive losses of the diode become more dominant.

7.2.5 Reverse Recovery (Q_{RR}) Losses

The body diode reverse recovery losses occur when the body diode transitions from the on-state to the off-state. Enhancement-mode GaN transistors, unlike standard power MOSFETs or cascode GaN devices, have no minority carriers to be stored in a junction, and therefore, have no reverse recovery charge and thus no associated losses.

7.2.6 Total Hard-Switching Losses

The total dynamic power loss is the sum of the individual components:

$$P_{DYN} = (P_{sw} + P_{OSS} + P_G + P_{SD} + P_{RR}) \tag{7.41}$$

Owing to the GaN transistor's lower Q_{SG2} and Q_{GD}, P_{sw} is much lower than a comparable power MOSFET. The output capacitance for all types of GaN transistors is smaller than MOSFETs of comparable $R_{DS(on)}$, making P_{OSS} relatively low. Both gate drive voltages and gate charge are also lower, making P_G lower. Finally, due to the reverse current conduction mechanism, enhancement-mode GaN transistors have a higher V_{SD} when compared with the body diode forward voltage of a MOSFET. This characteristic of an enhancement-mode GaN transistor has the potential to increase the power loss P_{SD} and is influenced by the total reverse conduction time, a condition that can be controlled by the time that the rectifier switch is acting like a diode [5]. Enhancement-mode GaN transistors have zero reverse recovery, making the final term in Eq. (7.41) zero (or small in the case of a cascode device).

Overall, the dynamic power losses of GaN transistors are significantly lower than power MOSFETs and enable power converters using hard-switching topologies to be more efficient and smaller. Now let's look at a simple figure of merit that can be used to estimate circuit performance and compare expected results between technologies and products within the same technology [6–9].

7.2.7 Hard-Switching Figure of Merit

Several figures of merit have been proposed for evaluation of transistor losses. The preceding analysis indicates several factors that should be included in the hard-switching figure of merit. One such FOM_{HS} is given in Eq. (7.42) and discussed in [10]:

$$\text{FOM}_{HS} = (Q_{GD} + Q_{GS2}) \cdot R_{DS(on)} \tag{7.42}$$

For a given on-resistance and application, selecting a technology with a lower value of FOM_{HS} indicates lower power loss proportional to

$$P_{TOTAL} \alpha \sqrt{\text{FOM}_{HS}} \tag{7.43}$$

The FOM_{HS} can be plotted on a graph with $R_{DS(on)}$ on the x-axis and the charge-related terms $(Q_{GS2} + Q_{GD})$ on the y-axis as shown in Figure 7.21.

From Figure 7.21, 200 V GaN transistors have a similar FOM_{HS} to 40 V Si MOSFETs, and 600 V GaN transistors have a similar FOM_{HS} to 100 V MOSFETs.

7.3 External Factors Impacting Hard-Switching Losses

In Section 7.2, a detailed analysis of the derivation of hard-switching losses was presented. In practical applications, this picture is somewhat incomplete, as there are additional factors that can further impact hard-switching losses such as common-source inductance (CSI) denoted by (L_{CS}) and power loop inductance (L_{Loop}). These factors appear in practical circuits with physical limitations brought on by device size, packaging parasitics, and circuit layout parasitics.

7.3.1 Impact of Common-Source Inductance (L_{CS})

The two main inductances, common-source inductance and high-frequency power loop inductance for a half-bridge configuration, are shown in Figure 7.22.

The impact of common-source inductance on gate drive performance has been discussed in Chapters 3 and 4. In this section, its impact will be quantified for the hard-switching current transition.

During a current transition event, the voltage generated across the common-source inductance will oppose the gate drive voltage, thereby reducing the gate current used to charge the gate capacitance. This effectively lengthens the current transition period as shown in Figure 7.23.

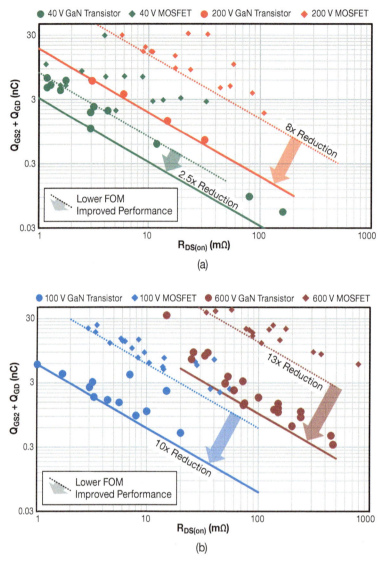

Figure 7.21 Comparison of hard-switching FOM$_{HS}$ between GaN transistors and Silicon MOSFETs at various voltages: (a) 40 V and 200 V and (b) 100 V and 600 V.

An analysis of the gate circuit shown in Figure 7.24 can be used to determine the amount of time the current transition is lengthened due to its common-source inductance. Because a full analysis reveals terms with exponential and sinusoidal components, some simplifying assumptions can be made.

The first simplifying assumption is that the voltage induced across the common-source inductance can be regarded as a voltage source in phase with the gate voltage, and thus will only impact the magnitude of the voltage in the gate circuit. The second assumption ignores the impact of the gate circuit inductance. In Chapter 4 it was shown that this inductance contributes negligibly to circuit switching performance. The third assumption is that the external drain current is constant during the transition.

Hard-Switching Topologies | 219

Figure 7.22 Power loop (L_{LOOP}) and common-source inductance (L_{CS1} and L_{CS2}) in a half-bridge configuration.

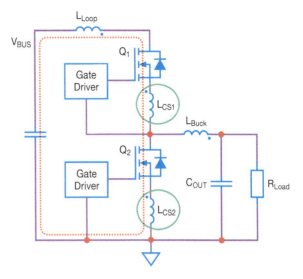

Figure 7.23 Effect of common-source inductance on the gate voltage.

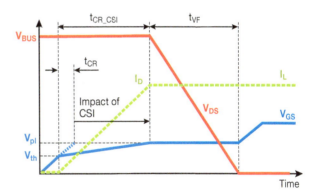

Figure 7.24 Gate circuit loop including common-source inductance.

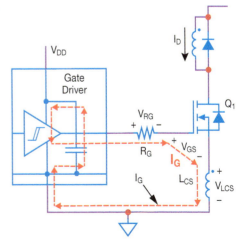

Neglecting the gate driver voltage drop, as it can be included as part of the gate resistance, the Kirchhoff voltage loop in the gate circuit is given by

$$V_{DD} = V_{RG} + V_{GS} + V_{LCS} \qquad (7.44)$$

The gate and drain currents can be added to yield:

$$V_{DD} = I_G \cdot R_{G\_on} + V_{GS} + \frac{L_{CS} \cdot I_L}{t_{CR\_CSI}} \qquad (7.45)$$

Where the gate circuit current is given by

$$I_G = \frac{Q_{GS2}}{t_{CR\_CSI}} = \frac{C_{GS} \cdot I_L}{t_{CR\_CSI} \cdot g_m} \qquad (7.46)$$

where t_{CR_CSI} is the new current rise interval time, accounting for common-source inductance and g_m is the transconductance during this period, which can be estimated as the slope of the transfer function between $V_{GS(th)}$ and V_{PL}.

Combining Eqs. (7.45) and (7.46), t_{CR_CSI} can be determined as follows:

$$t_{CR\_CSI} = \left(\frac{I_L}{V_{DD} - V_{GS}}\right) \cdot \left(\frac{C_{GS}}{g_m}\right) \cdot \left(\frac{L_{CS} \cdot g_m}{C_{GS}} + R_{G\_on}\right) \qquad (7.47)$$

From Eq. (7.47), the equivalent common-source inductance resistance (R_{CSI}) can be extracted as

$$R_{CSI} = \frac{L_{CS} \cdot g_m}{C_{GS}} \qquad (7.48)$$

The impact of common-source inductance on the gate circuit resistance can be significant, given that the device input capacitance is already small. The value of L_{CS}, therefore, will need to become very small to minimize the impact of the common-source inductance. For example, 100 pH of common-source inductance in a MOSFET circuit with C_{GS} of 2900 pF, and $g_m = 60$ Siemens results in 2 Ω equivalent resistance. The same 100 pH common-source inductance in an equivalent GaN transistor circuit with C_{GS} of only 850 pF, and $g_m = 60$ Siemens, results in a 7 Ω equivalent resistance.

Equation (7.48) can be used to estimate the current transition time, including the influence of the CSI. It can simply be added to the R_G term in Eqs. (7.11) and (7.12) for the Q_{GS2} component only.

As an example, a 1 MHz converter was experimentally evaluated with a fixed-loop inductance and the common-source inductance varied, with the results shown in Figure 7.25. It is notable how quickly the losses increase as a function of common-source inductance.

7.3.2 Impact of Power Loop Inductance on Device Losses

Another factor that impacts hard-switching losses at higher frequencies is the power loop inductance that affects the commutation of voltage and current between the switching devices. This is the inductance encompassed by the bus supply as well as the devices connected to this bus, as shown in Figure 7.22, and discussed in Chapters 3 and 4. Component parasitic inductance and physical layout inductance elements all contribute to this total loop inductance.

The power loop inductance has two main negative effects on the switch during turn-off: (i) it slows the transition and (ii) it increases the voltage across the drain and source. During device turn-on, the loop inductance reduces the device drain-to-source voltage, which decreases

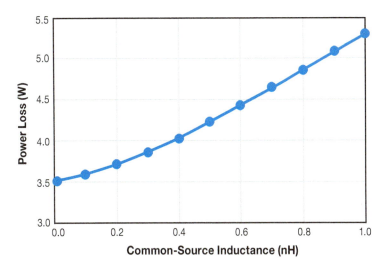

Figure 7.25 Effect of common-source inductance on power loss [11–13] (V_{BUS} = 12 V, V_{OUT} = 1.2 V, I_{OUT} = 20 A, f_{SW} = 1 MHz, control transistor is the EPC2015C [14], synchronous rectifier transistor is the EPC2015C.

losses. However, the sum of the two negative effects and the positive effect has a net negative result, which means that the loop inductance will increase losses in the circuit as can be seen in example shown in Figure 7.26. An experimental example is shown in which the loop inductance was gradually increased in a 12–1.2 V converter to measure its impact on total power loss. Adding 3 nH of loop inductance increases loss by 30% over the ideal case.

With the fast switching speeds of GaN transistors, even small values of high-frequency loop inductance can increase the voltage overshoot. Decreasing this inductance, therefore, results in lower voltage overshoot, increased input voltage capability, and reduced electromagnetic interference (EMI). Figure 7.27 shows the switch-node voltage waveforms for a design with

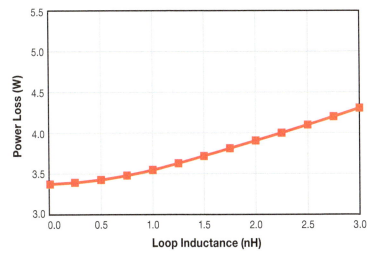

Figure 7.26 Effect of loop Inductance on power loss [11–13] (V_{BUS} = 12 V, V_{OUT} = 1.2 V, I_{OUT} = 20 A, f_{SW} = 1 MHz, control transistor is the EPC2015C [14], synchronous rectifier transistor is the EPC2015C). *Source:* Adapted from Efficient Power Conversion Corporation [14].

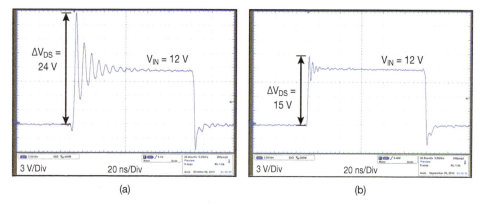

Figure 7.27 Synchronous rectifier switching waveforms of designs with (a) $L_{LOOP} \approx 1.6$ nH and (b) $L_{LOOP} \approx 0.4$ nH ($V_{BUS} = 12$ V, $V_{OUT} = 1.2$ V, $I_{OUT} = 20$ A, $f_{SW} = 1$ MHz, $L = 150$ nH, control transistor is the EPC2015C [14] synchronous rectifier transistor is the EPC2015C). *Source:* Adapted from Efficient Power Conversion Corporation [14].

a high-frequency loop inductance of 1.6 nH compared with 0.4 nH. The voltage overshoot on the synchronous rectifier switch is reduced from 100% of the input voltage to 25%, respectively. This overshoot occurs during the turn-on transient of the control switch.

During the turn-on transient of Q_1, the overshoot voltage on Q_2 was found to primarily be a function of the loop inductance and switching speed (i.e. gate driver strength). However, during the turn-off transient of Q_1, the overshoot voltage on Q_1 was found to depend more heavily on the inductive load current than gate driver strength. In some very high current designs, this turn-off overshoot voltage can exceed the voltage rating of the device, imposing an inherent limit on the current-handling capability of the converter at a given operating voltage [15]. In this way, lower power loop inductance can potentially increase the current-handling capability of the converter, in addition to input voltage capability. In Chapter 5 the potential impact of this overshoot voltage on reliability was discussed.

To calculate the power losses due to the loop inductance, one must determine the magnitude of the inductance and the energy stored in the inductance at the switching event. This energy must be dissipated at completion of the switching transition, which occurs in the form of voltage-overshoot ringing. The challenge is to determine the inductance, and this can be done by determining the ringing frequency (f_{ring}) at the end of the switching transition and using the output capacitance of the FET that is off at the bus voltage ($C_{OSSQ2}(V_{Bus})$), using the following equation:

$$L_{Loop} = \frac{1}{C_{OSSQ2}(V_{Bus}) \cdot \left(2 \cdot \pi \cdot f_{ring}\right)^2} \quad (7.49)$$

The energy stored in the loop inductance at the instant of switching is thus:

$$E_{Loop} = \frac{L_{Loop} \cdot I_{turn\_on}^2}{2} \quad (7.50)$$

The power loss for the loop inductance can then be calculated as follows:

$$P_{Loop} = E_{Loop} \cdot f_{sw} \quad (7.51)$$

7.4 Frequency Impact on Magnetics

Magnetic components, such as transformers and inductors, account for the other large contribution to power loss in switching power converters.

7.4.1 Transformers

For a magnetic core with a specific cross-sectional core area and specific winding window area, the core-area product (cross-sectional area multiplied by specific winding window area) is commonly used to design magnetic structures [16] and directly relates to the volume of the core. A constant core-area product results in similar losses, and consequently, similar converter efficiencies for a given operating frequency.

For a given design, as the switching frequency is increased over a substantial range for a given material, the core losses will decrease with frequency due to the decrease in flux density amplitude faster than they increase due to frequency [17, 18], an effect that can be used to an advantage in a converter using GaN transistors compared with one using Si MOSFETs. Eventually, as frequency increases beyond the design range for the core material, core losses will start to increase again. At this point, it may be beneficial to consider magnetic materials with lower core loss density as the frequency is increased.

As an example, consider what happens when the switching frequency is increased from 300 to 500 kHz. The core cross-sectional area of the 300-kHz design can be decreased to the point where there is the same flux density as the 500 kHz design. This results in a core cross-sectional area that is 60% of the 300 kHz design, as shown in Figure 7.28. Additional effects from this new core design are:

a) Magnetic core volume decreased to approximately 60% of the original value.
b) Core losses per unit volume may have increased, but this depends upon the core material and switching frequency.
c) Winding volume and mean length-per-turn also has been reduced to approximately 85–90% (depending on the length(l)-to-width(w) ratio). This results in a lower DC-winding resistance and resulting copper wire conduction losses.
d) The AC-winding resistance per unit length increased, due to reduced skin depth, depending on the design and conductor thickness. Furthermore, the AC-winding resistance is proportional to the decrease in DC-winding resistance.

Typically, (a) is greater than (b) and (c) is greater than (d) and, therefore, the transformer will yield a higher efficiency at 500 kHz compared with 300 kHz. The extent that frequency can be

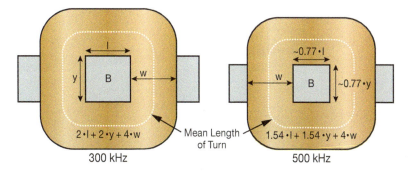

Figure 7.28 Cross-sectional view of two equivalent transformer structures (constant flux density) for different switching frequencies.

increased is material-dependent; as the material is pushed beyond its intended operating frequency range, any real benefit from increasing the frequency will become negated. An alternative core material may result in higher frequency capability, but there are tradeoffs such as reduced permeability and increased cost. In the multi-MHz frequency range, many core materials are operating at their upper limit, and in some cases, air–core approaches may need to be investigated.

7.4.2 Inductors

The higher switching frequency capability of GaN transistors enables much lower filter inductances than comparable Si-based designs. For a given ripple current, doubling the switching frequency allows for half the filter inductance to achieve the same ripple current. This lower inductance means that the filter inductor can be re-optimized for smaller size, lower power loss, or a combination of both. Power loss models for commercially available inductors are typically available from their manufacturers [19, 20].

The impact of the change in magnetic size is like the transformer, but due to a slightly different mechanism. The core material in a transformer will experience a positive and negative magnetic flux swing due to the voltage excitation but, in the case of an inductor the current in the winding has a DC component to it and thus the magnetic flux is either positive or negative only. This means that the flux excitation and associated losses are proportionately lower than those of a transformer for the same frequency. However, the inductor conduction losses will be higher due to the DC component. Using the same analysis as for the transformer, a more efficient inductor will again result at higher frequencies because (a) is greater than (b) and (c) is greater than (d). The same upper frequency operating limits apply to the inductor as for the transformer.

7.5 Buck Converter Example

The analyses in this chapter can now be applied to an actual converter design example. A buck converter was chosen because it provides a simple circuit that includes a hard-switching device and a transistor acting as a synchronous rectifier, as shown in Figure 7.29. Since most of the

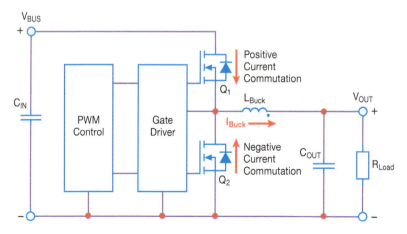

Figure 7.29 Basic buck converter circuit for example.

dynamic losses are related to the input voltage and to the ratio between the input voltage and output voltage, the following basic rule applies: *The higher the input voltage and the greater the ratio between the input and output voltage, the greater the benefits derived from using GaN transistors.*

In the following example a buck converter is operating at 1 MHz and delivering up to 15 A to a 12 V load, with a supply voltage of 48 V. The hard-switching loss analysis is based on the EPC2619 [4] for both the control switch Q_1 and the synchronous rectifier Q_2. The output inductor is 2.2 μH, with a specified DC series resistance of 770 μΩ. Both devices are driven by a 5 V supply, with 1.8 Ω of external gate resistance for turn-on and 470 mΩ for turn-off. The gate driver [21] has pull-up and pull-down resistances of 700 and 400 mΩ, respectively. The dead time was set to 10 ns for each switching edge. As an example, the calculation of overall losses at 15 A will be reviewed here using the equations previously presented. For this initial estimation, the impact of parasitic inductances was neglected; however, the temperature-dependent FET characteristics will be adjusted to 115 °C and the gate drive voltages reduced from 5 to 4.7 V and 4.9 V for the upper and lower devices to match those observed in the experimental data.

The buck converter has two switching events, defined relative to the control switch, as turn-on (where the switch-node voltage will rise to the bus voltage) and turn-off (where the switch-node voltage falls to zero). Since there are two transistors involved in this design, four conditions need to be analyzed: the turn-on and turn-off events for each of the transistors, as shown in Figure 7.30. The turn-off switching transient is defined as self-commutation. The current (I_{Buck}) in the buck inductor (L_{Buck}) at the time Q_1 is turned off will discharge the output capacitance, without Q_2 being turned on, and hence reduces the switch-node voltage on its own. The turn-on switching transient is defined as forced commutation. Regardless of the current in the buck inductor at the time Q_2 is turned off, the switch-node voltage must be forced to the bus voltage when Q_1 turns on.

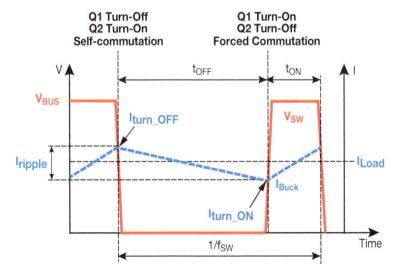

Figure 7.30 Basic buck converter waveforms.

An analysis of the buck converter is required before calculating device losses. The control switch duty cycle (D) is given by [22–24]:

$$D = \frac{V_{\text{OUT}}}{V_{\text{BUS}}}$$

$$= \frac{12\,\text{V}}{48\,\text{V}} \tag{7.52}$$

$$= 0.25 = 25\%$$

Using this duty cycle, the peak-to-peak ripple current in the inductor can be calculated:

$$I_{\text{ripple}} = \frac{(V_{\text{BUS}} - V_{\text{OUT}}) \cdot D}{f_{\text{sw}} \cdot L_{\text{OUT}}}$$

$$= \frac{(48\,\text{V} - 12\,\text{V}) \cdot (0.25)}{(1\,\text{MHz}) \cdot (2.2\,\mu H)} \tag{7.53}$$

$$= 4.09\,\text{A}$$

The inductor currents at the moment of turn-on can be calculated as follows:

$$I_{\text{Turn-ON}} = I_{\text{OUT}} - \frac{I_{\text{ripple}}}{2}$$

$$= 15\,\text{A} - \frac{4.09\,\text{A}}{2} \tag{7.54}$$

$$= 12.95\,\text{A}$$

The corresponding turn-off (falling transition of the switch node) current of the control switch is given by

$$I_{\text{Turn-OFF}} = I_{\text{OUT}} + \frac{I_{\text{ripple}}}{2}$$

$$= 15\,\text{A} + \frac{4.09\,\text{A}}{2} \tag{7.55}$$

$$= 17.05\,\text{A}$$

The operating mode of the buck converter results in positive current commutation for the control switch and negative current commutation for the synchronous rectifier (reverse conduction operation). The dead time is set to 10 ns for both transitions. Due to the ripple current, the current-related losses for the control switch will need to be determined independently from the synchronous rectifier.

Given all these conditions, Table 7.1 highlights parameters that need to be calculated for a complete loss analysis.

7.5.1 Output Capacitance Losses

P_{OSS} losses will be analyzed first. Since C_{OSS} is only dependent on the input voltage, both the control switch and synchronous rectifier can be analyzed simultaneously. As was shown in Section 7.2.2, the C_{OSS} loss is dependent on Q_{OSS} and not E_{OSS} because the circuit uses the same device for the control switch and the synchronous rectifier. From the EPC2619 [4] datasheet, the C_{OSS} as function of drain-to-source voltage can be used to determine $E_{\text{OSS\_SHB}}$ as

Table 7.1 Loss analysis parameters.

FET function	Control switch		Synchronous rectifier	
Transition state	Turn-off	Turn-on	Turn-on	Turn-off
Commutation	Self	Forced	Self	Self
Effective dead-time	Fixed at 10 ns	Fixed at 10 ns	Fixed at 10 ns	Fixed at 10 ns
P_{oss}	No	Yes	Minimal	Induced in control FET
P_G	Yes (½ · P_G)	Yes (½ · P_G)	Yes (½ · P_G)	Yes (½ · P_G)
P_{SD}	No	No	Reverse conduction	Reverse conduction
P_{RR}	None	None	None	None
P_{on}	N/A	Yes	Small (from reverse)	N/A
P_{off}	Yes	N/A	N/A	Small (to reverse)
P_{cond}	$I_{RMS}^2 \cdot R_{Dson}$		$I_{RMS}^2 \cdot R_{DSon}$	

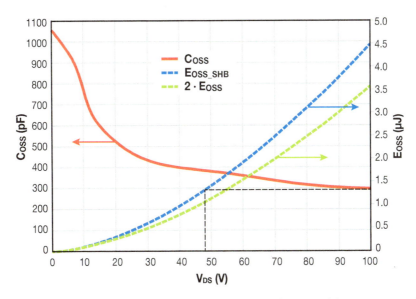

Figure 7.31 C_{OSS}, Symmetrical half bridge (E_{OSS_SHB}) and $2 \cdot E_{OSS}$ as function of drain-to-source voltage for the EPC2619. *Source:* Adapted from Efficient Power Conversion Corporation [4].

function of drain-to-source voltage using Eq. (7.21), and is plotted in Figure 7.31 together with the classic C_{OSS} loss calculation method using the sum of each device's E_{OSS}. It is clear from Figure 7.31 that the higher the operating voltage becomes, the larger the error will be for the C_{OSS} loss calculation using the classic approach.

Using Eqs. (7.21) and (7.15), the symmetrical half bridge $E_{\text{OSS\_SHB}}$ can be calculated as follows:

$$\begin{aligned} E_{\text{OSS\_SHB}} &= v_{\text{BUS}} \cdot \int_0^{V_{\text{BUS}}} C_{\text{OSS}}(v_{\text{DS}}) \cdot dv_{\text{DS}} \\ &= 48 \cdot \int_{0\,\text{V}}^{48\,\text{V}} C_{\text{OSS}}(v_{\text{DS}}) \cdot dv_{\text{DS}} \\ &= 1.325\,\mu\text{J} \end{aligned} \quad (7.56)$$

Despite there being two devices in the circuit, there is only one transition that creates C_{OSS} losses. Those losses will be dissipated in the control switch only, and at 1 MHz the P_{OSS} is 1.325 W.

7.5.2 Gate Losses

The magnitude of the gate losses is small in comparison to switching losses, but the values for Q_{GS2} and Q_{GD} need to be determined for later use in the switching loss calculations.

Both devices incur gate losses. The drain current will affect the total gate charge such that the values differ for the control switch and synchronous rectifier. In the case of the control switch, the total gate charge includes Q_{GD} as the device experiences a voltage transition. The synchronous rectifier, however, has no Q_{GD} component as it always switches from reverse conduction.

Before Q_{GS} can be determined for the specific operating condition, the value of the plateau voltage must be read from the device transfer characteristic graph (extracted from the EPC2619 [4] datasheet) for the specific operating condition, compensated for the elevated temperature, and as shown in Figure 7.32, with $V_{\text{pl}} = 1.9$ V at $I_{\text{DS}} = 29$ A for the datasheet value, and $V_{\text{pl(op)H}} = 1.8$ V at $I_{\text{DS}} = 17$ A and 1.71 V at $I_{\text{DS}} = 13$ A which represent the high load current ripple current range calculated using Eqs. (7.54) and (7.55).

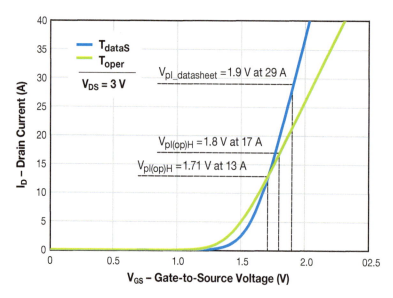

Figure 7.32 Transfer characteristic for EPC2619 [4] with $V_{\text{pl\_ds}}$ (at 29 A), V_{plH} (at 17 A) and, V_{plL} (at 13 A) shown. Source: Adapted from Efficient Power Conversion Corporation [4].

The value for $Q_{GS(op)}$ can now be determined for each of the current values using Eq. (7.7) and the EPC2619 [4] datasheet, giving $Q_{GS} = 2.2$ nC:

$$
\begin{aligned}
Q_{GS(op)H} &= \left(\frac{Q_{GS}}{V_{pl}}\right) \cdot V_{pl(op)H} \\
&= \left(\frac{2.2 \text{ nC}}{1.9 \text{ V}}\right) \cdot 1.8 \text{ V} \\
&= 2.08 \text{ nC}
\end{aligned}
\tag{7.57}
$$

$$
\begin{aligned}
Q_{GS(op)L} &= \left(\frac{Q_{GS}}{V_{pl}}\right) \cdot V_{pl(op)L} \\
&= \left(\frac{2.2 \text{ nC}}{1.9 \text{ V}}\right) \cdot 1.71 \text{ V} \\
&= 1.98 \text{ nC}
\end{aligned}
\tag{7.58}
$$

Using the same method and Eq. (7.6), and the typical value given in the datasheet for V_{th} as 1.1 V, Q_{GS1} results in

$$
\begin{aligned}
Q_{GS1} &= \left(\frac{Q_{GS}}{V_{pl}}\right) \cdot V_{th} \\
&= \left(\frac{2.2 \text{ nC}}{1.9 \text{ V}}\right) \cdot 1.1 \text{ V} \\
&= 1.27 \text{ nC}
\end{aligned}
\tag{7.59}
$$

Q_{GS2} can then be calculated using Eq. (7.8), which yields:

$$
\begin{aligned}
Q_{GS2H} &= Q_{GS(op)H} - Q_{GS1} \\
&= 2.08 \text{ nC} - 1.27 \text{ nC} \\
&= 0.81 \text{ nC}
\end{aligned}
\tag{7.60a}
$$

$$
\begin{aligned}
Q_{GS2L} &= Q_{GS(op)L} - Q_{GS1} \\
&= 2.0 \text{ nC} - 1.27 \text{ nC} \\
&= 0.71 \text{ nC}
\end{aligned}
\tag{7.60b}
$$

The value of Q_{GD} cannot be linearly approximated, so in this case, it needs to be determined from the C_{RSS} capacitance. From the EPC2619 [4] datasheet, the C_{RSS} as a function of drain-to-source voltage can be used to determine Q_{GD} as a function of drain-to-source voltage using Eq. (7.3). This result has been plotted in Figure 7.33, and the value of Q_{GD} at 48 V is 0.9 nC.

Knowing both Q_{GS} and Q_{GD}, it is possible to estimate $Q_{G(op)}$. Once the device has fully turned on, the slope of the Q_G graph will largely be the same, regardless of the voltage or current to which the device is switching. The slope ($m_{QGslope}$) can be determined from the datasheet for the region between the plateau voltage and the final gate voltage (5 V) using Eq. (7.61) for the datasheet-provided conditions:

$$
\begin{aligned}
m_{QGslope} &= \frac{Q_G - (Q_{GS} + Q_{GD})}{V_{DR} - V_{pl}} \\
&= \frac{8.5 - (2.2 + 0.9)}{5 - 1.9} \\
&= 1.742 \text{ nC}/_V
\end{aligned}
\tag{7.61}
$$

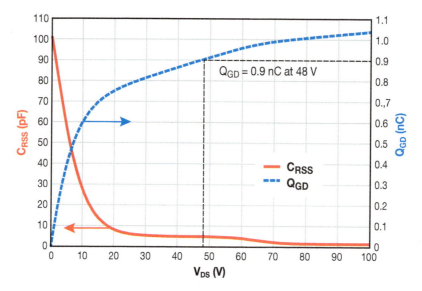

Figure 7.33 C_{RSS} and Q_{GD} as a function of drain-to-source voltage for the EPC2619 [4] plotted with the value of the working voltage in this example. *Source:* Adapted from Efficient Power Conversion Corporation [4].

Using the slope ($m_{QGslope}$) the $Q_{G(op)}$ for each of the operating conditions can be determined. For the control switch turning on at the lower inductor current value:

$$Q_{G(op)sw} = \left(Q_{GS(op)L} + Q_{GD}\right) + \left(m_{QGslope} \cdot (V_{DR} - V_{pl})\right)$$
$$= (1.98 + 0.9) + (1.742 \cdot (4.7 - 1.71)) \quad (7.62)$$
$$= 8.09 \text{ nC}$$

And for the synchronous rectifier turning on at the higher inductor current value, but with "zero" voltage across it:

$$Q_{G(op)rc} = \left(Q_{GS(op)H} + Q_{GD}\right) + \left(m_{QGslope} \cdot (V_{DR} - V_{pl})\right)$$
$$= (2.08 + 0) + (1.742 \cdot (4.9 - 1.8)) \quad (7.63)$$
$$= 7.48 \text{ nC}$$

The new calculated values for Q_{GS}, Q_{GD}, and Q_G are plotted as a function of gate-to-source voltage in Figure 7.34 together with the plot given in the datasheet. Figure 7.34 shows the gate-charge difference between the control switch and synchronous rectifier.

The gate power now can be determined using Eq. (7.22). For the control switch:

$$P_{Gcs} = Q_G \cdot V_{DR} \cdot f_{sw}$$
$$= 8.09 \text{ nC} \cdot 4.7 \text{ V} \cdot 1 \cdot 10^6 \quad (7.64)$$
$$= 38 \text{ mW}$$

and for the synchronous rectifier:

$$P_{Gsr} = Q_G \cdot V_{DR} \cdot f_{sw}$$
$$= 7.48 \text{ nC} \cdot 4.9 \text{ V} \cdot 1 \cdot 10^6 \quad (7.65)$$
$$= 36.7 \text{ mW}$$

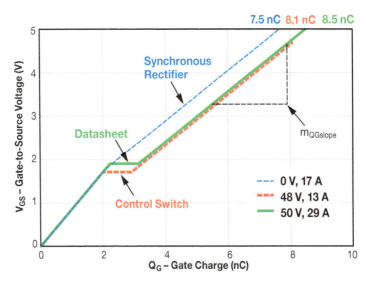

Figure 7.34 Q_G as function of gate-to-source voltage for the EPC2619 [4] plotted with the datasheet value and re-plotted for the two cases of the example. *Source:* Adapted from Efficient Power Conversion Corporation [4].

7.5.3 Reverse Conduction Losses

In this buck converter example and operating conditions, only the synchronous rectifier device experiences reverse conduction. The reverse conduction time needs to be determined using the selected dead time of 10 ns. The two cases that need to be analyzed are for turn-off (falling switching-node voltage) and turn-on (rising switch-node voltage).

7.5.3.1 Turn-Off Transient Reverse Conduction Losses (t_{rc_on})

Since the turn-off transition is self-commutating, the buck inductor current at the time of turn-off and the total output charge for both devices need to be determined. The turn-off current already has been calculated using Eq. (7.55), $I_{turn-off}$ = 17.05 A.

Next, the Q_{OSS} for the devices needs to be determined to determine which transition condition exists such as ZVS or partial ZVS of reverse conduction following a self-commutated transition. Using Eq. (7.15), the Q_{OSS} of the switching device is plotted together with C_{OSS} and shown in Figure 7.35. The value for Q_{OSS} at 48 V is 27.6 nC.

The Q_{OSS} for the converter is used to determine the ZVS self-commutating time using Eqs. (7.31), (7.32a), and (7.32b):

$$t_{ZVS} = \frac{Q_{OSSQ1} + Q_{OSSQ2}}{I_{turn\_off}}$$
$$= \frac{2.76 \text{ nC} + 27.6 \text{ nC}}{17.05 \text{ A}} \qquad (7.66)$$
$$= 3.24 \text{ ns}$$

It is important to include both devices' Q_{OSS} in Eq. (7.66), hence Q_{OSStot}. If both devices are the same, Q_{OSS} simply doubles. If both devices are not the same, each device's Q_{OSS} needs to be determined independently for the same voltage condition and added together for the total Q_{OSStot} to calculate the fall-time. Furthermore, if the circuit includes a Schottky diode across the synchronous rectifier, it too must be included in the Q_{OSS} calculation.

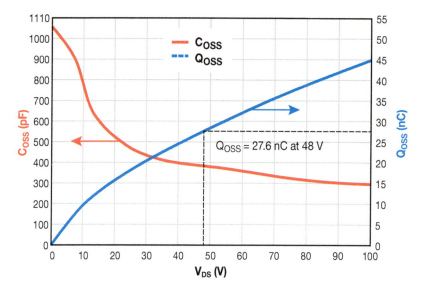

Figure 7.35 C_{OSS} and Q_{OSS} as a function of drain-to-source voltage for the EPC2619 [4] plotted with the value for working voltage of this example. *Source:* Adapted from Efficient Power Conversion Corporation [4].

The fall-time is determined to be 3.24 ns. Having chosen a 10 ns dead time, it appears that the circuit will switch well after the switch-node has completed its self-commutation transition. This means Q_2 incurs reverse conduction prior to turn-on.

There are three time-intervals that affect the turn-on reverse conduction time that was discussed in Section 7.2.4.2. The determination of each of those intervals will be calculated next. Q_1 current turn-off can be calculated using Eq. (7.30), including the gate driver turn-off drive resistance (R_{DRoff}), and using the value determined in Eq. (7.60a):

$$t_{Q1Ioff} = (R_{Goff} + R_{DRoff}) \cdot \frac{Q_{GS2\_Q1}}{\frac{V_{pl} + V_{th}}{2}}$$

$$= (0.47\,\Omega + 0.4\,\Omega) \cdot \frac{0.81\,\text{nC}}{\frac{1.8\,\text{V} + 1.1\,\text{V}}{2}} \quad (7.67)$$

$$= 0.486\,\text{ns}$$

Similarly, for Q_2 turn-on can be calculated using Eq. (7.35), including the gate driver turn-on drive resistance (R_{DRon}), and using the value determined in Eq. (7.60a):

$$t_{Q2Ion} = (R_{Gon} + R_{DRon}) \cdot \frac{Q_{GS2\_Q2}}{V_{DR} - \left(\frac{V_{pl} + V_{th}}{2}\right)}$$

$$= (1.8\,\Omega + 0.7\,\Omega) \cdot \frac{0.81\,\text{nC}}{4.9\,\text{V} - \left(\frac{1.8\,\text{V} + 1.1\,\text{V}}{2}\right)} \quad (7.68)$$

$$= 0.587\,\text{ns}$$

And finally, the reverse conduction voltage transition time using Eq. (7.33):

$$t_{\text{Q2Vrcz}} = \frac{Q_{\text{OSSQ1 rc}} + Q_{\text{OSSQ2 rc}}}{4 \cdot I_{\text{turn\_off}}}$$

$$t_{\text{Q2Vrcz}} = \frac{0.67 \text{ nC} + 1.82 \text{ nC}}{4 \cdot 17.05}$$

$$= 0.04 \text{ ns}$$

(7.69)

where Eq. (7.34a) is used for Q_1 output charge change:

$$Q_{\text{OSSQ1rc}} = \int_{V_{\text{BUS}}}^{V_{\text{BUS}} + V_{\text{PL}}} C_{\text{OSSQ1}}(v_{\text{DS}}) \cdot dv_{\text{DS}}$$

$$= \int_{48 \text{ V}}^{48 \text{ V} + 1.79 \text{ V}} C_{\text{OSSQ1}}(v_{\text{DS}}) \cdot dv_{\text{DS}}$$

$$= 0.67 \text{ nC}$$

(7.70a)

and Eq. (7.34b) is used for Q_2 output charge change:

$$Q_{\text{OSSQ2rc}} = \int_{0}^{V_{\text{PL}}} C_{\text{OSSQ2}}(v_{\text{DS}}) \cdot dv_{\text{DS}}$$

$$= \int_{0}^{1.79 \text{ V}} C_{\text{OSSQ2}}(v_{\text{DS}}) \cdot dv_{\text{DS}}$$

$$= 1.82 \text{ nC}$$

(7.70b)

The total turn-on reverse conduction time ($t_{\text{rc\_on}}$) can now be calculated using Eq. (7.29):

$$t_{\text{rc\_on}} = \left(t_{\text{dt}_{\text{dis}}} + t_{\text{Q2Ion}}\right) - \left(t_{\text{ZVS}} + t_{\text{Q1Ioff}} + t_{\text{Q2Vrcz}}\right)$$

$$= \left(10 \text{ ns} + 0.587 \text{ ns}\right) - \left(3.24 \text{ ns} + 0.486 \text{ ns} + 0.04 \text{ ns}\right)$$

$$= 6.82 \text{ ns}$$

(7.71)

Comparing this to the simpler method that only accounts for the ZVS transition time results in 6.76 ns with difference of 61 ps.

Next, the reverse-conduction voltage drop needs to be determined. Again, the datasheet is referenced for the drain current value. From Figure 7.36, at 17.05 A the voltage drop for diode conduction is 1.8 V. This value should be similar to the plateau voltage in the case of an enhancement-mode GaN transistor only.

With the reverse conduction time and the reverse voltage, the reverse conduction loss can be determined using Eq. (7.23):

$$P_{\text{SD\_on}} = V_{\text{SD}} \cdot I_{\text{DS}} \cdot t_{\text{rc}} \cdot f_{\text{sw}}$$

$$= 1.8 \text{ V} \cdot 17.05 \text{ A} \cdot 6.82 \text{ ns} \cdot 1 \text{ MHz}$$

$$= 209.3 \text{ mW}$$

(7.72)

7.5.3.2 Turn-on Transient Reverse Conduction Losses ($t_{\text{rc\_off}}$)

Since this transition is forced-commutating, the buck inductor current at the time of turn-on needs to be determined. In this case, the diode will conduct after the synchronous rectifier Q_2 is turned off and will keep conducting until the control switch is turned on. This is due to the buck inductor's current keeping the diode in the conduction state. Thus, the diode conduction time is nearly equal to the dead time, which is 10 ns for this example. However, a more detailed

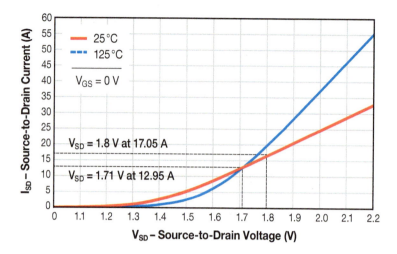

Figure 7.36 Reverse current conduction for the EPC2619 [4] plotted with the value for working currents of this example.

timing analysis will be calculated to improve accuracy. The turn-on current already has been calculated using Eq. (7.54), $I_{\text{turn-on}} = 12.95$ A.

There are three time-intervals that affect turn-off reverse conduction time that was discussed in Section 7.2.4.1. The determination of each of those intervals will be calculated next. First the current turn-off of Q_2 can be calculated using Eq. (7.25), including the gate driver turn-on drive resistance (R_{DRoff}), and the value determined in Eq. (7.60b):

$$t_{Q2\text{Ioff}} = (R_{\text{Goff}} + R_{\text{DRoff}}) \cdot \frac{Q_{\text{GS2\_Q2}}}{\frac{V_{\text{pl}} + V_{\text{th}}}{2}}$$

$$= (0.47\,\Omega + 0.4\,\Omega) \cdot \frac{0.71\text{ nC}}{\frac{1.71\text{ V} + 1.1\text{ V}}{2}} \quad (7.73)$$

$$= 0.437\text{ ns}$$

Similarly, for Q_1 turn-on can be calculated using Eq. (7.28), including the gate driver turn-on drive resistance (R_{DRon}), and using the value determined in Eq. (7.60b):

$$t_{Q1\text{Ion}} = (R_{\text{Gon}} + R_{\text{DRonn}}) \cdot \frac{Q_{\text{GS2\_Q1}}}{V_{\text{DR}} - \left(\frac{V_{\text{pl}} + V_{\text{th}}}{2}\right)}$$

$$= (0.47\,\Omega + 0.4\,\Omega) \cdot \frac{0.71\text{ nC}}{4.7\text{ V} - \left(\frac{1.71\text{ V} + 1.1\text{ V}}{2}\right)} \quad (7.74)$$

$$= 0.536\text{ ns}$$

Followed by the voltage transition time for Q_2 using Eq. (7.26):

$$t_{Q2\,\text{Vrc}} = \frac{Q_{\text{OSSQ1 rc}} + Q_{\text{OSSQ2 rc}}}{2 \cdot I_{\text{turn\_on}}}$$

$$= \frac{0.57\text{ nC} + 1.56\text{ nC}}{2 \cdot 12.95\text{ A}} \quad (7.75)$$

$$= 0.04\text{ ns}$$

where Eq. (7.27a) is used for Q_1 output charge change:

$$
\begin{aligned}
Q_{\text{OSSQ1rc}} &= \int_{V_{\text{BUS}}}^{V_{\text{BUS}}+V_{PL}} C_{\text{OSSQ1}}(\nu_{\text{DS}}) \cdot d\nu_{\text{DS}} \\
&= \int_{48\,\text{V}}^{48\,\text{V}+1.71\,\text{V}} C_{\text{OSSQ1}}(\nu_{\text{DS}}) \cdot d\nu_{\text{DS}} \\
&= 0.57\,\text{nC}
\end{aligned}
\tag{7.76a}
$$

and Eq. (7.27b) is used for Q_2 output charge change:

$$
\begin{aligned}
Q_{\text{OSSQ2rc}} &= \int_{0}^{V_{PL}} C_{\text{OSSQ2}}(\nu_{\text{DS}}) \cdot d\nu_{\text{DS}} \\
&= \int_{0}^{1.71\,\text{V}} C_{\text{OSSQ2}}(\nu_{\text{DS}}) \cdot d\nu_{\text{DS}} \\
&= 1.56\,\text{nC}
\end{aligned}
\tag{7.76b}
$$

The total turn-off reverse conduction time ($t_{\text{rc\_off}}$) can now be calculated using Eq. (7.24):

$$
\begin{aligned}
t_{\text{rc\_off}} &= \left(t_{\text{dt\_chrg}} + t_{\text{Q1lon}}\right) - \left(t_{\text{Q2loff}} + t_{\text{Q2Vrc}}\right) \\
&= (10\,\text{ns} + 0.536\,\text{ns}) - (0.437\,\text{ns} + 0.04\,\text{ns}) \\
&= 10.059\,\text{ns}
\end{aligned}
\tag{7.77}
$$

Comparing this to the simpler method that only accounts for the dead time only time results in a difference of 59 ps.

Next, the reverse-conduction voltage drop needs to be determined. Again, the datasheet is referenced for the drain current value. From Figure 7.36, at 12.95 A the voltage drop for diode conduction is 1.71 V. This value should be similar to the plateau voltage in the case of an enhancement-mode GaN transistor only.

With the reverse conduction time and the reverse voltage, the reverse conduction loss can be determined using Eq. (7.23):

$$
\begin{aligned}
P_{\text{SD\_off}} &= V_{\text{SD}} \cdot I_{\text{DS}} \cdot t_{\text{rc}} \cdot f_{\text{sw}} \\
&= 1.71\,\text{V} \cdot 12.95\,\text{A} \cdot 10.059\,\text{ns} \cdot 1\,\text{MHz} \\
&= 222.8\,\text{mW}
\end{aligned}
\tag{7.78}
$$

7.5.4 Control Switch Dynamic Losses

The control switch experiences both hard turn-on and some turn-off losses. Since all the components have already been determined, Eq. (7.11) can be used to determine the turn-on switching power losses where the blocking voltage prior to turn-on is adjusted to include reverse conduction.

$$
\begin{aligned}
P_{\text{on}} &= \frac{(V_{\text{Bus}} + V_{\text{SD}}) \cdot I_{\text{DS}} \cdot f_{\text{sw}} \cdot (R_{\text{Gon}} + R_{\text{DRon}})}{2} \cdot \left[\frac{Q_{\text{GD}}}{V_{\text{DR}} - V_{\text{pl}}} + \frac{Q_{\text{GS2}}}{V_{\text{DR}} - \left(\frac{V_{\text{pl}} + V_{\text{th}}}{2}\right)}\right] \\[2ex]
&= \frac{(48\,\text{V} + 1.71\,\text{V}) \cdot 12.95\,\text{A} \cdot 10^{6}\,\text{Hz} \cdot (1.8\,\Omega + 0.7\,\Omega)}{2} \cdot \left[\frac{0.9\,\text{nC}}{4.7\,\text{V} - 1.71\,\text{V}} + \frac{0.71\,\text{nC}}{4.7\,\text{V} - \left(\frac{1.71\,\text{V} + 1.1\,\text{V}}{2}\right)}\right] \\[2ex]
&= 401\,\text{mW}
\end{aligned}
\tag{7.79}
$$

The analytical determination of turn-off losses for the control switch under self-commutation is a highly complex process due to the interaction between the non-linearity of the drain current as function of the gate voltage (transconductance) and non-linearity of the output capacitance. For this example and given the low turn-off gate resistance it is assumed that the transition between the channel carrying the inductor current to the output capacitance carrying the inductor current within the device occurs very quickly, and thus any voltage, current, and time exposure will be very small, and thus the losses will be small enough to be ignored.

7.5.5 Conduction Losses

The conduction time for each device is based on the duty cycle. The duty cycle (D) for the control switch already has been calculated using Eq. (7.52) and is 25%. The duty cycle for the synchronous rectifier (D_{Sync}) can be determined from the control switch duty cycle as follows:

$$
\begin{aligned}
D_{\text{Sync}} &= 1 - D \\
&= 1 - 0.25 \\
&= 0.75 \\
&= 75\%
\end{aligned}
\tag{7.80}
$$

The conduction losses ($P_{\text{Conduction}}$) for the control switch then can be calculated. The equation for the RMS of the current is given in [1] and adjusted to the elevated temperature:

$$
\begin{aligned}
P_{\text{Conduction}} &= \left(I_{\text{load}}^2 + \frac{I_{\text{ripple}}^2}{12} \right) \cdot R_{\text{DS(on)}} \cdot D \\
&= \left(15\,A^2 + \frac{4.09\,A^2}{12} \right) \cdot 4.85\,\text{m}\Omega \cdot 0.25 \\
&= 274.5\,\text{mW}
\end{aligned}
\tag{7.81}
$$

and for the synchronous rectifier:

$$
\begin{aligned}
P_{\text{Conduction}} &= \left(I_{\text{load}}^2 + \frac{I_{\text{ripple}}^2}{12} \right) \cdot R_{\text{DS(on)}} \cdot D_{\text{Sync}} \\
&= \left(15\,A^2 + \frac{4.09\,A^2}{12} \right) \cdot 4.85\,\text{m}\Omega \cdot 0.75 \\
&= 824.5\,\text{mW}
\end{aligned}
\tag{7.82}
$$

The total conduction loss is thus the sum of the switch and rectifier conductions losses and is 1099 mW.

7.5.6 Total Hard-Switching Losses

The total device switching losses can be calculated by adding the various components. For the control switch:

$$
\begin{aligned}
P_{\text{HSQ1}} &= P_{\text{OSS}} + P_{\text{G}} + P_{\text{on}} + P_{\text{Conduction}} \\
&= 1.325\,\text{W} + 38\,\text{mW} + 401\,\text{mW} + 274.5\,\text{mW} \\
&= 2.04\,\text{W}
\end{aligned}
\tag{7.83}
$$

and for the synchronous rectifier:

$$P_{\text{HSQ2}} = P_{\text{SDoff}} + P_{\text{SDon}} + P_{\text{G}} + P_{\text{Conduction}}$$
$$= 222.8\,\text{mW} + 209.3\,\text{mW} + 36.7\,\text{mW} + 824.5\,\text{mW} \qquad (7.84)$$
$$= 1.293\,\text{W}$$

7.5.7 Inductor Losses

The final loss component is the inductor loss (P_{L}). The inductor used for this example is the IHTH-1125KZ-5A-2R2 [25] and has a DC resistance (DCR) = 0.7 mΩ and inductance of 2.2 µH. The manufacturer also provides a loss calculator on its website [20], which was utilized in this example. Alternative methods can be used to determine inductor losses that include referencing the manufacturers' core losses as function of flux density and frequency. Using the loss calculator, the core loss (P_{core}) was given as 947 mW, DCR loss (P_{DCR}) as 182 mW, and AC winding loss (P_{AC}) as 123 mW with total inductor losses of 1.251 W.

7.5.8 Total Buck Converter Estimated Losses

Adding all the loss components together yields the total estimated power loss for the buck converter example:

$$P_{\text{Total}} = P_{\text{control}} + P_{\text{Sync}} + P_{\text{L}}$$
$$= 2.04\,\text{W} + 1.293\,\text{W} + 1.251\,\text{W} \qquad (7.85)$$
$$= 4.584\,\text{W}$$

This results in an estimated efficiency for the converter of 97.5%, which excludes control circuit power.

7.5.9 Buck Converter Losses Accounting for Common Source Inductance

Ignoring the effect of common-source inductance can lead to an artificially low loss prediction. Therefore, in this section, the switching losses will be recalculated to include the effect of CSI. By the very nature of CSI, it is impossible to measure without significant perturbation of the circuit.

CSI can be estimated using a commercial parametric extraction simulation program [26] that can compute the inductance from the layout and device design. This would require knowledge of the internal design of the device that would seldom be made available. Alternatively, CSI can be estimated using circuit simulation software and using an ideal switch in the simulation. In the simulation, CSI can be added, and waveforms compared with measured waveforms until enough correlation is found. For this calculation, a value of 110 pH will be used as a reasonable approximation.

Equation (7.48) can be used to determine the equivalent gate circuit resistance introduced by the CSI. This resistance will be used in conjunction with Eqs. (7.11) and (7.12) to determine the CSI-impacted losses. Note that CSI only impacts the current transition interval, and therefore, Eqs. (7.11) and (7.12) need to be adjusted so that R_{CSI} is only added to R_{G} for the Q_{GS2} interval.

To calculate R_{CSI}, the value of C_{GS} and transconductance (g_m) at the operating conditions is required. First, the transconductance can be determined using the Shichman–Hodges model or small-signal model for a MOSFET [27, 28]:

$$
\begin{aligned}
g_{mL} &= \frac{2 \cdot I_{DS}}{V_{pl} - V_{th}} \\
&= \frac{2 \cdot 12.95 \text{ A}}{1.71 \text{ V} - 1.1 \text{ V}} \\
&= 42.5 \text{ S}
\end{aligned}
\tag{7.86a}
$$

$$
\begin{aligned}
g_{mH} &= \frac{2 \cdot I_{DS}}{V_{pl} - V_{th}} \\
&= \frac{2 \cdot 17.05 \text{ A}}{1.8 \text{ V} - 1.1 \text{ V}} \\
&= 48.7 \text{ S}
\end{aligned}
\tag{7.86b}
$$

Next C_{GS} needs to be determined for of the operating currents. These values can be derived from Q_{GS}, which yields a time-equivalent capacitance, by reading off the values at the plateau voltage and which we determined using either Eq. (7.57) or (7.58), respectively:

$$
\begin{aligned}
C_{GSL} &= \frac{Q_{GS}}{V_{pl}} \\
&= \frac{2.0 \text{ nC}}{1.71 \text{ V}} \\
&= 1.17 \text{ nF}
\end{aligned}
\tag{7.87a}
$$

$$
\begin{aligned}
C_{GSH} &= \frac{Q_{GS}}{V_{pl}} \\
&= \frac{2.08}{1.8 \text{ V}} \\
&= 1.16 \text{ nF}
\end{aligned}
\tag{7.87b}
$$

as they should result in nearly the same value since the difference in gate voltage is very small.

The equivalent common-source inductance impedance (R_{CSI}) is then calculated for the low current transient as follows:

$$
\begin{aligned}
R_{CSIL} &= \frac{L_S \cdot g_m}{C_{GS}} \\
&= \frac{110 \cdot 10^{-12} \cdot 42.5 \text{ S}}{1.17 \cdot 10^{-9}} \\
&= 4.0 \, \Omega
\end{aligned}
\tag{7.88a}
$$

and for the high current transient:

$$
\begin{aligned}
R_{CSIH} &= \frac{L_S \cdot g_m}{C_{GS}} \\
&= \frac{110 \cdot 10^{-12} \cdot 48.7 \text{ S}}{1.16 \cdot 10^{-9}} \\
&= 4.62 \, \Omega
\end{aligned}
\tag{7.88b}
$$

The control switch turn-on losses, adjusted for CSI, can be determined by updating Eq. (7.11) as follows:

$$P_{\text{on}} = \frac{(V_{\text{Bus}} + V_{\text{SD}}) \cdot I_{\text{DS}} \cdot f_{\text{sw}}}{2} \cdot \left[\frac{Q_{\text{GD}} \cdot (R_{\text{Gon}} + R_{\text{DRon}})}{V_{\text{DR}} - V_{\text{pl}}} + \frac{Q_{\text{GS2}} \cdot (R_{\text{Gon}} + R_{\text{CSIL}} + R_{\text{DRon}})}{V_{\text{DR}} - \left(\dfrac{V_{\text{pl}} + V_{\text{th}}}{2}\right)} \right]$$

$$= \frac{(48\,\text{V} + 1.71\,\text{V}) \cdot 12.95\,\text{A} \cdot 10^6\,\text{Hz}}{2} \cdot \left[\frac{0.9\,\text{nC} \cdot (1.8\,\Omega + 0.7\,\Omega)}{4.7\,\text{V} - 1.71\,\text{V}} + \frac{0.71\,\text{nC} \cdot (1.8\,\Omega + 4.0\,\Omega + 0.7\,\Omega)}{4.7\,\text{V} - \left(\dfrac{1.71\,\text{V} + 1.1\,\text{V}}{2}\right)} \right]$$

$$= 693\,\text{mW}$$

$$(7.89)$$

This result represents an increase of 292 mW in turn-on loss resulting from CSI.

The case for the synchronous rectifier is more complicated due to the change in timing effects, so many calculations need to be updated. For the synchronous rectifier turn-on case, starting with Q_1 current turn-off can be re-calculated using Eq. (7.67):

$$t_{\text{Q1Ioff}} = (R_{\text{Goff}} + R_{\text{DRoff}} + R_{\text{CSIH}}) \cdot \frac{Q_{\text{GS2\_Q1}}}{\dfrac{V_{\text{pl}} + V_{\text{th}}}{2}}$$

$$= (0.47\,\Omega + 0.4\,\Omega + 4.62\,\Omega) \cdot \frac{0.81\,\text{nC}}{\dfrac{1.8\,\text{V} + 1.1\,\text{V}}{2}}$$

$$(7.90)$$

$$= 3.07\,\text{ns}$$

Similarly, for Q_2 turn-on can be re-calculated using Eq. (7.68):

$$t_{\text{Q2Ion}} = (R_{\text{Gon}} + R_{\text{DRon}} + R_{\text{CSIH}}) \cdot \frac{Q_{\text{GS2\_Q2}}}{V_{\text{DR}} - \left(\dfrac{V_{\text{pl}} + V_{\text{th}}}{2}\right)}$$

$$= (1.8\,\Omega + 0.7\,\Omega + 4.62\,\Omega) \cdot \frac{0.81\,\text{nC}}{4.9\,\text{V} - \left(\dfrac{1.8\,\text{V} + 1.1\,\text{V}}{2}\right)}$$

$$(7.91)$$

$$= 1.67\,\text{ns}$$

The total turn-on reverse conduction time ($t_{\text{rc\_on}}$) can now be re-calculated using Eq. (7.29):

$$t_{\text{rc\_on}} = \left(t_{\text{dt}_{\text{dis}}} + t_{\text{Q2Ion}}\right) - \left(t_{\text{ZVS}} + t_{\text{Q1Ioff}} + t_{\text{Q2Vrcz}}\right)$$

$$= (10\,\text{ns} + 1.67\,\text{ns}) - (3.24\,\text{ns} + 3.07\,\text{ns} + 0.04\,\text{ns})$$

$$(7.92)$$

$$= 5.32\,\text{ns}$$

The shift in reverse time with the addition of CSI effects stems from the change in gate drive strength between Q_1 turn-off and Q_2 turn-on being more sensitive to drive voltage with higher gate resistances.

The reverse conduction losses can be updated to:

$$P_{\text{SD\_on}} = V_{\text{SD}} \cdot I_{\text{DS}} \cdot t_{\text{rc}} \cdot f_{\text{sw}}$$

$$= 1.8\,\text{V} \cdot 17.05\,\text{A} \cdot 5.32\,\text{ns} \cdot 1\,\text{MHz}$$

$$(7.93)$$

$$= 163.3\,\text{mW}$$

For the synchronous rectifier turn-off case:

$$t_{Q2Ioff} = (R_{Goff} + R_{DRoff} + R_{CSIL}) \cdot \frac{Q_{GS2\_Q2}}{\frac{V_{pl} + V_{th}}{2}}$$

$$= (0.47\,\Omega + 0.4\,\Omega + 4.0\,\Omega) \cdot \frac{0.71\,\text{nC}}{\frac{1.71\,\text{V} + 1.1\,\text{V}}{2}} \tag{7.94}$$

$$= 2.45\,\text{ns}$$

Similarly, for Q_1 turn-on can be calculated using Eq. (7.28), including the gate driver turn-on drive resistance (R_{DRon}), and using the value determined in Eq. (7.60b):

$$t_{Q1\,Ion} = (R_{Gon} + R_{DRonn} + R_{CSIL}) \cdot \frac{Q_{GS2\_Q1}}{V_{DR} - \left(\frac{V_{pl} + V_{th}}{2}\right)}$$

$$= (0.47\,\Omega + 0.4\,\Omega + 4.0\,\Omega) \cdot \frac{0.71\,\text{nC}}{4.7\,\text{V} - \left(\frac{1.71\,\text{V} + 1.1\,\text{V}}{2}\right)} \tag{7.95}$$

$$= 1.39\,\text{ns}$$

The total turn-off reverse conduction time (t_{rc_off}) can now be re-calculated using Eq. (7.24):

$$t_{rc\_off} = (t_{dt\_chrg} + t_{Q1Ion}) - (t_{Q2Ioff} + t_{Q2Vrc})$$

$$= (10\,\text{ns} + 1.39\,\text{ns}) - (2.45\,\text{ns} + 0.04\,\text{ns}) \tag{7.96}$$

$$= 8.9\,\text{ns}$$

$$P_{SD\_off} = V_{SD} \cdot I_{DS} \cdot t_{rc} \cdot f_{sw}$$

$$= 1.71\,\text{V} \cdot 12.95\,\text{A} \cdot 8.9\,\text{ns} \cdot 1\,\text{MHz} \tag{7.97}$$

$$= 197.1\,\text{mW}$$

The total adjusted power loss can now be determined which accounts for CSI-induced losses. For this example, the total losses increase by 245.6 mW to 4.83 W, which decreased the efficiency to 97.4%.

7.5.10   Experimental Comparison

An experimental buck converter that mimics the example was tested using the EPC90153 [29] development board. The buck converter was operated and measured at the same operating voltages and load current as the example. The dead time was set to 10 ns.

Before making a comparison between the experimental evaluation and the analysis, one must understand the difference between them. There are many additional factors that need to be accounted for and a few are listed as follows:

1) PCB traces have resistance and thus losses.
2) Capacitances used for the DC input and output have ESR and the ripple current will induce loss.
3) Gate driver bootstrap circuit has reverse recovery and thus introduces loss in the upper FET of a buck converter.
4) Device and inductor will increase in temperature and loss mechanisms need to be temperature compensated. The analysis has already accounted for this mechanism.

5) Gate driver's drive voltages are not exactly 5 V but will decrease for various reasons. The analysis has already accounted for this mechanism.
6) Printed circuit board introduces capacitance in parallel with the switch-node and ground or supply bus. That capacitance appears in parallel with the lower FET output capacitance and thus increases losses.
7) Measurement system accuracy is extremely important, and one must perform a thorough calibration of the entire setup to ensure it results in zero loss or at least only have resistive losses for the connections and PCB traces.

The analysis did not account for all mechanisms listed above, so some need to be determined based on the experimental system and are shown in Figure 7.37. These differences need to be excluded from the analytical comparison.

7.5.10.1 PCB Trace Resistance Losses

The first additional loss mechanism is the copper traces of the board under test. These are identified as R_{hs}, R_{ls}, and R_{sw} in Figure 7.37 and were measured to be 1.97, 360, and 490 µΩ, respectively. Equations (7.81) and (7.82) can be used to calculate the losses for R_{hs} and R_{ls}, respectively, using the appropriate resistance. The result is P_{Rhs} = 111.5 mW and P_{Rls} = 61.1 mW. For the switch-node current the RMS calculation uses the same equation as Eq. (7.81) or (7.82) except the duty cycle becomes 1 with result of P_{Rsw} = 110.9 mW. The total PCB copper losses is then P_{PCB} = 283.5 mW.

7.5.10.2 Bus and Output Capacitance ESR Losses

Next the contribution of the capacitors needs to be determined represented by ESR_{Cin} and ESR_{Cout} in Figure 7.37. On the EPC90153, the input bus comprised four sets of capacitors:

1) Seven paralleled 220 nF, 100 V rated, X7S, size 0603 ceramic capacitor with equivalent ESR = 2.143 mΩ and C = 413.2 nF adjusted for bias voltage,
2) Ten paralleled 1 µF, 100 V rated, X7R, size 0805 ceramic capacitor with equivalent ESR = 1.712 mΩ and C = 2.5 µF adjusted for bias voltage,

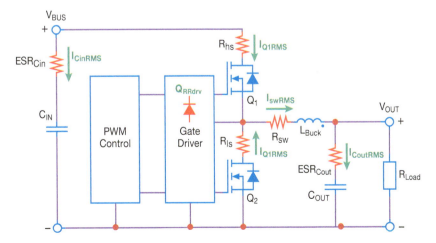

Figure 7.37 Updated buck converter schematic showing locations of additional loss mechanisms present in the experimental setup.

3) Five paralleled 1 µF, 250 V rated, X7R, size 2220 ceramic capacitor with equivalent ESR = 1.12 mΩ and 4.15 µF adjusted for bias voltage,
4) One 470 µF, 250 V rated electrolytic capacitor with equivalent ESR = 338.6 mΩ.

For the output, there are five paralleled 4.7 µF, 100 V rated, X7R, size 1210 ceramic capacitor with equivalent ESR = 430 µΩ and 22 µF adjusted for bias voltage. The methodology used to determine the capacitor ESR losses is given in [30] resulting in 60.7 mW of loss for the input capacitors and 600 µW for the output capacitor with total of 61.3 mW.

7.5.10.3 Power Loop Losses

Figure 7.38 shows the measured switch-node waveform of the experimental converter operating from 48 V and delivering 15 A into a 12 V load. The waveform shows the zoomed-in portion of the ringing voltage with cursors marking the peak of each cycle for three cycles. Within the cursors, the time difference between the cursors is 9.05 ns which gives a ringing frequency of 331.5 MHz.

The loop inductance can be calculated using the method presented in Section 7.3.2. and starting with Eq. (7.49). Since this example is a buck converter, the off-device is the synchronous rectifier device Q2, so only its output capacitance at the bus voltage is used:

$$L_{\text{Loop}} = \frac{1}{C_{\text{OSSQ2}}(V_{\text{Bus}}) \cdot \left(2 \cdot \pi \cdot f_{\text{ring}}\right)^2}$$

$$L_{\text{Loop}} = \frac{1}{383 \text{ pF} \cdot (2 \cdot \pi \cdot 331.5 \text{ MHz})^2}$$

$$L_{\text{Loop}} = 1.25 \text{ nH}$$

(7.98)

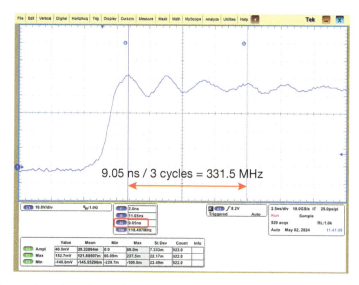

Figure 7.38 Measured switch-node waveform of the experimental converter operating with 48 V input and delivering 15 A into a 12 V load and switching at 1 MHz with the ringing period for three cycles highlighted. *Source:* With permission of EPC.

Using the loop inductance and Eq. (7.50), the energy stored in the inductor prior to the switching event can be determined:

$$E_{\text{Loop}} = \frac{L_{\text{Loop}} \cdot I_{\text{turn\_on}}^2}{2}$$

$$E_{\text{Loop}} = \frac{1.25\,\text{nH} \cdot 12.95\,\text{A}^2}{2} \tag{7.99}$$

$$E_{\text{Loop}} = 105.1\,\text{nJ}$$

Finally, the power loss for the loop inductance can then be calculated using Eq. (7.51)

$$P_{\text{Loop}} = E_{\text{Loop}} \cdot f_{\text{sw}}$$

$$P_{\text{Loop}} = 105.1\,\text{nJ} \cdot 1\,\text{MHz} \tag{7.100}$$

$$P_{\text{Loop}} = 105.1\,\text{mW}$$

7.5.10.4 Gate Driver Bootstrap Diode Reverse Recovery Losses

The loss contribution of the gate driver bootstrap diode Q_{RRdrv}, shown in Figure 7.37, needs to be determined as it increases losses in the upper FET when that FET hard-switches. The contribution of this loss is often overlooked due to a lack of information and may be considered negligible. However, GaN devices enable higher switching frequency making this contribution important enough to calculate. The reverse recovery charge of the gate driver operating at 48 V is 2.2 nC. At 1 MHz and 48 V, the power loss it induces in the upper FET is 52.8 mW.

7.5.10.5 PCB Switch-Node Capacitance Losses

The EPC90153 converter was operated without any external impedance connection to the switch-node and allowed a separate measurement of the contribution of the output capacitance, gate driver bootstrap diode reverse recovery allowing extraction of PCB switch-node capacitance losses. In this test there are two forced commutated transitions, and thus the P_{OSS} loss calculation given in Section 7.2.2 needs to be doubled. In this test the power delivered to the board at 48 V input was measured at 2.793 W and the calculated value of two times the result from Section 7.5.1 is 2.65 W plus the gate driver bootstrap diode yields 2.703 W resulting in a difference of 90.2 mW, which is the contribution of the PCB capacitance to the switch-node ($P_{\text{Sw\_PCB}}$).

Using this loss, the value of the PCB capacitance on the switch-node can be determined as follow:

$$C_{\text{Sw\_PCB}} = \frac{2 \cdot P_{\text{Sw\_PCBOSS}}}{V_{\text{Bus}}^2 \cdot f_{\text{Sw}}}$$

$$C_{\text{Sw\_PCB}} = \frac{2 \cdot 90.2\,\text{mW}}{48\,\text{V}^2 \cdot 1\,\text{MHz}} \tag{7.101}$$

$$C_{\text{Sw\_PCB}} = 78.3\,\text{pF}$$

7.5.10.6 Total Experimental Loss Breakdown

The total experimental adjustment is thus 487.6 mW. A breakdown of the total losses in the experimental converter is shown in Figure 7.39. Figure 7.39 highlights the two largest contributors to losses are output capacitance and the filter inductor. These account for approximately half the losses. Also shown in Figure 7.39 is the contribution of PCB losses (Joule heating was discussed in Chapter 6) in comparison to the overall losses, and as devices tend

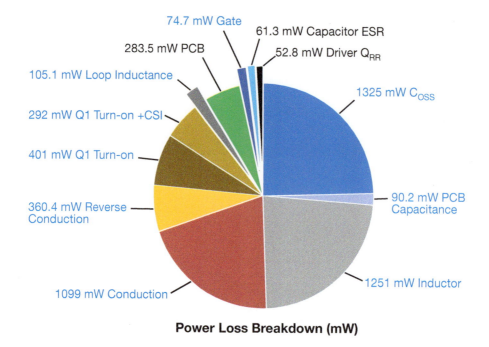

Figure 7.39 Breakdown of the losses in the experimental converter with blue text showing the calculated losses and the black italics text showing the additional losses from the experimental setup.

to conduct higher currents, those losses will increase proportionally and will drive layout techniques to minimize those losses through design.

The measured losses for the buck converter when operating with 48 V input and delivering 15 A into a 12 V load was 7.25 W. Subtracting out the adjustment yields 6.762 W compared to the total calculated value of 4.83 W, giving a difference of 1.85 W.

It is noteworthy to mention that should a current meter be under-reading by just 1%, a converter with 12 V load and delivering 180 W can result in an additional 1.8 W of loss that is not real. The problem increases as converter efficiency increases requiring creative and alternative measurement techniques to measure power loss accurately. Every effort was taken in the experimental section to ensure this error did not affect the results. Additional errors that can arise include:

- The inductor loss is probably underestimated, and it is difficult to accurately separate out device losses from the inductor loss in an experimental setup; thus, the loss numbers from the manufacturer need to be relied on for the analysis.
- The capacitors ESR are also underestimated given those are determined using small-signal techniques and the converter operates with large signals. Furthermore, there is a frequency element to the losses that make it difficult to determine the correct value to be used.
- Losses in the PCB used are based on DC measurement, but the converter operation includes high AC current magnitudes that can increase losses, particularly around the device itself where the conductors tend to neck down leading to additional skin and proximity effect losses. In addition, copper thickness has a high tolerance and can be a source of error relied on for multiple experimental boards.

- The gate driver output stage is a miniature switching converter in its own right and is also subject to capacitance effects. These capacitance effects further slow the gate voltage transitions and thus increase losses in the main power converter.

It is also important to note that an analysis is a tool to predict the performance of using a device in a converter. The complexity to achieve ever higher accuracy may become too complex to be useful or manageable and other techniques, such as circuit simulation tools, may be more valuable. In this example we could look at the time after the gate has reached the end of the Q_{GD} time, but that is not fully on. In the period where the gate voltage continues to rise to the final on-state may still take 10s of nano-seconds to complete and during that time the on-resistance of the device transitions from a high value to the value used in this analysis. Additional effects not included are loop inductance losses as those have not been analytically quantified.

7.6 Summary

In this chapter, the mechanism and key factors that contribute to hard-switching losses have been discussed. The analytical tools needed to calculate losses from the greatest contributors were developed and used in a concrete example. The calculations in large part agreed with the actual measured results from a circuit built to these same specifications. It is important to note that such a loss analysis would need to account for many experimental factors, such as contribution by the PCB and support components, to ensure accuracy and for which designers may not necessarily have available at the onset of a design. This means the analysis is a tool to be used to guide the designer into making good component choices that will have a high probability of achieving the desired specifications.

References

1 Erickson, R.W. and Maksimović, D. (2001). *Fundamentals of Power Electronics*, 2e. Springer.
2 Glaser, J.S. and Reusch, D. (2016). Comparison of deadtime effects on the performance of DC-DC converters with GaN FETs and silicon MOSFETs. *IEEE Energy Conversion Congress and Exposition (ECCE) Conference*. Milwaukee, WI, (September 18–22, 2016).
3 Vishay (2004). Power MOSFET basics: understanding gate charge and using it to assess switching performance. Application note AN608.
4 Efficient Power Conversion Corporation (2023). EPC2619 – Enhancement-mode Power Transistor. EPC2619 datasheet, [Revised November 2023]. https://epc-co.com/epc/Portals/0/epc/documents/datasheets/EPC2619_datasheet.pdf.
5 Texas Instruments (2019). Does GaN have a body diode? – Understanding the third Quadrant operation of GaN. Application report SNOAA36. https://www.ti.com/lit/an/snoaa36/snoaa36.pdf.
6 Huang, A.Q. (2004). New unipolar switching power device figures of merit. *IEEE Electron Dev. Lett.* 25: 298–301.
7 Kim, I.-J., Matsumoto S., Sakai, T. et al. (1995). New power device figure-of-merit for high frequency applications. *Proceedings of International Symposium on Power Semiconductor Devices and IC's (ISPSD)*, (May 1995), 309–314.
8 Baliga, B.J. (1989). Power semiconductor device figure-of-merit for high frequency applications. *IEEE Electron Dev. Lett.* 10 (10): 455–457.

9 Ying, Y. (2008). Device selection criteria – based on loss modeling and figure of merit. M.S. thesis. Virginia Tech, Blacksburg, VA. https://vtechworks.lib.vt.edu/server/api/core/bitstreams/64268248-691d-41b5-8d85-873a2048da20/content.

10 Reusch, D., Strydom, J., and Lidow A., Improving system performance with eGaN® FETs in DC-DC applications. *46th International Symposium on Microelectronics, (IMAPS)*, Orlando, FL, October 2013. https://www.google.com/url?sa=t&source=web&rct=j&opi=89978449&url=https://meridian.allenpress.com/ism/article-pdf/2013/1/000764/2254424/isom-2013-wp63.pdf&ved=2ahUKEwjIlo_x_vuFAxVFOUQIHcrjACYQFnoECBIQAQ&usg=AOvVaw0FxeTNvwwmJNMzXF8ad4re.

11 Reusch, D. (2012). High frequency, high power density integrated point of load and bus converters. PhD dissertation. Virginia Tech, Blacksburg, VA. http://scholar.lib.vt.edu/ theses/available/etd-04162012-151740/.

12 Reusch, D. (2013). eGaN® FET-silicon power shoot-out, Vol. 13, Part 1: impact of parasitics. *Power Electronics Technology*. https://www.electronicdesign.com/technologies/eda/semiconductors/gan/article/21195695/egan-fet-silicon-power-shoot-out-vol-13-part-1-impact-of-parasitics.

13 Reusch, D.2013). eGaN® FET-silicon power shoot-out Vol. 13, Part 2: Optimal PCB Layout. *Power Electronics Technology*. https://www.electronicdesign.com/technologies/eda/semiconductors/gan/article/21195833/egan-fet-silicon-power-shoot-out-vol-13-part-2-optimal-pcb-layout.

14 Efficient Power Conversion Corporation (2021). EPC2015C – Enhancement-mode power transistor. EPC2015C datasheet, 2021 [Revised April 2021]. http://epc-co.com/epc/documents/datasheets/EPC2015C_datasheet.pdf.

15 Jones, E.A., Williford, P., Yang, Z. et al. (2017). Maximizing the voltage and current capability of GaN FETs in a hard-switching inverter. *Proceedings of the. IEEE International Conference on Power Electronics and Drive Systems (PEDS)*, (December 2017), 740–747.

16 McLyman, T. and Colonel, W. (2004). *Transformer and Inductor Design Handbook*, 3e. CRC Press.

17 Steinmetz, C.P. (1892). On the law of hysteresis. *AIEE Trans.* 9: 3–64. Reprinted under the title A Steinmetz contribution to the ac power revolution, introduction by J. E. Brittain (1984) in *Proceedings of the IEEE* 72 (2): 196–221.

18 Reinert, J., Brockmeyer, A., and De Doncker, R.W. (2001). Calculation of losses in ferro- and ferrimagnetic materials based on the modified steinmetz equation. *IEEE Trans. Ind. Appl.* 37 (4): 1055–1061.

19 Coilcraft. Power inductor finder and analyzer. http://www.coilcraft.com/apps/loss/loss_1.cfm.

20 Vishay Dale. IHLP® Inductor loss calculator tool. https://www.vishay.com/en/inductors/calculator/calculator/.

21 uPI Semi, Dual-channel gate driver for enhanced mode GaN transistors. uP1966E datasheet, August 2021. https://epc-co.com/epc/Portals/0/epc/documents/datasheets/uP1966E_datasheet.pdf.

22 Hauke, B. (2011). Basic calculation of a buck converter's power stage. Texas Instruments Application report SLVA477B. December 2011 [Revised August. 2015].

23 Ejury, J. (2013). Buck converter design. Infineon design note DN 2013-01, V1.0. https://cdn.badcaps-static.com/pdfs/2a997c023d0eda74b0a3c42d4b38ca9c.pdf.

24 Schelle, D. and Castorena, J. (2006). Buck-converter design demystified, *Electronics Design*. https://www.electronicdesign.com/technologies/power/power-supply/dc-dc-converters/article/21188914/buck-converter-design-demystified.

25 Vishay Dale (2020). High current through-hole inductor, high temperature series, IHTH-1125KZ-5A datasheet. [Revised January 1, 2024] https://www.vishay.com/docs/34349/ihth-1125kz-5a.pdf.

26 Ansys (2021). Ansys Q3D Extractor. http://www.ansys.com/Products/Simulation+Technology/Electromagnetics/Signal+Integrity/ANSYS+Q3D+Extractor.

27 Cartwright, K.V. (2009). Derivation of the exact transconductance of a FET without calculus. *Technol. Inter. J.* 10 (1): https://www.google.com/url?sa=t&source=web&rct=j&opi=89978449&url=https://tiij.org/issues/issues/fall09/Fall09/011.pdf&ved=2ahUKEwjqoubZl_yFAxVYle4BHdUoDIUQFnoECB4QAQ&usg=AOvVaw18mJ7 reaK808fwDmLaroLf.

28 Shichman, H. and Hodges, D.A. (September 1968). Modeling and simulation of insulated-gate field-effect transistor switching circuits, *IEEE J. Solid State Circuits*, SC-3: 285–289. https://ieeexplore.ieee.org/document/1049902.

29 Efficient Power Conversion Corporation (2022). 80 V, 20 A half-bridge development board featuring EPC2619, Development board EPC90153 quick start guide. https://epc-co.com/epc/products/evaluation-boards/epc90153.

30 Xie, M. (2016). How to select input capacitors for a buck converter. Texas Instruments Incorporated, *Analog Applications Journal*, 2016. https://www.google.com/url?sa=t&source=web&rct=j&opi=89978449&url=https://www.ti.com/lit/pdf/slyt670&ved=2ahUKEwjq7P_im_yFAxXYIUQIHSToBEwQFnoECCIQAQ&usg=AOvVaw30iu3U94CjSYMuFclEK8nx

8

Resonant and Soft-Switching Converters

8.1 Introduction

In the previous chapter, the application of GaN transistors in hard-switching power converters was addressed, and the benefits GaN transistors provided. In this chapter, the fundamentals of resonant and soft-switching applications will be discussed and the superior performance of GaN transistors over silicon MOSFETs will be evaluated. The chapter will conclude with a design example comparing GaN transistors and Si MOSFETs in an isolated, high-frequency 48 V intermediate bus converter (IBC) with a 12 V output, utilizing a resonant topology operating at 1.2 MHz

8.2 Resonant and Soft-Switching Techniques

Resonant and soft-switching techniques can improve performance in converters by reducing switching-related losses compared to conventional hard-switching converters. This is accomplished by creating operating conditions where the transistor does not encounter simultaneous high voltage and high current during the switching commutation. There are many different resonant and soft-switching techniques [1–4], and the two conditions in common use are zero-voltage switching (ZVS) and zero-current switching (ZCS).

8.2.1 Zero-Voltage and Zero-Current Switching

ZVS is used to eliminate the turn-on commutation losses in a switching device. ZVS is achieved when the transistor drain-to-source voltage is reduced to zero before turning on. To reduce the drain-to-source voltage across the transistor in the majority of resonant and soft-switching topologies, the device output charge (Q_{OSS}) must be removed by conducting current from source-to-drain through the output capacitance (C_{OSS}) until the drain-to-source voltage reaches zero. The switching trajectory of a traditional hard-switching and ZVS, soft-switching, turn-on transition is shown in Figure 8.1a, where the x-axis, V_{DS}, represents switch voltage, and the y-axis, I_{DS}, represents switch current. For the hard-switching transition, the current first rises to the load current, then the voltage falls to zero. This results in large values of voltage and current being commutated simultaneously in the transistor, generating loss as well as inducing C_{OSS} losses as discussed in Chapter 7. For the ZVS transition, before the transistor current rises to the load current, a negative current drives the drain-to-source voltage to zero, creating a

---

GaN Power Devices for Efficient Power Conversion, Fourth Edition. Alex Lidow, Michael de Rooij, John Glaser, Alejandro Pozo, Shengke Zhang, Marco Palma, David Reusch, and Johan Strydom.
© 2025 John Wiley & Sons Ltd. Published 2025 by John Wiley & Sons Ltd.

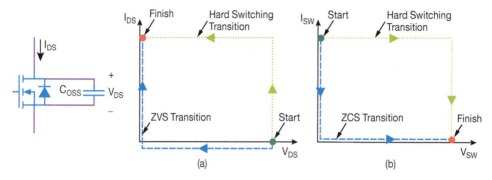

Figure 8.1 Ideal switching transition for (a) zero-voltage switching turn-on transition; (b) zero-current switching turn-off transition.

soft turn-on condition. The device does not commutate high voltage and high current simultaneously, reducing the turn-on commutation losses in the device to zero, and eliminating E_{OSS} losses.

With the mechanism described above, ZVS erases turn-on commutation losses, but has little effect on turn-off losses. A realistic transistor can also exhibit a form of quasi-ZVS switching, which occurs when the output capacitance of the transistor slows the voltage rate of rise enough that the transistor channel can be switched before the voltage rises to the full bus voltage. This is discussed in Chapter 7. ZCS provides true soft commutation during device turn-off, as shown in Figure 8.1b. For the hard-switching turn-off transition, also shown in Figure 8.1b, the voltage first rises to the bus voltage, the current then falls to zero. This results in large values of voltage and current being commutated simultaneously in the transistor, generating losses. ZCS occurs when the transistor drain-to-source current is reduced to zero and before turning off. To achieve ZCS, the current is shaped as a sinusoidal pulse using a resonant network. When the switch current resonates to zero, the device can turn off with ZCS. During a ZCS turn-off transition, large values of voltage and current are not commutated simultaneously in the transistor, virtually eliminating the turn-off losses in the transistor. For ZCS resonant converters, the hard-switching turn-on commutation remains, and the turn-on switching transition and C_{OSS} losses are dissipated in the device.

For resonant and soft-switching topologies to operate under ZVS and ZCS, reactive networks are required to shape the transistor's voltage and current. Often, the addition of physical components can be avoided by utilizing the parasitic components internal to the device and stray elements such as package and PCB inductances. This allows the parasitics that diminished hard-switching converter performance to be used effectively to achieve soft-switching commutations in resonant and soft-switching converters. For the majority of resonant and soft-switching DC–DC power converters, ZVS is preferred to ZCS due to the reduction of the C_{OSS} losses, which are incurred only during the turn-on switching transition.

8.2.2 Resonant DC–DC Converters

The traditional approach for resonant converters used for DC-DC power conversion is shown in Figure 8.2. The input voltage source, V_{IN}, connects to a switching network, which outputs a pulsed waveform to the resonant network. The resonant network then shapes the voltage or

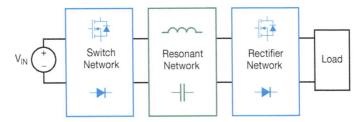

Figure 8.2 Resonant DC–DC converter block diagram consisting of switch network, resonant network, and rectifier network.

current to achieve soft switching in the switching network's power devices. Following the resonant network is the rectifier network that rectifies and filters the voltage and current to deliver DC power to the load.

8.2.3 Resonant Network Combinations

The most basic resonant networks consist of a series network, where the load is connected in series with the resonant capacitance (C_S), as shown in Figure 8.3a, a parallel network, where the load is connected in parallel with the resonant capacitance (C_P), as shown in Figure 8.3b, or a combination of the series and parallel networks, also known as a series–parallel network, as shown in Figure 8.3c. Another popular resonant network is the LLC, where the parallel capacitance (C_P) of the series–parallel network is replaced with a parallel inductor (L_P), and L_R and C_S are rearranged, as shown in Figure 8.3d. These different resonant networks can operate with ZVS or ZCS and provide unique benefits as well as disadvantages when compared to traditional hard-switching converters [1–3, 5, 6]. There are also less common resonant network topologies as well as multi-element resonant networks that can be employed as resonant tanks [5, 7, 8].

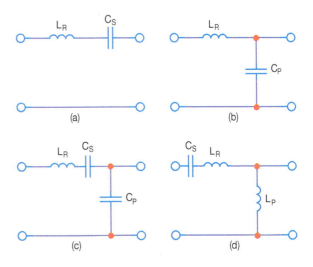

Figure 8.3 DC–DC converter resonant networks: (a) series LC, (b) parallel LC, (c) series–parallel LCC, and (d) LLC.

8.2.4 Resonant Network Operating Principles

In this section, the basic operation of a common resonant configuration, the series-resonant network, shown in Figure 8.3a will be discussed. This series-resonant network allows for either ZVS or ZCS to occur in the switching network transistors, reducing either turn-on or turn-off commutation losses. The magnitude of the impedance of the series network can be given by

$$|Z_{SRN}| = \sqrt{R^2 + (X_L - X_C)^2} = \sqrt{R^2 + \left(\omega \cdot L_R - \frac{1}{\omega \cdot C_S}\right)^2} \tag{8.1}$$

$$X_L = \omega \cdot L_R \tag{8.2}$$

$$X_C = \frac{1}{\omega \cdot C_S} \tag{8.3}$$

where X_L and X_C are the reactances of the inductor and capacitor, respectively, R is the equivalent load resistance, Z_{SRN} is the magnitude of the impedance of the series-resonant network, and ω is the angular frequency of the resonant network. The standard frequency common to DC–DC converters is given by

$$f = \frac{\omega}{2 \cdot \pi} \tag{8.4}$$

where f is frequency measured in Hertz.

The magnitude of the impedance plot for a series-resonant network is shown in Figure 8.4. The resonant frequency is the point where the network transitions from appearing capacitive to inductive. The resonant frequency provides the minimum impedance and is given by

$$f_0 = \frac{1}{2 \cdot \pi \cdot \sqrt{L_R \cdot C_S}} \tag{8.5}$$

The turn-on commutation will switch for a transistor connected to a series-resonant network operating above the resonant frequency under ZVS. Operation above the resonant frequency makes the resonant tank appear inductive, as shown in Figure 8.4. For an inductive load, the current lags the voltage, as shown in Figure 8.5a. The inductive-resonant tank produces a negative current in the device before turn-on, discharging the transistor output charge and allowing for a soft ZVS turn-on commutation. For operation above the resonant frequency, there will be voltage and current in the device at turn-off, leading to a hard turn-off commutation.

For operation below the resonant frequency, the transistor will turn off under ZCS. For a series-resonant converter, operation below the resonant frequency makes the resonant tank

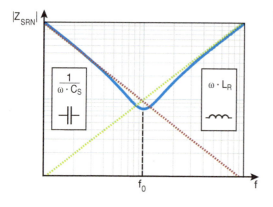

Figure 8.4 Impedance plot of a series-resonant network.

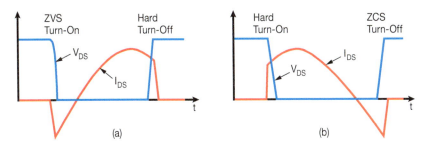

Figure 8.5 Operation of series-resonant converter transistor: (a) above the resonant frequency, and (b) below the resonant frequency.

appear capacitive, as shown in Figure 8.4. For a capacitive load, the current leads the voltage, as shown in Figure 8.5b. The capacitive-resonant tank produces a negative current in the device before turn-off, allowing for a soft ZCS turn-off commutation. For operation below the resonant frequency, there will be voltage and current in the device at turn-on, leading to a hard turn-on commutation and C_{OSS} losses.

8.2.5 Resonant Switching Cells

The best performance is achieved for traditional resonant converters operating close to the resonant frequency, yet output regulation is achieved by varying the switching frequency. The further the operating frequency is moved from the resonant point to maintain regulation, the more the performance of the resonant converter suffers from higher circulating energy and component stresses [1–8].

To apply the principles of resonant power conversion to pulse-width modulated (PWM) converters, another family of resonant converters was developed [1, 9]. These quasi-resonant (QR) cells commonly are seen in DC–DC power conversion and combine a resonant network with a single transistor to create a ZVS or ZCS device. They can be applied to traditional non-isolated topologies like the buck converter discussed in Chapter 7 [1, 3, 4], as well as various other topologies and applications. The resonant cells use the same concept of shaping of the voltage and/or current to achieve soft switching as the traditional resonant networks. (Later in this chapter there will be a design example to demonstrate the use of a QR cell in a GaN-based resonant converter.) The ZVS QR cell, shown in Figure 8.6a, places a resonant capacitor, C_R, in parallel with the transistor and a resonant inductor in series with the capacitor-switch network, while the ZCS QR cell, shown in Figure 8.6b, places the resonant capacitor in parallel with the series combination of the resonant inductor-switch network. In practice, the ZVS QR cell is

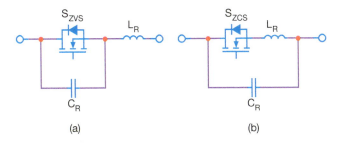

Figure 8.6 Quasi-resonant switching cells: (a) ZVS and (b) ZCS.

GaN Power Devices for Efficient Power Conversion

more common since the switch capacitance is absorbed into the ZVS network, but interferes with the ZCS network. As with the resonant converter, the QR converter has many variations.

8.2.6 Soft-Switching DC–DC Converters

Soft-switching converters can be seen as a hybrid between hard-switching PWM converters and frequency-controlled resonant converters. Soft-switching converters employ resonant techniques for a portion of the operating period to achieve a soft device commutation, with the remaining period operating as a PWM converter [2, 3]. This allows for transistor soft commutation while reducing the higher circulating energy and device stresses associated with resonant converters, as well as offering PWM control for output regulation.

8.3 Key Device Parameters for Resonant and Soft-Switching Applications

In resonant and soft-switching applications, the switching-related losses are minimized by using techniques to achieve ZVS and ZCS. With the reduction of switching losses, the Q_{GD} and Q_{GS2} terms that dominate losses in hard-switching applications are no longer the critical device parameters determining in-circuit performance. The two device parameters key to high performance in resonant and soft-switching applications are device output charge, Q_{OSS}, and gate charge, Q_G.

8.3.1 Output Charge (Q_{OSS})

Output charge has a large impact on performance of resonant and soft-switching converters as it directly impacts the energy required to achieve ZVS. A reduction in this energy can result in reduced ZVS transition times and currents, providing both a longer power delivery period and lower RMS currents in high-frequency resonant and soft-switching converters. In a ZCS topology, the energy of the output capacitance (E_{OSS}) is dissipated when the transistor turns on in the same manner as a hard-switching commutation.

Before a ZVS transition can occur, the output capacitance must be discharged, bringing the drain-to-source voltage of the transistor to zero volts before turning on the transistor. The time required to achieve ZVS is given by:

$$t_{ZVS} = \frac{C_{OSS(TR)} \cdot V_{DS}}{I_{ZVS}} = \frac{Q_{OSS}}{I_{ZVS}} \tag{8.6}$$

where t_{ZVS} is the time required to discharge the output capacitance, $C_{OSS(TR)}$ is the time-related output capacitance, V_{DS} is the transistor drain-to-source voltage, I_{ZVS} is the soft-switching current used to discharge the transistor's output capacitance, and Q_{OSS} is the output charge of the transistor.

8.3.2 Determining Output Charge from Manufacturers' Datasheet

To properly design a ZVS transition, the values of C_{OSS} and Q_{OSS} at the proper in-circuit operating conditions are critical. Values of C_{OSS} and Q_{OSS} generally are given for a single drain-to-source operating voltage in manufacturers' datasheets as shown in Table 8.1 for a 100 V enhancement-mode GaN transistor, EPC2204 [10], and in Table 8.2 for a 100 V silicon MOSFET, IQE065N10NM5SC [11].

Table 8.1 Data from an efficient power conversion 100 V EPC2204 enhancement-mode GaN transistor datasheet showing transistor capacitances and associated charges [10].

Parameter		Test conditions	Min.	Typ.	Max.	Unit
Dynamic characteristics (T_J = 25 °C unless otherwise stated)						
$R_{DS(on)}$	Drain-to-source on resistance	V_{GS} = 5 V, I_D = 16 A		4.4	6	mΩ
	Packaged device size		—	3.8	3.9	mm^2
C_{ISS}	Input capacitance	V_{DS} = 50 V, V_{GS} = 0 V	—	644	851	pF
C_{OSS}	Output capacitance		—	304	456	
C_{RSS}	Reverse transfer capacitance		—	2.3	—	
Q_G	Total gate charge	V_{DS} = 50 V, V_{GS} = 5 V, I_D = 16 A	—	5.7	7.4	nC
Q_{GD}	Gate-to-drain charge	V_{DS} = 50 V, I_D = 16 A	—	0.8	—	
Q_{GS}	Gate-to-source charge		—	1.8	—	
Q_{OSS}	Output charge	V_{DS} = 50 V, V_{GS} = 0 V	—	25	38	

All measurements were done with substrate connected to source.
Source: Adapted from Efficient Power Conversion Corporation [10].

Table 8.2 Data from an IQE065N10NM5SC 100 V silicon MOSFET datasheet showing transistor capacitances and associated charges.

Parameter		Test conditions	Min.	Typ.	Max.	Unit
Dynamic characteristics (T_J = 25 °C unless otherwise stated)						
$R_{DS(on)}$	Drain-to-source on resistance	V_{GS} = 10 V, I_D = 20 A		5.7	6.5	mΩ
	Packaged device size		—	10.9	11.6	mm^2
C_{ISS}	Input capacitance	V_{DS} = 50 V, V_{GS} = 0 V	—	2300	3000	pF
C_{OSS}	Output capacitance		—	340	440	
C_{RSS}	Reverse transfer capacitance		—	18	32	
Q_G	Total gate charge	V_{DS} = 50 V, V_{GS} = 10 V, I_D = 20 A	—	34	43	nC
Q_{GD}	Gate-to-drain charge		—	7.4	11.1	
Q_{GS}	Gate-to-source charge		—	10.1	—	
Q_{OSS}	Output charge	V_{DS} = 50 V, V_{GS} = 0 V	—	40	53	

Source: Adapted from Infineon [11].

The single output charge and capacitance point given in manufacturers' datasheets does not provide enough information to properly design for a wide range of operating conditions. The output capacitance of both GaN transistors and MOSFETs is highly non-linear, and the output charge varies with drain-to-source voltage. Figure 8.7 shows the capacitance curves for the 100 V GaN transistor [10] and 100 V Si MOSFET [11] where it can be seen that the output capacitance changes approximately by factors of three and six, respectively, from 0 to 50 V.

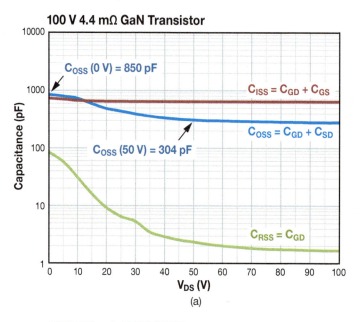

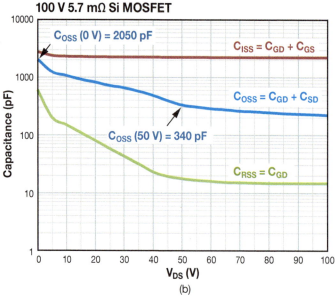

Figure 8.7 Capacitance curves of (a) 100 V, 4.4 mΩ GaN transistor. *Source:* Adapted from Efficient Power Conversion Corporation [10]. (b) 100 V, 5.7 mΩ Si MOSFET. *Source:* Adapted from Infineon [11].

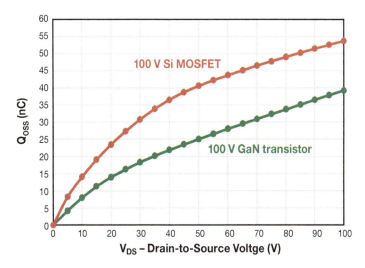

Figure 8.8 Output charge for varying drain-to-source voltages for 100 V, 4.4 mΩ GaN transistor. Source: Adapted from Efficient Power Conversion Corporation [10] and 100 V, 5.7 mΩ Si MOSFET. Source: Adapted from Infineon [11].

The output charge and effective time-related capacitance for a transistor at any given voltage can be calculated from the manufacturers' datasheets in the same manner as the hard-switching converters discussed in Chapter 7:

$$Q_{OSS}(V_{DS}) = \int_0^{V_{DS}} C_{OSS}(V_{DS}) \cdot dV_{DS} \tag{8.7}$$

$$C_{OSS(TR)}(V_{DS}) = \frac{Q_{OSS}(V_{DS})}{V_{DS}} \tag{8.8}$$

where V_{DS} is the drain-to-source operating voltage of the transistor.

The output charge for the EPC2204 [10] GaN transistor and IQE065N10NM5SC [11] Si MOSFET are calculated and plotted in Figure 8.8 using Eq. (8.7) for drain-to-source voltages varied from 0 V to the maximum voltage listed in the manufacturers' datasheets. The GaN transistor, with a similar on-resistance, offers a significant reduction in output charge over the entire voltage range.

8.3.3 Comparing Output Charge of GaN Transistors and Si MOSFETs

To compare the output charge figure of merit (FOM) between GaN and MOSFET technologies in resonant and soft-switching applications, the product of the on-resistance and output charge for state-of-the-art 40, 100, 200, and 650 V GaN and Si MOSFETs are plotted and shown in Figure 8.9. The GaN devices offer a significant reduction in output charge FOM, with the gains increasing as voltage increases. The reduction in FOM allows the circuit designer to reduce the transistor conduction losses, shorten the ZVS transition, or reduce the ZVS current, all of which would lower converter loss and improve efficiency. These benefits will be demonstrated experimentally in the design example to follow later in this chapter.

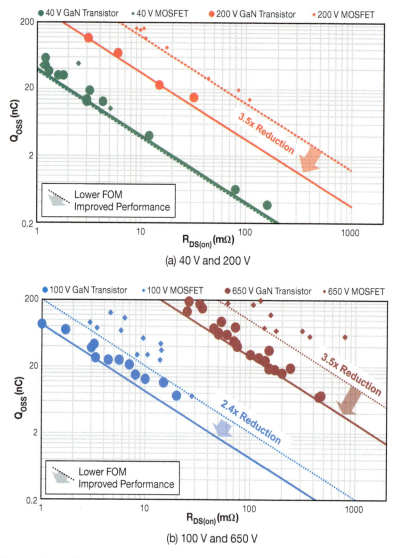

Figure 8.9 Output charge figure of merit comparison between GaN and Si devices. (a) 40 and 200 V, and (b) 100 and 600 V.

8.3.4 Gate Charge (Q_G)

The frequency capability of resonant and soft-switching topologies is also significantly impacted by the gate charge, Q_G. The gate charge is the amount of charge required to fully turn on or off the transistor. Voltage source drivers are employed for the vast majority of DC–DC converters. This voltage source appears in series with the input capacitance of the transistor and has an effective resistance equal to the sum of the gate driver circuit's internal and external resistance and the internal gate resistance of the transistor. The gate charge is dissipated each switching cycle, resulting in a gate drive loss equal to

$$P_G = Q_G \cdot V_{DR} \cdot f_{sw} \tag{8.9}$$

where V_{DR} is the gate drive voltage and f_{sw} is the switching frequency.

Looking beyond the gate drive power loss, the gate driving speed also has a large impact on the performance of high-frequency resonant and soft-switching converters. The switching period is inversely proportional to switching frequency and, as frequencies increase, the gate rising and falling speeds can become limitations to the minimum switching time. The design example at the end of this chapter will illustrate this issue and show how GaN technology can offer superior performance in high-frequency DC–DC converters than is possible with Si MOSFETs.

8.3.5 Determining Gate Charge for Resonant and Soft-Switching Applications

To enable users to calculate gate charge, manufacturers supply a gate charge curve similar to Figure 8.10 for the 100 V EPC2204 [10] GaN transistor. The gate charge, Q_G, is given for a hard-switching transition and not directly applicable to resonant and soft-switching applications. The definition of charges Q_{GS1}, Q_{GS2}, Q_G, Q_{GD}, and Q_{GS} is given in Chapters 3 and 7. For a ZVS application, the voltage commutation period occurs before the device turns on, and the Miller plateau region as well as the accompanying charge, Q_{GD}, are eliminated from the total gate charge. The gate charge for ZVS topology can be given by:

$$Q_{G\_ZVS} = Q_G - Q_{GD} \tag{8.10}$$

where Q_G is the gate charge of a hard-switching application and Q_{GD} is the gate-to-drain charge.

For a ZCS transition, the current commutation period occurs before device turn-off, and the Q_{GS2} period is eliminated. While this reduces switching commutation losses, it does not significantly impact the total gate charge as the slopes of the Q_{GS2} period and the Q_G period following the Q_{GD} region are generally similar.

$$Q_{G\_ZCS} = Q_G \tag{8.11}$$

8.3.6 Comparing Gate Charge of GaN Transistors and Si MOSFETS

The gate charge figure of merit comparison for state-of-the-art 40, 100, 200, and 650 V GaN and Si MOSFETs is plotted in Figure 8.11. The GaN devices offer a significant reduction in gate

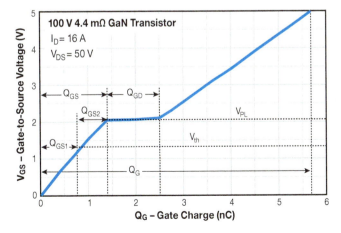

Figure 8.10 Gate charge curve for a 100 V 4.4 mΩ GaN transistor. *Source:* Adapted from Efficient Power Conversion Corporation [10].

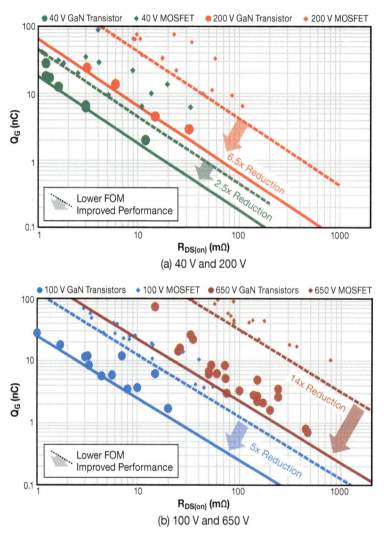

Figure 8.11 Gate charge figure of merit comparison between GaN and Si devices. (a) 40 and 200 V, and (b) 100 and 600 V.

charge FOM, with the gains increasing as voltage increases. The reduction in FOM allows the circuit designer to reduce the gate drive losses and shorten the gate drive transition period, leading to lower converter loss and improved efficiency.

8.3.7 Comparing Performance Metrics of GaN Transistors and Si MOSFETS

There are many different resonant and soft-switching techniques, and, therefore, distilling a single FOM for the wide variety of topologies into a simple metric is not practical. As discussed earlier, for resonant and soft-switching applications, the output charge, Q_{OSS}, and gate charge, Q_G, are the dominating device parameters. To allow designers to simply compare different devices to determine the technology providing the best relative in-circuit performance for resonant and soft-switching applications, a practical soft-switching FOM is

$$\text{FOM}_{SS} = (Q_{OSS} + Q_G) \cdot R_{DS(on)} \tag{8.12}$$

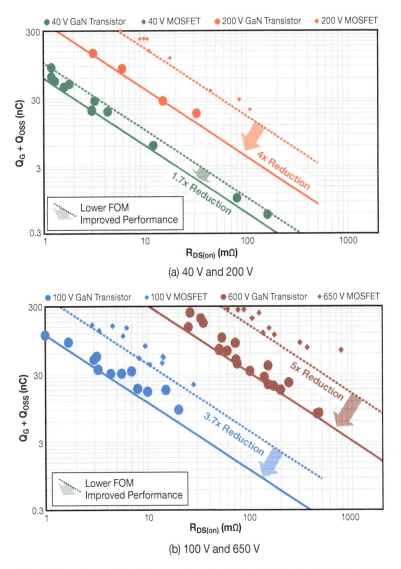

Figure 8.12 Soft-switching figure of merit comparison for GaN and Si devices. (a) 40 and 200 V, and (b) 100 and 600 V.

The comparison of this soft-switching FOM for 40, 100, 200, and 650 V GaN and Si MOSFETs is plotted and shown in Figure 8.12. The GaN technology offers a significant FOM reduction for all voltages, indicating significant performance improvements in high-frequency soft-switching applications. The benefits of replacing a Si MOSFET with an enhancement-mode GaN transistor in a high-frequency resonant converter will be quantified next.

8.4 High-Frequency Resonant Bus Converter Example

Distributed power systems are prevalent in telecommunications, networking, and high-end server applications, and generally utilize a 48 V bus voltage adopted from the telecom industry. The traditional distributed power architecture (DPA), shown in Figure 8.13a, employs a

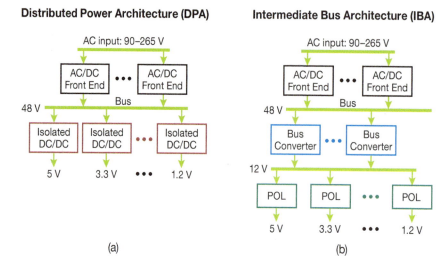

Figure 8.13 (a) Traditional distributed power architecture and (b) intermediate bus architecture.

number of 48 V isolated hard-switching point-of-load (POL) converters to power the end loads. Having a large number of regulated and isolated POLs, however, significantly increases the cost, volume, and complexity of the system. To simplify design and improve performance, the intermediate bus architecture (IBA) has been widely adopted [12, 13]. A popular IBA approach, shown in Figure 8.13b, employs a lower number of 48 V isolated bus converters that satisfy isolation requirements and supply an intermediate 12 V bus voltage. With the final regulation to the loads provided by smaller, more efficient, non-isolated POL buck converters, the bus converters can be operated as unregulated DC–DC transformers, improving efficiency and reducing cost.

The unregulated bus converter, also known as a DCX, or DC–DC transformer, can provide the highest efficiency by being designed to deliver power close to 100% of a switching period, something not possible for a regulated converter. The latter requires the converter to vary the fraction of a switching period during which power is delivered to the output to regulate output voltage against changes in input voltage or load. The majority of today's regulated bus converters use traditional hard-switching bridge topologies, which, due to large switching related losses, are forced to operate at lower frequencies where the bulky isolation transformer and output inductors occupy a large portion of board area. In an effort to improve power density and performance, unregulated bus converters can be used to shrink or eliminate the output inductors. The operating frequency of the unregulated bus converter can be increased through the use of resonant and soft-switching converters [14–17], further shrinking passive components and improving performance [18].

An experiment was undertaken to compare an enhancement-mode GaN transistors with a Si MOSFETs in this application. The subject design was a high-frequency resonant converter, 48–12 V unregulated isolated bus converter operating at a switching frequency of 1.2 MHz, and an output power of up to 400 W. The topology, shown in Figure 8.14, employs a soft-switching technique to achieve ZVS for the primary devices, and a resonant approach to achieve ZCS in the secondary devices as well as to limit the turn-off current in the primary devices [14].

Referring to Figure 8.14, it can be seen that the leakage inductance (L_{K1}) during the power delivery period, t_0–t_1, resonates with a small output capacitance (C_O). With proper timing, this

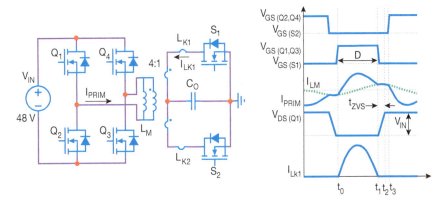

Figure 8.14 High-frequency resonant bus converter schematic and key waveforms.

results in ZCS for the secondary-side device (S_1), and significantly reduces turn-off current in the primary-side devices (Q_1, Q_3). Since the topology is an unregulated bus converter, the circuit can always operate at the optimal operating point (the resonant frequency), providing the highest efficiency. The ZVS transition begins at the end of the power delivery period. For t_1–t_2, the magnetizing current of the transformer is used to charge and discharge the output capacitances of the devices, setting up a ZVS turn-on transition for devices Q_2, Q_4, and S_2. If the ZVS transition period is too long, the body diode of the devices Q_2 and Q_4 will turn on and conduct current as seen in period t_2–t_3. At time t_3, this operation is repeated for the other switching leg with the current flowing through switches Q_2, Q_4, and S_2, and leakage inductance L_{K2}, delivering power to the load while providing flux balancing in the transformer.

8.4.1 Resonant GaN and Si Bus Converter Designs

To obtain a direct comparison in performance between GaN transistors and Si MOSFETs in an isolated converter, having identical layouts and using the same topology are critical. Isolated DC–DC converter performance is heavily dependent on topology selection, printed circuit board (PCB) layout, number of PCB inner layers, copper weight of inner layers, and the design of the transformer. To accurately compare the performance of GaN and Si in a high-frequency resonant bus converter application, the same circuit topology was used and a similar layout was maintained for both designs.

Two bus converters, shown in Figure 8.15, were built, based on the schematic in Figure 8.14, to operate at a switching frequency of 1.2 MHz. Both PCBs were constructed with 12 layers and two-ounce copper thickness for all layers. To accurately compare only device performance, these converters both had the same transformer core material, core shape, and winding arrangement, designed from [18]. The placement of the primary-side input capacitors and secondary resonant capacitors is similar for both designs to ensure similar parasitic inductances for the primary and secondary loops, with the only differences being those introduced by the different packages of the Si MOSFETs and GaN transistors. By using GaN transistors with smaller size for the same on-resistance the active footprint area shrank significantly, reducing the power stage size by 30% compared to the Si MOSFET design.

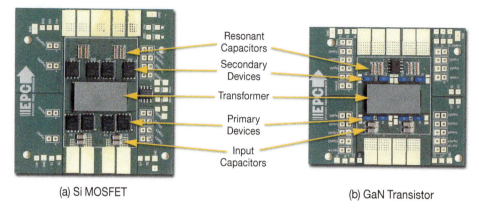

Figure 8.15 48–12 V bus converters operating at a switching frequency of 1.2 MHz constructed with (a) silicon MOSFETs and (b) GaN transistors. *Source:* With permission of EPC.

8.4.2 GaN and Si Device Comparison

To obtain a direct comparison in performance between GaN transistors and Si MOSFETs in the high-frequency resonant bus converter application, GaN and Si devices with similar on-resistance were selected. Comparisons of the key parameters for the GaN transistors and Si MOSFETs are shown in Tables 8.3 and 8.4 for the primary and secondary devices, respectively. The soft-switching FOM for the GaN devices (($Q_{OSS} + Q_G$) × $R_{DS(on)}$) is reduced by over a factor of two for both the primary and secondary devices, leading to proportionally shorter resonant transition periods and increased power delivery periods. The GaN transistor provides additional performance improvements in the form of reduced miller charge, Q_{GD}, further reducing the turn-off switching losses incurred in the primary devices. As a further advantage, the LGA packaging of the GaN transistor has lower parasitic package inductance compared to the traditional Si MOSFET package (TDSON-8). When putting all these benefits together, efficient multi-MHz switching can be obtained through the use of advanced topologies combined with low-loss GaN transistors.

Table 8.3 Device comparison between GaN and Si devices for primary devices for V_{IN} = 48 V, V_{OUT} = 12 V.

Parameter	GaN transistor	Si MOSFET [19]	FOM Ratio
Voltage rating (V_{DSS})	100 V	80 V	
$R_{DS(on)}$	5.6 mΩ at 5 V	5.2 mΩ at 8 V[a]	
Q_G	5.8 nC at 5 V	25.9 nC at 8 V[a]	
Q_{GD} at V_{IN}	2.2 nC	8.1 nC[a]	
Q_{OSS} at V_{IN}	35 nC	62 nC[a]	
$Q_G \times R_{DS(on)}$	32.5 pC·Ω	134.7 pC·Ω	4.14 × reduction
$Q_{OSS} \times R_{DS(on)}$	196 pC·Ω	322.4 pC·Ω	1.64 × reduction
FOM_{SS} ($Q_{OSS} + Q_G$) × $R_{DS(on)}$	228.5 pC·Ω	457.1 pC·Ω	2.00 × reduction

[a] Calculated from manufacturer's datasheet curves.

Table 8.4 Device comparison between GaN and Si for secondary devices for V_{IN} = 48 V, V_{OUT} = 12 V.

Parameter	EPC2015C GaN transistor [20]	BSC027N04LS G Si MOSFET [21]	FOM ratio
Voltage rating (V_{DSS})	40 V	40 V	
$R_{DS(on)}$	3.2 mΩ at 5 V	2.9 mΩ at 5 V[a]	
Q_G	8.7 nC at 5 V	27.5 nC at 5 V[a]	
Q_{GD} at 20 V	1.2 nC	6.5 nC	
Q_{OSS} at 20 V	19 nC	40 nC	
$Q_G \times R_{DS(on)}$	27.8 pC·Ω	79.8 pC·Ω	2.87 × reduction
$Q_{OSS} \times R_{DS(on)}$	60.8 pC·Ω	116 pC·Ω	1.91 × reduction
FOM_{SS} $(Q_{OSS} + Q_G) \times R_{DS(on)}$	88.6 pC·Ω	195.8 pC·Ω	2.21 × reduction

[a] Calculated from manufacturers' datasheet curves.

8.4.3 Zero-Voltage Switching Transition

The experimental waveforms for the ZVS transition period are shown in Figure 8.16 for the GaN and Si designs. By replacing a Si MOSFET with a GaN transistor, the ZVS transition period is reduced from 87 to 42 ns as a result of the reduced output charge enabled by GaN technology. Looking at the gate waveforms, it also can be seen that the gate drive speed for the GaN transistor is significantly faster than the Si MOSFET counterpart, even when

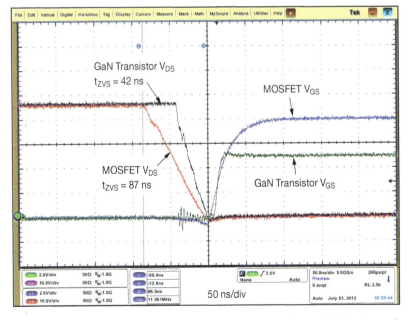

Figure 8.16 ZVS switching transitions for primary-side GaN transistor and Si MOSFET designs at f_{sw} = 1.2 MHz, V_{IN} = 48 V, and I_{OUT} = 26 A.

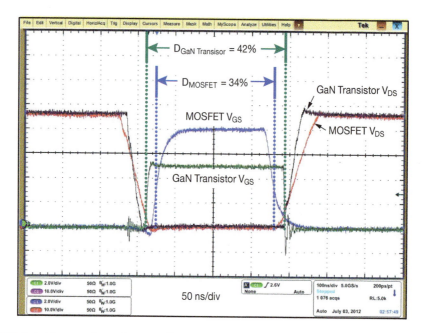

Figure 8.17 Switching waveforms showing effective duty cycle for primary-side GaN transistor and Si MOSFET designs at f_{sw} = 1.2 MHz, V_{IN} = 48 V, and I_{OUT} = 26 A.

driven with a lower gate drive voltage, providing both a longer power delivery period and reduced gate losses. The Si MOSFET takes almost 100 ns to reach its steady-state gate voltage; this is over 10 times longer than the GaN device and reflects the gate charge reduction enabled by GaN technologies.

The power delivery period for a half cycle is shown in Figure 8.17 for the GaN- and Si-based resonant bus converters. The effective duty cycles, D, which represent the power delivery period for each half cycle for the GaN and Si designs were measured to be 42 and 34%, respectively. The soft-switching FOM from Eq. (8.12) predicts the duty cycle gains, along with a 50% reduction of the soft-switching FOM, translate into a 50% reduction in the dead-time, including the ZVS transition and gate charging periods.

As the power delivery periods increase in duration, the circulating energy and resonant currents decrease, reducing the conduction losses in the resonant converter. For the resonant converter used in this design, the conduction losses, related to the resonant converter's RMS current, I_{RES}, are inversely proportional to effective duty cycle, D, by

$$I_{RES} \infty \frac{1}{\sqrt{D}} \tag{8.13}$$

For this design example, the increased duty cycle provided by GaN devices can reduce the conduction losses in the devices, transformer, PCB, and components by almost 20%.

8.4.4 Efficiency and Power Loss Comparison

The comparison in efficiency and power loss between the two designs operating at 1.2 MHz is shown in Figure 8.18. The GaN transistor-based converter offers a one-percentage point

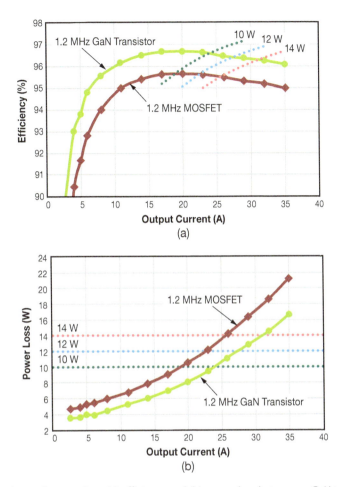

Figure 8.18 Experimental comparison (a) efficiency and (b) power loss between a GaN transistor and a Si MOSFET-based resonant bus converters (V_{IN} = 48 V, V_{OUT} = 12 V, f_{sw} = 1.2 MHz).

improvement in peak efficiency over its Si MOSFET counterpart, resulting in about a 25% decrease in power loss. Since it is typical for bus converter designs to be thermally limited by losses under full load for a fixed converter size, the reduction in power loss translates directly into higher output power handling capability. For a design capable of dissipating 14 W, the GaN transistor converter can increase the output power capability by up to 65 W while maintaining the same total converter loss when compared to the benchmark Si MOSFET design. Assuming a 12 W maximum power loss for both designs, the output power of the GaN transistor-based converter increases by 55 W, from 270 to 325 W.

The loss breakdown for the 1.2 MHz designs at output currents of 2.5 and 20 A is shown in Figure 8.19, and leads to the conclusion that the GaN technology improves efficiency for all load conditions. At lower currents, the gate driving losses dominate the transistor-related losses, and the GaN device's lower gate charge enables substantially reduced drive loss. At high currents, the conduction loss dominates total power loss, and the GaN-based converter's shorter ZVS dead-time and gate charging time provide lower conduction losses inversely proportional to the effective duty cycle. The one area where the Si-based design offers lower loss is

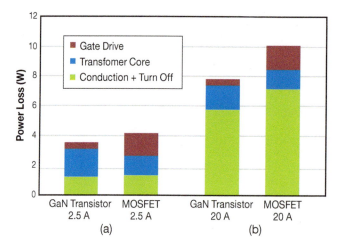

Figure 8.19 Power loss breakdown of a GaN transistor and Si MOSFET-based resonant bus converter at (a) 2.5 A, and (b) 20 A (V_{IN} = 48 V, V_{OUT} = 12 V, f_{sw} = 1.2 MHz).

in the transformer core. The longer power delivery period provided by the GaN-based design increases the transformer flux density, leading to higher core loss, but the increase in transformer core loss is more than offset at lower currents by the gate drive loss savings, and at high currents by the combination of conduction and gate loss savings.

From the results at 1.2 MHz, it can be seen that the Si MOSFET converter is approaching its frequency limit because the ZVS transition time and gate charging time are becoming a significant portion of the overall period. To compare the frequency improvements possible with GaN transistors over Si MOSFETs, the converter frequency was reduced to 800 kHz for the Si MOSFET design while increasing the GaN transistor design to 1.6 MHz. In both cases, the core structure remained the same and was not optimized for the different operating frequencies. The efficiency and loss comparisons between the designs are shown in Figure 8.20, with the GaN transistor-based design offering a 0.9% peak efficiency improvement and less power loss up to an output current of 29 A. The sharp drop-off in efficiency at currents above 20 A for the GaN transistor-based converter is a result of increased AC transformer winding losses, and an effective duty cycle reduction. Conversely, the flattening out of the efficiency for the 800 kHz Si MOSFET design was a result of reduced AC transformer winding losses and a higher effective duty cycle at the lower frequency.

8.4.5 Impact of Further Device Improvements on Performance

As discussed in Chapters 1 and 2, GaN technology is rapidly advancing compared to Si MOSFETs. Two critical figures of merit for higher density and higher performance in resonant and soft-switching applications, device size and output charge, are plotted in Figure 8.21a,b, respectively, for the last three generations of GaN transistors. Over the last three generations, both figures of merit have roughly halved for GaN transistors. The significantly denser and better performing devices translate directly into higher performance and power density in-circuit. Plotted in Figure 8.22 is the in-circuit performance comparison of a 48–12 V highly resonant converter based on circa 2012 and circa 2024 GaN transistors. With the latest GaN devices, the efficiency is increased by almost 2%, translating to total converter losses being

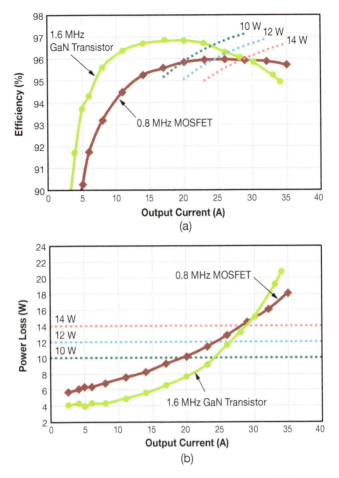

Figure 8.20 (a) Efficiency and (b) power loss comparison between f_{SW} = 1.6 MHz GaN transistor and f_{SW} = 800 kHz Si MOSFET-based V_{IN} = 48 V, V_{OUT} = 12 V resonant bus converter.

reduced by around 50%. With the lower losses and smaller power devices, the power density is also increased by over a factor of two. Further details on this design will be covered in depth in Chapter 10.

8.5 Summary

It was shown in previous chapters that enhancement-mode GaN transistors have a distinct advantage over silicon MOSFETS in hard-switching applications due to reduced Q_{GD} and Q_{GS2}, both of which are critical in hard-switching applications, but have little impact in resonant and soft-switching converters. In this chapter, the benefits of GaN technology as applied to resonant and soft-switching applications were discussed, and the in-circuit superiority of GaN transistors was demonstrated in a 48 V bus converter operating at 1.2 MHz. This chapter also discussed a simple soft-switching FOM that allows designers to quickly compare device technologies for use in resonant and soft-switching applications. The soft-switching FOM is

270 | *GaN Power Devices for Efficient Power Conversion*

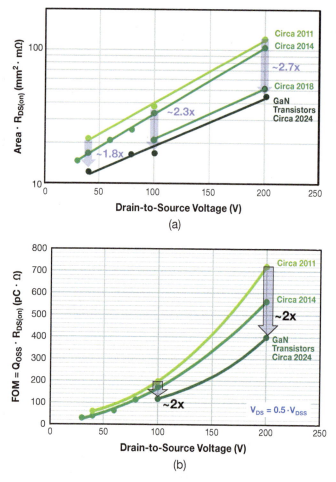

Figure 8.21 Historical plot of (a) size and (b) output charge (Q_{OSS}) performance figures of merit (FOM) for GaN transistors from 25 to 200 V from 2011 to 2024.

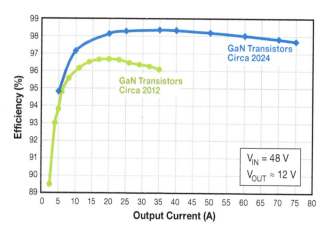

Figure 8.22 Experimental comparison between circa 2012 and circa 2024 GaN transistor-based resonant bus converters (V_{IN} = 48 V, V_{OUT} ≈ 12 V, f_{sw} ≈ 1–1.2 MHz).

made up of the two parameters most critical for resonant and soft-switching applications: output charge, Q_{OSS}, and gate charge, Q_G.

In Chapter 9 we will explore the performance of GaN transistors designed for power conversion at RF frequencies.

References

1 Lee, F.C. (1989). *High-Frequency Resonant, Quasi-Resonant, and Multi-Resonant Converters.* Blacksburg, VA: Virginia Power Electronics Center.

2 Lee, F.C. (1989). *High-Frequency Resonant and Soft-Switching PWM Converters.* Blacksburg, VA: Virginia Power Electronics Center.

3 Erickson, R.W. and Maksimovic´, D. (2001). *Fundamental of Power Electronics.* Norwell, MA: Kluwer.

4 Kazimierczuk, M.K. and Czarkowski, D. (2011). *Resonant Power Converters.* New Jersey: Wiley.

5 Yang, B. (2003). Topology investigation for front end DC/DC power conversion for distributed power system. Ph.D. dissertation. Virginia Tech, Blacksburg, VA.

6 Vorperian, V. (1984). Analysis of resonant converters. Ph.D. dissertation. California Institute of Technology, Pasadena, CA.

7 Severns, R.P. (1992). Topologies for three-element resonant converters. *IEEE Trans. Power Electron.* 7 (1): 89–98.

8 Fu, D. (2010). Topology investigation and system optimization of resonant converters. Ph.D. dissertation. Virginia Polytechnic Institute State University, Blacksburg, VA.

9 Liu, K. H. and Lee, F. C. (1984). Resonant switches – a unified approach to improved performances of switching converters. *IEEE International Telecommunications Energy Conference*, (November 1984), pp. 334–341.

10 Efficient Power Conversion Corporation (2022). EPC2204 – enhancement-mode GaN power transistor, 100 V, 125 A, EPC2022 datasheet, September 2020 [Revised March 2024]. https://epc-co.com/epc/Portals/0/epc/documents/datasheets/EPC2204_datasheet.pdf.

11 Infineon (2022). OptiMOS™ 5 power-transistor, 100 V, IQE065N10NM5SC datasheet, 2022 [Rev. 2.0]. https://www.infineon.com/dgdl/Infineon-IQE065N10NM5SC-DataSheet-v02_00-EN.pdf?fileId=8ac78c8c81ae03fc0181d247e8ad0a30.

12 Schlecht, M. (2007). High efficiency power converter. US Patent No. 7,269,034B2,11 September 2007.

13 White, R.V.(2003). Emerging on-board power architectures. *APEC '03, Eighteenth Annual IEEE* 2, (9–13 February 2003): 799–804.

14 Ren, Y., Xu, M., Sun, J., and Lee, F.C. (2005). A family of high power density unregulated bus converters. *IEEE Transactions, Power Electron.* 20 (5): 1045–1054.

15 Ren, Y. (2005), High frequency, high efficiency two-stage approach for future microprocessors. Ph.D. dissertation, Virginia Tech, Blacksburg, VA. April 2005.

16 Ren, Y., Lee, F.C. and Xu, M. (2008). Power converters having capacitor resonant with transformer leakage inductance. US Patent 7,196,914, 27 March 2008.

17 Vinciarelli, P. (2006), Point of load sine amplitude converters and methods. US Patent 7,145,786, 5 December 2006.

18 Reusch, D. (2012), High frequency, high power density integrated point of load and bus converters. Ph.D. Dissertation, Virginia Tech, Blacksburg, VA.

19 Infineon (2009). OptiMOS™3 Power-transistor, BSC057N08NS3 G datasheet, October 2009 [Rev. 2.4]. https://www.infineon.com/dgdl/Infineon-BSC057N08NS3G-DS-v02_04-en.pdf?fileId=db3a30431add1d95011ae803c9345616.

20 Efficient Power Conversion Corporation (2011). EPC2015C – Enhancement-mode power transistor, EPC2015C datasheet, March 2011 [Revised Jan. 2013]. http://epc-co.com/epc/Portals/0/epc/documents/datasheets/epc2015c_datasheet.pdf.

21 Infineon (2009). OptiMOS™3 Power-transistor, BSC027N04LS G datasheet, October 2009 [Rev. 104]. https://www.infineon.com/dgdl/Infineon-BSC027N04LSG-DS-v01_04-en.pdf?fileId=db3a30431689f4420116c4323646080c.

9

RF Performance

9.1 Introduction

The main focus to this point in the book has been the switching capabilities of GaN transistors. Now, the RF capabilities of these same GaN transistors and, in particular, enhancement-mode transistors will be examined, highlighting specific RF applications that can benefit from their adoption.

High electron mobility transistors (HEMT), using GaN as a semiconductor material, are available from Microchip, MACOM, NXP, Integra, and Qorvo, among others. These transistors are all depletion-mode, which is not as limiting for RF power amplifiers as it is for switching converters. This is because the potential of device failure upon power up, due to the short condition, can be mitigated easily in the RF design, which is not the case for a switching converter. Depletion-mode transistors require additional circuitry for gate circuit biasing, due to the negative gate voltage requirements to regulate the drain current, which is also viewed as a disadvantage in cases where enhancement-mode devices are an option.

The main alternative today for GaN RF transistors operating in the range of 500 MHz–3 GHz is the laterally diffused metal oxide semiconductor (LDMOS) FET made using silicon. Compared to LDMOS transistors, GaN is well suited for RF transistors for many of the same reasons as for switching applications [1–3]. GaN transistors exhibit superior RF performance over LDMOS in general, most notably with respect to power density [4], frequency range (bandwidth) [1, 5], and noise figure. This leads to improvements in RF power capability [2, 6] with transistors being specified over a very wide frequency range [7]. In addition, the lower input and output capacitances lead to higher impedances that allow for higher drain efficiencies and reduced impedance transformation ratios required for matching. Both these factors lead to improvements in amplifier efficiency, size reduction, and ultimately cost.

When used in pulsed-RF applications, the bias circuit will also need to be operational before main power and RF are applied, to prevent starting with very high currents that can potentially damage the circuit. Therefore, RF circuits can benefit from using enhancement-mode transistors for many of the same reasons as in switching converters.

The measurement and performance metrics used for RF transistors differ significantly from switching devices. These metrics will be presented, along with their relevance, and guidance on how to measure and use them (Table 9.1).

GaN Power Devices for Efficient Power Conversion, Fourth Edition. Alex Lidow, Michael de Rooij, John Glaser, Alejandro Pozo, Shengke Zhang, Marco Palma, David Reusch, and Johan Strydom.
© 2025 John Wiley & Sons Ltd. Published 2025 by John Wiley & Sons Ltd.

GaN Power Devices for Efficient Power Conversion

Table 9.1 Definitions of terms used in RF analysis and design.

V_{GSQ}	RF circuit gate voltage quiescent bias point
I_{DQ}	Drain current in the transistor at the quiescent operating point
P_{DQ}	Power losses at the quiescent operating point
P_{DC}	DC power delivered to the RF transistor
P_{RFout}	Output RF power
η_D	Drain efficiency – the ratio of P_{RFou}/P_{DC}
s_{11}	Input port reflection coefficient: the percentage of the input incident wave that is reflected back from the input port
s_{12}	Reverse gain: the percentage of the output port incident wave that is reflected to the input port
s_{21}	Forward gain: the percentage of the input port incident wave that is reflected to the output port
s_{22}	Output port reflection coefficient: the percentage of the output incident wave that is reflected back from the output port
K	Rollett stability factor
C_S	The source-side stability circle center on a Smith chart
C_L	The load-side stability circle center on a Smith chart
C_A	Constant available gain circle center on a Smith chart
R_S	The source-side stability circle radius on a Smith chart
R_L	The load-side stability circle radius on a Smith chart
R_A	The radius of the constant available gain circle on a Smith chart
Γ_{in}	Input reflection coefficient of the transistor
Γ_{out}	Output reflection coefficient of the transistor
Γ_S	Input-side matching reflection coefficient
Γ_L	Output-side matching reflection coefficient
G_T	Transducer power gain
G_{tu}	Unilateral transducer power gain
U	Unilateral figure of merit
g_u	Normalized unilateral transducer gain
G_A	Available gain
G_{msg}	Maximum stable gain of the transistor
X	Matching network series reactance
B	Matching network shunt susceptance

9.2 Differences Between RF and Switching Transistors

The main difference between switching transistors and RF transistors is that RF transistors are designed to work best in the linear region of the transfer characteristic to maximize RF power gain and minimize RF signal distortion. Switching devices are designed to work best in the on-state and transition states to minimize losses [8–14]. For performance comparisons RF devices have different evaluation metrics [1] from switching devices, among which are power gain, linearity or 1 dB compression [8], and drain efficiency. The power gain determines how much the transistor amplifies the input signal with power. The 1 dB compression point determines the maximum output power that the transistor can deliver without distorting the signal. The drain efficiency determines how efficient the transistor is at amplifying.

Typically, RF devices are biased to an operating point onto which the RF signal is superimposed. To bias the transistor, a drain-to-source current is established such that the voltage remains at the supply voltage. Figure 9.1 illustrates this concept, using the transistor transfer characteristic. The bias voltage and current are named quiescent points and denoted as V_{GSQ} and I_{DQ}, respectively, and typically are provided in datasheets for baseline performance reporting.

The bias point will be associated with losses ($P_{DQ} = I_{DQ} \cdot V_{Supply}$), and as such, RF devices have a higher operating power (P_{DQ}) loss ratio with respect to power delivered (P_{RFout}), compared to switching devices. The ratio P_{RFout}/P_{DC} is referred to as drain efficiency (η_D) where P_{RFout} is the output RF power and P_{DC} is the DC power delivered to the transistor. When employing an RF FET as a Class A amplifier, this power loss ratio reaches a theoretical maximum of 50%, whereas a switching converter can have efficiencies as high as 98%. This means that the thermal dissipation of an RF device is significantly higher than equivalently sized switching devices and necessitates the ability to efficiently dissipate heat to the environment. Figure 9.2 shows the difference between a packaged RF FET (left) and a chip-scale switching FET (right), both with similar drain voltage and current ratings.

Another key difference between RF devices and switching devices is in how they are characterized. RF transistors are characterized as part of a transmission line in terms of incident

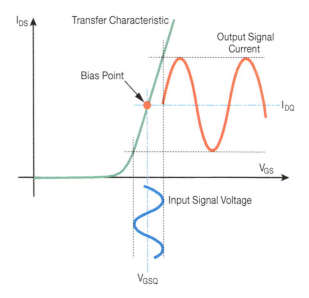

Figure 9.1 Transfer characteristic for an enhancement-mode transistor showing the bias point and input voltage signal with corresponding output current.

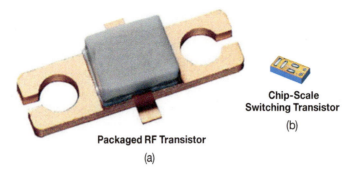

Figure 9.2 (a) An RF-packaged FET. *Source:* Ref. [15]/MACOM. (b) The equivalent chip-scale switching FET [16], with similar voltage and current rating. *Source:* Ref. [16]/Efficient Power Conversion Corporation.

and reflected waves, whereas switching devices are characterized in terms of energy commutation. This opens up many new terms and definitions unfamiliar to many power circuit designers. The common metrics used to characterize RF devices are the *s*-parameters, which represent a measure of incident, reflection, and transmission of electromagnetic waves. S-parameters will also be used in this chapter to characterize the enhancement-mode GaN transistor originally designed for switching converter applications.

9.3 RF Basics

Before delving into RF transistor measurement and analysis, some basics need to be reviewed. In this chapter, all discussions will be limited to two-port networks, as this is sufficient to describe a transistor. Figure 9.3 shows the basic diagram for a two-port network showing the incident waves, denoted a_1 and a_2, and reflected waves denoted b_1 and b_2. Any incident wave at a port can be reflected to either of the ports. For example, an incident wave on port 1 (a_1) can reflect from port 1 (b_1) and/or port 2 (b_2).

An *s*-parameter is defined as the ratio between reflected (b_1 and b_2) and incident (a_1 and a_2) waves and is given by Eq. (9.1).

$$s_{nm} = \frac{b_n}{a_m} \qquad (9.1)$$

and s_{nm} is a complex number with the general form:

$$s_{nm} = \mathfrak{R}(s_{nm}) + i \cdot \mathfrak{J}(s_{nm}) \qquad (9.2)$$

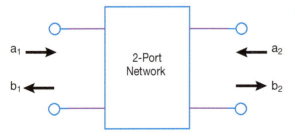

Figure 9.3 Two-port network with (a) incident waves and (b) reflected waves.

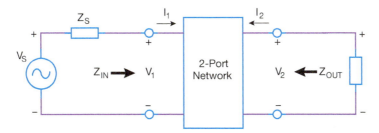

Figure 9.4 A two-port network with source and load.

In this chapter, port 1 will be designated as the input port, or gate, and port 2 as the output port, or drain.

From the *s*-parameters, it is possible to derive useful characteristics of the two-port network such as impedance, gain, and isolation. (The methodology is described in detail in [9].)

Figure 9.4 shows the two-port network connected to a load and source, and gives the input and output impedance for those ports.

The Smith chart is a useful tool [8] that simplifies the conversion of *s*-parameters into impedances and will be extensively used in this chapter.

9.4 RF Transistor Metrics

The main metric for RF transistor performance evaluation is RF power gain. Maximum power gain is defined by the limit of linear performance for a transistor. Figure 9.5 shows data from the MACOM NPT1012 datasheet [7] showing the key RF metrics for the depletion-mode GaN-on-silicon HEMT under certain operating conditions.

RF power gain is a measure of how much power is increased or decreased when incident on a port. Mathematically it is expressed as:

$$G = \frac{P_{out}}{P_{in}} \qquad (9.3)$$

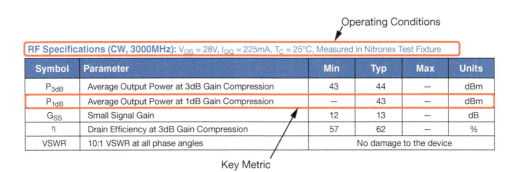

Figure 9.5 Data from MACOM NPT1012 datasheet [7] showing the Key RF metrics for the transistor. *Source:* Adapted from MACOM [7].

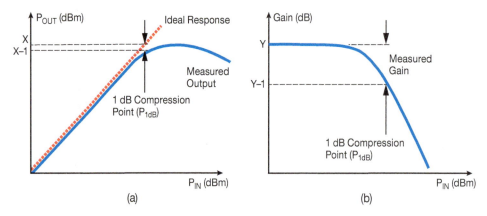

Figure 9.6 Graphical representation of the definition of linearity based on (a) the 1 dB compression point and (b) linearity on the gain graph.

Gain can also be expressed logarithmically, with units of decibel (dB) as follows:

$$G(\text{dB}) = 10 \cdot \log\left(\frac{P_{\text{out}}}{P_{\text{in}}}\right) \quad (9.4)$$

Using the definition of gain, linearity can be defined for the amplifier as having a fixed gain value and is characterized by the linear relationship between the input power and the output power. At its limit, there is a loss in gain due to amplification saturation. Linearity is also known as linear dynamic range with its limit being the 1 dB compression point [8]. For an RF transistor, the gain will be a constant as a function of input power until it exceeds a specific value. The 1 dB compression point is defined as the point when the measured amplifier output power (dB) deviates by 1 dB from the ideal predicted power, and is shown in Figure 9.6. Figure 9.6a is the output power as a function of the input power, and Figure 9.6b shows the same result for gain as a function of input power. Above a specific input power level, the gain will begin to decrease and the output power will no longer be a linear function of the input power.

Typically, the linearity of a transistor is provided in a datasheet at a specific frequency and bias condition. Power gain is given over a frequency range, based on a specific bias setting. Small-signal s-parameters can be used to design a Class A RF power amplifier, but large-signal s-parameters are needed to predict power performance.

9.4.1 Determining the High-Frequency Characteristics of RF FETs

To determine the RF characteristics of a transistor requires a few steps. The procedure starts with s-parameter measurements of the device itself. The s-parameters are used to test for stability and to identify if the device is unilateral (negligible reverse gain) or bilateral (reverse gain is high enough to affect stability). Once the stability criteria have been determined, an amplifier can be designed, typically a Class A or Class AB.

Small-signal s-parameters are measured using a vector network analyzer (VNA) with the transistor under specific bias conditions. Several measurements may be required to determine the bias conditions that yield the highest performance metrics.

To measure the RF characteristics requires that the device be mounted to a test fixture. The test fixture will need to be calibrated using a set of standards to obtain the device-only s-parameters. The thru-reflect-line (TRL) [17] and short-open-load-thru (SOLT) methods are the

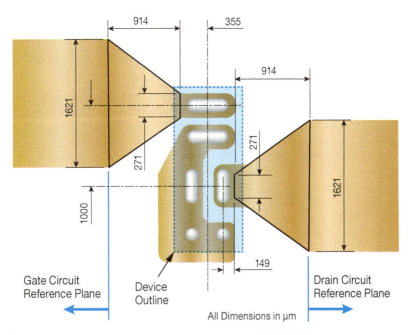

Figure 9.7 Reference plane design for RF connection to the EPC8009 [16] FET. *Source:* Adapted from Efficient Power Conversion Corporation [16].

most popular methods of calibration. The calibration process and accuracy between the two methods is well documented in [18, 19].

Figure 9.7 shows a reference plane design suitable for the EPC8009 [16] enhancement-mode switching FET that can be used to test its RF characteristics. Due to the small size of this device and its connections, a reducing taper microstrip is required to make the connection. The taper's interface to the 50 Ω microstrip transmission line in this example is based on a 30-mil thick, one-once copper, Rogers 4350 substrate [20].

9.4.2 Pulse Testing for Thermal Considerations

It has already been mentioned that RF devices have a high ratio of power dissipation to power output, and therefore require substantial cooling. In the case where power dissipation exceeds the capability of the device, pulsed-mode testing can be employed.

Most RF amplifiers are operated with a continuous RF signal and bias. This is known as continuous wave (CW) operation. Pulse testing is used to lower the average power dissipation in the device. The bias pulse and pulsed RF signal used in pulse testing are shown in Figure 9.8.

It is important to maintain the drain bias current stable during the on-state of the pulse as changes can cause inaccurate measurements. This is difficult to achieve, particularly for Class A amplifiers, as the drain current will increase in response to an increase in RF power without any means to differentiate between what is the bias component and what is the RF component. In addition, rapid changes in gate voltage can lead to unwanted ringing of the drain and can cause bias instability and oscillations. The choice of bias tees for pulse testing is also critical because the frequency response of the bias tees must satisfy both RF and bias requirements.

A dedicated pulse controller similar to the schematic block diagram shown in Figure 9.9 can be used for RF testing FETs. Using an isolation amplifier, the drain current is measured and

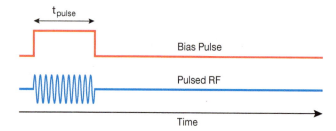

Figure 9.8 Bias pulse and pulsed RF signal for pulsed RF testing.

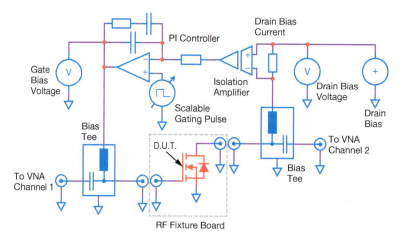

Figure 9.9 Schematic block diagram for transistor bias control under pulsed RF testing.

used with a proportional-integral (PI) controller to regulate the gate voltage to maintain the drain current. The PI controller is also controlled by the gate pulse to avoid rapid gate transitions.

For higher power RF measurements, the pulse controller can be adapted such that gate voltage can be fed to a second device while the drain bias current for the second device is not measured by the controller [21]. The first device will not be exposed to RF thereby providing a stable reference for the device under RF testing. This setup is shown in Figure 9.10.

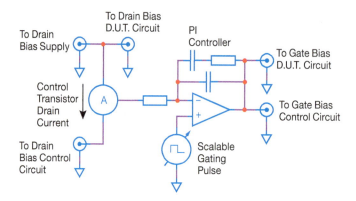

Figure 9.10 Power RF testing using the pulse controller having a reference device and a device under test.

9.4.3 Analyzing the s-Parameters

With the ability to measure the s-parameters, the data now needs to be analyzed so that an amplifier can be designed. The procedure involves checking for stability issues and determining the input and output reflection coefficients for the device.

9.4.3.1 Test for Stability

It is important to determine whether the device is conditionally or unconditionally stable. An unconditionally stable device will remain stable (will not oscillate) regardless of the impedance presented to its gate or drain. The test for unconditional stability is given by the Rollett stability factor (K) [22]:

$$K = \frac{1 - |s_{11}|^2 - |s_{22}|^2 + |\Delta|^2}{2 \cdot |s_{12} \cdot s_{21}|} \geq 1 \tag{9.5}$$

and

$$|\Delta| \leq 1 \tag{9.6}$$

where

$$\Delta = s_{11} \cdot s_{22} - s_{12} \cdot s_{21} \tag{9.7}$$

If this stability criterion for either K or $|\Delta|$ cannot be satisfied, then the transistor will be defined as conditionally stable. This means that any design using the transistor must avoid the unstable region. Stability circle plots on a Smith chart [8] are used to determine where the unstable regions are located. The stability circles are given by the following equations:

$$C_s = \left(\frac{s_{11} - \Delta \cdot s_{22}^*}{|s_{11}|^2 - |\Delta|^2} \right)^* \tag{9.8}$$

$$R_S = \frac{|s_{12} \cdot s_{21}|}{|s_{11}|^2 - |\Delta|^2} \tag{9.9}$$

$$C_L = \frac{(s_{22} - \Delta \cdot s_{11}^*)^*}{|s_{22}|^2 - |\Delta|^2} \tag{9.10}$$

$$R_L = \frac{|s_{12} \cdot s_{21}|}{|s_{22}|^2 - |\Delta|^2} \tag{9.11}$$

The superscript asterisk ($*$) denotes the complex conjugate of the parameter, also known as the reflection of the parameter. Using Eq. (9.2) as an example, the complex conjugate for s_{nm} is given as:

$$s_{nm}^* = \Re(s_{nm}) - i \cdot \Im(s_{nm}) \tag{9.12}$$

The unstable region falls inside the stability circles, as shown for example in Figure 9.11 (for a tutorial on the use of Smith Charts see reference [8]).

9.4.3.2 Transistor Input and Output Reflection

The RF transistor will ultimately be placed in an amplifier circuit with input and output matching networks that transform the standard source impedance (Z_0) and standard load

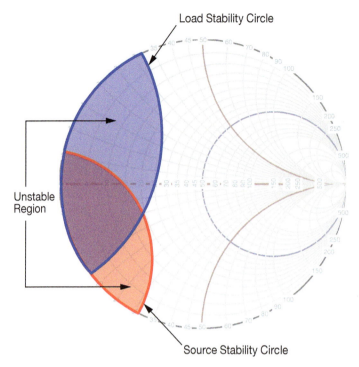

Figure 9.11 Stability circle plot showing unstable regions for both source and load ports.

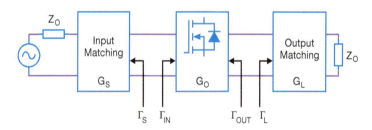

Figure 9.12 Basic amplifier structure with input and output matching with reflection coefficients for the FET and matching networks.

impedance (Z_0) to desired values (Γ_S and Γ_L), as shown in Figure 9.12. Matching networks will be discussed in detail later in this chapter.

Although they share the same nomenclature, the transistor input and output reflection coefficients are not simply given by s_{11} and s_{22} (reflection coefficients) as one might expect, but are given by Γ_{in} and Γ_{out} for the input (gate circuit) and output (drain circuit), respectively. This is due to the effect of the transmission coefficients s_{12} and s_{21} cross influencing the input and output by way of the load and source impedances. This can be seen in the equations for the transistor input and output reflection [8], where the load network impacts the input reflection, and the source network impacts the output reflection:

$$\Gamma_{in} = s_{11} + \frac{s_{12} \cdot s_{21} \cdot \Gamma_L}{1 - s_{22} \cdot \Gamma_L} \qquad (9.13)$$

$$\Gamma_{\text{out}} = s_{22} + \frac{s_{12} \cdot s_{21} \cdot \Gamma_{\text{S}}}{1 - s_{11} \cdot \Gamma_{\text{S}}} \tag{9.14}$$

9.4.3.3 Transducer Gain

Transducer power gain (G_T) is defined as the ratio of power delivered to the load to the power available from the source. From Figure 9.12, it can be seen that there are multiple gain components which are gain of the source side matching G_S, gain of the transistor G_0, and gain of the load side matching G_L. Together, these components make up the transducer gain (G_T), which can be determined using the following equations:

$$G_T = G_S \cdot G_0 \cdot G_L \tag{9.15}$$

and written using s-parameters:

$$G_S = \frac{1 - |\Gamma_S|^2}{|1 - \Gamma_{\text{in}} \cdot \Gamma_S|^2} \tag{9.16}$$

$$G_0 = |s_{21}|^2 \tag{9.17}$$

$$G_L = \frac{1 - |\Gamma_L|^2}{|1 - \Gamma_{\text{out}} \cdot \Gamma_L|^2} \tag{9.18}$$

9.4.3.4 Unilateral/Bilateral Transistor Test

A unilateral transistor is defined as one where s_{12} is very small relative to s_{21} [8, 23]. Since a value of zero for s_{12} is not physically possible, we can perform a test to determine if a transistor can be regarded as unilateral or bilateral. Using Eqs. (9.16) and (9.18) and substituting Γ_{in} and Γ_{out} with s_{11} and s_{22}, respectively, a unilateral transducer gain (G_{TU}) for Eq. (9.15) can be derived. The ratio of the transducer gain to the unilateral transducer gain can be used to determine the normalized unilateral transducer gain as follows:

$$g_u = \frac{G_T}{G_{TU}} \tag{9.19}$$

which is bounded as follows:

$$\frac{1}{(1 + U)^2} < g_u < \frac{1}{(1 - U)^2} \tag{9.20}$$

where U is the unilateral figure of merit and given by the following equation:

$$U = \frac{|s_{11}| \cdot |s_{12}| \cdot |s_{21}| \cdot |s_{22}|}{\left(1 - |s_{11}|^2\right) \cdot \left(1 - |s_{22}|^2\right)} \tag{9.21}$$

A transistor can be regarded as unilateral if g_u is within 10% of unity; otherwise, it is considered bilateral. Also, a unilateral transistor will always be stable by definition, as the device effectively does not have a feedback mechanism. A unilateral amplifier design is simplified greatly as Γ_{in} and Γ_{out} (Eqs. (9.13) and (9.14)) can be reduced to s_{11} and s_{22}, respectively. The unilateral solution will not be covered in this book, as the procedure to design an amplifier using a bilateral transistor also can be adapted for a unilateral transistor.

9.5 Amplifier Design Using Small-Signal *s*-Parameters

The essence of an amplifier design is to determine the input (Γ_S) and output (Γ_L) reflection coefficients for the matching networks of the transistor. The basis for the design can be maximum gain, or a specific gain that is within the capability of the transistor.

The example here assumes a bilateral transistor, and the procedure followed will be based on a specific gain requirement. The unique solution for an unconditionally stable transistor that yields the maximum transducer gain (G_{Tmax}) can also be determined using this procedure. The specific maximum transducer gain needs to be determined to know its gain limit. For an unconditionally stable transistor, the maximum transducer gain is given by:

$$G_{Tmax} = \frac{|s_{21}|}{|s_{12}|} \cdot \left(K - \sqrt{K^2 - 1} \right) \tag{9.22}$$

The unconditionally stable transistor maximum transducer gain amplifier solution is defined as a conjugately matched amplifier, where the matching networks are designed to have zero reflection. It will be the complex conjugate of the transistor ports and can be written in the form:

$$\Gamma_{in} = \Gamma_S^* \tag{9.23}$$

$$\Gamma_{out} = \Gamma_L^* \tag{9.24}$$

9.5.1 Conditionally Stable Bilateral Transistor Amplifier Design

The design of an amplifier using a conditionally stable bilateral transistor can involve plotting many gain circles to find an acceptable solution. In this section, we will present a simpler method based on conjugately matching one port of the transistor and mismatching the other port. The solution can then be adjusted to find a solution where both input (Γ_S) and output (Γ_L) reflection coefficients fall within the stable operating regions. The design procedure will make use of constant gain circles where an arbitrary gain value and reflection coefficient for that port are chosen, and the equations solved to find the other port's reflection coefficient. Amplifier design using feedback networks will not be covered.

9.5.1.1 Available Gain
The amplifier will be designed using the available gain (G_A) approach [11], so that the output network will be conjugately matched with the transistor and the input network mismatched with the transistor. The approach will reduce the amount of reflected power being transmitted back to the input via s_{12} due to any output mismatch. A mismatch on the input results in a significantly lower reflection magnitude than it would if the output were mismatched. Available gain (G_A) is defined as the ratio of the power available from the amplifier to the power available from the source. Substituting Eq. (9.24) into Eq. (9.18), Eq. (9.13) into Eq. (9.16), and solving for Eq. (9.15) under these conditions yields:

$$G_A = \frac{1 - |\Gamma_S|^2}{|1 - s_{11} \cdot \Gamma_S|^2} \cdot |s_{21}|^2 \cdot \frac{1}{1 - |\Gamma_{out}|^2} \tag{9.25}$$

where the unknown is Γ_S, which will be chosen based on a specific gain requirement for the amplifier. The output will be conjugately matched using the constant available gain circle, G_A. The normalized available power gain (g_A) is given by:

$$g_A = \frac{G_A}{|s_{21}|^2} \tag{9.26}$$

9.5.1.2 Constant Available Gain Circles

Using the available gain, constant available gain circles can be derived [11] and are summarized as follows:

$$C_A = \frac{g_A \cdot (s_{11} - \Delta \cdot s_{22}^{*})^{*}}{1 + g_A \cdot (|s_{11}|^2 - |\Delta|^2)} \tag{9.27}$$

$$R_A = \frac{\sqrt{1 - 2 \cdot K \cdot |s_{12} \cdot s_{21}| \cdot g_A + |s_{12} \cdot s_{21}|^2 \cdot g_A^2}}{1 + g_A \cdot (|s_{11}|^2 - |\Delta|^2)} \tag{9.28}$$

A specific available gain is chosen and the gain circle is plotted. The gain circle gives all possible values for the input reflection coefficient (Γ_S) that will yield this gain. For each value of (Γ_S), a value for (Γ_L) can be determined, and a circle of (Γ_L) can be calculated using Eqs. (9.14) and (9.24). A range of values where both (Γ_S) and (Γ_L) fall outside their respective stability circles, and that also lie inside the unity circle of the Smith chart, will yield a stable amplifier design. The best choice would be based on values that lie furthest from the stability circles.

9.6 Amplifier Design Example

Next, an amplifier will be designed using an enhancement-mode GaN transistor [16], and the available gain approach. In this example, the s-parameters have been measured using the reference planes shown in Figure 9.7 and will operate at 500 MHz.

At 500 MHz, $V_{DsQ} = 30$ V, and $I_{DQ} = 500$ mA, the enhancement-mode GaN transistor (EPC8009 [16]) has the following s-parameter values:

$$s_{11} = -0.926 \quad -i \cdot 0.157, \quad |s_{11}| = 0.939$$

$$s_{22} = -0.658 \quad -i \cdot 0.46, \quad |s_{22}| = 0.803$$

$$s_{12} = -0.002 \quad +i \cdot 0.013, \quad |s_{12}| = 0.013$$

$$s_{21} = 5.280 \quad +i \cdot 4.042, \quad |s_{21}| = 6.65$$

From the s-parameters, it can be determined whether the transistor is unilateral or bilateral using Eqs. (9.20) and (9.21).

$$U = \frac{|s_{11}| \cdot |s_{12}| \cdot |s_{21}| \cdot |s_{22}|}{\left(1 - |s_{11}|^2\right) \cdot \left(1 - |s_{22}|^2\right)}$$

$$= \frac{0.926 \cdot 0.013 \cdot 6.65 \cdot 0.803}{\left(1 - 0.926^2\right) \cdot \left(1 - 0.803^2\right)} \tag{9.29}$$

$$= 1.534$$

and

$$\frac{1}{(1+U)^2} < g_u < \frac{1}{(1-U)^2}$$

$$\frac{1}{(1+1.534)^2} < g_u < \frac{1}{(1-1.534)^2} \qquad (9.30)$$

$$0.156 < g_u < 3.5$$

From the result, it is clear that neither unilateral FOM boundary is within 10% of unity, and therefore, the transistor is considered bilateral.

Next, we need to decide whether the transistor is conditionally or unconditionally stable using Eqs. (9.5)–(9.7):

$$\Delta = s_{11} \cdot s_{22} - s_{12} \cdot s_{21}$$

$$= (-0.926 - i \cdot 0.157) \cdot (-0.658 - i \cdot 0.46) - (-0.002 + i \cdot 0.013) \cdot (5.280 + i \cdot 4.042)$$

$$= -0.6 + i \cdot 0.472$$

$$(9.31)$$

and

$$|\Delta| = 0.763 \qquad (9.32)$$

with the Rollett stability factor

$$K = \frac{1 - |s_{11}|^2 - |s_{22}|^2 + |\Delta|^2}{2 \cdot |s_{12} \cdot s_{21}|}$$

$$= \frac{1 - 0.939^2 - 0.803^2 + 0.763^2}{2 \cdot |(-0.002 + i \cdot 0.013) \cdot (5.280 + i \cdot 4.042)|} \qquad (9.33)$$

$$= 0.326$$

The unconditional stability of $K \geq 1$ and $|\Delta| \leq 1$ has not been met for this transistor due to K. Therefore, the device is conditionally stable and the stability circles can be plotted to determine the unstable regions.

The stability circles can be calculated using Eqs. (9.8)–(9.11) and are shown in Figure 9.13 with the unstable regions shaded. The amplifier design for (Γ_S) and (Γ_L) needs to avoid these regions. As the transistor is conditionally stable, we need to determine a suitable gain that will yield an amplifier that will not oscillate (is always stable). Before we can select a working gain, we need to determine the maximum stable gain of the transistor using the following equation:

$$G_{MSG} = \frac{|s_{21}|}{|s_{12}|}$$

$$= \frac{6.65}{0.013} \qquad (9.34)$$

$$= 520.3 = 27.2 \ \text{dB}$$

The maximum stable gain is a simple method to determine the stable gain limit of the transistor; however, it may not yield a workable solution. For this example, a design gain of $200 = 23$ dB is selected.

Next, using the gain value selected, the available gain circle is plotted as shown in Figure 9.13, which reveals all values of Γ_S that will yield a gain of 23 dB. Next, a specific value for (Γ_S) can be

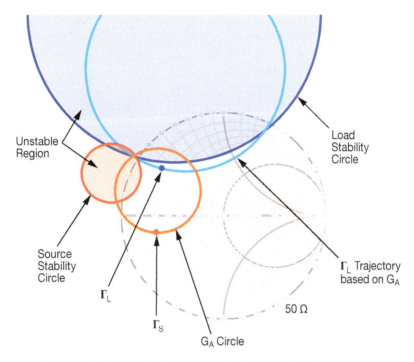

Figure 9.13 Stability circles for EPC8009 [16] at 500 MHz with V_{DSQ} = 30 V and I_{DQ} = 500 mA together with the available gain circle of 23 dB and load reflection coefficient Γ_L trajectory. *Source:* Adapted from Ref. [16].

selected and, using Eqs. (9.14) and (9.24), a value for Γ_L can be determined. Both Γ_S and Γ_L must lie outside the unstable regions of the stability circles. The trajectory of Γ_L, based on the available gain circle, has been plotted in Figure 9.13 to make it easier to observe if a solution exists. If there are points where both the G_A circle and the Γ_L trajectory lie outside the stability circles and also lie inside the unity Smith chart, then a workable solution exists.

The reflection coefficients at 500 MHz are:

$$\Gamma_S = -0.604 - i \cdot 0.167$$
$$\Gamma_L = -0.557 + i \cdot 0.458$$

Using these reflection coefficients, the amplifier matching networks can be designed. These reflection coefficients are usually provided in the datasheets of RF components over a range of frequencies.

9.6.1 Matching and Bias Tee Network Design

Figure 9.14 shows a block diagram for the RF amplifier design. To accommodate a small heatsink for the transistor, a 50 Ω transmission line 12.25 mm long for the gate circuit and 14 mm long for the drain circuit needs to be added to bring out the terminals before connection to the bias tees and matching networks. The impact of both the transmission lines and bias tees on the reflection coefficients will need to be calculated for adjustment prior to calculating the matching network.

The 50 Ω transmission line rotates the selected impedance (reflection coefficient) around the center of the Smith chart to a new location, since only the phase component changes and not the characteristic impedance. Given that this is the transmission line, the direction of rotation

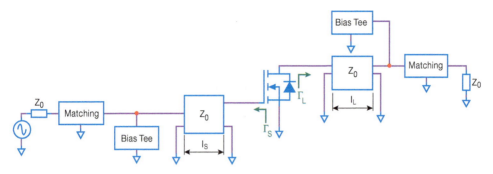

Figure 9.14 Block diagram of the complete amplifier design that includes transmission lines, bias tees, and matching networks.

is counter-clockwise. The angle of rotation will depend on the length of the transmission line, transmission line design, and the frequency of operation. The derivation for the electrical angle based on a microstrip is given in [24]. In this example, the electrical phase is 12.46° for the gate transmission line and 14.24° for the drain transmission line. The rotation on the Smith chart is twice the electrical angle.

The bias tee circuits are used to provide the quiescent supply to the transistor. It must be designed so as to not affect the RF characteristics of the circuit, yet it needs to adequately provide the required bias conditions. The bias tee circuit for the amplifier consists of a second-order passive filter (DC pass–AC block) as shown in Figure 9.15. The impact on the RF circuit can be determined as the series combination of the two passive elements, shunting the RF signal at the point of connection. Since the amplifier will be pulsed, additional design considerations need to be made to ensure stable pulse operation of the transistor with some useful design tips given in [25].

The selected bias tee components have the following electrical properties at 500 MHz:

$L_{Bin} = 48.4 \text{ nH, ESR} = 16 \, \Omega$

$C_{Bin} = 100 \text{ pF}$

$L_{Bout} = 240.8 \text{ nH, ESR} = 126 \, \Omega$

$C_{Bout} = 10 \text{ nF}$

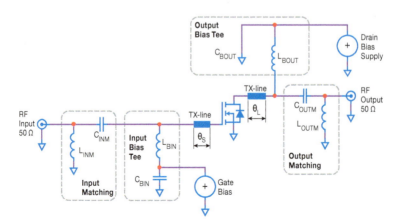

Figure 9.15 Amplifier schematic showing transmission lines, bias tees, and matching networks.

Using the bias tee network values, the calculated input and output reflection coefficients, and the effect of the transmission lines, the new reflection coefficients (Γ_S) and (Γ_L) can be determined. In this example, the results are:

$$\Gamma_S = -0.543 - i \cdot 0.414 = 10.45 \, \Omega, 19.65 \text{ pF (at 500 MHz)}$$
$$\Gamma_L = -0.707 + i \cdot 0.13 = 8.26 \, \Omega, 1.41 \text{ nH (at 500 MHz)}$$

The original reflection coefficients, impact of the transmission lines, and bias tees are shown in Figure 9.16, with the new reflection coefficients that are used to design the matching networks.

The design of the matching network is to convert the reflection coefficients to that of the load and the source impedances. In this design example, the source and load impedances are 50 Ω. For the case where the transistor resistance is less than the source impedance (Z_0), the matching network will take the form shown in Figure 9.17.

The matching network design has two solutions: (i) where X is capacitive and B is inductive, (ii) where X is inductive and B is capacitive. The preferred solution is (i), as it acts as a high-frequency pass with low-frequency filtering, and is desirable because the transistor has a higher gain in the lower frequencies and any unwanted signal can easily corrupt the amplifier. The

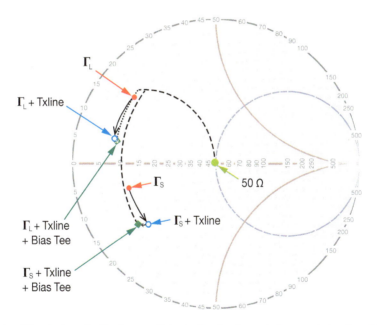

Figure 9.16 Amplifier design of reflection coefficients showing impact of transmission lines, bias tees, and matching network trajectories for each port are also shown.

Figure 9.17 Matching network configuration for the amplifier.

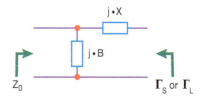

GaN Power Devices for Efficient Power Conversion

solution for the matching network design for the schematic of Figure 9.17 is given in [8] and repeated in Eqs. (9.35) and (9.36):

$$B = \pm \frac{\sqrt{\dfrac{Z_0 - R_L}{R_L}}}{Z_0} \tag{9.35}$$

$$X = \pm\sqrt{R_L \cdot (Z_0 - R_L)} - X_L \tag{9.36}$$

Since a DC block is required for both the RF input and output, and this function can be integrated into the matching network, the case with the negative values for X and B in Eqs. (9.35) and (9.36) can be calculated for the matching networks as follows.

For the gate circuit the matching network:

$$B_{in} = -\frac{\sqrt{\dfrac{50 - 10.45}{10.45}}}{50} \tag{9.37}$$
$$= -0.039 \cdot S = 8.18 \, nH = L_{inM}$$

$$X_{in} = -\sqrt{10.45 \cdot (50 - 10.45)} - 16.2 \tag{9.38}$$
$$= -36.53 \, \Omega = 8.71 \, pF = C_{inM}$$

and for the drain circuit the matching network:

$$B_{out} = -\frac{\sqrt{\dfrac{50 - 8.26}{8.26}}}{50} \tag{9.39}$$
$$= -0.045 \cdot S = 7.08 \, nH = L_{outM}$$

$$X_{out} = -\sqrt{8.26 \cdot (50 - 8.26)} + 4.42 \tag{9.40}$$
$$= -14.14 \, \Omega = 22.51 \, pF = C_{outM}$$

The matching network solution trajectories for these solutions have been plotted in Figure 9.16.

9.6.2 Experimental Verification

Based on the design example, a 500 MHz RF amplifier was designed and tested using the EPC8009 [16] eGaN FET. The drain bias current was set to 250 and 500 mA, and the drain bias voltage at 30 V. The amplifier was tested in pulse mode with a pulse duration of 240 μs and a repetition rate of 10 Hz. The amplifier was tested using a VNA with an additional RF amplifier to boost the input RF power to the amplifier, and loaded with a 20 W-capable, 30 dB RF attenuator. An RF power-in to power-out sweep was performed. Figure 9.18 shows the 1 dB compression point for this amplifier at both bias current settings.

Figure 9.19 shows the gain and drain efficiency for the amplifier as a function of output power.

It can be seen from the experimental results that the EPC8009 with 500 mA-drain bias current has a 1 dB compression point at 40.6 dBm (11.6 W) output power, where the power gain is 20.6 dB with drain efficiency of 57.4%. At a drain bias current of 250 mA, the device has a 1 dB

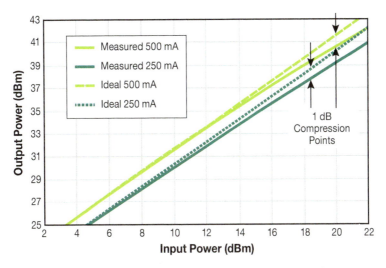

Figure 9.18 Measured 1 dB compression points for the EPC8009-based RF amplifier with 30 V drain bias voltage and 250 and 500 mA drain bias currents while operating at 500 MHz.

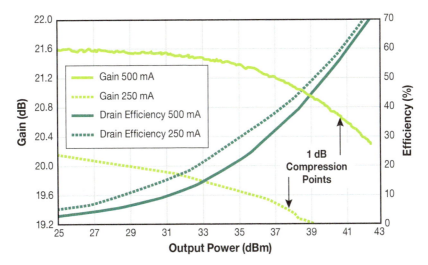

Figure 9.19 Measured gain and drain efficiency for the EPC8009-based RF amplifier with 30 V drain bias voltage and 250 and 500 mA drain bias currents while operating at 500 MHz.

compression at 38.4 dBm (6.96 W) output power, where the power gain is 19.3 dB with drain efficiency of 45.9%.

Having determined the RF performance of the GaN transistor, it may be compared to state-of-the-art LDMOS FETs with similar characteristics. Since the GaN FET was designed as a switching device, and not an RF device, the comparison focuses on the differences pertinent to RF designs. The metrics compared are power gain, linearity (1 dB compression,) and drain efficiency at the same frequency of operation. The LDMOS devices selected for comparison are [26, 27] based on comparable power capabilities at 500 MHz. The comparison data between the GaN transistor and LDMOS FETs are given in Table 9.2.

From Table 9.2, it can be seen that the GaN FET has a higher gain than the LDMOS FET while operating at a higher voltage with comparable drain efficiency despite the higher bias

Table 9.2 Comparison between GaN transistors and LDMOS FET at 500 MHz.

Parameter	GaN transistor [16] (500 mA)	GaN transistor [16] (250 mA)	LDMOS FET [26]	LDMOS FET [27]
Output power (W)	11.6	6.96	15	9
1 dB gain (dB)	20.6	19.3	14	13
Drain efficiency (%)	57.4	45.9	55	60
Rated voltage (V)	65	65	40	40
Bias voltage (V)	30	30	12.5	12.5
Bias current (mA)	500	250	150	150

power. The capacitances of the GaN transistor are also significantly lower than the LDMOS FETs, ensuring lower matching impedance conversion.

9.7 Summary

In this chapter, the key metrics and a methodology for an RF amplifier design were presented and compared against an actual amplifier. The design was based on the EPC8009 eGaN FET, which was not originally designed to be an RF device. Despite this, the results show excellent RF characteristics with stable gain in excess of 20 dB, and a drain efficiency approaching 60% at the 1 dB compression point. The GaN transistor was compared against two LDMOS FETs with similar RF characteristics, and it was seen that the GaN transistor yields a higher gain than the LDMOS FETs, as well as comparable drain efficiency. Many LDMOS devices are internally matched to enhance their RF performance around a specific frequency. This reduces the usable bandwidth to a few tens of MHz, whereas an LDMOS FET designed for broadband applications can have an operating bandwidth around 100 MHz. GaN transistors, designed as power switching transistors, are not internally matched and, as a result, the higher impedances enable the device to span a bandwidth of as much as 3 GHz.

It was also demonstrated that, despite being designed as a switching device, the GaN transistor can easily be connected to an RF circuit using reference planes and microstrip tapers. This allows for more compact designs than possible with packaged LDMOS devices. The lack of a package, however, may present thermal limitations that can be overcome by thermal design methods discussed in Chapter 6. The more compact layout and lack of package can also lead to cost reductions and system size reduction.

The enhancement-mode switching GaN transistor paves the way for reduced cost RF applications and is ideally suited for applications such as MRI systems. GaN transistors can also offer higher blocking voltage than LDMOS, which can increase the voltage standing wave ratio capability and increase an amplifier's ability to absorb RF energy due to impedance mismatching.

References

1 White, D. and Wilcox, G. (2012). *New GaN FETs, amplifiers and switches offer system engineers a way to reduce RF board space and system prime power*. White paper. TriQuint.

2 Inoue, K., Sano, S., Tateno, Y. et al. (2010). Development of gallium nitride high electron mobility transistors for cellular base stations. *SEI Tech. Rev.* 71: 88–93.

3 Aethercomm (2007). *Gallium nitride (GaN) microwave transistor technology for radar applications*. White paper. Aethercomm.

4 Murphy, M. (2011). NXP goes with GaN. Compound Semiconductor, August/September, 23–26.

5 GaN devices set benchmarks for power and bandwidth. *Microwave Product Digest,* February 2012. www.mpdigest.com.

6 Ishida, T. (2011). GaN HEMT technologies for space and radio applications. *Microwave J.* 54 (8): 57–63.

7 MACOM, NPT1012B – Gallium nitride 28V, 25W RF power transistor. NPT1012 datasheet, NDS-025 Revision 3. https://www.mouser.com/datasheet/2/249/NDS_025_Rev_3_NPT1012-693303.pdf.

8 Pozar, D.M. (2005). *Microwave Engineering*, 3e. Wiley.

9 Gonzales, G. (1997). *Microwave Transistor Amplifiers*, 2e. NJ: Prentice Hall.

10 Hejhall, R.C. (1993). RF small signal design using two-port parameters. Motorola, Application note AN215A.

11 Payne, K. (2008). Practical RF amplifier design using the available gain procedure and the advanced design system EM/circuit co-simulation capability. White paper, 5990-3356EN, Agilent Technologies. www.agilent.com.

12 Lidow, A., Strydom, J., de Rooij, M., and Ma, Y. (2012). *GaN Transistors for Efficient Power Conversion*, 1e. El Segundo: Power Conversion Press.

13 Strydom, J. (2012). eGaN® FET-silicon power shoot-out volume 11: optimizing FET on-resistance. *Power Electron. Technol.* 38: http://powerelectronics.com/discrete-semis/gan_transistors/egan-fet-silicon-power-shoot-out-volume-11-optimizing-fet-on-resistance-1001/.

14 de Rooij, M. and Strydom, J. (June 2012). eGaN® FET-silicon power shoot-out volume 9: low power wireless energy converters. *Power Electron. Technol.* 38: http://powerelectronics.com/discrete-power-semis/egan-fet-silicon-shoot-out-vol-9-wireless-power-converters.

15 MACOM. CGH55015 – GaN HEMT RF FET, CGH55015 datasheet. https://cdn.macom.com/datasheets/CGH55015.pdfhttp://www.cree.com/RF/Products/SBand-XBand-CBand/Packaged-Discrete-Transistors/CGH55015F2-P2.

16 Efficient Power Conversion Corporation (2013). EPC8009 – Enhancement-mode power transistor. EPC8009 datasheet, September. http://epc-co.com/epc/documents/datasheets/EPC8009_datasheet.pdf.

17 Engen, G.F. and Hoer, C.A. (1979). Thru-reflect-line: an improved technique for calibrating the dual six-port automatic network analyzer. *IEEE Trans. Microwave Theory and Tech.* 27: 987–993.

18 Agilent. Network analysis applying the 8510 TRL calibration for non-coaxial measurements. Product note 8510-8A.

19 Fleury, J. and Bernard, O. (2001). Designing and characterizing TRL fixture calibration standards for device modeling. *Appl. Microwave and Wireless Technical Note* 13: 26–55.

20 Rogers Corporation (2014). Rogers 4350 laminates. Datasheet. http://www.rogerscorp.com/acm/products/55/RO4350B-Laminates.aspx.

21 de Rooij, M.A. and Strydom, J.T. (September 2013). Method for bias control of a Class A power RF amplifier. US patent pending.

22 Rollett, J.M. (1962). Stability and power-gain invariants of linear two ports. *IRE Trans. Circuit Theory* 9 (1): 29–32.

23 Orfanidis, S.J. (2014). Electromagnetic waves and antennas. http://www.ece.rutgers.edu/~orfanidi/ewa/.

24 Bahl, I.J. and Trivedi, D.K. (1977). A designer's guide to microstrip line. *Microwaves* 1977: 174–182.

25 Baylis, C., Dunleavy, L., and Clausen, W. (2006). Design of bias tees for a pulsed-bias, pulsed-RF test system using accurate component models. *Microwave J.* 49 (10): 68–75.

26 STMicroelectronics (2011). PD55015 – RF power transistor. Datasheet, August 2011. http://www.st.com/web/en/resource/technical/document/datasheet/CD00128612.pdf.

27 NXP. RF power field effect transistor. MRF1518N datasheet [Rev. 11 June 2009]. https://www.nxp.com/docs/en/data-sheet/MRF1518N.pdf.

10

DC–DC Power Conversion

10.1 Introduction

This chapter presents DC conversion application examples using GaN transistors. As discussed in the previous chapters, the converter performance improvements stem directly from the GaN transistor's relative improvement in FOM over the silicon MOSFET, be it in hard-switching or soft-switching applications. GaN transistors offer the potential to improve performance over the aging population of Si MOSFETs, enabling a new generation of power converters offering higher frequencies, efficiencies, and densities than ever achievable before.

Examples that will be covered will show various design techniques and methods presented in earlier chapters used in complete solutions. The examples that will be presented are:

1) A complete DC–DC buck converter using a GaN IC with digital control.
2) A complete discrete GaN FET converter using an ASIC controller IC specifically designed for GaN FETs.
3) A demonstration design for fast transient paralleling of GaN FETs.
4) A complete bi-directional capable resonant DC converter.

10.2 DC–DC Converter Examples

DC–DC converters are found in servers, solar systems, robots, computers, telecommunication systems, handheld electronics, electric vehicles, and many other applications. With the ever-increasing power demands of modern technologies, combined with the desire for smaller size and lower power consumption, designs must drive toward higher power density and efficiency to meet these system demands. The most straightforward way to improve power density is to increase switching frequency, enabling a volume reduction in the passive components. The practical issue with increasing switching frequency is a decrease in efficiency because of higher losses, thus limiting silicon-based solutions ability to increase power density without significant efficiency penalty.

10.2.1 Complete DC–DC Buck Converter Using a Gan IC With Digital Control

In this example a GaN IC will be used to realize a buck converter design that need to meet the following specifications:

- Input voltage range of 24–60 V,
- output voltage of 12 V,

GaN Power Devices for Efficient Power Conversion, Fourth Edition. Alex Lidow, Michael de Rooij, John Glaser, Alejandro Pozo, Shengke Zhang, Marco Palma, David Reusch, and Johan Strydom.
© 2025 John Wiley & Sons Ltd. Published 2025 by John Wiley & Sons Ltd.

- output current capability of 15 A,
- maximum component thickness of 3.5 mm,
- fit in an area not to exceed 22 mm × 15 mm,
- maximize the efficiency and power density.

The featured GaN IC is the 100 V rated, 6.6 mΩ, EPC23102 [1], a monolithically integrated half bridge with half-bridge gate driver and is shown in Figure 10.1.

The EPC23102 features a near-zero common-source inductance that results in the shortest possible current transitions. It also has drivers well matched to the size of the FETs, yet switch-node transitions can still be adjusted using R_{Boot} and R_{DRV}. The monolithic integration also ensures thermal balancing where the hotter device is cooled by the cooler part of the IC. The simple design of the GaN IC also makes it easy to draft a layout.

The EPC23102 is not a complete DC–DC converter and it therefore requires some support circuits. These include a controller, voltage and current feedback, and a housekeeping power supply. The EPC23102 was specifically designed to work with various support circuits including micro-controllers that typically operate at 3.3 V. The input PWM ports of the EPC23102 can directly accept 3.3 V logic and require no level translation.

For this design a dSPIC33CK32MP102 [2] DSP controller was chosen for its small size, high performance PWM generator, and advanced compensator controller capability. Input and output voltage is fed back to the controller using simple voltage divider resistor networks and bidirectional current is measured using a 1 mΩ shunt and high-voltage (65 V) common-mode amplifier, MCP6C02T-50E [3] with a gain of 50. The board is powered using an external 5 V supply for the GaN IC and, via a 3.3 V LDO, powers the controller. This design was implemented on the EPC9177 [4] and a detailed diagram of the design is shown in Figure 10.2.

A comprehensive study was performed, and the optimal inductance was found to be 1.5 µH. The inductor chosen for testing was a custom design from SunLord WTXE1330T1R5MT with physical dimensions of 12.5 mm by 12.5 mm, and a thickness of 3 mm. The optimal switching frequency was found to be 720 kHz.

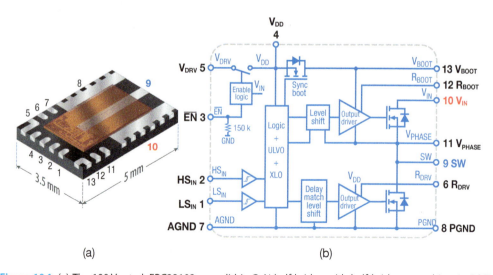

Figure 10.1 (a) The 100 V rated, EPC23102 monolithic GaN half bridge with half-bridge gate driver in PQFN package. *Source:* With permission of EPC. (b) Block diagram of key circuits monolithically integrated on GaN.

DC–DC Power Conversion | 297

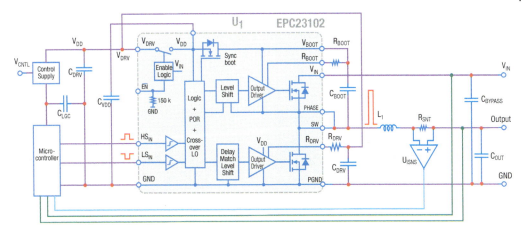

Figure 10.2 Detailed diagram of the EPC9177 [4] buck converter demonstration board that uses the EPC23102 monolithic GaN IC power stage. Source: Ref. [4]/with permission of Efficient Power Conversion Corporation.

With the key components chosen, a PCB layout needed to be captured. The design had to fit within an area of 22 mm by 15 mm. This restriction meant that both sides of the board had to be utilized for components. The top side of the board was designated for the main power circuit and the bottom side assigned for the controller. To accomplish this required blind and buried vias, and enhanced PCB technology such as high-density interconnect HDI [5]. Figure 10.3 shows how the critical power loop was designed to minimize the inductance using the top and inner layer 1.

Figure 10.3 also shows the simplicity of the layout design around the GaN IC. An experimental board was designed to evaluate the performance of the complete DC converter design using a GaN IC. The design is shown in Figure 10.4 for the EPC9177 [4] showing the location of the main components and circuits on both sides of the board.

Figure 10.5 shows the zoom in images of the EPC9177 [4] design showing just the solution area that fits within 21 mm by 13 mm, well within the given specifications. On the top side of the board are the main bus capacitors, filter inductor, and EPC23102 [1] ePower™ Stage. On the bottom side of the board is the controller [2], shunt with amplifier [3], and LDO for controller power and output capacitors.

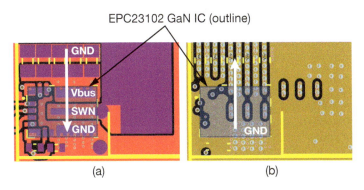

Figure 10.3 Top layer (a) and first inner layer (b) of the EPC9177 [4] design showing the minimization of the critical power loop inductance using the inner vertical design technique [6]. Source: Adapted from Refs. [4] and [6].

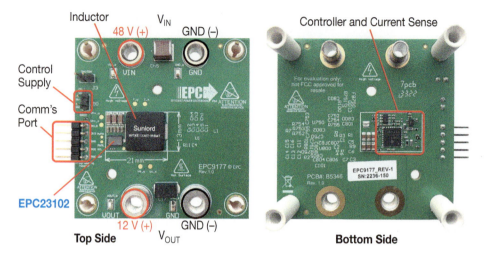

Figure 10.4 Photo of the EPC9177 [4] DC converter board showing the location and details of key components and circuits. *Source:* Ref. [4]/Efficient Power Conversion Corporation.

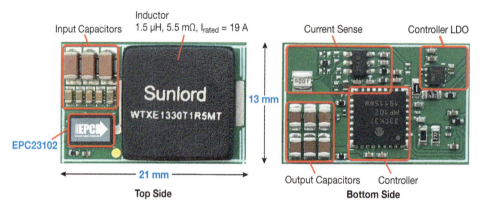

Figure 10.5 Zoomed-in photo of the EPC9177 [4] DC converter board showing only solution area and key components. *Source:* Ref. [4]/Efficient Power Conversion Corporation.

The EPC9177 [4] buck converter board was programmed with a type II compensator controller, where the desired crossover frequency is greater than both the buck converter filter resonance and output capacitor ESR frequency, set to 12 V output and tested under various input voltages while sweeping the load current to the maximum value or halted if thermal limit was reached.

Figure 10.6 shows the measured switch-node waveform of the EPC9177 [4] converter operating from a 48 V input and delivering 10 A into a 12 V load. From Figure 10.6 the switch-node ringing overshoot is just 15 V and was "tuned" using a value of R_{Boot} of 2.2 Ω.

Figure 10.7 shows the measured power loss for the EPC9177 [4] DC converter operating at 720 kHz from the various input voltages of 24, 36, and 48 V, and with output set to 12 V. The power losses include 480 mW of power consumption used by the EPC23102 gate driver, current sense amplifier, and controller. For the 48 V input case, a small heatsink was installed on the GaN IC and inductor that extended the load current range to 20 A.

DC–DC Power Conversion | 299

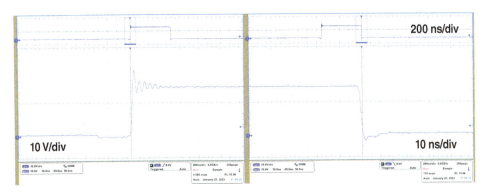

Figure 10.6 Measured waveforms of the switch-node for the EPC9177 [4] operating from 48 V and delivering 10 A into a 12 V load. *Source:* Ref. [4]/Efficient Power Conversion Corporation.

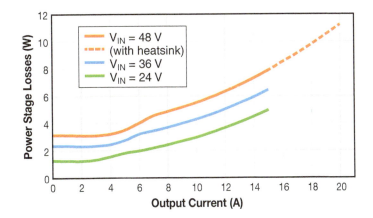

Figure 10.7 Measured power losses of the EPC9177 [4] DC converter for various input voltages and load current range. *Source:* Ref. [4]/Efficient Power Conversion Corporation.

The measured losses are plotted as efficiency in Figure 10.8 under the same operating conditions as Figure 10.7.

Figure 10.9 shows the thermal performance of the EPC9177 [4] DC converter operating at 720 kHz with 400 LFM airflow, from 48 V input and delivering 15 A into a 12 V load and the GaN IC is the hottest component at 86 °C which gives 14 °C margin before reaching an operating temperature limit of 100 °C.

10.2.2 Complete Discrete GaN FET Converter Using an ASIC Controller IC Specifically Designed For GaN FETs

The eco-system for GaN FET-compatible ASIC analog controllers with integrated gate drivers has evolved with many new controllers [7–11] to choose from. Such GaN-compatible controllers simplify the design of converters when using GaN FETs. There are several benefits of using ASIC controllers over digital, including that within the controller are, housekeeping power supply, current sense feedback circuits, and lower power consumption.

However, with integration comes challenges. A single IC solution can restrict the layout around the IC, and a designer will need to take advantage of the many design tools covered

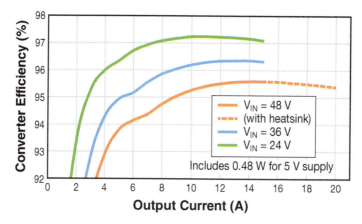

Figure 10.8 Measured power losses of the EPC9177 [4] DC converter for various input voltages and load current range. *Source:* Ref. [4]/Efficient Power Conversion Corporation.

Figure 10.9 Thermal performance of the EPC9177 [4] DC converter operating from 48 V and delivering 15 A into a 12 V load. Airflow used was 400 LFM in the direction shown. *Source:* Ref. [4]/Efficient Power Conversion Corporation.

earlier in this book to yield the best outcome. The example in this section will highlight the most important techniques.

In this example, an LTC7890 [7] controller will be paired with the 100 V rated, 3.2 mΩ, EPC2218 [12] in a two-phase synchronous buck converter that can operate from 48 V and deliver up to 50 A into a 12 V load. There are no additional support ICs required. Figure 10.10 shows a block diagram of the EPC9158 [13] two-phase buck converter where the LTC7890 [7] ASIC controller is highlighted in the green enclosing polygon.

Chapters 3 and 4 presented various layout designs to maximize the performance of the GaN FETs. When using an ASIC controller IC these challenges increase due to the proximity of two sets of gate drivers with respect to each other and the relative size of the power stage circuit. Figure 10.11 shows typical ASIC controller configurations with the physical location of the pins assigned to the output stage of the gate drivers within the physical controller. Red triangles represent the upper FET driver and black the lower FET driver. On the left, images of two examples among the many controllers listed in Chapter 3, Table 3.4, are provided. In the center

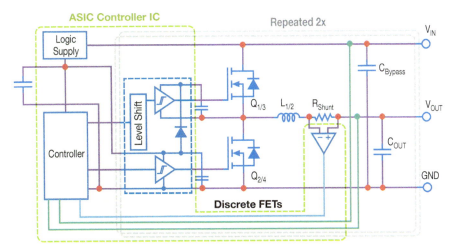

Figure 10.10 Detailed diagram of the EPC9158 [13] ASIC controller two-phase buck converter demonstration board that uses the EPC2218 [12] GaN FET for the power stage. *Source:* Adapted from Efficient Power Conversion [12] and [13].

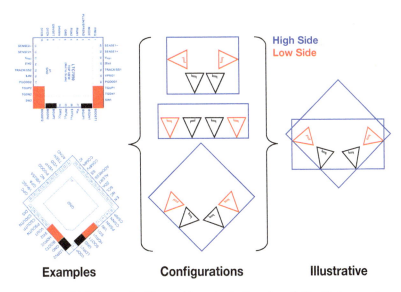

Figure 10.11 Examples of ASIC controller IC gate driver physical locations (left) with three main configurations shown (center), where the blue boxes exemplify the ASIC IC outline with the respective gate drivers shown as triangles, and illustrative drawing (right) that will be used in the layout discussion.

of the figure, three graphics are shown to display all possible configurations. And, on the right, the three graphics are combined into a single model representative of all configurations. This single model will be used to represent the controller for subsequent layout discussions.

Chapters 3 and 4 highlighted the importance of keeping common-source inductances power loop, and gate loop inductance to a minimum and in the order of priority listed. When working with ASIC ICs the same requirements apply, however, physical restrictions may require additional compromises. Figure 10.12 shows three layout configurations for a two-phase ASIC controller. Figure 10.12a,b shows vertical layouts with the bus capacitors below and above the

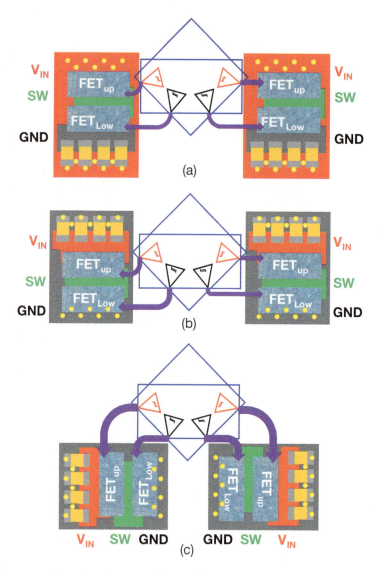

Figure 10.12 Layout configurations for a two-phase ASIC IC controller with (a) vertical layout with input bus capacitors below the controller, (b) vertical layout with the input bus capacitors above the controller, and (c) horizontal layout.

controller, respectively. Figure 10.12c shows a horizontal configuration and only one is possible due to physical size constraints. The choice of vertical or horizontal configurations is a designer's choice. The key compromise in any of the layouts stems from the converter operation.

In a buck converter, the upper FET acts as the control device, it performs a time-critical function, and it is subject to higher losses from the hard-switching transients. The lower FET behaves as the synchronous rectifier, experiencing near zero current switching, which is more forgiving should gate timing be affected by layout. This means that for a converter design, and considering physical layout limitations, one must prioritize the layout of the control FET over the synchronous rectifier.

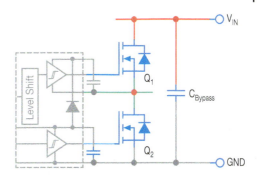

Figure 10.13 Simplified power schematic of the half-bridge power cell highlighting the main nodes with specific colors for corresponding identification in the layout image.

From Chapter 4, the gate loop inductance is proportional to the physical length; and hence, why a shorter path yields lower inductance. The priority compromise can be seen in Figure 10.12, where the gate path to the control device (high-side FET) is shorter than the synchronous rectifier (low-side FET). Figure 10.12 is for illustrative purposes and one should review the layout options when making a physical design.

Using the layout configuration options, the EPC9158 [13] was designed using the configuration shown in Figure 10.12b. To simplify the layout discussion, Figure 10.13 shows the important main half-bridge nodes with specific colors that can be seen in the corresponding layout in Figure 10.14.

From Figure 10.14a, the dark blue node gate of the control device (Upper FET) has been prioritized in layout, with the shortest gate path length and thus lowest inductance. The compromise for the low-side synchronous rectifier was minimal and is shown with the light blue gate node. In each case, the main return path for the gate is on the next layer down (inner layer 1) shown in Figure 10.14b with the large gray power ground (GND) and small green islands of the switch node (SWN). The layout in Figure 10.14 follows the internal vertical layout technique presented in Chapter 4.

The EPC9158 [13] used 2 μH inductors [14] and the switching frequency was set to 500 kHz. Current is measured using 1 mΩ shunts and the two outputs are tied together with the controller set to operate with current tracking between the two phases. The dead time was fixed to 10 ns rising switch-node and 6 ns falling switch node.

Figure 10.15 shows the measured power loss of the EPC9158 [13] as function of load current up to 50 A, operating from an input voltage range of 28–54 V and powering a 12 V load. As expected, higher input voltages incur higher loss. The losses include power consumed by the controller IC and inductor losses.

Figure 10.16 shows the corresponding EPC9158 [13] efficiency for the losses shown in Figure 10.17. At 48 V input, the peak efficiency is 96.5% and occurs when the output current is around 30 A (60% load).

The EPC9158 [13] was operated to thermal steady-state power from 48 V and delivering 50 A into a 12 V load with 400 LFM airflow and no heatsink installed, with the results shown in Figure 10.18. Under these operating conditions and 25 °C ambient, the maximum device temperature reached is 91 °C.

10.2.3 Demonstration Design for Fast Transient Paralleling of GaN FETs

In Chapter 4, techniques for effectively paralleling high-speed GaN transistors were discussed. In this section, the impact of in-circuit parasitic inductance on performance and a comparison of PCB layouts will be examined for a 48–12 V, 480 W, 40 A buck converter operating at a

304 *GaN Power Devices for Efficient Power Conversion*

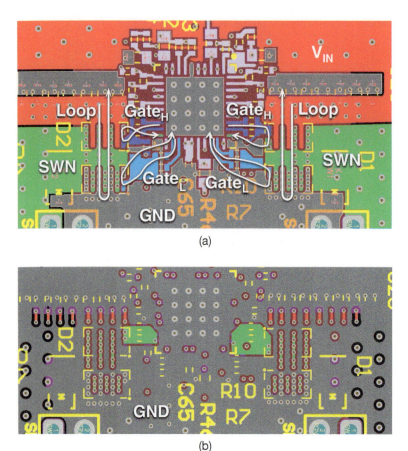

Figure 10.14 Layout of the EPC9158 [13] showing the (a) top layer and (b) inner layer 1, with color nodes; red for Vin, gray for GND, green for the switch-node, and blue for gate with dark blue for the upper FET. The looped arrows shows the current paths for the gate and power loops. *Source:* Adapted from Efficient Power Conversion [13].

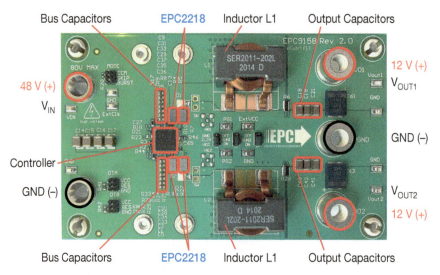

Figure 10.15 Photo of the EPC9158 [13] board showing the main components and input and output connections. *Source:* Ref. [13]/Efficient Power Conversion Corporation.

DC–DC Power Conversion | 305

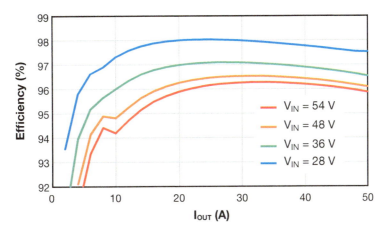

Figure 10.16 Measured efficiency as function of load current of the EPC9158 [13] operating from various input voltages and powering a 12 V load. *Source:* Ref. [13]/with permission of Efficient Power Conversion Corporation.

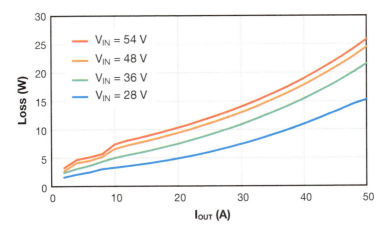

Figure 10.17 Measured power loss as function of load current of the EPC9158 [13] operating from various input voltages and powering a 12 V load. *Source:* Ref. [13]/with permission of Efficient Power Conversion Corporation.

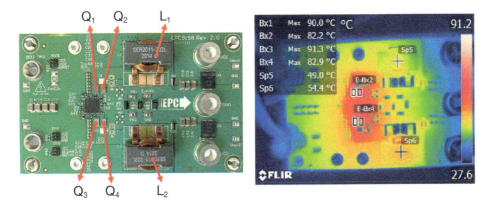

Figure 10.18 Measured thermal steady-state performance of the EPC9158 [13] board operating with 48 V input and delivering 50 A into a 12 V load with 400 LFM airflow and no heatsink installed.

switching frequency of 300 kHz using paralleled GaN transistors. High-performance GaN transistors operated in parallel can achieve higher frequency and power with substantially higher efficiency than Si MOSFETs. The ability to simply and effectively parallel high-performance GaN transistors enables a variety of high-current, high-frequency applications.

The objective of paralleling devices is to combine multiple devices so they may operate as a single device with lower on-resistance, enabling higher power-handling capability. To effectively parallel devices, each device in a switch cluster should dynamically share current equally, and equally divide switching-related losses in steady state. The introduction of imbalanced in-circuit parasitics between parallel devices leads to uneven power sharing and degraded electrical and thermal performance, limiting the effectiveness of paralleling [15]. For high-speed devices, such as GaN transistors, the increased switching speeds can amplify the impact of parasitic inductance mismatches [16].

In a GaN transistor-based buck converter, common source inductance (L_{CSI}) and high-frequency power loop inductance (L_{LOOP}) have been shown in Chapter 4 to impact switching speeds and performance significantly. For paralleling GaN transistors, these parasitics must be minimized and balanced to ensure proper parallel operation.

Figure 10.19a shows the impact of parasitic imbalance in the high-frequency loop inductance for two parallel GaN transistors operating at 48 V with various common source inductances. As the difference between the high-frequency loop inductance between the parallel devices increases, so does the difference in dynamic current between them, causing electrical and thermal performance degradation. As the common source inductance decreases, higher switching speeds can be achieved, and the sharing issues become more pronounced.

Figure 10.19b shows the dynamic current difference resulting from parasitic imbalance in the common source inductance for two parallel GaN transistors operating at 48 V with various high-frequency loop inductances. Similar to loop inductance imbalance, as common source inductance varies, current sharing worsens. This trend is magnified as loop inductance decreases and capable switching speeds increase.

To improve the parallel performance of high-speed GaN devices, the parasitic imbalance contributed by the PCB layout must be minimized. Two different parallel layouts will be

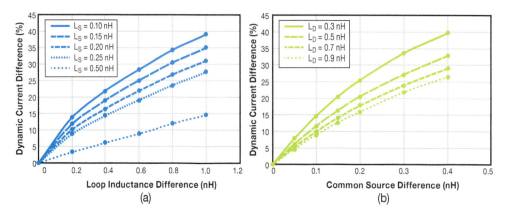

Figure 10.19 Impact of (a) high-frequency loop inductance (b) common-source inductance parasitic imbalance on transistor dynamic current sharing for a V_{IN} = 48 V, I_{OUT} = 25 A, GaN transistor-based buck converter with two devices operating in parallel (GaN transistors: EPC 2001C).

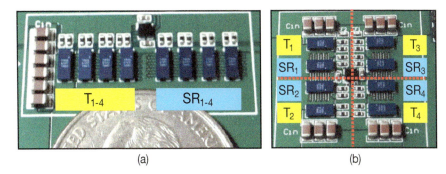

Figure 10.20 Four parallel GaN transistor layouts with (a) single high-frequency power loop (b) four distributed high-frequency power loops. *Source:* With permission of EPC.

reviewed, each based on the internal vertical layout, and an assessment given on their ability to provide parallel performance similar to the equivalent single transistor design. Each half-bridge design contains four devices in parallel for the control switch (T_{1-4}) and the synchronous rectifier (SR_{1-4}). The design was tested in a buck converter configuration operating with input of 48 V and output of 12 V with a switching frequency of 300 kHz. In total, eight 100 V EPC2001C GaN transistors were used to achieve an output power up to 480 W and an output current of up to 40 A.

The first parallel design is shown in Figure 10.20a using the single switch paralleling layout approach presented in Chapter 4.5.1. The second parallel design is shown in Figure 10.20b using the half-bridge paralleling layout approach presented in Chapter 4.5.2.

The switch-node voltage waveforms of the synchronous rectifiers for the two designs are shown in Figure 10.21. The switch-node waveforms for the single high-frequency power loop design are shown in Figure 10.21a, the voltage transitions for the inner-most and outer-most devices show an almost two nanosecond switching time difference, which equates to about 25% of the total switching time. This voltage difference demonstrates the parasitic inductance imbalance in this PCB layout, which leads to inefficient paralleling, resulting in current sharing and thermal issues.

The switch-node waveforms for the symmetrical four high-frequency power loop design are shown in Figure 10.21b. The voltage transitions for the devices are almost identical, demonstrating this layout's ability to balance the parasitic inductance well, thus improving overall performance by offering better electrical and thermal performance.

The thermal evaluation of the two designs, shown in Figure 10.22, demonstrates the thermal imbalance of the single high-frequency loop design. Figure 10.22a shows that a hot spot will develop on the devices handling a greater portion of the power as a result of the parasitic imbalance. The control switch located closest to the input capacitors, T_1, has a maximum temperature more than 10 °C higher than the control switch device located furthest away from the input capacitors, T_4. For the four distributed power loop design, shown in Figure 10.22b, there is a very good thermal balance, with negligible difference in temperature between devices. The average temperature of the structure is also lower for the four distributed power loop design.

By offering lower individual parasitic inductance and better parasitic inductance balance, the four high-frequency loop design is more effective at paralleling. This results in better thermal and electrical performance as shown in Figure 10.23. The distributed high-frequency loop

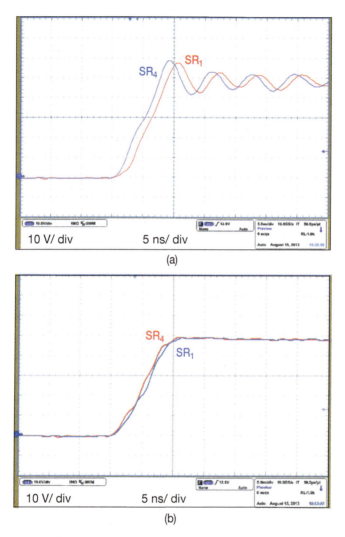

Figure 10.21 Switching waveforms of parallel GaN transistors for (a) single high-frequency power loop (b) four distributed high-frequency power loop designs (V_{IN} = 48 V, V_{OUT} = 12 V, I_{OUT} = 30 A, f_{sw} = 300 kHz, L = 3.3 μH, GaN transistors Control Device/Synchronous Rectifier: 100 V EPC2001C). *Source:* With permission of EPC.

design offers a 0.2% gain in efficiency at 40 A and has an almost constant 10 °C improvement in the maximum transistor temperature.

The switching waveforms for an optimal PCB design using a single GaN transistor, two parallel GaN transistors, and four parallel GaN transistors are shown in Figure 10.24. The parallel designs operate like a single device with lower resistance and slower switching speed in proportion to the number of devices connected in parallel.

10.2.4 Resonant LLC DC-to-DC Converter

This example demonstrates the use of GaN transistors in a 48 V to 12 V high-power-density resonant converter, widely used in data centers and server racks as intermediate bus converters (IBC). These converters enable system-level reduction of conduction losses by bringing the

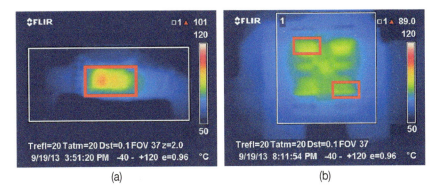

Figure 10.22 Thermal measurements of parallel GaN transistors layouts with (a) single high-frequency power loop (b) four distributed high-frequency power loops (V_{IN} = 48 V, V_{OUT} = 12 V, I_{OUT} = 30 A, f_{sw} = 300 kHz, L = 3.3 µH, GaN transistors: Control Device/Synchronous Rectifier: 100 V EPC2001C).

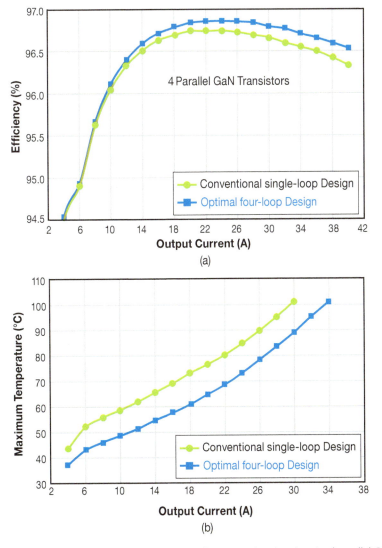

Figure 10.23 (a) Efficiency and (b) Thermal comparison for conventional and optimal parallel GaN transistor designs (V_{IN} = 48 V, V_{OUT} = 12 V, f_{sw} = 300 kHz, L = 3.3 µH, control switch: 4× EPC2001C, synchronous rectifier: 4× EPC2001C).

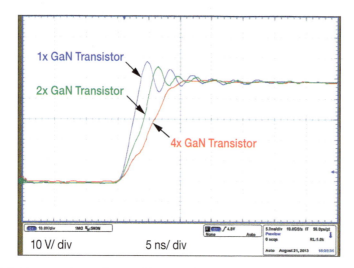

Figure 10.24 Switching waveforms of optimal layout with 1, 2, and 4 parallel GaN transistors (V_{IN} = 48 V, V_{OUT} = 12 V, I_{OUT} = 30 A/(Number of GaN transistors), f_{sw} = 300 kHz, L = 3.3 μH, GaN transistors: control device/synchronous rectifier: 100 V EPC2001C).

48 V bus used for distribution as close as possible to the load (CPU or GPU) [17]. Because of their location, close to the CPU or GPU, they require very high-power-density and low profile, only achievable with very high efficiency. The inductor–inductor–capacitor (LLC) [18–22] resonant converter, which is implemented in this example, is a popular candidate to meet these requirements. This solution can maintain a very high efficiency over a wide range of output power when operated as a DC transformer (DCX) with fixed conversion ratio, making it ideal for applications having relaxed output voltage regulation requirements. The LLC converter presented in this section uses GaN transistors in the primary and secondary sides of the transformer to achieve a power density that exceeds 5 kW/in^3 when processing 1 kW of power between the 48 V bus and 12 V load. This power density can be achieved in a form factor of 17.5 mm × 23 mm × 7.2 mm.

Four EPC2619 [23], a 100 V-rated GaN transistor with a typical $R_{DS(on)}$ of 3.3 mΩ, configure the primary full bridge; and six EPC2067 [24], a 40 V-rated GaN transistor with a typical $R_{DS(on)}$ of 1.3 mΩ, the synchronous rectifiers in the secondary. All devices are in CSP format for minimum size.

Figure 10.25 shows a simplified schematic of the converter. On the primary side a full bridge generates a bipolar square waveform from the input. After that, a resonant tank filters it before entering a 3:1 transformer with a center-tapped full wave synchronous rectifier in the secondary. To maximize efficiency, the switching frequency of the primary and secondary switches is matched to the resonant frequency. This, combined with a carefully tuned dead time, minimizes switching losses by ensuring ZVS in the primary FETs and ZCS in the secondary.

In addition, the isolation provided by the transformer permits the use of a partial power technique to enable higher efficiency and current capability [25, 26]. Figure 10.25 shows the two possible configurations of the converter, Partial power or Through power. In both cases, the LLC converter can process up to 750 W, but owing to the different configuration of the input with respect to the output, the two converter specifications summarized in Table 10.1 are realized.

Figure 10.26 shows the hardware implementation of the EPC9159 [27]. The power converter is separated into two boards, EPC9556P, containing the primary full-bridge; and EPC9551T, a

DC–DC Power Conversion

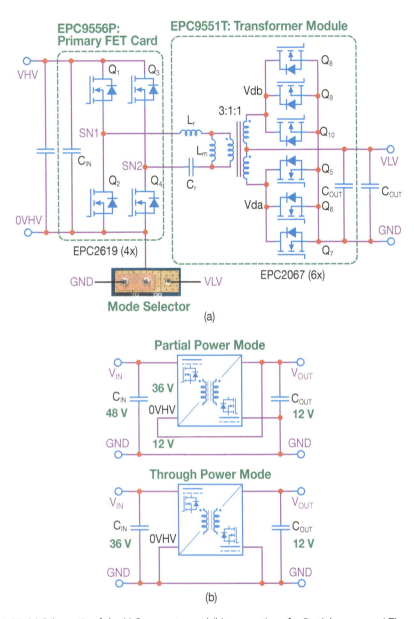

Figure 10.25 (a) Schematic of the LLC converter and (b) connections for Partial power and Through power modes. *Source:* With permission of EPC.

Table 10.1 Specifications of the LLC converter when configured in Partial power and Through power modes.

Mode	Partial power	Through power	Units
Input/output voltage ratio	4:1	3:1	
Nominal input voltage	48	36	V
Nominal output voltage	12	12	V
Maximum output power	1000	750	W
Maximum output current	83	62.5	A

312 *GaN Power Devices for Efficient Power Conversion*

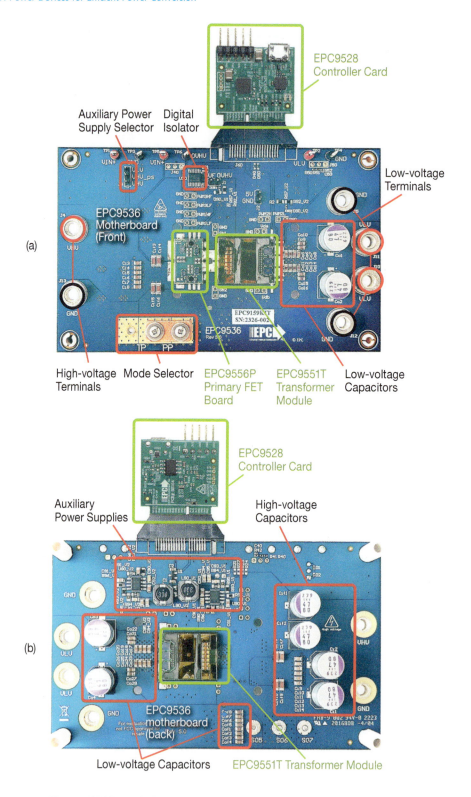

Figure 10.26 Photos of EPC9159 [27]: (a) top view, and (b) bottom view. *Source:* Ref. [27]/Efficient Power Conversion Corporation.

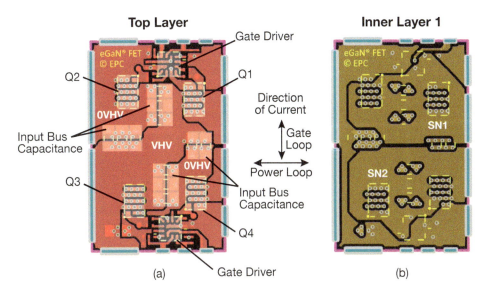

Figure 10.27 Example of internal-vertical layout used in the primary full bridge of an LLC converter: (a) top layer, and (b) inner layer 1. *Source:* With permission of EPC.

module with the resonant tank, transformer, and rectifier. Both boards are mounted onto a motherboard, EPC9536, with additional input/output bus capacitance, auxiliary power supplies, the mode selector, connections for power and controller, and probe points for waveform captures [27].

The primary full bridge is composed of two half bridges, each one using two EPC2619 GaN transistors and a half-bridge gate driver, uP1966E [28]. It is implemented in a six-layer board using an internal-vertical layout with the input bus capacitors located between the FETs, as shown in Figure 10.27. This layout was previously discussed in Chapter 4, Figure 4.11b. In this particular example, the power loop is closed by placing the switch nodes (SN1 and SN2) in Inner Layer 1. A symmetrical layout with the capacitors between FETs was chosen for this design because both high-side and low-side FETs in each half bridge of the LLC see the same RMS currents. Figure 10.27 shows images of the top two layers, the most critical ones in determining the power loop and common source inductances. Following the recommendations from Chapter 4, orthogonality between power and gate loops was maintained to minimize CSI.

The resonant tank, transformer, and rectifier FETs are all located on the same module, labeled as EPC9551T in Figure 10.26. The module is mostly occupied by the transformer, with a custom magnetic core, and the windings built into two PCBs with 12 layers each. Both PCBs are soldered together in the middle to form a 24-layer stack. Each PCB uses high-density-interconnect (HDI) technology for the outer two layers, with a copper thickness of 1.5 oz. All the remaining layers have a 2 oz. thickness. A total of twenty layers are used to construct the transformer windings by implementing concentric turns around the core's center post, as shown in Figure 10.28. These primary and secondary turns are interleaved to alternate magnetomotive force and minimize proximity losses. The magnetic core is split into two halves with a center leg and four small satellite legs for the return flux path. The magnetic path is closed with two end plates whose geometry was designed for uniform and adequate flux density while freeing enough board space for the resonant capacitors and rectifier FETs [29]. The cores are manufactured with ML91S [30] and the airgap is set to 7.5 mil between the core halves to control the magnetizing inductance to be approximately 1.8 uH.

314 | *GaN Power Devices for Efficient Power Conversion*

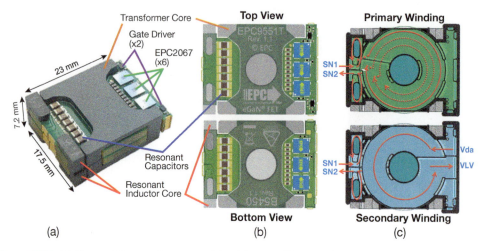

Figure 10.28 (a) Photo of the transformer module EPC9551T, (b) top and bottom views of the module. *Source:* With permission of EPC. (c) Example layout of the primary and secondary windings with a description of the current flow.

The second element on the transformer module is the GaN FETs used as synchronous rectifiers. Since a center-tapped full-wave rectifier only requires low-side switches, the single switch approach discussed in Chapter 4, Section 4.5.1, can be followed for paralleling GaN transistors. In this example, three EPC2067 GaN transistors are paralleled to form a single switch with an equivalent 0.433 mΩ of typical $R_{DS(on)}$. Each group of paralleled GaN transistors is controlled with a single LMG1020 low-side gate driver [31].

The third element on the module is the resonant tank. It is formed by fourteen 22 nF resonant capacitors connected in parallel, and two external resonant inductors of approximately 12 nH each. These, combined with a leakage inductance of under 2 nH, make up the 26 nH of resonant inductance needed to create a resonant frequency of nearly 1.8 MHz. The external inductors are also built into the transformer module and consist of a single turn. The core material is a soft-ferrite, ML95S [30].

As discussed previously, the LLC converter in EPC9159 [27] is designed to operate at resonant frequency. As a result, switching losses, both in primary and secondary FETs, can be eliminated. To achieve this, the actual switching frequency of the FETs (1.67 MHz) needs to be slightly lower than the resonant frequency (approximately 1.8 MHz). The difference is caused by the dead time, which is the time needed by the magnetizing current to charge and discharge the output capacitance (C_{OSS}) of the GaN FETs in the primary full bridge. Note that for a given C_{OSS}, the dead time can be decreased by increasing the magnetizing current. This is desirable because shorter dead time increases the percentage of time the converter transfers power from input to output. However, the additional magnetizing current also increases conduction losses in the primary FETs and transformer windings. From this trade-off stems GaN technology's superiority for this application. As shown previously in Chapter 8, Figure 8.9, for the same $R_{DS(on)}$, a 100 V-rated GaN transistor, like the EPC2619, has a Q_{OSS} close to 2.5 times lower than an equivalent silicon MOSFET, which means dead times can be decreased by 2.5× without increasing magnetizing current. Moreover, a 100 V-rated GaN transistor like EPC2619 offers outstanding improvements in gate driver power owing to the five times lower Q_G than a similar silicon MOSFET.

Figures 10.29 and 10.30 show the waveforms of the LLC converter operating in Through power and Partial power modes, respectively. For each case, the primary switch node waveforms (SN1, SN2), the secondary drain voltages (Vda and Vdb), and the resonant current

DC–DC Power Conversion | **315**

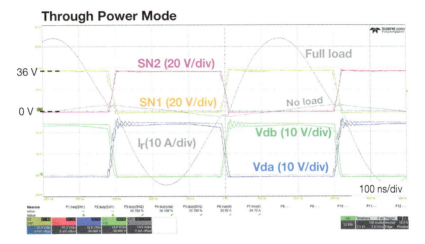

Figure 10.29 Converter waveforms operating in Through power mode at no load and full load.

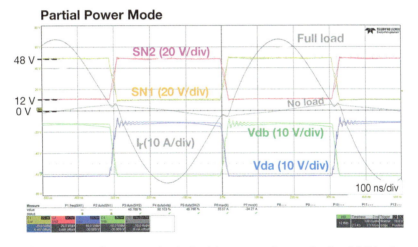

Figure 10.30 Converter waveforms operating in Partial power mode at no load and full load.

(I_r), are shown with no load and full load. The converter is operating at resonance because the resonant portion of I_r starts and ends at the same point regardless of the load current.

An additional observation is that the shape and amplitude of the resonant current and secondary drain waveforms are independent of the operating mode. In contrast, the primary switch node waveforms differ between modes, even though the differential voltage applied to the primary side of the transformer is the same, 36 V. In Through power mode, a trapezoidal waveform oscillates between 0 V and 36 V, whereas in Partial power mode it does so between 12 V and 48 V. This is because in the latter, 0VHV is floating above the output voltage as opposed to GND, as shown in the schematics of Figure 10.25b.

The performance curves in Figure 10.31 show the efficiency and power losses measured for EPC9159 [27] operated in both modes up to full power, 1000 W in Partial power mode, and 750 W in Through power mode. In Partial power mode, peak efficiency is above 97% and full load efficiency remains above 95.5%. Because the power processed by the module in both

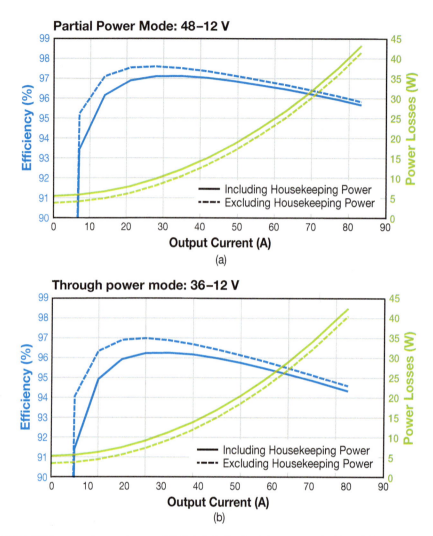

Figure 10.31 Efficiency and power losses of EPC9159 [27] in Partial power mode (a) and Through power mode (b). *Source:* Ref. [27]/with permission of Efficient Power Conversion Corporation.

modes is the same, so are the power losses. However, since the output power in Partial power mode is 33% higher, its efficiency is higher. Similarly, the difference in power losses with and without housekeeping power is independent of the mode. At 1.67 MHz of switching frequency, this housekeeping power is dominated by the power required to switch all the GaN transistors in EPC9159 [27].

To operate EPC9159 [27] at full power, a heatsink and forced air are required. Figure 10.32 shows a picture of EPC9159 with the heatsink installed operating in Partial power mode at full power in steady state. The corresponding performance curves for this scenario are shown in Figure 10.31 (top), and the waveforms in Figure 10.30. The heatsink surface temperature reaches 61 °C.

If the heatsink is not an option, the converter can be operated with some power derating, depending on the maximum desired increase in temperature. Figure 10.33 shows this scenario with EPC9159 [27] operating at 64% of full power and 400 LFM of forced air. With an ambient temperature of 25 °C, the highest temperature reached was approximately 85 °C.

DC–DC Power Conversion | 317

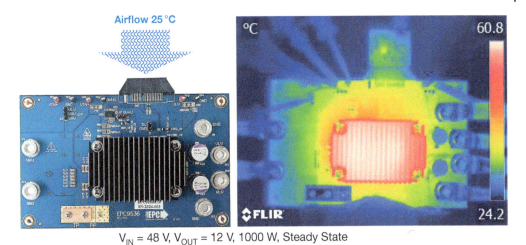

V_{IN} = 48 V, V_{OUT} = 12 V, 1000 W, Steady State

Figure 10.32 Thermal performance of EPC9159 with a heatsink installed at full load and 400 LFM of forced air. *Source:* With permission of EPC.

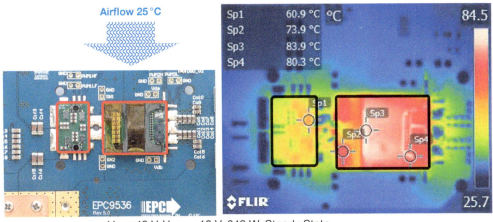

V_{IN} = 48 V, V_{OUT} = 12 V, 640 W, Steady State

Figure 10.33 Thermal performance of EPC9159 without a heatsink at 64% of full load and 400 LFM of forced air. *Source:* With permission of EPC.

10.3 Summary

Highlighted in this chapter were some of the DC–DC converter applications that benefit from the superior performance of GaN transistors compared with their silicon MOSFET counterparts. Improvements in switching losses are complemented by improvements in gate drive losses and layout parasitics to demonstrate that GaN can outperform silicon in almost any power conversion application.

In Chapter 11, multilevel converters that have enabled high-voltage power conversion voltages using relatively low-voltage GaN transistors will be explored.

References

1 Efficient Power Conversion (2024). EPC23102 – Empower™ Stage IC. EPC23102 datasheet, March 2024 [Revised April 2024]. https://epc-co.com/epc/Portals/0/epc/documents/datasheets/EPC23102_datasheet.pdf.

2 Microchip (2020). dSPIC33CK64MP105 Family – 16-Bit digital signal controllers with high-speed ADC, op amps, comparators and high-resolution PWM. dSPIC33CK32MP102 datasheet, February 2020. https://www.microchip.com/en-us/product/dspic33ck32mp102.

3 Microchip Technology (2020). MCP6C02 – Zero-drift, 65V high-side current sense amplifier. MCP6C02T-50E datasheet, February 2020. https://www.microchip.com/en-us/product/mcp6c02#document-table.

4 Efficient Power Conversion (2023). EPC9177: Small area, low-profile, synchronous buck converter, EPC9177 quick start guide. https://epc-co.com/epc/products/evaluation-boards/epc9177.

5 Printed circuit board. Wikipedia [edited 4 April 2024]. https://en.wikipedia.org/wiki/Printed_circuit_board.

6 Reusch, D. and Strydom J. (2013). Understanding the effect of PCB layout on circuit performance in a high frequency gallium nitride based point of load converter. *Applied Power Electronics Conference (APEC)*, Long Beach, CA, (March 17–21, 2013), 649–655.

7 Analog Devices (2022). LTC7890 – Low I_Q, dual, 2-phase synchronous step-down controller for GaN FETs. LTC7890 datasheet [Updated March 16, 2023]. https://www.analog.com/en/products/ltc7890.html.

8 Texas Instrument (2016). LM5140-Q1 Wide input range dual synchronous buck controller. SNVSA02A datasheet [Revised December 2016]. https://www.ti.com/lit/gpn/lm5140-q1.

9 Renesas Electronics (2022). ISL81806 – 80V Dual synchronous buck controller optimized to drive e-mode GaN FET. ISL81806 datasheet [Revised January 28, 2022]. https://www.renesas.com/us/en/document/dst/isl81806-datasheet?r=1488361.

10 Analog Devices (2022). LTC7891 – 100 V, Low I_Q, synchronous step-down controller for GaN FETs. LTC7891 datasheet, March 2022. https://www.analog.com/media/en/technical-documentation/data-sheets/ltc7891.pdf.

11 Analog Devices (2022). LT8390A – 60V 2MHz synchronous 4-switch buck-boost controller with spread spectrum. LT8390A datasheet [Revised C, July 2022]. https://www.analog.com/media/en/technical-documentation/data-sheets/lt8390a.pdf.

12 Efficient Power Conversion (2024). EPC2218 – Enhancement mode power transistor. EPC2218 datasheet. March 2024. https://epc-co.com/epc/Portals/0/epc/documents/datasheets/EPC2218_datasheet.pdf.

13 Efficient Power Conversion (2022). EPC9158: 48V/54V – 12 V, 50 A, synchronous buck converter. EPC9158 quick start guide [Revised 2.0]. https://epc-co.com/epc/products/evaluation-boards/epc9158.

14 CoilCraft (2023). Shielded power inductor – SER2000. SER2011-202 datasheet [Revised February 16, 2023]. https://www.coilcraft.com/getmedia/56db8f04-6d7a-4e93-ad3f-acf7c2a14b74/ser2000.pdf.

15 Forsythe, J. B. (n.d.). Paralleling of power MOSFETs for high power output. http://educypedia.karadimov.info/library/para.pdf.

16 Wu, Y. F. (2013). Paralleling high-speed GaN power HEMTs for quadrupled power output. *Applied Power Electronics Conference (APEC)*, Long Beach, CA, (March 17–21, 2013), 211–214.

17 Gendron, R. (2016). 48 V: The new standard for high-density, power efficient data centers. *Power Electronic News*, August 2016. https://www.powerelectronicsnews.com/48-v-the-new-standard-for-high-density-power-efficient-data-centers/.

18 Mammano, B. (2021). Resonant mode converter topologies. Unitrode design seminar, 1985, TI Literature No. SLUP085. https://www.google.com/search?client=firefox-b-1-e&q=Resonant+mode+converter+topologies.

19 Huang, H. (2010). Designing an LLC resonant half-bridge power converter. Reproduced from 2010 Texas Instruments Power Supply Design Seminar SEM1900, Topic 3, TI Literature No. SLUP263. https://bbs.dianyuan.com/upload/community/2013/12/01/1385867010-65563.pdf.

20 Fei, C., Ahmed, M.H., Lee, F.C., and Li, Q. (2017). Two-stage 48 V-12 V/6 V-1.8 V voltage regulator module with dynamic bus voltage control for light-load efficiency improvement. *IEEE Trans. Power Electron.* 32 (7): 5628–5636.

21 Reusch, D. (n.d.). High frequency, high power density integrated point of load and bus converters. Ph.D. dissertation. Department of ECE, Virginia Tech, Blacksburg, VA.

22 Ahmed, M.H., Fei, C., Lee, F.C., and Li, Q. (2017). 48V voltage regulator module with PCB winding matrix transformer for future data centers. *IEEE Trans. Ind. Electron.* 64 (12): 9302–9310.

23 Efficient Power Conversion (2023). EPC2619 – Enhancement mode power transistor. EPC2619 datasheet, November 2023. https://epc-co.com/epc/Portals/0/epc/documents/datasheets/EPC2619_datasheet.pdf.

24 Efficient Power Conversion (2021). EPC2067 – Enhancement mode power transistor. EPC2067 datasheet, October 2021. https://epc-co.com/epc/Portals/0/epc/documents/datasheets/EPC2067_datasheet.pdf.

25 Zhao, J., Yeates K., and Y. Han (2013). Analysis of high efficiency DC/DC converter processing partial input/output power. *14th IEEE Workshop on Control and Modeling for Power Electronics (COMPEL)*, (June 2013), 1–8.

26 Itoh, J.-I. and Fujii, T. (2008). A new approach for high efficiency buck-boost DC/DC converters using series compensation. *IEEE Power Electronic. Specialists Conference*, (June 2008), 2109–2114.

27 Efficient Power Conversion (2023). EPC9159: 1 kW, 48V/12V, LLC converter. EPC9159 quick start guide. https://epc-co.com/epc/Portals/0/epc/documents/guides/EPC9159_qsg.pdf.

28 uPI Semi (2021). uP1966E: Dual-channel gate driver for enhancement mode GaN transistors. uP1966E datasheet, July 2021. https://www.upi-semi.com/en-article-upi-718-2279

29 de Rooij, M. and Negahdari, A. (2022). Beyond 4 kW/in3 power-density for 48 V to 12 V conversion using eGaN FETs in an LLC DC-DC bus converter. *PCIM Europe 2022; International Exhibition and Conference for Power Electronics, Intelligent Motion, Renewable Energy and Energy Management*, Nuremberg, Germany, May 10–12, 2022, 1–9, https://doi.org/10.30420/565822013.

30 Proterial Ltd. (n.d.) Mn-Zn soft ferrite cores for high frequency power supplies MaDC-F^{TM}. https://www.proterial.com/e/products/soft_magnetism/pdf/PR-EM13_MaDC-F.pdf.

31 Texas Instruments (2018). LMG1020: 5-V, 7-A, 5-A Low-side GaN and MOSFET driver for 1-ns pulse width applications. LMG1020 datasheet [Revised October 2018]. https://www.ti.com/lit/gpn/LMG1020.

11

Multilevel Converters

11.1   Introduction

In the past decade, applications such as cloud computing and big data processing have placed strenuous demands on the electric energy consumption of data centers worldwide. The annual electricity report from the International Energy Agency (IEA) states data centers consumed 460 TWh in 2022, which could rise to more than 1000 TWh by 2026 in a worst-case scenario [1]. In the US, where 33% of the world's data centers are located, consumption is expected to rise from 200 TWh in 2022 to 260 TWh in 2026, roughly 6% of the US total electric energy consumption.

It is well understood that a significant portion of this consumption is due to losses resulting from the use of inefficient power delivery architectures that require significant improvements [2, 3]. Meanwhile, in higher-voltage applications, such as laptop adapters, telecom rectifiers, and inverter-fed drives, there are also urgent needs to improve efficiency and reliability [4, 5]. In response to these trends, the majority of mobile phone chargers and laptop power adapters are now based on GaN technology.

Since GaN transistors are smaller and faster than MOSFETs, and offer a significant reduction in PCB real estate, topologies that require a greater number of active GaN devices as a tradeoff for reduced passive component size have become attractive alternatives for realizing higher power density and efficiency. Multilevel converters can utilize the benefits of wide bandgap transistors and provide a significantly reduced amount of space required for passive components [6]. However, to realize the promise of GaN-based multilevel converters, many challenges need to be overcome and are discussed in this chapter.

11.2   Benefits of Multilevel Converters

11.2.1   Transistor Figures of Merit

Chapter 3 introduced various Figures of Merit (FOM) as a method for comparing the performance of different transistor technologies within a given converter. In hard-switching converters, such as buck, boost, and voltage source inverters, the most important parameters to consider are the on-resistance $R_{DS(on)}$, the gate charge Q_G, the output charge Q_{OSS}, and the reverse recovery charge Q_{RR}.

GaN Power Devices for Efficient Power Conversion, Fourth Edition. Alex Lidow, Michael de Rooij, John Glaser, Alejandro Pozo, Shengke Zhang, Marco Palma, David Reusch, and Johan Strydom.
© 2025 John Wiley & Sons Ltd. Published 2025 by John Wiley & Sons Ltd.

322 | GaN Power Devices for Efficient Power Conversion

When doing a comparison, in hard-switching converters, the most used FOMs are:

- Conduction: $FOM_1 = R_{DS(on)} \times Q_G$
- Switching: $FOM_2 = R_{DS(on)} \times Q_{OSS}$
- Reverse Recovery: $FOM_3 = R_{DS(on)} \times Q_{RR}$

The lower the FOM number, the more superior the device contribution in the considered application. For GaN FETs, FOM_3 is null because Q_{RR} is zero. When comparing devices with different maximum drain-source (V_{DS}) breakdown voltage ratings and from different technologies, Table 11.1 shows that higher-voltage-rated devices have higher FOMs.

Table 11.1 also shows, in the FOM columns, that GaN transistors outperform silicon and silicon carbide MOSFETs given the same breakdown voltage. Considering that the lower the switch voltage, the better all FOMs are, finding a way to use low-voltage devices is a viable solution to increasing efficiency and power density in high-voltage converters.

Multilevel converters comprise power semiconductors and voltage sources that generate a stepped voltage waveform with variable and controllable frequency, phase, and amplitude. The multiple-step waveform can be synthesized by properly switching the power semiconductors, selecting the appropriate voltage level, and connecting it to the load. The number of levels is the number of steps, or constant voltage values, generated between the output terminal and any arbitrary internal reference node within the converter. Figure 11.1 shows a single-phase example of a classic two-level converter and a nine-level multilevel converter.

Historically, multilevel converters appeared as a promising alternative to stacked switches in high-voltage applications [6], which were used to achieve high-voltage without using high-voltage rated semiconductors. The simple stacked switch topology requires tight driving synchronization and perfect voltage balancing, which are hard to realize, particularly with fast switching wide bandgap transistors. Therefore, they were not favored.

Many multilevel topologies use high-voltage switches or diodes, and three topologies use only low-voltage switches:

1) cascaded half bridge (also known as modular multilevel converter [MMC])
2) cascaded full-bridge (CHB) converter
3) flying capacitor multilevel converter (FCML).

Table 11.1 Hard-switching figure of merit.

Device	$V_{DS(max)}$ (V)	$R_{DS(on)}$ (mΩ)	Q_G (nC)	Q_{OSS} (nC)	Q_{RR} (nC)	FOM_1	FOM_2	FOM_3
EPC2302[a]	100	1.4	23	85	0	32.2	119.0	0
IPTG014N10[b]	100	1.4	169	214	316	236.6	299.6	442.4
EPC2304[a]	200	3.1	24	116	0	74.4	359.6	0
IPT067N20[b]	200	6.7	72	227	364	482.4	1520.9	2438.8
IMBG65R007[c]	650	6.7	179	337	~0	1199.3	2257.9	0
GS66516T[a]	650	25	14.2	134	0	355.0	3350.0	0

[a] GaN FET.
[b] Si MOSFET.
[c] SiC MOSFET.

Multilevel Converters | 323

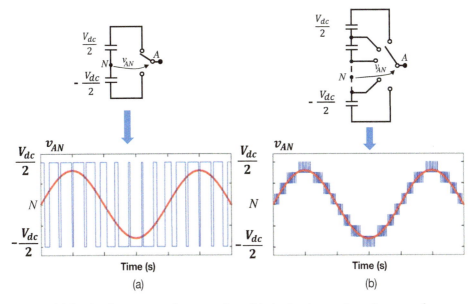

Figure 11.1 (a) Two-level converter voltage waveform, (b) nine-level converter voltage waveform.

The cell configurations are shown in Figure 11.2 for these three topologies that use low-voltage switches only. The flying capacitor converter (FCML) is the most popular of the three designs because it can be operated from a single DC source. The MMC and the CHB topologies are less prevalent because they require multiple DC sources with isolated power supplies or batteries at each level.

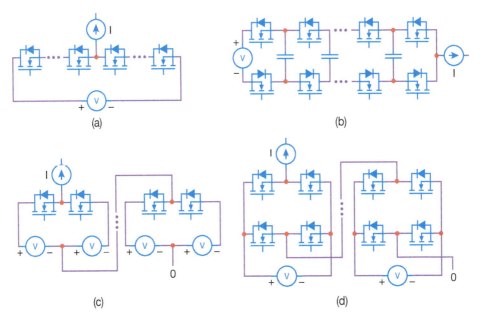

Figure 11.2 (a) Stacked switch configuration, (b) flying capacitor multilevel converter FCML, (c) cascaded half-bridge or modular multilevel converter MMC, (d) cascaded full-bridge converter CHB.

Table 11.2 Pros and Cons of converters' cell configuration in Figure 11.2 (b)–(d).

Cell configuration	Pros	Cons
MMC	Modular High reliability Does not require high-voltage switches	Large number of capacitors complex control and precharging total capacitance independent of switching frequency
CHB	Modular No floating capacitors High reliability Does not require high-voltage switches	Requires isolated DC sources More power switches Common mode present at 0 V state
FCML	Modular Cost efficient at high level number Does not require high-voltage switches	High number of floating capacitors Many voltage sensors Complex control

Table 11.2 summarizes the pros and the cons of the three topologies. The FCML topology can be generalized and used in various power converter circuits, such as buck, boost, inverter, or rectifier [6]. In the flying capacitor multilevel converter, the load current passes through the flying capacitors; hence, a higher current rating means a higher number of capacitors are needed at each level.

11.2.2 The Flying Capacitor Multilevel Converter Basics

The flying capacitor multilevel converter (FCML) is a topology that uses capacitors as DC sources and is scalable to high-level counts. The capacitance required at each level is inversely proportional to the converter switching frequency, as derived in Eqs. (11.5)–(11.10) below. Therefore, the total energy storage in the FCML converter can be reduced by increasing the switching frequency. As described in the following paragraphs, the FCML, using phase-shift PWM, has the property of capacitor voltage self-balancing. More accurate voltage control of the flying capacitors can be achieved using an active balancing method (by active modulation of the switches control signals) instead of passive balancing. This characteristic can extend to higher-level counts and bus voltages using low-voltage gallium nitride (GaN) devices.

11.2.3 Transistor Figures of Merit in FCML Converters

The different types of Figures of Merit introduced previously must be properly combined to compare multilevel converters. In the case of FCML, it is essential to include the number of levels (N) in determining the optimum number of levels and the associated device.

The example in Figure 11.3 shows the difference between a classical two-level converter and a three-level ($N = 3$) converter when the switching node is connected to the highest voltage. In the two-level converter case, given in Figure 11.3a, with switch S1 turned on, the complementary switch S2 must withstand the full V_{IN} voltage. The total resistance from V_{IN} to the load is $R_{DS(on)}$, where the gate, the reverse recovery, and the output charges involved are Q_G, Q_{RR}, and Q_{OSS} (excluding load parasitic capacitance).

Following the similar three-level case example, in Figure 11.3b, with both S_1 and S_2 turned on, the $V_{IN}/(N-1) = V_{IN}/2$ voltage is applied to the series of S_3 and S_4 (which are both off).

Multilevel Converters | 325

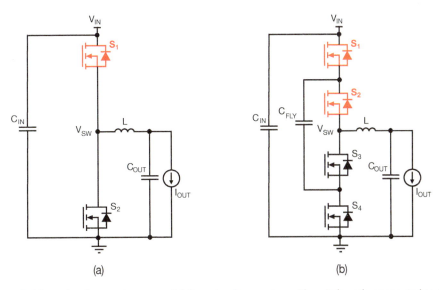

Figure 11.3 (a) two-level converter versus (b) three-level converter, with switch-node connected to V_{IN}.

The total resistance is $(N-1) \times R_{DS(on)} = 2 \times R_{DS(on)}$, the gate charge involved is $(N-1) \times Q_G$, the reverse recovery charge is $(N-1) \times Q_{RR}$, and the output charge is $Q_{OSS}/(N-1) = Q_{OSS}/2$, because the two output capacitors (C_{OSS}) of the switches are connected in series behave as a single capacitor $C_{OSS}/(N-1) = C_{OSS}/2$. In the case of GaN FETs or SiC MOSFETs, the effect of the reverse recovery charge is negligible.

Given the above considerations, a figure of merit for the multilevel converter can be obtained by calculating the modulus of the vector with components FOM_{1N}, FOM_{2N}, and FOM_{3N}, which are obtained from FOM_1, FOM_2, and FOM_3 by including the effect of the N number of levels.

$$FOM_{1N} = FOM_1 \cdot (N-1)^2 = R_{DS(on)} \cdot Q_G \cdot (N-1)^2 \quad (11.1)$$

$$FOM_{2N} = \frac{(N-1) \cdot FOM_2}{(N-1)} = R_{DS(on)} \cdot Q_{OSS} = FOM_2 \quad (11.2)$$

$$FOM_{3N} = FOM_3 \cdot (N-1)^2 = R_{DS(on)} \cdot Q_{RR} \cdot (N-1)^2 \quad (11.3)$$

$$FOM_{ML} = \sqrt{FOM_{1N}^2 + FOM_{2N}^2 + FOM_{3N}^2} \quad (11.4)$$

Consider as an example a V_{IN} voltage of 400 V and the devices given in Table 11.1; the FOM_{ML} is calculated and shown in Table 11.3 for two-level, four-level, and seven-level converters. Table 11.3 reveals that a multilevel topology has no practical value when using Si MOSFETs, due to the significant adverse effect of reverse recovery charge. The four-level converter based on 200 V GaN FETs shows a tangible improvement over a conventional two-level topology made with 650 V GaN FETs and 650 V SiC MOSFETs. It is worth noting that the FOM_{ML} does not consider the number of capacitors and gate drivers that must be added per each level.

11.2.4 FCML Basic Switching Power Cell and Self-Balancing Characteristics

The FCML converter is shown in Figures 11.2b and 11.3b. A closer look at two adjacent basic switching cells is illustrated in Figure 11.4, where the flying capacitor in the middle is charged when $S1_H$ and $S2_L$ are on, discharged when $S1_L$ and $S2_H$ are on, and bypassed in the remaining two states (either $S1_H$ and $S2_H$ are on, or $S1_L$ and $S2_L$ are on).

Table 11.3 Hard switching figure of merit for a multilevel converter with N levels (Table 11.1 has been re-ordered by semiconductor technology dependency).

Device	$V_{DS(max)}$ (V)	$R_{DS(on)}$ (mΩ)	Q_G (nC)	Q_{OSS} (nC)	Q_{RR} (nC)	N	FOM_{ML}
EPC2302[a]	100	1.4	23	85	0	7	1,165.3
EPC2304[a]	200	3.1	24	116	0	4	760.1
GS66516T[a]	650	25	14.2	134	0	2	3,368.8
IPTG014N10[b]	100	1.4	169	214	316	7	18,063.5
IPT067N20[b]	200	6.7	72	227	364	4	22,426.1
IMBG65R007[c]	650	6.7	179	337	~0	2	2,556.6

[a] GaN FET.
[b] Si MOSFET.
[c] SiC MOSFET.

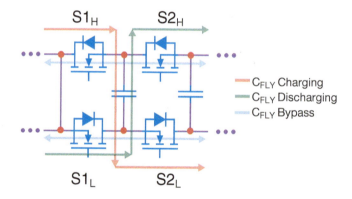

Figure 11.4 Charging and discharging paths for flying capacitor.

In normal operation within a PWM cycle, $S1_H$ and $S1_L$, as well as $S2_H$ and $S2_L$, are complementary switched, and their control commands can be operated to ensure the same time for charging and discharging of the capacitor. One common technique for the modulation is the phase-shifted PWM (PSPWM), which can be demonstrated to allow the same charging and discharging time regardless of the duty cycle (D) applied at the two adjacent switching cells. When using the PSPWM, the signal with duty cycle applied to each cell is shifted by $T_{PWM}/(N-1)$, thus ensuring a symmetrical overlap of charging, discharging, and bypass conditions.

Figure 11.5 shows an example of the modulation of the two adjacent cells of Figure 11.4 in the case of a five-level converter, where the phase shift is $T_{PWM}/(N-1) = 25\%$ of the PWM period. The three cases of Figure 11.5 are the following: (i) D < phase shift; (ii) D > phase shift and 1−D > phase shift; (iii) 1−D < phase shift. In all three cases, the charging time (in red) always equals the discharging time (in green).

In most designs, no active methods are used to balance the capacitor voltages, thanks to this self-balancing property. In some designs, however, it is necessary to further minimize the ripple and the unbalance due to non-ideal switching. As an example, the gate driver integrated circuit propagation delay distributions cause a variation of the PWM signals timing between

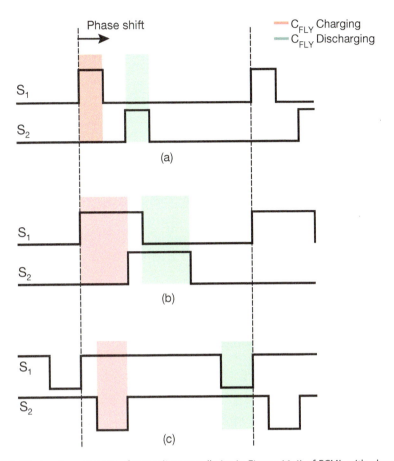

Figure 11.5 Self-balancing property of two adjacent cells (as in Figure 11.4) of FCML with phase-shifted PWM. The 25% phase shift in the diagram is for N = 5 levels. (a) D < phase shift; (b) D > phase shift and 1−D > phase shift; (c) 1−D < phase shift.

the different gate drivers in the converter. This delay mismatch at the different levels distorts the applied pulse lengths and the duties between adjacent cells. In these cases, active methods are used based on measuring each flying capacitor's voltage and on the instantaneous modification of one of the duties (charging or discharging).

11.2.5 Level Selection and Capacitance Sizing

In the N-level flying capacitor converter shown in Figure 11.2b, the voltage rating of each switch is equal and limited to the voltage difference between two capacitors per Eq. (11.5), while each capacitor voltage, and hence its rating, varies with the position in the multilevel structure per Eq. (11.6).

$$V_{SW} = \frac{V_{IN}}{N-1} \tag{11.5}$$

$$V_C(m) = \frac{m \cdot V_{IN}}{N-1}, \quad m = 1, 2, \ldots, (N-2) \tag{11.6}$$

Figure 11.5b illustrates the condition that imposes the longest discharge time that occurs when both D and $1-D$ are larger than the phase shift. In this case, the ripple on the flying capacitor is maximum when D = phase shift. Introducing Δt_{PS} as the phase shift time:

$$\Delta t_{\mathrm{PS}} = \frac{T_{\mathrm{PWM}}}{N-1} \tag{11.7}$$

the charge stored in the flying capacitor varies during Δt_{PS} as:

$$\Delta Q_{\mathrm{CFLY}} = i_{\mathrm{load}} \cdot \frac{T_{\mathrm{PWM}}}{N-1} \tag{11.8}$$

Given the desired ripple voltage percentage (α_{ripple}), the voltage ripple across the capacitor can be expressed as:

$$\Delta V_{\mathrm{CFLY}} = 2 \cdot \alpha_{\mathrm{ripple}} \cdot \frac{V_{\mathrm{IN}}}{N-1} \tag{11.9}$$

By combining Eqs. (11.8) and (11.9), the flying capacitor sizing equation can be derived:

$$C_{\mathrm{FLY}} = \frac{\Delta Q_{\mathrm{CFLY}}}{\Delta V_{\mathrm{CFLY}}} = \frac{i_{\mathrm{load}}}{2 \cdot \alpha_{\mathrm{ripple}} \cdot V_{\mathrm{IN}} \cdot f_{\mathrm{PWM}}} \tag{11.10}$$

Equation (11.10) is independent of the level number, so it is important to consider that the total converter capacitance increases with each additional level. The number of discrete capacitors increases more than linearly because capacitors have lower capacitance density at higher voltage. Increasing the number of levels also means linearly increasing the number of gate drivers (one per switch) that must provide functional or galvanic isolation and the floating power supplies for each of them. The choice of levels is a tradeoff between the benefits on the inductor (or on the motor) and the increased complexity of the design.

11.2.6 Flying Capacitors Technology Selection

Class II ceramic capacitor energy storage density is significantly higher than other non-electrolytic dielectric technologies. However, the main drawback of ceramics is the dielectric dependence on voltage bias. This means that the effective capacitance of the device operated at a DC bias results in a fraction of the nominal capacitance. As a result, flying capacitors must be sized, based on their de-rated capacitance value for the voltage at which they will be operating. Many capacitor producers offer tools to select and analyze the capacitor derating with applied voltage and temperature conditions.

A good design practice considers the de-rated value when simulating the converter. Below 450 V, the X6S dielectric capacitor shows a good energy density; between 450 and 650 V, the X7T is a good choice. Above 650 V, it is possible to use X7T capacitors in series. Figure 11.6 shows the outputs for an X7T 450 V rated, 2.2 μF capacitor obtained with SimSurfing design support software from muRata [7] that aids with capacitor selection. Figure 11.6a shows that the effective capacitance drops to 880 nF when the capacitor is used to store charge at 300 V_{DC}. Figure 11.6b shows the temperature dependence of the same capacitor. It is worth noting that in some cases the change in capacitance as function of temperature decreases to very low values as function of DC voltage bias. This needs to be checked in the specific capacitor datasheet. Figure 11.7 shows that the de-rated capacitance of an X6S 450 V rated, 2.2 μF capacitor is quite stable with temperature.

From Eq. (11.10), the capacitance increases with the load current. Thus, many parallel discrete capacitors are required, especially at higher-voltage levels. With the actual ceramic

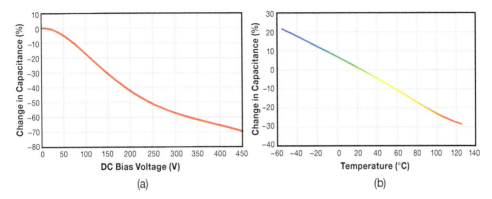

Figure 11.6 muRata 450 V, 2.2 μF X7T (a) capacitance derating versus DC bias, and (b) capacitance versus temperature.

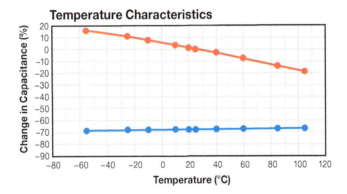

Figure 11.7 TDK C5750X6S2W225KT 450 V 2.2 μF X6S, capacitance derating versus temperature at 0 V bias (red) and 225 V bias (blue).

technology, the FCML converter is a practical solution for currents below 100 A_{RMS}. It is also clear that above a certain number of levels, the benefits of using a FCML converter diminish, and the specific level number depends on application, bus voltage, and load power.

11.2.7 Effect of Multilevel on the Inductor Sizing

In a buck or boost converter, the multilevel topology allows the reduction of the required filtering inductance for a given ripple specification. The maximum voltage step applied to the inductor is the voltage across one level, $V_{IN}/(N-1)$, and, thanks to the phase-shifted PWM modulation, the effective switching frequency seen by the inductor is equal to $(N-1) \times f_{PWM}$, where f_{PWM} is the switching frequency of the converter as applied to the switching devices. The combined effects reduce the required inductance to meet a given ripple by a factor of $(N-1)^2$ compared to a two-level converter switching at f_{PWM}. This can be observed by looking at the peak-to-peak inductor current ripple that for a two-level buck converter is given by:

$$\Delta i_{pp} = \frac{V_{IN} \cdot (1-D) \cdot D}{L \cdot f_{PWM}} \qquad (11.11)$$

Δi_{pp} has a maximum when $D = 50\%$. In a multilevel converter, the formula is slightly more complex because the voltage V_{IN} is divided into $(N-1)$ steps, the frequency is multiplied by the factor $(N-1)$, and the effective duty D_{eff} depends on the difference between the control signals of two adjacent switching cells. It can be demonstrated that when the duty is between these values:

$$\frac{m-1}{N-1} < D_{eff} < \frac{m}{N-1} \qquad m = 1, \ldots, (N-1) \tag{11.12}$$

the current peak is reached when:

$$D_{eff,pk} = \frac{1}{2} \cdot \frac{2m-1}{N-1} \qquad m = 1, \ldots, (N-1) \tag{11.13}$$

Given the above equations, Eq. (11.11) can be generalized as:

$$\Delta i_{pp} = \frac{V_{IN} \cdot (1 - D_{eff}) \cdot D_{eff}}{L \cdot f_{PWM} \cdot (N-1)^2} \tag{11.14}$$

Equation (11.14) can be used to derive the inductance L given the desired ripple, the number of levels, and the PWM frequency. For the same ripple current, a four-level converter requires an inductance nine times smaller than a two-level converter.

In a PFC for a 150 W TV adapter, where $V_{IN} = 400$ V to satisfy the PFC requirement for universal voltages (85–265 V_{AC} rms) and $f_{PWM} = 100$ kHz to avoid the differential EMI filter requirement [8], it can be shown that $L_{two\text{-}level} = 1000$ µH while the $L_{three\text{-}level} = 250$ µH, illustrating how the three-level converter offers a four times inductance reduction.

The impact on the core volume reduction is shown in Figure 11.8. A core volume reduction of approximately a factor of two is obtained with the three-level design. The height reduction is particularly significant for low-profile TV and laptop adapters.

11.2.8 Effect of Multilevel Inverters on Motor Drives

In a motor drive inverter, the inductance is generally part of the motor design and may not be reduced as in a buck or boost converter. Regardless, the multilevel topology still brings several benefits, even if the motor inductance does not change. Given the optimal frequency operation,

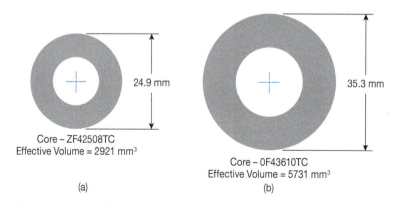

Figure 11.8 Inductor area and volume comparison between (a) three-level boost PFC, and (b) two-level boost PFC.

which varies with motor construction, and the $(N-1) \times f_{\mathrm{PWM}}$ frequency multiplication factor of the multilevel inverter, there are two possible benefits to motor drives:

- Reduced PWM frequency so that the overall inverter switching dissipation is lower and the motor is operated at its optimal frequency.
- Higher PWM frequency in conjunction with LC filtering at the inverter's output to operate the motor with an almost ideal sinusoidal voltage waveform.

In both cases, multilevel inverters lower the total harmonic distortion (THD) of the applied voltage to the motor, making the motor more mechanically efficient, as described in detail in Chapter 14.

In servo drive inverters, there are cases where the motor is located physically far from the inverter. When using high-voltage two-level inverters, the motor cables, due to their length, act as transmission lines and generate many reflections, which cause dangerous over-voltage spikes for the power switches and motor windings, notwithstanding detrimental EMI emissions. With two-level inverters, the only chance to minimize these transmission line effects is to use expensive shielded cables or to reduce the dv/dt to a point where the switching power dissipation becomes the dominant factor. On the other hand, multilevel inverters, with their multiple voltage steps approach, reduce the magnitude of the voltage transitions, and hence, the reflections associated with over-voltage spikes.

11.2.9 Effect of Multilevel Designs on Electromagnetic Interference (EMI) Filter

In power converters, the dv/dt of the switches is responsible for the common mode noise due to the capacitive coupling to the Earth, while the di/dt in the switches and the current ripple in the inductors are responsible for the differential mode noise. These noise sources must be minimized at the source or attenuated to comply with emissions regulations.

In applications connected to the grid, an input EMI filter is used to attenuate the interference, and depending on the power, it can be the bulkiest part of the converter. Multilevel topologies reduce the voltage steps at every level, thus reducing the switching dv/dt. Also, these topologies can reduce the ripple in the inductors, thus reducing the differential noise.

To achieve an EMI filter benefit with the multilevel boost topology, it is generally required to go to many levels, as shown in a similar seven-level topology in [9]. To avoid a complex high number of levels and still reduce the EMI filter size, a totem-pole topology (TP) [10–12], which exploits the zero reverse recovery losses of GaN transistors and common-mode symmetry, can be used. The availability of GaN transistors has made topologies like Totem-Pole Boost Rectifiers viable and efficient [10–12]. Such topologies enable high efficiencies, exceeding 98%, and power densities exceeding 40 W/in^3 for laptop adapters, telecom rectifiers, and onboard hybrid electric vehicle (HEV) chargers.

11.2.10 FCML Converter Simulation

Different tools and methods are used to simulate a power converter operation, and the choice of the proper tool depends on the type of analysis that must be carried out for a specific design. Generally, good design practices involve two distinct simulation types: a system-level simulation to define the converter's main components and a circuit-level simulation to refine circuit design and layout.

The most common system-level simulation tools are based on simplified piece-wise linear switch simulation. The switches are modeled as ideal and are either ON or OFF. Electrical analysis of each configuration can be made by obtaining a set of differential equations that describe the dynamics of the converter for every switching state. These tools are useful for performing frequency analysis and determining the controller strategy at the system level. Moreover, by knowing the switch-on (E_{ON}), switch-off (E_{OFF}), and recovery (E_{rec}, if there are p-n diodes) dissipated energies per device, it is possible to derive the converter efficiency and thermal behavior. The energy tables can be derived experimentally or using powerful circuit simulation software such as Spice. They define the switch energies E_{ON}, E_{OFF}, and E_{rec} at different voltages, temperatures, and currents.

11.2.11 Gate Driver Implementation

There are challenges to implementing multilevel converters with GaN devices because most switches are not ground-referenced. Each switch in the FCML requires a dedicated gate driver referenced to its source. Hence, each gate driver must be able to level shift the controller command PWM signal to the appropriate voltage level, as depicted in Figure 11.9.

As a practical rule, reverse junction [13] functional-isolated high-side gate drivers can be used when the nominal V_{IN} voltage is less than 400 V. Above 400 V, it is better to use galvanic isolated gate drivers. The use of reverse junction gate drivers below 400 V, due to their low quiescent current on the switch side, allows the use of bootstrapping techniques. Galvanic

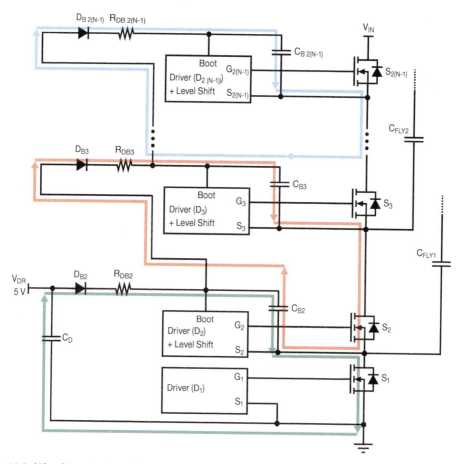

Figure 11.9 N-level inverter leg with conventional cascaded diode bootstrap circuit.

isolated gate drivers require a higher current to supply the switch side, which can be achieved only with isolated floating power supplies.

11.2.12 Bootstrap Power Supply Solutions for GaN Transistors

For GaN transistor-based multilevel converters, the common Schottky diode bootstrapping technique must be adapted to ensure proper gate voltage (between 4.75 and 5.5 V) is applied at each switch gate. To understand this point, consider the generic N-level inverter leg circuit shown in Figure 11.9, along with the switches and the bootstrap circuit. The gate drive supply is obtained with a conventional cascaded diode bootstrap circuit.

In Figure 11.9, blue denotes the charging path for bootstrap capacitor $C_{B\,2(N-1)}$ corresponding to the uppermost device $S_{2(N-1)}$, red is used to denote the charging path for an intermediate bootstrap capacitor C_{B3}, and green is used to show the charging path of the lowest bootstrap capacitor C_{B2}. Each bootstrap capacitor is charged by the bootstrap capacitor at its immediate lower level through the corresponding bootstrap diode and resistor: C_{B4} charges from C_{B3} through D_{B4} and R_{DB4} when S_3 is on, C_{B3} charges from C_{B2} through D_{B3} and R_{B3} when S_2 is on, and so on.

In this scheme, the lower levels bootstrap capacitors must store the charge for the subsequent levels, i.e. in a three-level FCML, the charges for the bootstrap capacitors C_{B4}, C_{B3}, and C_{B2} must be 3Q, 2Q, and Q, respectively; the local 5 V capacitor, that supplies the lowest gate driver, must store the charge 4Q. The higher-level bootstrap capacitors suffer from an undercharging problem due to the addition of subsequent bootstrap diode drops. They can also overcharge during dead time because of the higher reverse conduction voltage of GaN transistors [14]. These problems are series (or cascaded) effects, and they multiply with a higher number of levels.

Several solutions have been researched in the past to drive GaN devices in multilevel topologies [15–20]. For most of the proposed solutions [15–19], several extraneous on-chip isolated DC–DC converters are utilized to power the gate drive of the floating switches. In [16], a complex double-charge pump that utilizes the power side flying capacitors (C_{FLY} in Figure 11.2a) is used for level-shifting and bootstrapping. This leads to layout and design complexities. In all these implementations, the complicated drive circuitry makes for very specialized solutions requiring tight component tolerances as well as added manufacturing difficulties, apart from limiting power densities.

In this section, simpler ways of implementing the gate drive scheme will be discussed, along with actual implementations.

To remedy the issues of the cascaded diode approach in multilevel converters, a cascaded synchronous bootstrapping method was introduced in [21]. The corresponding circuit is shown in Figure 11.10 for a generic N-level case. This method does not require the extraneous voltage regulators in [16], does not utilize power circuitry for charging the bootstrap capacitors, and is independent of the switching sequence of the converter. It only requires that each switch, other than the topmost switch, turns on at some point in the switching cycle. This scheme improves upon the cascaded diode scheme by replacing the large forward drop bootstrap diode with a much lower dropout synchronous GaN transistor (Q_{BST}) during the intended charging period and by preventing overcharging of the bootstrap capacitors during dead-time conduction periods by operating as a voltage balancing high-forward-drop GaN diode.

Referencing Figure 11.10, it works as follows: when the gate signal G_1 is low, C_{BST1} charges to $V_{DR} - V_{BST1} \approx 3.5$–$4.0$ V. When G_1 is high, the gate of Q_{BST2} is pushed to $V_{DR} + V_{CBST1}$ while its source is at V_{DR} (5 V). This reverse biases D_{BST1} while Q_{BST2} is turned on. With the low

334 | *GaN Power Devices for Efficient Power Conversion*

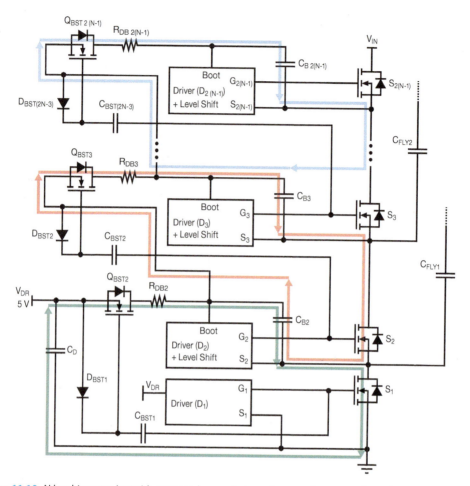

Figure 11.10 *N-level inverter leg with proposed cascaded synchronous bootstrapping scheme.*

dropout of Q_{BST2}, C_{B2} charges to a value very close to V_{DR} through the charging path shown in green. This also results in low losses in the charging circuit since the GaN transistor Q_{BST2} (high voltage, low current) is devoid of reverse recovery losses [14]. Similarly, C_{B3} charges off C_{B2}, C_{B4} charges off C_{B3}, and so on.

The gate drive waveforms of a three-level buck converter for the conventional diode approach (refer to Figure 11.9) are shown in Figure 11.11, along with the corresponding drive voltages at no load and no power. The gate drive voltages decrease with the higher levels as expected, with $V_{GS3} = 3.51$ V (drive voltage for S_3 with reference to Figure 11.9), which is well below the desired GaN transistor drive voltage [22]. V_{GS4} (drive voltage for S_4) is a diode drop below V_{GS3} at 2.81 V, which triggers the under-voltage lockout (UVLO) function in the topmost gate driver, thus making the converter non-functional. Hence, the classic scheme cannot be used for GaN transistor-based FCML (three and higher-level) converters.

The waveforms with the synchronous bootstrapped approach are shown in Figure 11.12a for a three-level version of the schematic proposed in Figure 11.10. The topmost (S_4) and bottom-most (S_1) switches' gate voltage waveforms are shown at no load in Figure 11.12a, which represents the worst-case condition for gate voltage variation. The difference is around 0.5 V, within the drive voltage range requirements for the GaN transistors [22].

Multilevel Converters | **335**

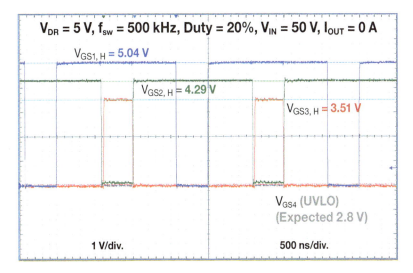

Figure 11.11 Gate drive waveforms for conventional bootstrap method using GaN transistors.

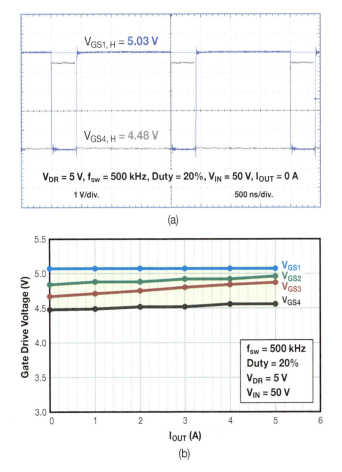

Figure 11.12 (a) Gate drive waveforms for cascaded synchronous bootstrap technique and (b) drive voltage variation overload current when using GaN transistors.

For the cases with load, the gate voltages are plotted in Figure 11.12b. The region in green represents a deviation of 0.5 V from 5 V. There is an insignificant drop between all GaN transistors for all operating conditions. The low-voltage difference between the bootstrap capacitors demonstrates that this technique can be implemented for a GaN-based multilevel converter with a higher number of levels.

The cascaded synchronous bootstrap method discussed can use a small GaN transistor in the bootstrap circuit (Q_{BSTx}). In the 48 V design example described later in this chapter, the 100 V-rated EPC2038 [23] was used. Such low-profile, low-current discrete GaN transistors are also available at higher voltages for higher-voltage implementations (e.g. EPC2054 rated 200 V and EPC2050 rated 350 V). The synchronous bootstrap circuit layout needs to have low inductance and be compact to be adequately implemented and provide the high power density characteristic of power converters designed with GaN transistors.

Another simpler discrete solution for a three-level converter is shown in Figure 11.13 [24]. In this solution, gate drive voltage regulation relies on the Zener diode reverse clamping mechanism and can be extended to four or more levels. The scheme works as follows. A voltage higher than 5 V is applied to the gate drive circuit as the single external power source. In this

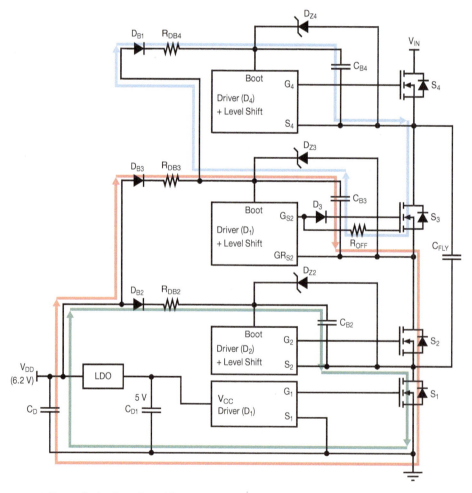

Figure 11.13 Zener diode clamp-based bootstrap scheme for three-level buck converter.

example, 6.2 V (V_{DD}) was used. A low dropout regulator (LDO) generates a 5 V ground-referenced supply (V_{CC}) for the bottom switch (S_1) gate drive. The LDO may be internal in the half-bridge driver ICs, such as in the LMG1210 from Texas Instruments [25]. The bootstrap diode D_{B2} has a forward drop of 1 V, which causes the bootstrap capacitor C_{B2} for switch S_2 to eventually charge to a value of $V_{DD} - V_{DB2}$ (6.2 V minus 1 V ≈ 5.2 V) when S_1 is on. This charging path is shown in green. The Zener Diode D_{Z2} has a reverse voltage rating of 5.1 V and hence clamps C_{B2} to 5.1 V. When the body diode of S_1 conducts during dead time intervals, the Zener diode also prevents overcharging of C_{B2}.

Using the same technique, the bootstrap capacitor C_{B3} for the switch S_3 charges when S_1 and S_2 are on simultaneously during the freewheeling period [21]. This path is shown in red. Since the body diode drops of S_1 and S_2 add in series, if a 5.1 V Zener diode is used for D_{Z3}, the clamping stress on it during dead-time conduction periods is more severe than D_{Z2}. To reduce this stress, a 5.7 V Zener diode is used to clamp the voltage of C_{B3} to 5.7 V. However, the gate of S_3 is still not in an over-voltage state since a diode, D_3, with a forward drop of 0.5 V, is inserted into the forward path of the gate charging circuit so that S_3 experiences a gate voltage of 5.2 V.

The top-level bootstrap capacitor C_{B4} for the switch S_4 charges from C_{B3} when S_3 is on. This path is shown in blue. Since D_{Z3} will clamp V_{CB3} to 5.7 V, V_{CB4} charges to $V_{CB3} - V_{DB4}$ (5.7 V − 1 V ≈ 4.7 V). Hence all the GaN transistors S_1 to S_4 are driven with gate drive voltages between 4.7 and 5.2 V, which provides satisfactory tight regulation with a relatively simple scheme. The analysis here assumes that the resistance drops in the bootstrap charging paths are zero.

The Zener-based gate drive scheme can also be used at higher V_{IN} voltage. The gate drive waveforms for the bottom switch (V_{GS1}) and the topmost switch (V_{GS4}) are shown in Figure 11.14 for a three-level converter with V_{IN} = 400 V at full load. S_4 is driven at 4.51 V, while S_1 is driven at 4.83 V. The offsets are due to finite values of resistances in the bootstrap charging paths.

The one drawback of this method is the extra power loss incurred due to the clamping action of the Zener switches. In this example, the total power loss due to the gate drive circuit is 0.36 W at full load, while the total gate drive loss in [21] is only 0.065 W at full load. Hence, for

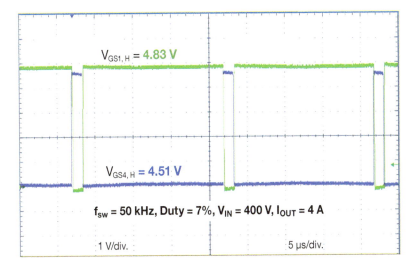

Figure 11.14 Gate drive waveforms for the Zener-clamp-based gate-drive scheme when using GaN transistors.

11.3 Experimental Examples

11.3.1 Multilevel Converters for 48 V Applications

With the advent of the 48 V server rack power delivery architecture in data centers [26–28], there has been a renewed interest in improving power efficiency through various circuit topological approaches. These designs range from hard-switching [29–32] to highly resonant [33–35], fully regulated to unregulated, and fully isolated to non-isolated. Due to the high power density requirements, multilevel converters have become attractive as well.

Switched capacitor circuits are good examples of multilevel topologies that can effectively reduce inductor size [6, 14, 15, 36–39]. As an interesting variant, the switched resonant tank and other hybrid converters have also become popular [40–42]. A three-level buck converter (shown in Figure 11.15a is built with GaN transistors as a design example. The primary timing diagrams and idealized waveforms of this design are shown in Figure 11.15b. As explained before, the inductance in a three-level buck is reduced by four times compared to a conventional buck, since the inductor experiences an effective doubling of switching frequency and at half the input voltage. This is evident by applying Eq. (11.14) to a three-level converter ($N = 3$) to obtain the following equations for the inductance:

$$L_{\text{three-level}} = \frac{V_{\text{IN}}(0.5 - D)D}{\Delta I_{\text{pp}} f_{\text{PWM}}}, D < 0.5 \tag{11.15}$$

$$L_{\text{three-level}} = \frac{V_{\text{IN}}(D - 0.5)(1 - D)}{\Delta I_{\text{pp}} f_{\text{PWM}}}, D > 0.5 \tag{11.16}$$

where V_{IN} is input voltage, D is duty ratio, f_{PWM} is switching frequency, ΔI_{pp} is peak-to-peak inductor current ripple.

These equations (normalized to $V_{\text{IN}}/\Delta I_{\text{pp}} f_{\text{PWM}}$) are plotted in Figure 11.16, demonstrating the inductance reduction gained with a three-level topology over the entire duty ratio range. It is to be noted that a 50% duty cycle is the operating condition of a conventional inductor-less switched capacitor circuit.

To illustrate the performance benefit, an ultra-thin three-level buck converter was compared to a conventional two-level converter. The converters' power schematic circuits are shown in Figure 11.17a,b, respectively.

Figure 11.18a shows a photo of the two-phase converter configured with four 100 V rated, 5.6 mΩ EPC2045 [43] FETs. Figure 11.18b shows a photo of the three-level converter configured with three 40 V rated, 2.8 mΩ EPC2055 [44] FETs and one 100 V rated, 2.8 mΩ EPC2218 [45] for the uppermost device, as this FET needs to block the full input voltage for a short period during start-up.

For the comparison between the two-phase and three-level designs, the same switching frequency of 400 kHz and the same inductor height of 3.5 mm were used. The 2.4 μH Würth 7443762504022 inductor was selected for the multilevel converter, and the 10 μH Vishay IHLP5050CE-01 inductor was selected for the two-phase converter.

The measured efficiency comparison between the two-phase and three-level converter is shown in Figure 11.19. The multilevel topology offers a 1.2% higher efficiency at 12.5 A load

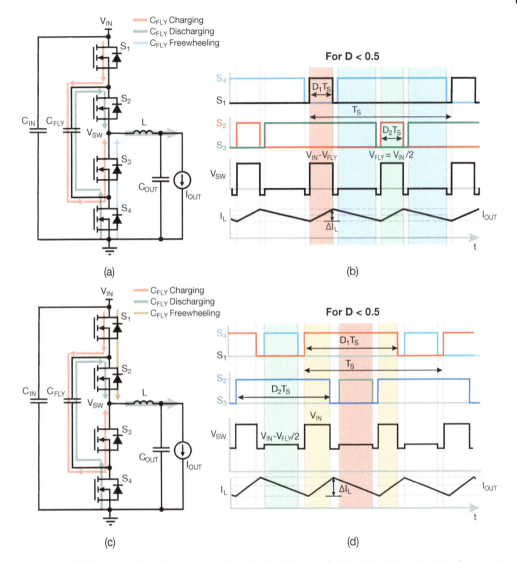

Figure 11.15 (a) Three-level buck converter and (b) timing diagrams for $D < 0.5$, (c) three-level buck converter and (d) timing diagrams for $D > 0.5$.

current than the two-phase configuration. Figure 11.20 illustrates that the multilevel converter operates cooler than the two-phase converter when delivering the same load power of 12.5 A into 20 V from a 48 V supply, reaching a thermal steady state with 800 LFM airflow.

11.3.2 A 300 V Bi-Directional Buck-Reverse Boost Converter

The second example of the benefit of multilevel converters is a high-voltage application with a three-level, two-phase, bi-directional buck-reverse boost converter.

The power schematic circuit for the bi-directional, two-phase, three-level converter is shown in Figure 11.21. Figure 11.22 shows a top-view photo of the three-level converter, and Figure 11.23 shows the zoomed-in area of the gate driver section on the top side of the board

Figure 11.16 Inductance reduction as a benefit of using a three-level converter.

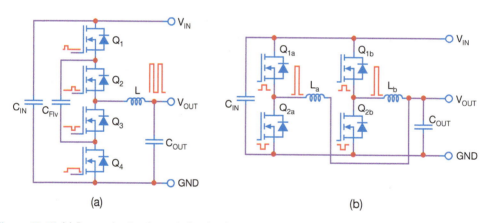

Figure 11.17 (a) Power circuit schematic for the three-level converter, (b) power circuit schematic for the two-phase converter.

and the GaN FETs on the bottom side. The converter is configured with two 200 V rated, 8 mΩ EPC2215 [46] FETs in parallel. This converter is designed to operate only with less than 50% duty cycle, allowing simplification of the bootstrap circuit and using the configuration shown in Figure 11.9 with minimal loss of voltage on the uppermost device.

The measured efficiency of the three-level converter operating with one phase in buck mode is shown in Figure 11.24 when operating from a 200 V supply and load of 40 V up to 29 A. From Figure 11.24, in the range from 15–20 A load, the converter efficiency exceeds 97.8%. Figure 11.25 shows the three-level converter thermal performance when operating from 200 V and delivering 24 A into 40 V, which has only a 30 °C temperature rise with 1700 LFM airflow. The converter was fitted with a small aluminum heatsink measuring just 50 × 30 mm and has 12 mm tall fins on a 3 mm base.

Multilevel Converters | 341

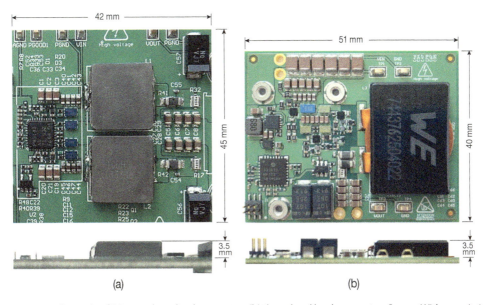

Figure 11.18 Example of (a) two-phase buck converter, (b) three-level buck converter. *Source:* With permission of EPC.

Figure 11.19 Efficiency comparison between the three-level buck converter and the conventional two-level buck converter.

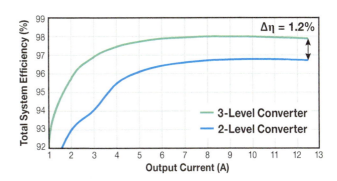

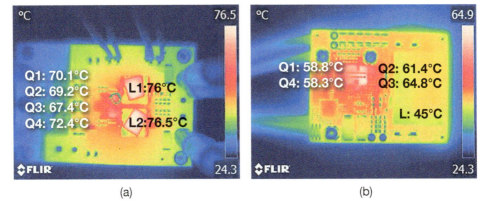

Figure 11.20 Thermal comparison between the three-level buck converter (b) and the conventional two-level buck converter (a) when powered from 48 V and delivering 12.5 A into a 20 V load reaching thermal steady state in 800 LFM airflow. *Source:* With permission of EPC.

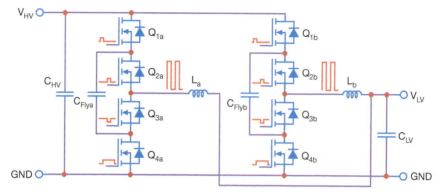

Figure 11.21 Power schematic circuit for the bi-directional three-level, two-phase converter.

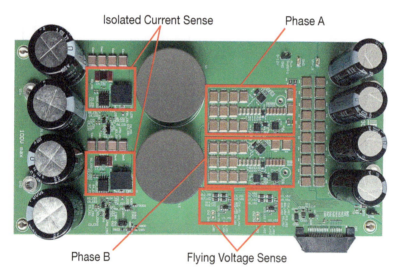

Figure 11.22 The experimental unit image with identification of the various function circuits. *Source:* With permission of EPC.

11.3.3 A 270 V_{AC}–400 V_{DC} PFC Totem Pole Boost Converter

The third example of the benefit of multilevel converters is a high-voltage application with a four-level totem-pole PFC boost converter designed for 3 kW load power. The schematic circuit for the PFC four-level converter is shown in Figure 11.26. The transistors Q_1–Q_6 GaN FETs are in the high-frequency leg, while Q_7 and Q_8 are responsible for the synchronous rectification at the mains line low frequency. Two MOSFETs in parallel per switch have been used for Q_7 and Q_8.

Figure 11.27 shows two pictures of the compact four-level converter, highlighting the main sections. The converter is an assembly of multiple PCBs, and it achieved a power density of

Multilevel Converters | 343

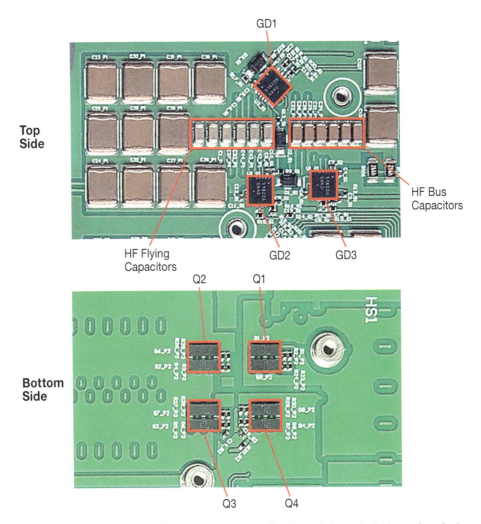

Figure 11.23 A zoomed-in layout of the power stage area showing switches (Q1–Q4 located on the bottom side of the PCB), gate drivers (GD1-3), and capacitor placement. *Source:* With permission of EPC.

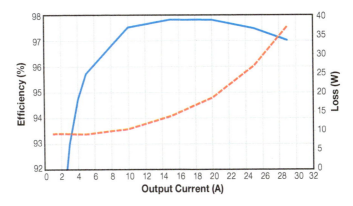

Figure 11.24 Measured efficiency (left axis) and power losses (right axis) with V_{OUT} = 40 V and V_{IN} = 200 V.

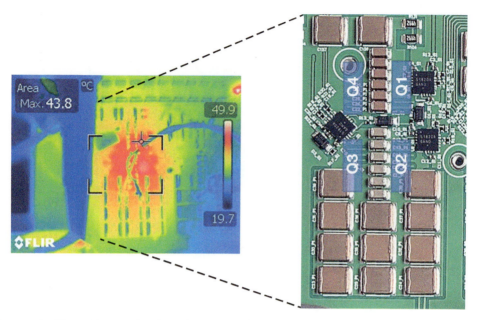

Figure 11.25 Thermal image taken from the top side of the PCB when operating from 200 V input and delivering 24 A into 40 V load in an ambient of 20 °C. *Source:* With permission of EPC.

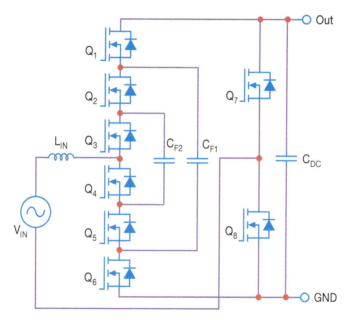

Figure 11.26 Power schematic circuit for the four-level flying capacitor totem pole PFC converter.

129 W/in^3. Figure 11.28 shows the details of the top and bottom of the GaN daughter card used in the circuit. The converter is configured with 200 V rated, 8 mΩ EPC2215 [46] FETs. The boost inductor is visible in Figure 11.28b and comprises four 3.3 μH low-profile inductors connected in series, for a total of 13.2 μH, that also reduces the total parasitic capacitance.

Multilevel Converters | 345

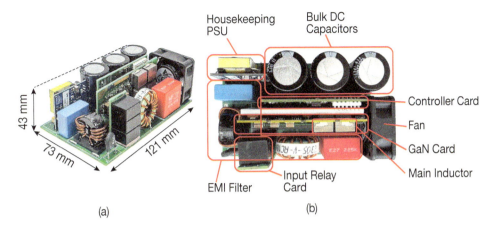

Figure 11.27 (a) Four-level, totem pole PFC converter dimensions. (b) Converter top view with main sections. *Source:* With permission of EPC.

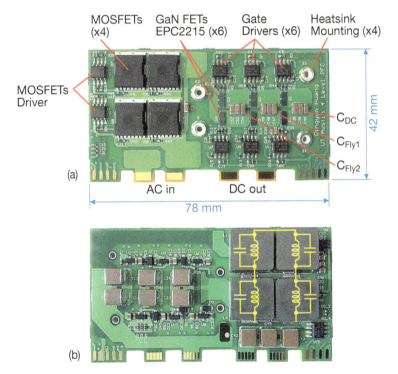

Figure 11.28 (a) GaN card top side, showing the low-frequency MOS and the high-frequency GaN FETs, and (b) GaN card bottom side showing the boost inductor made of four inductors in series. *Source:* With permission of EPC.

The multilevel topology brings an evident advantage in terms of boost inductor size reduction, as shown in Figure 11.28b, thanks to the frequency multiplication factor of the flying capacitors scheme. The frequency at the inductor is three times higher than the switching frequency of the GaN FETs. Given the same desired ripple current, the

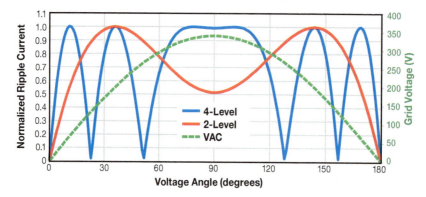

Figure 11.29 Normalized current ripple at the inductor. The ripple of a two-level converter is compared to that of a four-level converter with a 9× smaller PFC inductance and 3× higher frequency at the inductor.

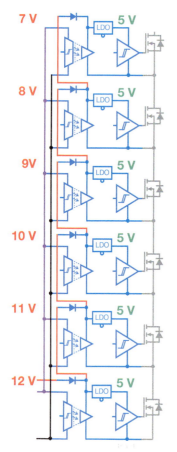

Figure 11.30 Bootstrap architecture with low drop regulators for 5 V stable supply at each switch in the four-level PFC converter.

designer can reduce the inductance value by a factor of nine because, compared to a two-level converter, the frequency seen by the inductor is three times higher, and the voltage drop is three times lower. The ripple diagram of a conventional two-level converter with an inductor L, compared to a four-level converter with an inductor $L/9$ as a function of the grid voltage angle, is shown in Figure 11.29.

The floating voltages for all gate drivers in the converter are generated as depicted in Figure 11.30. There is a series of bootstrap diodes with a low drop regulator following each to ensure a 5 V stable supply is applied to each GaN FET gate driver. To account for the voltage drop at each stage, the initial voltage has been set to 12 V, following a similar principle described in Figure 11.13.

The four-level totem pole PFC was tested up to 3 kW output power. The operation is depicted in Figure 11.31, which shows the regulated voltage V_{OUT} and the input current I_{LIN}.

Figure 11.32 shows the converter's efficiency curve compared to two other solutions based on two-level converters. From 1 to 3 kW power, the efficiency of the four-level totem pole PFC with EPC2215 GaN FETs is above 99%. The total harmonic distortion (THD) was measured in the same range at 3%.

Finally, Figure 11.33 shows the thermal behavior of the GaN daughter card at 3 kW operation. The converter was tested with 200 LFM airflow without a ducting box to convey the airflow. The maximum observed temperature was 71 °C.

Multilevel Converters | 347

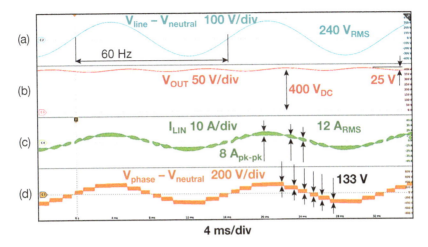

Figure 11.31 Converter operation at 2 kW load. (a) ac input line voltage 240 V_{RMS}, (b) output PFC voltage 400 V, (c) ac input line current, and (d) high-frequency leg switching node voltage with respect to ac neutral voltage. All graphs have 4 ms/div timescale.

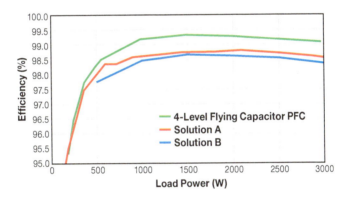

Figure 11.32 Four-level PFC converter efficiency compared to two other solutions.

Figure 11.33 Four-level PFC converter thermal image at 3 kW operation with 200 LFM airflow. *Source:* With permission of EPC.

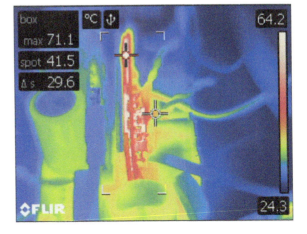

11.4 Summary

This chapter discussed the benefits of multilevel converters based on GaN transistors. Apart from topological benefits such as passive component size reductions, the benefits of using lower FOM devices were explored. To implement wide bandgap devices in these topologies, suitable gate drive schemes have been discussed for low- and high-voltage cases. The suitability of these topologies, including variants such as totem pole for PFC applications, is also briefly demonstrated. Finally, the electrical performance comparison of both the three-level prototypes has shown that they offer improvements over their two-level counterparts at low and high voltages.

In Chapter 12, the benefits of using GaN transistors in high-quality class D audio amplifiers will be discussed.

References

1 International Energy Agency (iea) (2024). Electricity 2024, Analysis and Forecast to 2026. https://iea.blob.core.windows.net/assets/6b2fd954-2017-408e-bf08-952fdd62118a/ Electricity2024-Analysisandforecastto2026.pdf.

2 Pratt, A., Kumar, P., and Aldridge, T. V. (2007). Evaluation of 400V DC distribution in telco and data centers to improve energy efficiency. *Proceedings of the IEEE International Telecommunications Energy Conference (INTELEC)*, October 2007, 32–39.

3 Ahmed, M.H., Fei, C., Lee, F.C., and Li, Q. (2017). 48-V voltage regulator module with PCB winding matrix transformer for future data centers. *IEEE Trans. on Ind. Electron.* 64 (12): 9302–9310.

4 Meaney, P. (2017). PFC for not dummies. High Volt Interactive Training, Texas Instruments.

5 Lu, B. (2006). Investigation of high-density integrated solution for AC/DC conversion of a distributed power system. Ph.D. dissertation. Virginia Tech.

6 Meynard, T.A. and Foch, H. (1992). Multi-level conversion: high voltage choppers and voltage-source inverter. PESC '92 Record. *23rd Annual IEEE Power Electronics Specialists Conference*, Toledo, Spain (1992), vol. 1, 397–403.

7 muRata. SimSurfing design support tool. https://www.murata.com/en-eu/tool/simsurfing.

8 IEC International Standard. (2018). IEC61000-3-2 Electromagnetic compatibility (EMC) – Part 3-2: Limits – Limits for harmonic current emissions (equipment input current ≤16 A per phase). https://webstore.iec.ch/publication/62553.

9 Lei, Y., Barth, C., Qin, S. et al. (2017). A 2 kW, single phase, 7-level flying capacitor multilevel inverter with an active energy buffer. *IEEE Trans. Power Electron.* 99: 1–1.

10 Liu, Z., Lee, F.C., Li, Q., and Yang, Y. (2016). Design of GaN-based MHz totem-pole PFC rectifier. *IEEE J. Emerg. Sel. Top. in Power Electron.* 4 (3): 799–807.

11 Texas Instruments (2018). TIDA-01604, 98.5% efficiency, 6.6-kW totem-pole PFC reference design for HEV/EV onboard charger. Texas Instruments Reference Design.

12 NXP Semiconductors (2016). Totem-pole bridgeless PFC design using MC56F82748. NXP Semiconductors Reference Design, Document: DRM174.

13 Texas Instruments (2017). Fundamentals of MOSFET and IGBT gate driver circuits. Application report SLUA618. (March 2017).

14 Reusch D. and de Rooij, M. (2017). Evaluation of gate drive overvoltage management methods for enhancement mode gallium nitride transistors. *IEEE Applied Power Electronics Conference and Exposition (APEC)*, Tampa, FL, 2017, 2459–2466.

15 Stillwell, A., Pilawa-Podgurski, R.C.N. (2017). A 5-level flying capacitor multi-level converter with integrated auxiliary power supply and start-up. *IEEE Applied Power Electronics Conference and Exposition (APEC)*, Tampa, FL, 2017, 2932–2938.

16 Ye, Z., Lei, Y., Liu, W. C., et al. (2017). Design and implementation of a low-cost and compact floating gate drive power circuit for GaN-based flying capacitor multi-level converters. *IEEE Applied Power Electronics Conference and Exposition (APEC)*, Tampa, FL, 2017, 2925–2931.

17 Moon, I., Carl F. Haken, Erik K. Saathoff, et al. (2017). Design and implementation of a 1.3 kW, 7-level flying capacitor multilevel AC-DC converter with power factor correction. *IEEE Applied Power Electronics Conference and Exposition (APEC)*, Tampa, FL 2017, 67–73.

18 Barth, C., Juan Colmenares, Thomas Foulkes, et al. (2017). Experimental evaluation of a 1 kW, single-phase, 3-level gallium nitride inverter in extreme cold environment. *IEEE Applied Power Electronics Conference and Exposition (APEC)*, Tampa, FL, 2017, 717–723.

19 Modeer, T., Barth, C.B., Pallo, N. et al. (2017). Design of a GaN-based, 9-level flying capacitor multilevel inverter with low inductance layout. *IEEE Applied Power Electronics Conference and Exposition (APEC)*, Tampa, FL, (2017), 2582–2589.

20 Chen, M. (2015). Merged multi-stage power conversion: A hybrid switched-capacitor/magnetics approach. Ph. D. dissertation. Massachusetts Institute of Technology.

21 Biswas S. and Reusch, D. (2018). GaN based switched capacitor three-level buck converter with cascaded synchronous bootstrap gate drive scheme. *IEEE Energy Conversion Congress and Exposition (ECCE)*, Portland, OR, 2018, 3490–3496.

22 Lidow, A., Strydom, J., de Rooij, M., and Reusch, D. (2014). *GaN Transistors for Efficient Power Conversion*, 2e. Hoboken, NJ: Wiley.

23 Efficient Power Conversion Corporation (2018). EPC2038 – enhancement-mode power transistor. EPC2038 datasheet. https://epc-co.com/epc/Portals/0/epc/documents/datasheets/EPC2038_datasheet.pdf.

24 Biswas S. and Reusch, D. (2018). Evaluation of GaN based multilevel converters. *IEEE Workshop on Wide Bandgap Devices and Applications (WiPDA)*, 2018, Atlanta, GA, 212–217.

25 Texas Instruments. LMG1210 – 200-V, 1.5-A, 3-A half-bridge GaN driver with adjustable dead time. http://www.ti.com/lit/ds/symlink/lmg1210.pdf.

26 Li, X. and Jiang, S. (2017). Google 48V power architecture. *IEEE Applied Power Electronics Conference and Exposition (APEC)*, Plenary. Tampa, FL, 2017.

27 Tung, L. (2016). Google, Facebook pause rivalries: Here's their 48V power-saving rack spec for Open Compute Project. ZDNET, August 2016. http://www.zdnet.com/article/google-facebook-pause-rivalries-heres-their-48v-power-saving-rack-spec-for-open-compute-project/.

28 Taranovich, S. (2016). Data center next generation power supply solutions for improved efficiency. EDN network, April 2016. https://www.edn.com/design/power-management/4441840/Data-center-next-generation-power-supply-solutions-for-improved-efficiency-.

29 Flex Power Modules, P.K.B. (2008). 4204B PI datasheet. https://flexpowermodules.com/resources/fpm-techspec-pkb4204.

30 General Electric (2021). EBDW025 – EBDW025A0B Barracuda∗ Series; DC-DC Converter Power Modules. EBDW025A0B datasheet. https://www.google.com/url?sa=t&source=web&rct=j&opi=89978449&url=https://www.mouser.com/datasheet/2/167/EBDW025A0B_DS-1920305.pdf&ved=2ahUKEwivvoqY16aGAxXQIUQIHfdXAv0QFnoECBMQAQ&usg=AOvVaw0YLkXYMTAKHQwjYW7eFxMd.

31 Ericsson. BMR457 – Advanced bus converter industry-standard Eighth-brick. BMR457 datasheet. www.ericsson.com.

32 Glaser, J. Strydom, J. and Reusch, D. (2015). High power fully regulated eighth-brick DC-DC converter with GaN FETs. *International Exhibition and Conference for Power Electronics, Intelligent Motion, Renewable Energy and Energy Management (PCIM Europe)*, 2015, 406–413.

33 Delta Electronics (2018). E54SJ12040 – 480W DC/DC Power Modules. E54SJ12040 datasheet. https://www.google.com/url?sa=t&source=web&rct=j&opi=89978449&url=https://filecenter. deltaww.com/Products/download/01/0102/datasheet/DS_E54SJ12040.pdf&ved=2ahUKEwig xKmZxqaGAxVFHUQIHa5FAi0QFnoECBAQAQ&usg=AOvVaw1hak_i0penKRH9JkIhyvZ9.

34 Vicor (2023). PI3546-00-LGIZ – ZVS Buck Regulator. PI3546-00-LGIZ evaluation board. https://www.vicorpower.com/products?productType=cfg&productKey=PI3546-00-LGIZ.

35 Vicor (2021). BCM48Bx120y300A00 – BCM® Bus Converter. BCM48Bx120y300A00 datasheet. https://www.google.com/url?sa=t&source=web&rct=j&opi=89978449&url=https:// www.vicorpower.com/documents/datasheets/BCM48B_120_300A00.pdf&ved=2ahUKEwj G08SByaaGAxXKOUQIHfuaBTgQFnoECBAQAQ&usg=AOvVaw06Kwt7dw-dhjsUtbWMIzqS.

36 Lei, Y., Barth, C., Qin, S. et al. (2016). A 2 kW, single-phase, 7-level, GaN inverter with an active energy buffer achieving 216 W/in^3 power density and 97.6% peak efficiency. *IEEE Applied Power Electronics Conference*, Long Beach, CA (2016).

37 Yousefzadeh, V., Alarcon, E., and Maksimovic, D. (2006). Three-level buck converter for envelope tracking applications. *IEEE Trans. Power Electron.* 21 (2): 549–552.

38 Reusch, D., Lee, F.C., and Xu, M. (2009). Three level buck converter with control and soft startup. *2009 IEEE Energy Conversion Congress and Exposition*, (September 2009), 31–35.

39 Rentmeister, J.S. and Stauth, J.T. (2017). A 48V:2V flying capacitor multilevel converter using current-limit control for flying capacitor balance. *IEEE Applied Power Electronics Conference and Exposition (APEC)*, Tampa, FL (2017), 367–372.

40 Li, Y., Lyu, X., Cao, D., Jiang S., and Nan, C. (2017). A high efficiency resonant switched-capacitor converter for data center. *IEEE Energy Conversion Congress and Exposition (ECCE)*, Cincinnati, OH, (2017), 4460–4466.

41 Jiang, S., Nan, C. Li, X., Chung, C. and Yazdani, M. (2018). Switched tank converters. *IEEE Applied Power Electronics Conference and Exposition (APEC)*, San Antonio, TX, (2018), 81–90.

42 Suo, G., Das R., and Le, H. (2018). A 95%-efficient 48-V to 1-V/10 A VRM hybrid converter using interleaved dual inductors. *IEEE Energy Conversion Congress and Exposition (ECCE)*, Portland, OR, (2018), 3825–3830.

43 Efficient Power Conversion Corporation (2018). EPC2045 – enhancement-mode power transistor. EPC2045 datasheet. http://epc-co.com/epc/Portals/0/epc/documents/datasheets/ EPC2045_datasheet.pdf.

44 Efficient Power Conversion Corporation (2022). EPC2055 – enhancement-mode power transistor. EPC2055 datasheet. https://epc-co.com/epc/Portals/0/epc/documents/datasheets/ EPC2055_datasheet.pdf.

45 Efficient Power Conversion Corporation (2024). EPC2218 – enhancement-mode power transistor. EPC2218 datasheet. http://epc-co.com/epc/Portals/0/epc/documents/datasheets/ EPC2218_datasheet.pdf.

46 Efficient Power Conversion Corporation (2023). EPC2215 – enhancement-mode power transistor. EPC2215 datasheet. http://epc-co.com/epc/Portals/0/epc/documents/datasheets/ EPC2215_datasheet.pdf.

12

Class D Audio Amplifiers

12.1 Introduction

High-fidelity audio amplifiers are a power application with demanding precision requirements. As power supplies and motor drives, starting in the 1970s, have improved in efficiency and power density by using switching converters instead of linear dissipative circuits, audio amplifiers moved from linear class A or class AB to switching class D [1, 2]. The precision requirement for audio amplifiers to achieve high fidelity delayed the wide adoption of class D, as the first amplifiers with acceptable performance were developed only in the late 1990s, and only very recently have the best class D amplifiers achieved the same performance of linear amplifiers [3].

The power ratings of audio amplification systems are continually increasing. These higher power ratings are needed to achieve a deeper bass in a smaller package and to cover large areas, such as public buildings, venues, arenas, and outdoor concerts and events spaces with improved audio quality. In some cases, the requirement for these large spaces exceeds a megawatt system power level [4]. In addition, the amplification channels count is growing, both in home, professional, and in-car immersive systems, calling for high power density [5]. Therefore, amplifiers with four or more channels, as shown in Figure 12.1, are becoming common [6]. Since the introduction of Class D helped this growth, better semiconductor switches such as GaN FETs can be used for lower loss, higher power density, multi-channel amplifiers of the next decade [7].

The processor for a class D amplifier creates a high-frequency, pulse-width-modulated small signal that represents the audio signal. Power transistors in either a half-bridge or full-bridge configuration convert the small signal to a large signal that drives the speakers through a filter. A diagram of a single-channel, bridge-tied load (BTL), class D amplifier with a split supply ($\pm$HV) is shown in Figure 12.2. Since each pulse is a square wave, increasing frequency gives a better representation of the audio signal. With each switching cycle, power is dissipated through both switching losses and conduction losses, creating a tradeoff between sound quality, operating frequency, and power dissipation.

The objective of the power stage in a class D amplifier is to create an exact large-signal replication of the small-signal source, while dissipating the least heat. Chapter 7 details the theory and practice of hard switching commutation with a buck converter example, which is relevant to the efficiency and linearity of a class D amplifier. A significant difference is that each transistor in a class D amplifier spends half of its time functioning as a control switch and half of its time operating as a rectifier switch. In a buck converter each transistor can be optimized for its function, while a transistor in a class D amplifier is required to perform both functions well [8].

GaN Power Devices for Efficient Power Conversion, Fourth Edition. Alex Lidow, Michael de Rooij, John Glaser, Alejandro Pozo, Shengke Zhang, Marco Palma, David Reusch, and Johan Strydom.
© 2025 John Wiley & Sons Ltd. Published 2025 by John Wiley & Sons Ltd.

352 | *GaN Power Devices for Efficient Power Conversion*

Figure 12.1 High power density 32 channels, 280 W per channel, GaN-based commercial amplifier in two rack units.

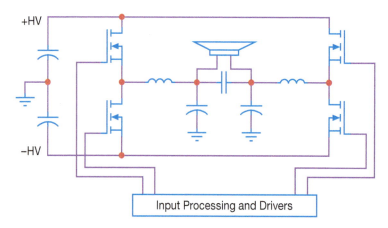

Figure 12.2 Diagram for a single channel of a BTL class D amplifier with a split supply (±HV).

There is no room for functional compromises as the transistor must have both outstanding on-state and switching characteristics.

Each parasitic element of the transistor that is dissipative creates distortion. For example, $R_{DS(on)}$ prevents the output waveform from reaching either voltage rail while dissipating power proportional to the square of the current. Input charge and gate resistance slow the switching transition, while applying both current and voltage to the transistor simultaneously. Stray inductance and reverse recovery charge cause voltage-overshoot and ringing while the associated energy is dissipated in each cycle [9]. Figure 12.3 compares an ideal waveform with measured waveforms of both a GaN transistor and a silicon MOSFET.

12.1.1 Total Harmonic Distortion

Total Harmonic Distortion (THD) is defined as the ratio of the equivalent root mean square (RMS) voltage of all the harmonic frequencies (beginning with the second harmonic) over the RMS voltage of the fundamental frequency [13] and is given by

$$\text{THD} = \frac{\sqrt{\sum_{n=2}^{\infty} V_{n\_RMS}^2}}{V_{Fund\_RMS}} \quad (12.1)$$

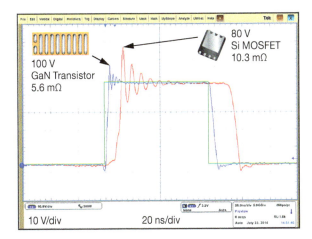

Figure 12.3 EPC2001C [10] GaN transistor (blue) and BSC123N08NS3 [11] MOSFET (red) waveforms compared with an ideal waveform (green) [12].

where V_{n_RMS} is the RMS voltage of the nth harmonic and V_{Fund_RMS} is the RMS voltage of the fundamental frequency. In practice, the number of harmonics to be summed is limited by the specified measurement bandwidth, which is in most cases up to 20 kHz, and often not exceeding the 20th harmonic.

Another major contributor to THD is dead-time distortion. Dead time is the time where both high-side and low-side transistors of the class D amplifier are commanded off between changing states. During dead time, turn on is delayed, preventing shoot-through of current caused by differences in propagation delay between high side and low side. The propagation delay differences are due to the variation in delay of the gate drivers along with variation of transistor turn-on and turn-off times. Turn-on and turn-off times are dependent on gate charges, and on common-source inductance (CSI) (see Chapter 3 and 4). GaN transistors in LGA or BGA chip-scale packages (CSP), and in power quad flat no-lead (PQFN), eliminate most CSI, while the wire bonds, clips, and cans of Si-MOSFETs have significant CSI. Low common-source inductance, fast switching due to small parasitic capacitances, and the availability of gate drivers with low propagation delay mismatch allow dead time to be reduced and related distortion to be minimized.

Referring to Figure 12.4, during most of the half-cycle, current flows in a single direction because output current is higher than ripple current. In this case, one transistor acts as the control switch and the other acts as the rectifier, similar to a buck converter at high current. When the control switch turns off, the switch-node voltage transition starts immediately at the beginning of the dead time. The reverse sequence is followed, when the rectifier turns off and it starts to freewheel by reverse conduction in the third quadrant, as explained in Chapter 7. This is followed by the switch-node voltage transition which does not happen until the control switch is turned on, at the end of the dead time. The result is that the effective on-time of the control switch is reduced, compared to the PWM command, as shown in Figure 12.4a. When the output current is near zero, the ripple current magnitude is sufficient to reverse the direction of the inductor current during half a cycle. In this case, a switch turn-off is followed immediately by the voltage commutation, at the beginning of dead time. Therefore, at low current, switch-node voltage on-time nearly matches PWM command on-time. Figure 12.4b shows detailed commutation when inductor current changes direction. Half-bridge commutation is described in much more detail in Chapter 7.

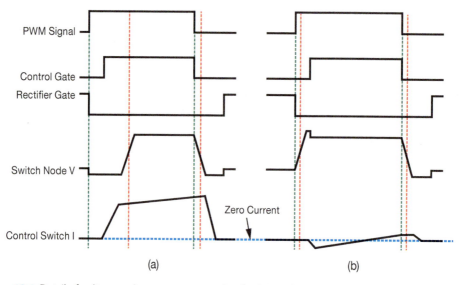

Figure 12.4 Detail of voltage and current commutation for (a) unidirectional current and (b) where current reverses direction.

The impact of reduced effective duty cycle due to dead time is a reduction of the output voltage as the PWM is averaged over many of its cycles. Near the zero crossing, the inductor current reverses direction, greatly reducing dead-time error. Figure 12.5 shows the error due to dead time in large signal compared with the theoretical waveform without dead time.

Many of the small-signal analog and digital components produce low levels of distortion at zero input, which is defined as noise. A common measure of sound quality is THD plus noise (THD + N). THD + N is typically measured with a 1 kHz signal, with target performance varying by product grade and cost. Also, signal-to-noise ratio or dynamic range are used to specify how quiet the amplifier output is, with no input signal, in comparison to the loudest signal the amplifier can reproduce. Usually, 100 dB is a sufficient value, while 120 dB is required only in

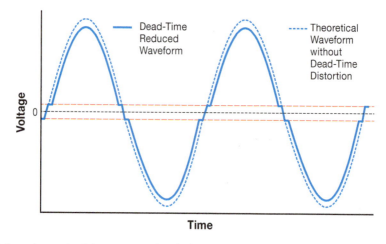

Figure 12.5 Error due to dead time compared with theoretical.

the most demanding applications, since it is approximately the dynamic range of human hearing.

In addition to the sources of noise in the analog path, such as thermal noise of resistors, and shot and flicker noise in operational amplifiers, in class D amplifiers care should be taken to avoid added noise in the digital path, by selecting components with short and consistent propagation delays (low jitter), starting from the comparator, up to and including the gate driver. Also, the gain of the class D power stage is proportional to the power supply voltage, so any variation of, or noise in the power supply is directly translated to the output. Feedforward and feedback control loops are used to provide sufficient power supply rejection ratio (PSRR) and achieve a low noise amplifier.

12.1.2 Intermodulation Distortion

A flaw of using THD + N as a sound quality metric is that the signal is constant. With a constant signal, feedback can be used to reduce THD + N to a level that meets design requirements but produces other undesirable side effects. In closed-loop systems negative feedback is used to correct for open-loop nonlinearity. The flaw in feedback is that it mixes the old sound with the new sound. This is not a problem at constant frequency, but music is dynamic with much color and richness in the intended harmonics [14], e.g. its wide range of frequencies and magnitudes. The distortion caused by the frequencies of the old sound mixing with the new sound is Dynamic Inter-Modulation (DIM), also called Transient Inter-Modulation Distortion (TIM), and our sense of hearing is very sensitive to it [15]. Intermodulation distortion (IMD) is thus a measure of signal errors due to two dissimilar but simultaneous signals.

IMD is measured in several ways. One method (DIM) combines a low-frequency, low-pass filtered square wave with a 15 kHz sine wave. The single-pole, low-pass filter frequencies are 30 kHz for DIM 30 and 100 kHz for DIM 100. DIM is calculated as the ratio of the sum of the RMS levels of the intermodulation components to the level of the sine wave [16].

Another method to measure IMD is based on the International Telecommunication Union (ITU) standard ITU-R, known also as CCIF [17]. Two sinusoidal signals of equal amplitude, at the upper extreme of the audio band, and spaced by 1 kHz, are summed and used as a stimulus. Usually 19 kHz/20 kHz or 18 kHz/19 kHz signals are used. Nonlinearity in the amplifier causes intermodulation products between the two signals. The distortion is evident in the 1 kHz component (second-order distortion, d2), as well as in the sidebands 1 kHz above and below the two tones (third-order distortion, d3). Higher-order modulation components can also appear, but they are not included in the CCIF IMD measurement. The second- and third-order distortions are summed up and expressed as a percentage of the upper frequency level [18–20].

12.2 GaN Transistor Class D Audio Amplifier Example

A two-channel GaN transistor-based class D audio amplifier capable of driving each channel at 350 W into 8 Ω speaker or 700 W into 4 Ω speaker is presented as an example [21]. It is configured as two single-ended (SE) channels and can also be used in a bridge-tied load for up to 1400 W into 8 Ω, or 70V/100V distribution line in constant voltage systems [22]. The PWM frequency in the system is set to about 600 kHz. The design uses the 200 V rated, 10 mΩ (max) EPC2307 GaN transistor [23], with an NCP51820 driver IC from onsemi as the half-bridge gate driver [24]. A photograph of the power stage of the amplifier, including the output filter and electrolytic bulk capacitors, is shown in Figure 12.6a, and measures about 50 mm × 50 mm.

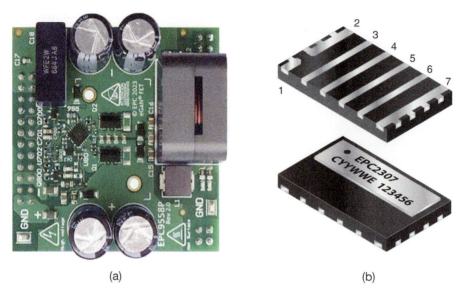

Figure 12.6 (a) Single class D amplifier power stage, and (b) EPC2307 GaN transistor. *Source:* Ref. [23]/Efficient Power Conversion Corporation.

Visible in the center of the board is the gate driver with the GaN FETs half bridge (U80, Q1, Q2), which fits within an area of approximately 15 mm × 12 mm. The EPC2307 GaN transistor in power quad flat no-leads (PQFN) package is shown in Figure 12.6b.

Each amplifier channel comprises a power stage board (EPC9558P) and a PWM modulator board (EPC9558L). The motherboard hosts two boards of each kind, as well as the housekeeping power supplies, protection, and ancillary function circuits. The complete system is shown in Figure 12.7.

Figure 12.7 EPC9192 class D amplifier can deliver 700 W into 4 Ω for each of the two channels. *Source:* Ref. [23]/Efficient Power Conversion Corporation.

The system is powered by dual split supply, within a range of ±42–85 V. Its 96% efficiency allows operation with no heatsink and no fan up to ±48 V, with heatsink and no fan up to ±70 V, and with heatsink and 200 LFM up to ±85 V.

A class D amplifier design is a tradeoff between efficiency, compactness, and outstanding audio performance. The capability of GaN transistors to switch at over 600 kHz with low losses allows simultaneous achievement of all of the above requirements.

With 600 kHz switching frequency, an off-the-shelf, compact (23 mm × 15 mm × 23 mm) inductor can be used that exhibits reasonably low losses, due to the limited ripple current, even at ±85 V supply. This allows low idle losses, which can be reduced further with a custom-designed inductor using Litz wire.

Operation without a heatsink is possible at reduced supply voltage and output power. The optional heatsink measures only 27 mm × 18 mm × 23 mm.

The PWM modulator is based on a self-oscillating closed-loop feedback design, with an inner loop based on AC inductor current and an outer loop based on output voltage and implemented as second-order integrator. The feedback is taken after the output filter, directly at the output terminals. The design is inspired by previous work from Williamson (used in Carver Corporation amplifier designs) [25], and Veltman [26, 27]. The self-oscillating frequency is variable and drops with modulation index, but it is kept above 70 kHz at onset of clipping, because of the high initial switching frequency. For the same reason, the feedback loop can be tuned to provide a loop gain of more than 40 dB across the whole audio band. Such a loop gain corrects for power stage and inductor non-linearities (improves SNR, THD, and IMD), rejects noise from power supply (improves PSRR), rejects coupling from nearby channels or other switching converters (reduces crosstalk and spurious noise), and lowers the output impedance, making the amplifier a more ideal voltage source, independent from the load.

12.2.1 Audio Performance

Audio measurements have been carried out using Audio Precision APx515 and APx525 audio analyzers with balanced connections for the inputs and outputs. The high-voltage PSU is a regulated LLC with +85 V and −85 V nominal outputs specifically designed for powering audio amplifiers.

The measured frequency response of the amplifier is shown in Figure 12.8. Note the full vertical scale span is only 2 dB. From open load to 2 Ω load, the gain at 1 kHz drops by only 0.04 dB.

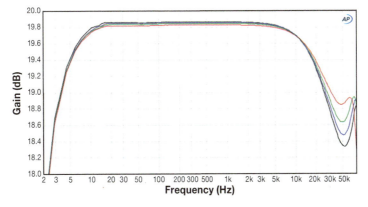

Figure 12.8 Measured EPC9192 GaN transistor class D amplifier frequency response at 2 V_{RMS} output, with open load (black), 8 Ω load (blue), 4 Ω load (green), 2 Ω load (red).

This corresponds to an output impedance of 8 mΩ, rising to about 25 mΩ at 20 kHz. Also, the voltage gain deviation from 20 Hz to 20 kHz is only +0.0 dB/−0.5 dB referenced to 1 kHz gain, regardless of the load.

The amplifier has a maximum of 0.017% THD + N from 20 Hz to 20 kHz (at 10 W output, into a 4 Ω speaker load) as shown in Figure 12.9.

Figure 12.10 shows an output power of 750 W at 1% THD + N at 1 kHz, less than 0.02% at −1 dB at full power (600 W), and approximately 0.007% at 10 W output.

Frequency analysis was conducted by applying a Fast Fourier Transform (FFT) to a 1 kHz, 1 W into a 4 Ω load, and yielded the result shown in Figure 12.11. The FFT was done with 32k samples, 48 kHz sample rate, AP-Equiripple window, and 8 averages. It is worth noting that sample size of the FFT, sampling frequency, and window function have an impact on the noise floor and need to be specified. For an incoherent signal such as noise, the spectrum level depends on the resolution bandwidth. Figure 12.12 shows the FFT of the output when no input signal is applied.

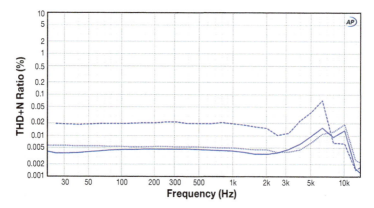

Figure 12.9 Measured EPC9192 GaN transistor class D amplifier THD + N as function of frequency, at 1 W (solid), 10 W (dotted), and 100 W (dashed) into 4 Ω loads.

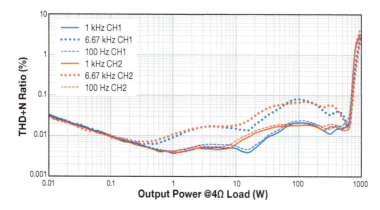

Figure 12.10 Measured EPC9192 GaN transistor class D amplifier THD + N as function of output power driving a 4 Ω load.

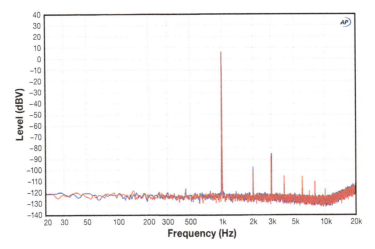

Figure 12.11 Spectrum of the output of the EPC9192 GaN transistor class D amplifier with a 1 kHz sine wave delivering 1 W into a 4 Ω load. Blue CH1, red CH2.

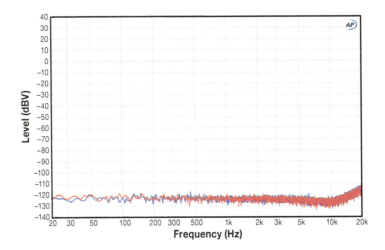

Figure 12.12 Spectrum of the output of the EPC9192 with no signal applied (residual noise), driving a 4 Ω load (blue for CH1, red for CH2). This corresponds to 38 μV$_{RMS}$ (A-weighted), 20 Hz to 20 kHz.

The noise floor corresponds to 38 μV (A-weighted), 20 Hz to 20 kHz. An A-weighing filter is commonly used by acoustical engineers to model the responsiveness of the human ear at varying frequencies [28]. Given the 55 V$_{RMS}$ maximum output, this amplifier yields a dynamic range of about 123 dB. Also, the low magnitude of all frequencies outside of the fundamental frequency indicates good suppression of harmonics.

Figure 12.13 shows the measured noise floor of the class D amplifier, from full power to low power, at 1 kHz driving a 4 Ω load. The measurement is obtained by removing the 1 kHz and its harmonics from the output spectrum. The RMS value of the remaining signal, the noise floor, is then computed. The procedure is repeated for a stepped amplitude signal input. The measured noise floor is less than −119 dBr (Ref = 700 W) unweighted and −122 dBr (Ref = 700 W) A-weighted.

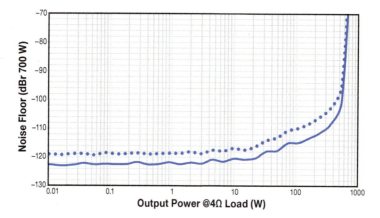

Figure 12.13 Measured EPC9192 GaN transistor class D amplifier noise floor as function of output power. (Dotted line is unweighted and line is solid A-weighted.)

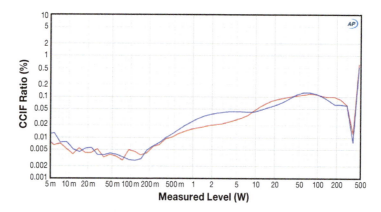

Figure 12.14 Measured CCIF Intermodulation Distortion (IMD) of the EPC9192 GaN transistor class D amplifier as a function of output power using 18 and 19 kHz, 1:1 ratio, d2 + d3, driving a 4 Ω load. Both channels are shown.

Figure 12.14 shows the measured CCIF IMD, from low power to full power, with the amplifier driving a 4 Ω load. IMD stays below 0.1% up to full power.

12.2.2 TradeOffs and Optimization

The efficiency of a class D amplifier is great at high power, but for most of its time, an audio amplifier is operating at mid-to-low output power, due to the dynamics of the audio signal and the typical mission profile, characterized by full power usage only for a small fraction of the lifetime. To improve the efficiency in the lowest range of the output power, the key is to minimize the losses at idle with the amplifier operating, switching, and no audio signal output.

For a half-bridge output stage like the class D amplifier, the idle losses are minimized by operating with zero voltage switching (ZVS) by using the ripple current in the inductor. In these conditions, it can be shown that most of the losses are caused by the inductor AC losses.

Gate drive losses can be significant for silicon, but they are very small for GaN FETs, due to the very low gate charge. Finally, conduction losses are usually small, due to the limited ripple current of a few A_{RMS} at most.

To achieve full charging of the switching node parasitic capacitance, mostly made up by the transistors C_{OSS}, a sufficient dead time, shown in Figure 12.4b, needs to be determined as follows:

$$DT_{ZVS} \geq \frac{2 \cdot Q_{OSS}}{I_{Lpk}} = \frac{2 \cdot Q_{OSS}}{\frac{V_{BUS}}{4 \cdot L \cdot f_{SW}}} = \frac{8 \cdot Q_{OSS} \cdot L \cdot f_{SW}}{V_{BUS}} \quad (12.2)$$

where Q_{OSS} is for a single FET at V_{BUS} and V_{BUS} is the supply voltage across the half bridge, i.e. $V_{POS} - V_{NEG}$ in case of dual split supply (see Chapter 7, Section 7.2.2). In the EPC9192 design, with ±85 V and 650 kHz f_{SW}, the Q_{OSS} of the EPC2307 at 170 V is 75 nC. A 10 µH inductor requires a dead time of about 45 ns, which is long enough to cause significant distortion. It should be noted that for an 8.2 mΩ, 200 V transistor, a GaN FET achieves one-third of the Q_{OSS} compared to a state-of-the-art equivalent silicon MOSFET, which enables a significantly shorter dead time. To further reduce the dead time, a lower inductance can be used.

Due to the lack of availability of off-the-shelf inductors with low enough losses, a custom 6.3 µH inductor based on gapped ferrite and Litz wire winding was tested. This allowed reduction of the dead time down to 30 ns, while maintaining the same power losses. The idle losses are about 3 W, at ±70 V supply, and about 4 W at ±85 V supply, and are mostly from the inductor. The shorter dead time greatly improved the THD at 6.67 kHz and mid–high power, as can be seen in Figure 12.15.

The effect of increasing the loop gain by 12 dB at low frequency (DC) was also evaluated. The original voltage control loop integrator DC-gain limiting resistor, setting a DC loop gain of about 50 dB, has been increased by a factor of 4. This proved to reduce the THD + N at 1 kHz, but to increase slightly both the THD + N at 6.67 kHz and the IMD.

To reduce distortion further, even shorter dead times can be used forcing partial ZVS operation, described in Chapter 7, Section 7.2.4.3, and trading off with additional losses. In addition, other techniques are possible, such as reducing the dead time at high output current, requiring even more advanced control techniques to manage.

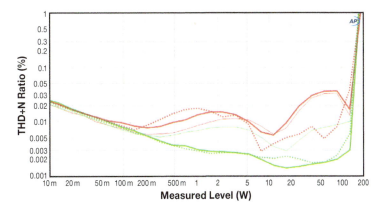

Figure 12.15 Measured THD + N with additional 12 dB loop DC gain, 50 ns dead time (solid) and 30 ns dead time (dotted) of the EPC9192 GaN transistor class D amplifier. Thin lines are the baseline design with 50 dB loop DC gain, 50 ns dead time. 1 kHz (green) and 6.67 kHz (red). 8 Ω load.

12.2.3 Efficiency and Thermal Management

The experimental EPC9192 class D audio amplifier efficiency was measured using a sine wave signal of 1 kHz, driving an 8 Ω load, and with three different supply voltage settings, corresponding to 125, 250, and 350 W maximum power into an 8 Ω load. Efficiency is about 96% at full load as shown in Figure 12.16.

The cooling system has been optimized for both PCB-based cooling and device back-side cooling, as described in Chapter 6, and is shown in Figure 12.17. PCB cooling is enhanced by utilizing the following design criteria:

- 2 oz copper thickness in the 4 layers of the PCB
- center high-frequency capacitor layout configuration that spreads out the FETs and reduces co-heating of the FETs.
- vias in pad plated over (VIPPO), 48 for each FET. About half of them are connected to copper polygons in the first inner layer, crossing only 0.2 mm in the thickness direction. The other half are crossing the full PCB thickness of 1.6 mm and connected to wide copper polygons in the bottom side.
- 1206 copper jumper (Keystone 5109) soldered very close to drain terminal of each FET. It helps to spread the heat on PCB and increase the contact surface to the top mounted heatsink.

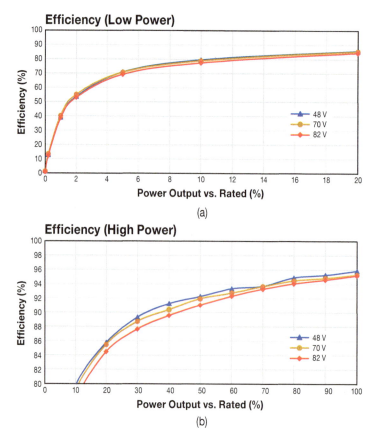

Figure 12.16 Measured efficiency as function of output Power of the EPC9192 GaN transistor class D amplifier in the low power range (a) and high power range (b) with ±48 V, ±70 V, and ±82 V supply voltages.

Class D Audio Amplifiers | 363

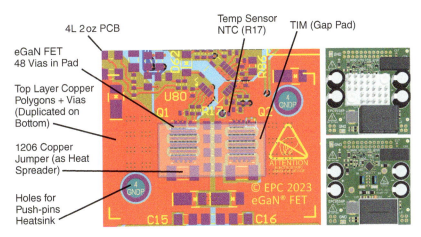

Figure 12.17 Implemented cooling solution for the power board of the EPC9192 amplifier. *Source:* With permission of EPC.

Back-side cooling is provided by installing a small heatsink on top of the devices and following the guidelines given in Chapter 6:

- high conductivity TIM gap pad, with 0.5 mm thickness, to minimize FET to heatsink thermal resistance. Suggested materials are T-Global TG-A1780 (17.8 W/mK) and Shiu Li Technology T-work9000 (20.0 W/mK). Size is 15 mm × 10 mm.
- custom heatsink by Alpha Novatec S08FKE03, size 27 mm × 18.5 mm × 21.3 mm. Heatsink is compressed on TIM by spring-loaded push-pins. Pressure on TIM is about 50 psi.

Thermal images have been captured for various operating conditions of the EPC9192 amplifier and are shown in Figure 12.18:

a) operation with no heatsink and convection cooling, using ±48 V supply and 4 A_{RMS} load current, corresponding to 125 W into 8 Ω, continuous sine wave, 3 dB Crest Factor (CF), or 250 W into 4 Ω with 9 dB CF signal, which is typical of music content.

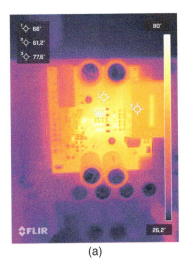
(a)

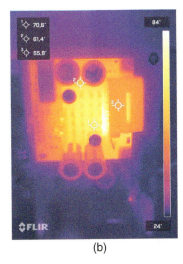
(b)

Figure 12.18 Measured thermal image of GaN power stage of the EPC9192 class D amplifier operating with (a) 4 A_{RMS} into a 8 Ω load and (b) 10 A_{RMS} into a 2 Ω load. *Source:* With permission of EPC.

b) operation with heatsink and 200 LFM airflow, using ±82 V supply and 10 A_{RMS} load current, corresponding to more than 700 W into 4 Ω with 6 dB CF signal. This is a very dense signal that is sometime found in the most demanding subwoofer applications.

12.3 Summary

Gallium nitride transistors lead to a superior tradeoff between compactness, audio performance, and efficiency for class D audio amplifiers. The low switching losses allow the use of higher switching frequency, which in turn reduces the losses and the size of the output filter inductor. Higher switching frequency makes practical the implementation of advanced control loops that corrects the non-linearities, rejects the noise from the PSU and the other interferences, and leads to achieving very low noise and THD.

With the use of GaN, it is possible to implement a high-power amplifier which retains outstanding performance, in a very compact size and with low losses. Professional audio and audiophile requirements can be met simultaneously in a single GaN-based amplifier design.

References

1 Self, D. (2013). *Audio Power Amplifier Design*, 6e, 80–85. Burlington, MA and Oxon, UK: Focal Press.
2 Electronics Tutorials. Amplifier classes. https://www.electronics-tutorials.ws/amplifier/amplifier-classes.html.
3 Hypex Electronics BV (2011). Ncore technology. https://www.hypex.nl/media/41/a4/c1/1697639958/Ncore%20Technology.pdf.
4 RCF Arena. https://www.rcf.it/en/-/welcome-to-rcf-arena-the-largest-outdoor-music-venue-in-europe.
5 Sphere at The Venetian Resort. https://en.wikipedia.org/wiki/Sphere_(venue).
6 Innosonix GmbH. MAxx-Series, A completely new experience of multi-channel-amplifier. https://innosonix.de/maxxSeries.html.
7 ICEpower (2024). Is GaN worth it? https://icepoweraudio.com/is-gan-worth-it/.
8 Colino, S. and Taylor, S. (2017). GaN FETs drive fidelity and efficiency in class-D audio amplifiers. *AES Convention 142*, May 11, 2017. http://www.aes.org/e-lib/browse.cfm?elib=18612.
9 Pavier, M., Sawle, A., Woodworth, A. et al. (2003). High frequency DC:DC power conversion: the influence of package parasitics. *Applied Power Electronics Conference and Exposition, 2003. APEC '03. Eighteenth Annual IEEE* 2 (February 2003), 9–13 and 699–704.
10 Efficient Power Conversion Corporation (2021). EPC2001C – Enhancement-mode power transistor. EPC2001C datasheet [Revised 2020]. http://epc-co.com/epc/Portals/0/epc/documents/datasheets/EPC2001C_datasheet.pdf.
11 Infineon Technologies AG (2009). OptiMOS™ 3 power transistor. BSC123N08NS3 datasheet, November 2009 [Revision 2.5] https://www.infineon.com/dgdl/Infineon-BSC123N08NS3G-DS-v02_05-en.pdf?fileId=db3a30431add1d95011ae80eb8555625.
12 Efficient Power Conversion Corporation (2018). eGaN® FETs and ICs for class-D audio applications. Efficient Power Conversion application brief AB003. http://epc-co.com/epc/Portals/0/epc/documents/briefs/AB003%20eGaN%20FETs%20for%20Class-D%20Audio.pdf.

13 Williams, D. (2017). Understanding, calculating, and measuring total harmonic distortion (THD). http://AllAboutCircuits.com and https://www.allaboutcircuits.com/technical-articles/the-importance-of-total-harmonic-distortion/.

14 Self, D. (2013). *Audio Power Amplifier Design*, 6e, 25–71. Burlington, MA and Oxon, UK: Focal Press.

15 Otala, M. and Leinonen, E. (1977). The theory of transient intermodulation distortion. *IEEE Trans. on Acoust., Speech Signal Process* 25 (1): 2–8.

16 Audio Precision (2009). DIM 30 and DIM 100 measurements per IEC 60268-3 with AP2700. https://www.ap.com/technical-library/dim-30-and-dim-100-measurements-per-iec-60268-3-with-ap2700/.

17 Bohn, D. (2003). Audio specifications. RaneNote 145. https://www.rane.com/note145.html.

18 Mathew, D. (2010). How to write (and read) audio specifications. Audio Precision white paper. https://www.ap.com/fileadmin-ap/technical-library/How_to_write_and_read_audio_specs.pdf?force.

19 Audio Precision, Inc. (2023). APx500 measurement software user's manual, version 8.1.0 [8211.0238 rev 041]. https://www.ap.com/fileadmin-ap/technical-library/APx500_User_Manual_v8-1-0.pdf?force.

20 Sound system equipment – Part 3: Amplifiers. International Electrotechnical Commission Standard IEC 60268-3:2018, Edition 5.0. April 4, 2018. https://www.en-standard.eu/iec-60268-3-2018-sound-system-equipment-part-3-amplifiers/.

21 Efficient Power Conversion Corporation (2024). Evaluation board EPC9192KIT user guide. https://epc-co.com/epc/documents/guides/EPC9192_qsg.pdf.

22 Constant-voltage speaker system available from: https://en.wikipedia.org/wiki/Constant-voltage_speaker_system.

23 Efficient Power Conversion Corporation (2023). EPC2307 – Enhancement mode power transistor. EPC2307 datasheet [Revised March 2023]. http://epc-co.com/epc/Portals/0/epc/documents/datasheets/EPC2307_datasheet.pdf.

24 onsemi (2022). NCP51820 – High speed half-bridge driver for GaN power switches. Onsemi datasheet [Revision 6 March 2022]. https://www.google.com/url?sa=t&source=web&rct=j&opi=89978449&url=https://www.onsemi.com/pdf/datasheet/ncp51820-d.pdf&ved=2ahUKEwiv5dH_p−FAxV9HkQIHXI8CY8QFnoECAgQAQ&usg=AOvVaw2JmjzceGvTHsgHjEwskaq2

25 Williamson, R. (1997). Audio frequency power amplifiers with actively damped filter. US Patent No. 5,606,289, 25 February 1997.

26 Veltman, A. and Domensino, H. (2003). Amplifier circuit having output filter capacitance current feedback. US Patent No. 6,552,606, 22 April 2003.

27 Van der Hulst, P., Veltman, A., and Groenenberg, R., (2002). An asynchronous switching high-end power amplifier. AES Convention, Paper 5502.

28 Vernier Science Education (2018). What is the difference between "A weighting" and "C weighting?" July 12, 2018. https://www.vernier.com/til/3500/.

13

High Current Nanosecond Laser Drivers for Lidar

13.1 Introduction to Light Detection and Ranging (Lidar)

Lidar is a form of radar where the electromagnetic radiation happens to be in the optical band [1, 2]. A typical lidar transmits light into a transmission medium such as the atmosphere, water, or space and detects reflected light from one or more targets located some distance from the lidar unit. A conceptual lidar system is shown in Figure 13.1. Comparison of the transmitted and reflected light makes it possible to determine target properties [2]. The target could be solid, a car for example; liquid, such as raindrops; or even gaseous, like a layer of Earth's atmosphere. One can measure obvious properties like distance, or far more subtle properties, such as the chemical content and particle size of pollution over a city. In the last few years, one form of lidar, time-of-flight (TOF) distance measurement, has become popular. Using a laser as the optical source, one can accurately and precisely measure the distance of a small target even at a long range. As with radar, it is possible to illuminate a solid angle and measure the distance to multiple points within the angle, producing an array of distance points that can provide 3-D spatial information about a target. The 3-D target data is often represented in the form of a point cloud (Figure 13.1). This measurement ability is extremely useful for 3-D mapping, surface measurements, obstacle detection, and autonomous navigation.

There are two main laser-based approaches to lidar distance measurement, based either on amplitude or frequency modulation (AM or FM) of the laser light output [4, 5]. Most AM systems are designed to turn the laser completely on or off; hence, they are frequently referred to as pulse-based lidar systems. Most FM systems operate as continuous-wave (CW) systems, also known as FMCW lidar systems. The latter operates the laser continuously at relatively low power while modulating the optical wavelength. While FMCW lidar has some attractive properties, it depends greatly on superior laser beam quality and requires substantial maturation of currently nascent photonic integrated circuit (PIC) technology to be cost-effective [6, 7]. On the other hand, pulsed-laser lidar systems can use low-cost lasers operated at high peak power and relatively low duty cycle, so that the average power is still relatively low. The peak power levels required are extremely high and the pulse widths very short.

Although TOF lidar systems have been in use for decades, until recently they have been expensive and bulky. In the case of FMCW lidars, PICs are still quite immature, and progress has been slow, so affordable systems have not reached mass production. On the other hand, pulsed TOF lidars have reached volume production [8]. This has been enabled in part by the commercialization of GaN power semiconductor technology. The high performance of GaN transistors has been a key factor in the development of modern, cost-effective lidar systems by

---

GaN Power Devices for Efficient Power Conversion, Fourth Edition. Alex Lidow, Michael de Rooij, John Glaser, Alejandro Pozo, Shengke Zhang, Marco Palma, David Reusch, and Johan Strydom.
© 2025 John Wiley & Sons Ltd. Published 2025 by John Wiley & Sons Ltd.

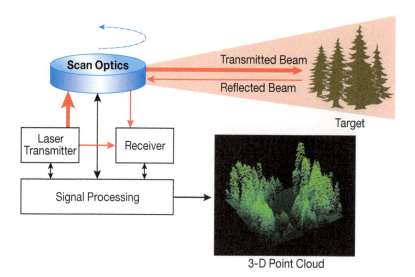

Figure 13.1 Basic lidar system measures distances to details of a target and generates a point cloud. Source of point cloud: Ingram [3]/Courtesy of Kim Ingram, University of California, Agriculture and Natural Resources.

enabling laser drivers with extremely fast, high-current pulses. As a result, the use of pulsed TOF lidar has rapidly expanded and the market continues to experience tremendous growth.

This chapter will discuss pulsed TOF lidar transmitters based on the use of GaN power semiconductors and semiconductor laser diodes. It will first explain some of the key specifications of time-of-flight pulsed laser transmitters. Next, it will provide a detailed explanation of laser driver operation and design. Examples of hardware implementation and measured results will be discussed. Finally, some additional topics useful for lidar laser driver design will be presented.

13.2 Pulsed Laser Driver Overview

Figure 13.2 shows a simplified pulsed laser driver. It comprises a loop with an energy source V_{IN}, an impedance Z_{PL} that represents all parasitic and intentional impedances in the loop, a laser diode D_L with a fixed forward voltage drop V_{DLF}, and a transistor Q_1, where Q_1 is controlled by a voltage v_{GS}. This voltage is an amplified version of $v_{command}$, where the amplifier is usually called a gate driver (GD_1). The key item of interest is the current i_{DL} since this controls the light output of the laser. Assume V_{IN}, Z_{PL}, and V_{DLF} characteristics to be fixed so that i_{DL} can only be controlled through Q_1. For an ideal transistor Q_1, we can control the impedance of the drain to source anywhere from zero ohms to nearly infinite via adjustment of v_{GS}, in which case the laser current, i_{DL}, will range from 0 A to the maximum of $(V_{IN} - V_{DLF})/Z_{PL}$. However, the real world is not so simple. V_{DLF} and the parameters of Q_1 have large temperature

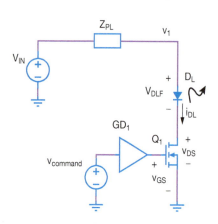

Figure 13.2 Simplifed pulse laser driver.

coefficients and exhibit substantial variability between devices. Furthermore, if Q_1 is not fully off or fully on, it will dissipate power, and for high-power laser drivers this is a problem. Hence, most high-power laser drivers turn Q_1 fully on or fully off, i.e. it is a switch operating in a pulsed mode. Given this, i_{DL} is determined by V_{IN}, V_{DLF}, and Z_{PL}, and if current control is needed, it is usually better to control V_{IN}. As we will show for very short pulses and high currents, Z_{PL} is dominated by parasitic inductance and this plays a major role in design.

In this chapter, we are only considering pulse laser drivers where Q_1 is turned fully on or fully off. We begin with the assumption that the switch Q_1 is fast enough that we can neglect the turn-on and turn-off times of the FET. Later in this chapter, we will show that this is possible with GaN power devices and discuss how to design a circuit and layout so that this can be achieved in practice. With this assumption, the pulse characteristics are nearly independent of the switch Q_1 and depend primarily on the power supply and the parasitic circuit components of the physical design.

The remainder of this section will present a high-level discussion of basic pulse requirements and key components in a laser transmitter. The desired pulse properties vary greatly depending on the type of application, distance range, frame rate, and eye safety limits. Thus, they drive the requirements of the pulse generator design. However, physical hardware limits, as well as size and cost requirements, constrain what is possible. For a designer to make appropriate trade-offs, they must first understand what drives the desired pulse properties and how this affects key components. This section provides the background necessary for this understanding.

13.2.1 Types of Pulsed Lidar Systems

There are two main methods of implementing a pulsed TOF lidar system, direct TOF (DTOF) and indirect TOF (ITOF). Simplified implementations of these are illustrated in Figure 13.3. Simple DTOF systems work by sending out an optical pulse from a transmitter, where the time of transmission t_0 is recorded. The pulse travels to a target, which reflects a fraction of the pulse back toward the lidar receiver. The latter determines the time of arrival t_1 to the receiver. Knowledge of the speed of light along with the light travel time yields the distance to the target. By contrast, simple ITOF systems sent a long burst of pulses at a particular frequency. With this method, a portion of the burst is reflected back to the lidar receiver, and the system measures the phase difference between the transmitted and received pulse bursts. Knowledge of the phase difference and burst frequency is used to determine the time delay, which is then used to calculate the distance.

These are descriptions of the simplest usable systems, and many variations are possible. The purpose of the variations is to tradeoff various system properties such as minimum and maximum range, point or frame acquisition time, resolution, accuracy, operating environment, qualification to various standards, safety, cost, and so forth. A thorough discussion of these can be found in [3]. However, the simple systems outlined above are enough to understand the fundamental needs of the laser transmitter.

Like radar, a basic knowledge of the effect of pulse shape on TOF lidar system performance clarifies the challenges in laser driver design for lidar applications [9]. The key factors are pulse amplitude, pulse timing, and pulse shape. Pulse amplitude is straightforward. All else being equal, the greater the amplitude, the longer the range of the lidar. However, several things can limit the pulse amplitude, such as transmitter hardware limitations, detector saturation, and eye safety. Since lidar systems produced in high volumes generally operate in the same environments as people, eye safety requirements generally limit the maximum pulse amplitude [10]. Pulse timing and pulse shape will be discussed in Sections 13.2.2–13.2.4.

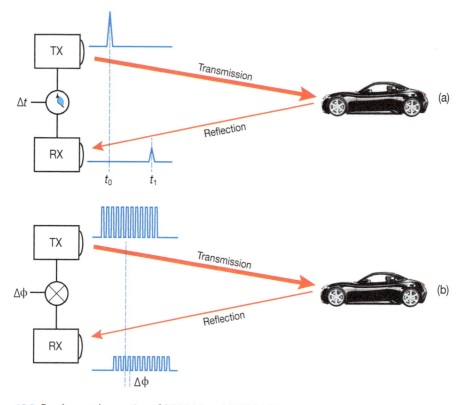

Figure 13.3 Fundamental operation of DTOF (a) and ITOF (b) lidar systems.

13.2.2 Pulse Requirements for DTOF

In simple DTOF lidar systems, pulses of width t_p are transmitted repeatedly at fixed intervals, with interval length T_p between pulses. This inter-pulse spacing is often specified by the equivalent pulse repetition frequency (PRF), where PRF = $1/T_p$. Such a system is shown in Figure 13.4.

Consider first a single ideal light pulse with infinitesimal width t_p. In Figure 13.4a, the pulse is transmitted, travels to the target, and some portion of the pulse reflects back toward the detector. The time Δt between the transmission and detection of the pulse is measured, and the distance to the target is given by:

$$d = \frac{c \Delta t}{2} \tag{13.1}$$

where c is the speed of light (approximately 30 cm/ns in air), and the factor of ½ results from the light traveling the round-trip distance $2d$. If there are multiple targets in the path of the beam, the resulting multiple reflections can be used to compute the different distances and resolve the targets as separate objects.

In practice, t_p is not infinitesimal. The finite speed of light dictates that the pulse occupies a distance l_p in space, where

$$l_p = c \cdot t_p \tag{13.2}$$

A wider pulse makes it difficult to resolve distances much shorter than l_p, especially if the target is not ideal. With multiple objects and corresponding reflections, Figure 13.4b shows

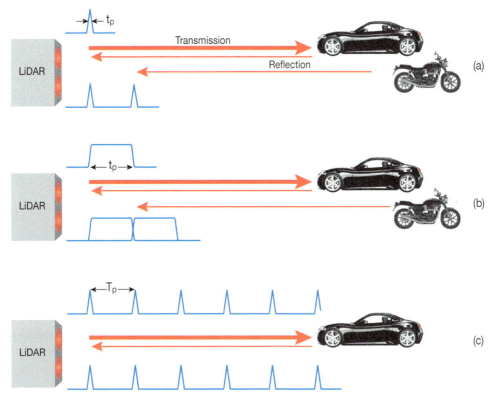

Figure 13.4 Effect of pulse width and frequency for DTOF lidar. (a) Narrow pulses allow reflections to be easily distinguished. (b) Wider pulses can overlap, making them harder to distinguish from each other and reducing distance resolution. (c) High PRF limits range because transmitted pulses cannot be distinguished from each other.

that the reflected pulses can overlap, making it difficult to distinguish them from each other. For autonomous vehicle navigation, the system must be able to identify and track objects around the vehicle. The required distance resolution is on the order of a few centimeters, so Eq. (13.2) implies a pulse width t_p on the order of 100–200 picoseconds for a rectangular pulse. This is an oversimplification, since the theoretical distance accuracy increases in proportion to the maximum frequency content of the signal, which depends greatly on the shape of the pulse. In practice, signal processing methods including variable pulse sequences and correlation techniques allow one to extract greater resolution than predicted by Eq. (13.2), but the cost is paid in time (latency) and computing power. The longer it takes to extract the required precision, the slower the frame rate of the lidar. Nevertheless, since shorter pulses have a greater proportion of their energy concentrated at higher frequencies, it is generally true that a shorter pulse width means faster measurement for a desired accuracy, and therefore better system performance.

The PRF must also be considered. The higher the PRF, the faster data can be generated, with obvious benefits for resolution and frame rate. Care needs to be taken with truly periodic pulses. For a target farther than $c \cdot T_P/2$, the receiver cannot distinguish between reflected pulses and the correct distance cannot be calculated, as shown in Figure 13.2c example. This is essentially an aliasing problem. If necessary, this can be mitigated in a similar manner to enhancing resolution of wide pulses. Pulse spacing can be made variable, with a common method being the use of pulse bursts. Furthermore, many DTOF systems only illuminate a small fraction of the field of view at any given time and use optical beam steering to scan the entire field of view. This results in a lower effective PRF for any given location in the FOV and thus reducing the distance aliasing problem.

Finally, consider the pulse shape. The ideal pulse is an impulse, i.e. "zero" width and finite area. For an impulse, shape is a meaningless concept. Any real pulse has non-zero width and a shape can be defined, but it is not clear what the latter should be. The pulse width is generally defined as the full-width-half-maximum (FWHM) because it is well defined for pulses with a single maximum and simplifies measurement. However, this obscures another requirement, namely that the pulse have a well-defined beginning and end, beyond which the value is effectively zero, i.e. below the detection limit. This is to prevent overlap of received pulses.

13.2.3 Pulse Requirements for ITOF

The requirements for ITOF are considerably different than those for DTOF. An ITOF detector is sensitive to phase differences, and phase is not defined for single pulses. Hence, the transmitted and reflected signals take the form of a pulse burst at frequency f_{burst} of a long enough duration that the phase may be detected (Figure 13.3b). Typically, the detector integrates over an entire burst, so for a given burst duration, a higher frequency f_{burst} will improve the signal-to-noise ratio (SNR) [11, 12]. The maximum burst duration is often limited by the system needs. For example, a full-frame detector with full-frame illumination operating at 30 frames/s means that a single burst can be no longer than 16.6 ms. Usually there are other limitations that reduce this time dramatically, such as an imager with a rolling shutter which requires at least one burst per scan line, as well as thermal limitations of the laser.

In the simple case where waveform within the burst is a square wave, we can detect the phase by computing the product of the transmitted and reflected signals, and calculating a measure of the product, e.g. the average or some other norm. Once the phase shift $\Delta\phi$ is determined, we can calculate the equivalent time delay:

$$\Delta t = \frac{\Delta\phi}{2\pi f_{burst}} \tag{13.3}$$

and the distance d can be determined using Eqs. (13.1) and (13.2).

The effects of pulse features are not identical for ITOF and DTOF. The key timing parameter for ITOF is the burst frequency, because, as the frequency increases the system signal-to-noise ratio (SNR) increases as well. However, frequency is limited by two things. First, there are hardware limits throughout the lidar system, including the detector, the signal processing, the laser driver, and even the printed circuit board (PCB). Second, ITOF suffers from worse distance aliasing effect compared with DTOF because the burst frequency tends to be much higher. This is because in a simple multiplier system phase detector the detected phase shift, $\Delta\phi$, can only occur between zero and π, so the maximum unambiguous detection range occurs when $\Delta\phi = \Delta\phi_{max} = \pi$. Thus,

$$d_{max} = \frac{ct_{dmax}}{2} = \frac{c}{4f_{burst}} \tag{13.4}$$

As with DTOF, there are ways to mitigate this through more complex modulation schemes, but this is beyond the scope of this chapter.

While edge speed is important if square waves are desired, the effect of waveshape on ITOF system performance is complex. While square waves seem desirable for simplicity, they contain higher frequency harmonics that can contribute to the distance aliasing [13]. In addition, there is the practical problem that square waves are very difficult to obtain at the frequency and power levels desired, as will be discussed in Section 13.5.4.

The role of pulse amplitude is similar for ITOF and DTOF, namely higher amplitude means more power and a higher SNR. Since both the burst length and average laser duty cycle of ITOF are usually a few orders of magnitude longer than the single pulse of DTOF, a much lower amplitude is needed to transmit the same total average optical power. Additional factors for determining optical power are that the range tends to be less due to the aliasing problem, and ITOF detectors are often full-frame imagers so that the entire field of view must be illuminated rather than one small spot. The former tends to reduce the necessary amplitude and the latter tends to increase the amplitude. Because of these factors, it is difficult to make simple statements based on analysis. However, Section 13.2.4 discusses typical numerical values seen based on the current designs in the market.

13.2.4 Typical Pulse Parameter Values for DTOF and ITOF

The previous discussion provides the basic background for desired optical pulses for DTOF and ITOF cases. These are quite idealized and general, so, while they provide a high-level view of what is important and why they don't cover realistic systems. This section will provide some numerical ranges of parameters for both types of systems.

First, it is a good idea to understand some system characteristics. DTOF systems tend toward high power and long range, though there are some notable exceptions [14]. These systems tend to use a low number of lasers (1–128) and combine them with beam-steering optics, so that within the field of view only a small number of spots, or in some cases a single line, are illuminated. This allows one to make simultaneous measurements of up to approximately a hundred points. Hence, it takes multiple sweeps to build up a whole frame covering the entire field of view. However, each sample point can have a high pulse energy, meaning the range of such systems can extend to beyond 300 m. Due to the long range, the distance resolution on these systems can be somewhat relaxed, though it is still often on the order of 10 cm.

ITOF systems tend to have short range and high distance resolution. Due to the long bursts on the order of microseconds, illumination of only a single point to a few hundred points would make the acquisition time of a 100 kpixel or 1 Mpixel frame prohibitively long. Hence, many ITOF systems use a CMOS focal plane array detector that can capture an entire frame at one time similar to a digital camera. As a result, the laser optics are designed such that the entire field of view is illuminated, and no beam steering is necessary. However, the optical energy now fills a large area so that the pulse energy density is low. The resulting small signal is one of the factors that limits the range. ITOF systems can have high frame rates with high pixel counts, and the distance resolution can range from ~1 cm to <1 mm, even for low-cost systems, and these capabilities are very compelling in applications such as blind spot detection in cars or robot vision systems.

Table 13.1 shows some typical numbers. These are not hard boundaries, and modern lidar system design continues to evolve. They do, however, represent the typical state of the art for systems either in, or near, commercial production. Real pulse shapes are generally quite different from impulses and square waves and are generally dominated by circuit parasitics. These will be discussed in Section 13.3 in conjunction with driver circuit design.

13.2.5 Semiconductor Optical Sources

The most common laser source for TOF lidar is a near-infrared (NIR) semiconductor laser diode. There are two main types, edge-emitting lasers (EELs) or vertical cavity surface-emitting

Table 13.1 Main specifications for lidar transmitters based on laser diodes.

	DTOF	ITOF	Units
Laser peak optical power	100–1000	2–20	W
Laser peak current	30–500	2–20	A
Pulse width	2–10	2–100	ns
PRF	0.01–1	10–250	MHz
Laser duty cycle	0.02–1	5–25	%
Range	50–>300	0.5–50	m

Figure 13.5 Some typical laser diodes used for TOF lidar, all from ams OSRAM. SPL S1L90A_3 A01 single-channel triple junction EEL (a), SPL S4L90A_3 A01 quad channel triple junction EEL (b), EGA2000-940-W VCSEL with diffusor intact (c) and removed (d). More details are shown in Table 13.2.

lasers (VCSELs). In some cases, LEDs may be used instead of lasers, but the basic design of the driver is similar, so in the remainder of this chapter only the term laser will be used. Some typical laser diodes are shown in Figure 13.5 [15–17].

Whether EEL or VCSEL, the basic laser diode is a form of junction diode. When forward biased above a certain threshold current a laser diode emits laser optical radiation at a certain wavelength. When driven with a pulse of current, a pulse of laser light whose instantaneous optical power is approximately proportional to the current is generated [6]. It is the job of a laser driver to generate the current pulse, and the driver must control two main parameters: pulse width and amplitude. These two factors have a large effect on the distance resolution and the range, respectively, as discussed in the previous section.

Some typical laser diode specifications are shown in Table 13.2. Laser diodes datasheets normally present a reasonably comprehensive view of optical performance, but electrical parameters are generally poorly specified compared to more conventional diodes. This is made more challenging by the lack of simulation models, and that pulsed laser diodes are often operated under conditions that are not addressed in typical datasheets. Laser diode manufacturers are working to remedy this situation, but they face several challenges. First, the state of the art is advancing rapidly. Second, the required testing methods are difficult or non-existent, and are in the process of being developed. Third, there has been explosive growth in production volume, which consumes much of a company's limited resources. Finally, the developers of laser diodes generally come from an optical or optoelectronic background and have little familiarity with the needs of circuit design. At this time, only a few laser manufacturers have usable electronic simulation models.

A laser diode die may comprise multiple laser diodes in series, parallel, or both. For example, both EELs and VCSELs have multiple stacked laser junctions grown on top of each other to increase total optical output capability, with as many as four or more junctions in series available in commercial lasers. Both EELs and VCSELs also allow fabrication of multiple parallel lasers on a single die. In such cases the lasers may all be connected in parallel to act as one light source, or they may be grouped in blocks that can be driven individually. The latter case is useful to form part of a beam-steering system. Multiple lasers on a single die can also simplify optical design because the lasers on a single die are precisely aligned by the semiconductor fabrication process.

Finally, note that most laser diodes used for lidar operate in the range of 900–950 nm wavelength, part of the near-infrared (NIR) band. There are several benefits to this. First, silicon photodiode-based detectors have theoretical peak responsivity for wavelengths in the range of 900–1000 nm, which is a tremendous benefit in the development of cost-effective detectors [18]. Second, the most common basis for such lasers is a gallium arsenide/gallium aluminum arsenide (GaAs/GaAlAs) heterostructure, which is also quite mature from a semiconductor process standpoint [19]. This fact also contributes to the relatively high electrical-to-optical conversion efficiency, which is often in the range of 20%–40% in application. Third, there is a notch in the solar spectrum in the same range, which helps reduce solar interference in lidar systems [18]. Finally, many optical techniques and materials for visible light are simple to apply to NIR optics, reducing cost and design time.

Table 13.2 Laser diode specifications.

Part number	λ(nm)	I_{Fmax}(A)	V_{Fmax}(V)	$P_{opt,max}$(W)	Package	L(nH)
SPL PL90_3	905	30	9	75	Through hole	5
TPGAD1S09H	905	30	12.5	75	Surface mount	2
SPL S1L90A_3 A01	915	40	11	120	Surface mount	<0.5
SPL S4L90A_3 A01	915	160	11	480	Surface mount	<0.5
EGA2000-940-W	940	10	3.3	7.5	Surface mount	<0.5

GaN Power Devices for Efficient Power Conversion

There are some drawbacks to NIR lasers, most critical is the concern for eye safety. NIR light is close enough to visible light that it can be focused onto the retina in the human eye. This places strict limits on peak and average optical power density, and these limit SNR. Some lidar systems use lasers in the range of 1400–1550 nm, which is considered much safer. However, high-power laser diodes in that wavelength range are not mature; hence, they are expensive, and the electrical to optical efficiency is low compared to 900–950 nm laser diodes [20]. Another solution is to use pumped fiber lasers, but the size, complexity, and cost are a barrier to high volume consumer or automotive applications.

Laser diodes are available in a variety of packaging options, ranging from bare die to complex metal–ceramic hermetic enclosures. The package design considers thermal, environmental, electrical, and optical performance, and has a large impact on lidar performance. From an electrical point of view, the inductance of the laser package is the dominant factor in the design of high current, nanosecond pulse lasers. This has driven laser manufacturers to move from through-hole to surface-mount packages.

13.2.6 Driver Switch Properties

The discussion above assumes that the transistor Q_1 used in Figure 13.4 is an ideal switch. Practical semiconductor switches are usually implemented with field effect transistors (FETs), and though they are fast, they have non-zero switching time and can limit the current due to saturation and resistive effects. In addition, switches and their packages can have significant inductance which not only increases the required voltage for a given pulse shape but can significantly limit turn-on speed.

The electrical performance advantages of GaN transistors are discussed in detail in many of the previous chapters, primarily in the context of power conversion. Lidar applications have many similarities, and a few key differences that are highlighted here. The benefits of high switching speed and low inductance packaging are even more important for lidar than for power conversion. On the other hand, the key specification for a lidar application is the maximum pulse current, I_{Dsat}, rather than $R_{DS(on)}$. The voltage drop due to the laser diodes is generally an order of magnitude higher than the transistor; hence, the conduction loss of the transistor is a secondary consideration. Finally, the small physical transistor size turns out to be crucial in many lidar applications because it allows many transmitter channels to be closely spaced, which reduces the overall lidar size, simplifies the optical hardware, and avoids additional parasitic inductance.

The development of GaN transistors and integrated circuits has made far better laser driver performance possible than with even the best silicon MOSFET technology. Table 13.3 shows a comparison of two automotive qualified (AEC-Q101) FETs. One is a GaN FET (EPC2252) and the second is a state-of-the-art silicon MOSFET (IAUZ40N08S5N100) [21, 22]. The components are chosen, based on voltage and pulse current ratings suitable for a typical laser driver for an automotive lidar. The required gate charge to turn the transistor on is 3.6 times smaller for the GaN device, and the internal gate resistance is 2.0 times smaller. Based on the gate figure of merit (FOM) $R_G \cdot Q_{Gtot}$, an ideal gate drive would allow the GaN transistor to switch 7 times faster. We also observe that the GaN transistor area is 6.2 times smaller. The inductances that control switching speed, namely L_G and L_S, are 2.5–10 times smaller due to the wafer-level chip-scale package, and this has a major benefit in achieving the maximum possible switching speed (see Chapter 4). Finally, it should be noted that high-performance gate drivers needed for lidar applications usually have an absolute maximum voltage rating of <6 V; hence, the silicon MOSFET parameters listed are higher than what is obtainable in a real design.

Table 13.3 Comparison of GaN transistor and MOSFET parameters relevant for lidar applications.

Parameter	EPC2252	IAUZ40N08S5N100	Benefit of GaN over Si
Technology	GaN transistor	Si MOSFET	
Voltage rating $V_{DS,max}$ (V)	80	80	1
Gate voltage V_{GS} (V)	5	6	1.2
On-state resistance $R_{DS(on)}$ (mΩ)	8	12	1.5
Max pulse current $I_{pulse,max}$ (A)	75	85	0.9
Total gate charge Q_{Gtot} (nC)	3.5	12.5	3.6
Internal gate resistance R_G (Ω)	0.6	1.2	2.0
Gate FOM $R_G \cdot Q_{Gtot}$ ($\Omega \cdot$ nC)	2.1	15	7.1
Gate inductance L_G (nH)	<0.2	2.0	>10
Source inductance L_S (nH)	<0.05	0.25	>5
Drain inductance L_D (nH)	<0.1	0.1	1
Package (mm × mm)	LGA 1.5 × 1.5	DFN 3.3 × 3.3	6.2

The greatly improved performance of GaN transistors over the aging silicon MOSFET technology translates to much faster switching for a given peak current capability, yielding higher image resolution at longer distances. In fact, GaN power semiconductors are a key enabling technology for cost-effective lidar systems.

13.2.7 Gate Drivers and Laser Driver Integrated Circuits

Up to this point, there has been an implicit assumption that switch Q_1 can be turned on and off instantaneously. Clearly, this can only be an approximation. The first consideration is the raw semiconductor capability of Q_1 itself, and this is captured by the gate FOM in Table 13.2, namely $R_G \cdot Q_{Gtot}$, which is a measure of how fast the gate can be brought to its required voltage. We see that GaN transistors offer greater than a sevenfold improvement over silicon. Another Q_1-based factor that affects the speed is the source inductance, and here the chip-scale GaN package offers a fivefold, or better, improvement. However, there are other factors external to Q_1 that influence the speed, and the main factor is the gate driver that triggers the GaN transistor.

The gate driver output impedance appears in series with the transistor's R_G and thus will slow the transition speed. Some of this impedance is due to the gate driver IC, but in addition, the physical separation of the gate driver IC from Q_1 causes additional inductance in the gate drive loop. This acts as an additional impedance which further slows Q_1 switching and has the potential to cause additional problems like gate voltage overshoot. A good gate driver IC should have low output impedance, small size, fast intrinsic switching, and very low inductance. Recently, such gate drivers have become commercially available [23–25], but they can be expensive compared to more conventional gate drivers. While good performance can be achieved with these gate driver ICs, the fact that they are separate from Q_1 means that

GaN Power Devices for Efficient Power Conversion

there will be additional inductance in the gate loop that cannot be eliminated. Additionally, they consume additional area which can be a challenge for multichannel lidar applications. Finally, the limited selection of gate drivers means that for many applications, the gate driver will be more powerful than needed, and for some high-power applications, they will be too weak, and one must then consider running gate drivers in parallel, bringing an accompaniment of addition challenges. In short, the gate drive is optimal for only a small set of FETs.

These challenges can be addressed via the inherent integration capability of lateral GaN power semiconductors. As shown in Chapters 1, 3, and 10, gate drivers, logic, and other functions can be monolithically integrated with a power device. In this way, performance can be improved by matching the gate driver size to the power FET and reducing the gate loop inductance. In addition, this reduces PCB area, parts count, and cost. Three examples of this are the EPC21601, EPC21603 (each 40 V, 15 A, 150 MHz), and EPC21701 (80 V, 15 A, 100 MHz) laser driver ICs [26–28]. Experimental results with these ICs are given in Section 13.5. In the future, the majority of laser driver applications will be met with monolithic GaN ICs.

13.3 Basic Design Process

In this section, we consider the hardware design of laser drivers that can achieve single-digit nanosecond pulses from tens to hundreds of amperes. Circuit design will be discussed, along with physical layout using common printed circuit board techniques. The latter is particularly important because the intrinsic speed of GaN power semiconductors is so fast that it is the parasitic components that dominate the design, with layout being a significant contributor.

Most of the discussion will focus on the design of the capacitive discharge resonant driver [29–31]. A key reason is that the rectangular pulse driver implicitly assumes that inductance is not a dominant factor in the design, but this assumption is generally not correct. Note that the underlying circuit design methods and layout methods are the same, so most of the discussion also applies to non-resonant drive circuits. Specific details that apply to the latter will then be discussed.

13.3.1 Basic Driver Circuits

There are many switch-based pulsed laser diode driver circuit topologies, but they all share a loop which includes a power source, an impedance, a switch, and the laser diode. We consider the two main methods for high current short-pulse laser drivers, both shown in Figure 13.6. This will cover the key functionality of all switched pulsed laser drivers. Figure 13.6a shows the resonant capacitive discharge laser driver, and Figure 13.6b shows the non-resonant pulse driver. These topologies are nearly identical, the main differences being the relative values of the various components and the drive requirements.

In the capacitive discharge circuit, capacitor C_1 is charged from a high impedance source (the current source in Figure 13.6a) during the off-time of transistor Q_1. In the normal case, Q_1 is turned on and the capacitor discharges through the laser diode, the switch, and inductance L_1, which includes stray inductance in the power loop. This forms an RLC circuit in which the current i_{DL} rings up and discharges the capacitor through the laser diode D_L. The resulting current pulse generates the desired optical pulse from D_L. After this point, the transistor is turned back off, and the capacitor C_1 is recharged for the next pulse. Thus, in the ideal, loss-less case the laser diode current is a half-sign pulse.

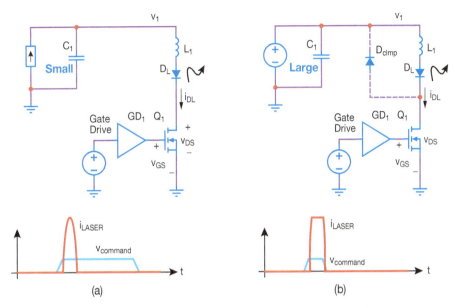

Figure 13.6 Most common laser diode driver circuits for pulsed lidar applications. (a) capacitive discharge driver; (b) rectangular pulse driver. For (b), an optional clamping diode D_{clmp} for laser protection is shown.

There are several benefits to the capacitive discharge approach. Stray inductance is incorporated into the circuit operation. Laser pulse energy is well controlled, and the minimum pulse width of the gate drive can be substantially longer than the desired diode current pulse. This is useful because many gate drivers have a minimum pulse width specification greater than the desired pulse width. The transistor and gate drive are ground-referenced, simplifying the circuit. Another benefit is that only the timing of the turn-on signal must be tightly controlled. This is highly beneficial as turn-on and turn-off propagation delays are not perfectly matched, nor are their temperature and power supply voltage coefficients.

However, the resonant discharge circuit has some drawbacks. Since the pulse shape is largely fixed for a given design, it places limitations on the types of performance-enhancing signal processing discussed previously. In addition, the time to recharge the capacitor limits the maximum pulse repetition frequency and the lidar sample rate. A high-voltage supply is needed, which adds expense and takes up valuable space within the system. Laser diodes often have relatively low reverse breakdown voltages; hence, an anti-parallel diode may be needed to protect the laser from reverse voltage. Despite the disadvantages, the simplicity and the incorporation of stray inductance make this approach popular.

In the rectangular pulse driver, a large C_1 acts as a bypass capacitor for voltage source V_1. A gate pulse width that simply produces the desired laser pulse width can then be used. Laser diode current is limited by the combination of the laser diode voltage drop and the transistor series resistance or saturation current.

The advantages of the rectangular pulse driver include control over pulse width and higher maximum pulse repetition frequency. The main drawbacks of this approach are threefold. First, laser diode current is difficult to control, and current sensing and voltage bus control are often necessary to ensure pulse-to-pulse repeatability. The second challenge is stray inductance. For this approach to work, the circuit must switch slowly enough that stray inductance does not significantly degrade the waveforms. With presently achievable values of stray inductance, the rectangular pulse approach is generally limited to lower peak amplitudes. Finally, the third problem is that under a transistor short failure, the laser can operate at full power for an indefinite period, and this is a possible eye safety hazard. This is a serious design concern and must be addressed with suitable protection circuitry.

Practical systems may share properties of both capacitive discharge and rectangular pulse drivers, and many variations are possible. At the time of this writing, high-power lidar systems favor the capacitive discharge approach since it can achieve stable, high-power, short, and well-controlled pulses with reduced need for complex compensation methods.

13.3.2 Resonant Capacitive Discharge Laser Driver Design

A laser driver starts with a set of specifications. For a resonant capacitive discharge driver, the usual starting information is:

I_{DLpk}: laser diode peak current
t_w: pulse width (FWHM)

From these values, one must pick the following:

C_1: resonant capacitor value
Q_1: GaN FET

There is one missing piece of information, namely the inductance L_1. As will be seen, L_1 dominates the design, but its value is not straightforward to determine. Some guidance will be given later in this section, but it is necessary to understand the design equations first, so let us start there.

Figure 13.7 shows a simplified schematic of a resonant capacitive discharge laser driver, and Figure 13.8 shows the main waveforms.

Assuming that Q_1 is an ideal switch and D_L an ideal diode with a fixed forward voltage drop, V_{DLF}, the driver works as follows: Q_1 starts in the off-state, so $i_{DL} = 0$. The capacitor voltage $v_1 = V_{IN}$, has been charged through R_1, where R_1 is chosen to be large enough that it has negligible effect on the resonant discharge (see Section 13.6.2). At $t = t_0$, $v_{command}$ triggers the gate drive, turning Q_1 fully on at $t = t_1$, and discharges C_1 through the laser diode D_L and inductor L_1. C_1 and L_1 form a resonant network, hence i_{DL} and v_{C1} ring sinusoidally. The effective initial capacitor voltage is $V_{C1,0} = V_{IN} - V_{DLF}$ due to the laser diode forward drop. At $t = t_2$, i_{DL} returns to zero and $v_{C1} = 2 \cdot V_{DLF} - V_{IN}$. At this point D_L prevents the current from reversing, and C_1 recharges via R_1. Switch Q_1 is turned off before v_1 crosses zero at $t = t_3$.

The capacitor charging time constant, τ_{chrg}, and the resonant period, t_{res}, are:

$$\tau_{chrg} = R_1 \cdot C_1 \tag{13.5}$$

$$t_{res} = 2\pi\sqrt{L_1 \cdot C_1} = 2t_{wb} \tag{13.6}$$

where t_{wb} is the width at the base of the pulse, i.e. half the resonant period. For high-power resonant pulse drivers, the laser duty cycle is usually <1% due to thermal limitations of the laser. Hence $\tau_{chrg} \gg t_{res}$, so R_1 has negligible effect on the L_1–C_1 resonance. The

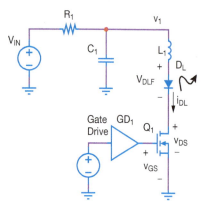

Figure 13.7 Capacitive discharge resonant driver.

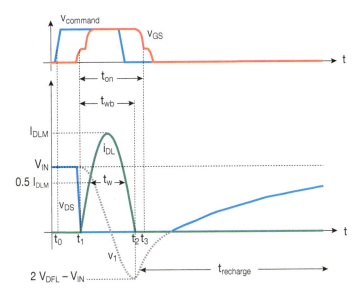

Figure 13.8 Key waveforms for capacitive discharge resonant driver of Figure 13.7.

resonant characteristic impedance R_0 and the full-width-half maximum (FWHM) pulse width t_w are given by:

$$R_0 = \sqrt{\frac{L_1}{C_1}} \tag{13.7}$$

$$t_w = t_{res} \cdot \frac{\pi - 2\sin^{-1}\frac{1}{2}}{2\pi} = \frac{t_{res}}{3} \tag{13.8}$$

The I–V curve of the laser diode and the inductance of L_1 determine the remainder of the design. Normally, the diode has been selected at the beginning of the design based on its optical power capability. Once the diode and optical power are known, the datasheet will provide the necessary peak current, I_{DLpk}. For design purposes, it is usually sufficient to approximate the diode forward voltage drop, V_{DLF}, based on the I–V curve of the laser diode. This will result in a slightly conservative design, and the error can easily be compensated for by reducing V_{IN} if necessary. The inductance, L_1, will be a function of the power loop components, C_1, Q_1, and D_1, *and the physical arrangement and connection of these components.* The inductance, L_1, cannot be zero because real parts have non-zero dimensions. Once all the components are chosen, a preliminary PCB layout can be done, and the inductance estimated via approximate analysis, FEM analysis, or experimentally. This means that some design iteration may be necessary to optimize the design.

We can calculate the peak laser diode current, I_{DLpk}, using the following equation:

$$I_{DLpk} = \frac{V_{IN} - V_{DLF}}{R_0} \tag{13.9}$$

From Eqs. (13.6)–(13.9), we solve for V_{IN} to find

$$V_{IN} = \frac{2\pi L_1}{3t_w} \cdot I_{DLpk} + V_{DLF} \tag{13.10}$$

$$C_1 = \frac{9t_w^2}{4\pi^2 L_1} \tag{13.11}$$

Once we know I_{DLpk}, t_w, and L_1, we can determine the required charging voltage, V_{IN}, and resonant capacitance, C_1.

From the above discussion, the following values are known: capacitance, C_1, peak current, I_{DLpk}, and the required voltage, V_{IN}. C_1 and V_{IN} determine the capacitor(s), and selection of Q_1 is done by choosing the smallest FET that meets the voltage and pulse current rating. This FET will have the lowest capacitance and thus be the fastest and usually the lowest cost component.

Now is an appropriate point to look further at L_1. From Eq. (13.10), V_{IN} is a decreasing function of L_1. Hence, a smaller L_1 means a lower voltage rating for C_1 and Q_1, and this generally means smaller, lower cost components, and potentially simpler layout. It is therefore desirable to minimize L_1 in any design. However, the designer would benefit from a starting point, so it is worth looking at some typical values for L_1. Based on several designs, it has been found that for an optimized design Figure 13.9 can be used to select a starting point.

Figure 13.9 shows ranges of some typical values of L_1 for designs in the range of I_{DLpk} of 10–200 A, with PCB designs suitable for V_{IN} values up to 200 V [32]. These ranges are based on a PCB with an optimized low inductance layout and typical laser diodes. Further details on this will be presented in Section 13.4.2. In most cases, it is the laser diode that contributes to the major portion of L_1, primarily due to the use of bond wires and in the case of through-hole inductors' leads. If we look at the latter, we see that the lowest achievable value is in the vicinity of 3 nH. Typical values tend to be in the range of 5–20 nH for through-hole packaged lasers mounted to the PCB with leads as short as possible. Surface mount lasers inherently have much lower values of L_1. Figure 13.9 also shows an inductance range based on PCB designs with a foil short circuit in place of a laser to use as a baseline minimum. The inductance ranges are based on experience with standard four-layer PCBs following IPC-2221 [33] recommended clearances and using via-in-pad construction (IPC4761 Type VII) but excluding blind vias, buried vias, or micro-vias [34].

Figure 13.10 shows the calculated voltage V_{IN} needed using Eq. (13.10) for 2, 5, and 10 ns pulse widths as function of L_1 for a 75 A pulse, and driving a laser diode with a 12 V forward voltage drop. The required V_{IN} increases linearly as function of L_1 for a given laser, pulse magnitude, and pulse width. Any parasitic resistance will drive the necessary V_{IN} higher, though this is not accounted for here. For example, a 2 ns pulse will need $L_1 < 1$ nH to allow for $V_{IN} < 100$ V. Section 13.4 will show how this can be accomplished.

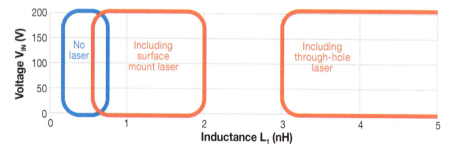

Figure 13.9 Typical ranges of total power loop inductance L_1 for various lasers, along with a baseline with laser replaced by copper foil.

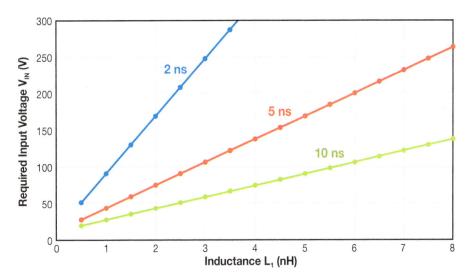

Figure 13.10 Bus voltage V_{IN} as function of inductance, L_1, for I_{DLpk} = 75 A, V_{DLF} = 12 V, and pulse widths of 2, 5, and 10 ns.

The resonant laser driver topology has the following benefits:

- The topology utilizes stray inductance.
- It has a stable pulse shape (L_1 is very stable).
- Pulse energy is set by the supply voltage V_{IN}.
- The switch is ground-referenced for simple drive.
- Only gate turn-on needs precise control (single edge control).
- Laser current pulse width can be shorter than gate drive minimum pulse width.
- Zero-current turn-off means there is no inductive spike, reducing voltage stress and EMI.

13.3.3 Non-Resonant Laser Driver Design

While it might seem that a non-resonant design would be simpler, this is not the case for high-current nanosecond pulses. From the resonant design process, we realize that it is important to minimize L_1 to minimize the required voltage needed to overcome the inductance. This is also applicable for a non-resonant design; hence, the only real change for a non-resonant design is that the power source impedance is low, i.e. C_1 is large and R_1 is small. This results in two problems. First, at the moment the switch Q_1 turns on, the rate of rise in current (di_{LD}/dt) is the same for both resonant and non-resonant cases. In the resonant case, it decreases as the capacitor is discharged, but it remains constant for the non-resonant case. To obtain a flat-topped pulse, something must limit the current. Second, at the end of the pulse, this current is interrupted by turning off the switch. At this point, the inductor current must rapidly decrease to zero, and this means a very high voltage v_{DS} appears at the drain of Q_1 unless there is an alternative path that either absorbs or diverts the inductive energy.

Consider the current limiting problem first. Current can be limited by resistance R_{stray} in the components or PCB, or by saturating Q_1. When the sum of v_{DS}, V_{LD}, and V_{Rstray} equals V_{IN}, then the current will stop increasing and become constant. This must be limited by resistance in the circuit or saturation of Q_1. The laser diode junction voltage drop is nearly constant, and it has a negative temperature coefficient, so it is not a good candidate for current limiting. However, there is a problem because it was shown in Section 13.3.2 that a relatively large V_{IN} is needed to turn the laser on quickly. This means that once the current is limited, the large V_{IN} means that a significant amount of power is dissipated, generating high heat flux. Thus, thermal management becomes a challenge. Since non-resonant pulses are usually used in ITOF systems where the laser duty cycle can be as high as 25%, this limits the design. For low-power systems, current control is achieved using saturation current limiting, but for medium- or high-power systems, the thermal problem becomes extremely challenging to manage.

Now consider the turn-off problem. The current in the laser and L_1 will be at the maximum. When Q_1 is turned off, this results in a large voltage spike and ringing across at the drain terminal of Q_1 until all the energy stored in L_1 is either stored elsewhere or dissipated. Some of the energy is stored in the output capacitance of Q_1, but this gets dissipated when Q_1 is turned on for the next pulse. There are various means to clamp the overshoot, but there is an inherent challenge as any clamping circuit will have its own inductance contribution. Since the required power loop inductances have already been minimized, the loop created by the clamp will always have the same or greater inductance. This greatly reduces the effectiveness of the clamp. The use of GaN power devices is of great benefit here because of the higher voltage rating of GaN FETs compared to equivalent speed Si MOSFETs. Hence, clamping is often not necessary to protect Q_1 when using a GaN FET. Note that EMI may still need to be addressed with non-resonant designs with careful attention to layout and, where necessary, shielding.

13.4 Hardware Driver Design

This section presents a design example to illustrate the effectiveness of the design procedure and to highlight key aspects of the physical design necessary to achieve the desired performance.

13.4.1 Resonant Pulse Hardware Design Example

This example is based on the EPC9179 laser driver demonstration board [35], shown in Figure 13.11. It uses the EPC2252 80 V, 75 A automotive qualified GaN FET as the main switch [21]. The design target is I_{DLpk} = 65 A and t_w = 2 ns.

The layout was designed to minimize the total inductance in keeping with the principles presented in Chapter 4. The design also includes high bandwidth voltage test points and a current measurement shunt. Details of the design can be found in [35], including Gerber layout files and a schematic.

An expanded view of the key section of the design is also shown in Figure 13.11, with the main parts labeled. To minimize inductance, both the energy storage capacitor and the current measurement shunt comprise five 0402 size surface mount packages connected in parallel. A set of pads to facilitate mounting of a laser are provided at the edge of the PCB. If possible, the laser should be mounted directly to the PCB, but this is not always possible. For lowest inductance, an interposer should be used to mount lasers that do not fit the pad layout. Note that a set of interposers is available for a small set of commercially available lasers [36].

Since inductance has a strong effect on design, the SPL S1L90A_3 A01 surface mount laser diode was used since it has the lowest known inductance for a high-power packaged laser diode able to handle the pulse current. Figure 13.12 shows the laser mounted using an interposer.

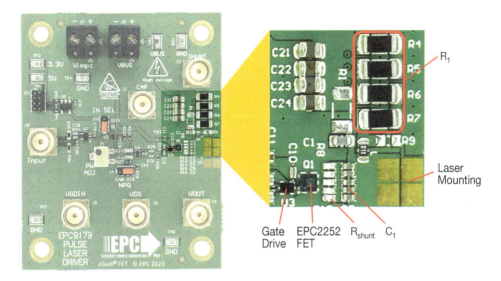

Figure 13.11 EPC9179 [36] laser driver with EPC2252 GaN FET [21] for experimental validation. *Source:* Refs. [36] and [21]/Efficient Power Conversion Corporation.

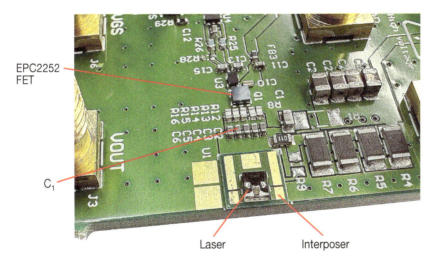

Figure 13.12 EPC9179 laser driver with laser mounted using an interposer. *Source:* Ref. [36]/Efficient Power Conversion Corporation.

While the target current is higher than the laser diode nominal rating of 40 A for 100 ns, the resulting pulses are short enough that the higher pulse current does no apparent harm to the laser (this is not recommended practice unless approved by the laser manufacturer).

The power loop inductance with mounted laser has an initial estimated value of 800 pH, based on experimental measurements. From the data sheet of the laser diode and knowledge of the diode junction drop for the laser, the voltage drop of the laser at the target current is estimated at $V_{DLF} = 12$ V. The ideal design based on Eqs. (13.10) and (13.12) yields $V_{IN} = 66.5$

V and C_1 = 1.14 nF, respectively. The capacitor value used was C_1 = 1.1 nF since this was the closest that could be achieved with standard component values. NPO/C0G ceramic capacitors were used due to their stable capacitance as function of voltage and temperature, and low loss.

13.4.2 Power and Gate Loop Layout

Layout plays a dominant role in laser driver performance because even the best layout contributes significantly to inductance in the design. A less-than-optimal layout will severely limit the performance. Fortunately, it is possible to achieve excellent results using standard PCB design methods. This section provides guidelines to achieve the desired results.

The first consideration is the board stackup, and the general principles of inductance cancellation are discussed in Chapter 4. Minimizing power loop inductance is of prime importance, but it is not enough. In lidar applications, turn-on speed is of utmost importance. Hence the gate loop inductance, and particularly the common source inductance (CSI), must be minimized. To understand why, consider Figure 13.13 where the CSI is shown explicitly. In a lidar application, a typical turn-on current slope is between 50 and 100 A/ns. A CSI of 50 pH will result in a voltage of 2.5–5 V appearing across the CSI, which subtracts from the gate voltage as explained in Chapter 4. Given that the gate turn-on voltage is 5 V, this clearly cannot work, and a CSI value in the range of 10–20 pH or less is a necessary target. If we consider that the rule-of-thumb for inductance is ~1 nH/mm, this might seem impossible, but we must consider that this number represents the coupling of the power loop and the gate loop and is not a standalone inductance. Hence, we need to take a careful look at how achieving this ultra-low CSI might be accomplished.

To understand how to decouple the gate and source loops from each other, one must first become familiar with Figure 13.14, which shows the physical layout of the key laser driver components of Figure 13.7 in the context of a PCB.

Figure 13.15 shows how the current flows when the switch Q_1 is turned on, with the left-hand side (a) showing the components and the PCB Layer 1 (top layer) and the right-hand side (b) with the Layer 1 removed to make Layer 2 visible. The red arrows show the gate current and the drain current for Q_1, whereas the blue arrows show the source current components.

Figure 13.15 shows how relevant currents flow on Layer 1. When the gate driver turns on, the gate driver bypass capacitor discharges through the gate driver and into the gate of Q_1. This transient turn-on gate current must then return into the negative terminal of the bypass capacitor by flowing left from the source of Q_1. As the gate is charged, Q_1 turns on and current flows out of the capacitor C_1, through the laser, and then down through the vias to Layer 2. Figure 13.15b shows the continuation of this current on Layer 2 back toward the FET, and then up through the vias into the drain terminals of Q_1. Similar to the gate loop, current flows through Q_1, and then out the source terminals. Returning

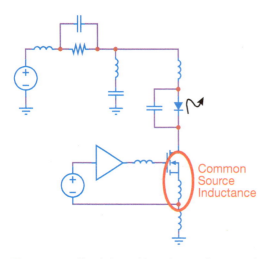

Figure 13.13 Simple laser driver showing location of multiple parasitic inductors, with common source inductance CSI highlighted.

High Current Nanosecond Laser Drivers for Lidar | 387

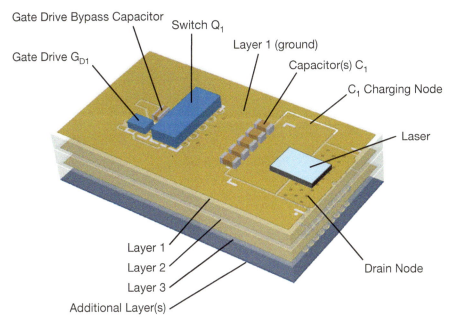

Figure 13.14 High-performance, low CSI PCB layout with main power loop components identified.

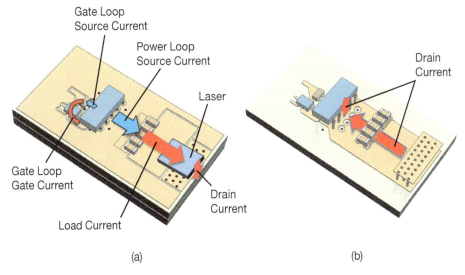

Figure 13.15 High-performance, low CSI PCB layout showing current flow during Q_1 turn-on. Layer 1 currents (a) and Layer 2 currents (b) are highlighted.

to Figure 13.15a, the gate return current is shown flowing out of the source and back into the negative terminals of the bus capacitor C_1, completing the loop.

There are two key points to understand. First, the gate loop and power loop source currents are almost completely separated, with the only common current path being the source bars on Q_1. As shown in Chapter 4, the source bars are large, interdigitated, and distributed over the entire die, which greatly reduces their contribution to inductance. Secondly, by moving the power loop drain current path to Layer 2 and the power loop source current path to Layer 1, we move the via inductance near the FET from the source loop to the drain loop. This does not change the total power loop inductance; however, it trades reduced source inductance for increased drain inductance. This in turn reduces ground bounce and can further reduce CSI with no negative effects on the main laser pulse.

13.5 Experimental Results

Experimental results show both the effectiveness of the design procedure and the performance that can be obtained. There will be three examples shown for resonant discharge laser drivers, and one example for a non-resonant monolithic GaN driver suitable for DTOF.

- A short resonant pulse (the design example of Section 13.4) DTOF design
- A high resonant current DTOF design
- A high resonant current DTOF design using a monolithic GaN IC
- A high-frequency ITOF design using monolithic GaN IC

13.5.1 Resonant Discharge Laser Driver with Short Pulse

Figure 13.16 shows the results for the EPC9179 [36] design example of the previous section. The results are very close to the values calculated, and the waveforms look clean and as expected. The waveforms are measured with probes built into the board, which allows high bandwidth and low noise measurement. Section 13.6 gives more detail on the design of these probes.

The behavior of the laser driver has a few key features worthy of discussion. First, the drain voltage fall time is <400 ps, so the overlap with the rising current is very small; hence, the switch behaves close to an ideal switch. Both the pulse width and current measure within 5% of the target value. Second, there are very few artifacts in the waveforms, though some ringing is visible after the current descends through zero near the end of the pulse. This additional ringing is most likely due to reverse recovery of the laser diode yet does not result in detectable optical emission. Unfortunately, this remains speculation for now since this cannot be confirmed with

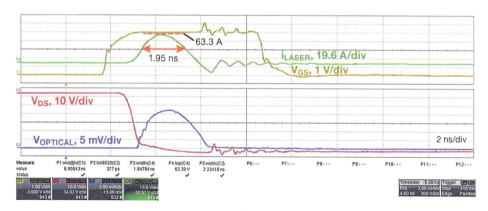

Figure 13.16 Experimental results for EPC9179 [36] laser driver fitted with an EPC2252 GaN transistor [20] with V_{IN} = 70 V and delivering a peak current I_{DLpk} = 63.8 A with t_w = 1.95 ns. The drain voltage exhibits a maximum dV_{ds}/dt = 132 V/ns and the drain current a maximum dI_d/dt = 79 A/ns. *Source:* Adapted from Glaser [36].

laser diode manufacturers. Third, the design assumes a fixed voltage drop across the laser, but a substantial proportion of the laser diode voltage drop is due to resistance, which slows and damps the resonance. There is additional resistance in the current measurement shunt which further slows and damps the resonance. In this case, these additional losses are small and hence the deviation from a half-sine is also small. This effect does not appear to be significant in lidar applications.

13.5.2 Resonant Discharge Laser Driver with High Current

For long-range sensing applications, a high maximum current must be achieved, but often the resolution, and thus pulse width requirement can be relaxed. This requires the selection of a larger transistor. For a good comparison, the same design as presented in Section 13.5.1 was used, except the FET is replaced with EPC2218A, which is an automotive qualified 80 V FET with a 230 A pulse current rating [37]. The EPC2218A uses the same process technology as the EPC2252 and has the same voltage rating, so the comparison starts on an equal footing. It is used in the EPC9180 laser driver [38], which is identical to the EPC9179 except for the FET, the resonant capacitor value (increased from 1.1 to 10.8 nF), and the current measurement shunt, decreased by a factor of three. In addition, the laser has been changed from the ams OSRAM SPL S1L90A_3 A01 to the SPL S1L904_3 A01, with all channels connected in parallel. The latter is electrically equivalent to four SPL S1L90A_3 A01 connected in parallel.

Starting from the assumption of fixed inductance and input voltage, it can be determined from the design equations that current and pulse width will both increase by a factor of $(10.8 \text{ nF}/1.1 \text{ nF})^{1/2} = 3.13$ with a new expected peak current of 198 A and pulse with of 6.1 ns. The measured results are shown in Figure 13.17, with a peak current of 219 A and a pulse width of 5.6 ns, exceeding initial expectations. The main reason is that the new laser is wider and has a factor of four increase in the number of bond wires, both factors that reduce inductance. In addition, the FET is also wider, and this will reduce the inductance slightly, though the impact is less than the contribution of the bond wires.

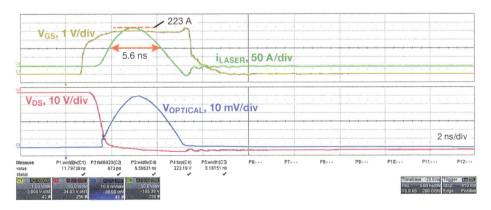

Figure 13.17 Experimental results for EPC9180 laser driver [38] fitted with an EPC2218A GaN transistor [37] with V_{IN} = 70 V and delivering a peak current I_{DLpk} = 219 A with t_w = 5.5 ns. The drain voltage exhibits a maximum dV_{ds}/dt = 60 V/ns and the drain current a maximum dI_d/dt = 112 A/ns. Source: Adapted from Efficient Power Conversion Corporation [38] and [37].

13.5.3 Resonant Discharge, High-Power Laser Driver IC Performance

As discussed in Section 13.2.7, the trend for lidar laser drivers is toward integration. Here, the performance of a monolithically integrated GaN laser driver is compared with a comparable discrete design. The IC is the EPC22704 [39], which will be compared to the EPC2218A of the previous example. The EPC22704 is tested on an EPC91100 design, which has the exact same layout and components as the EPC9180, with the exception that the IC driver replaces the discrete FET and its associated separate gate driver IC. The results are shown in Figure 13.19. The resulting peak current and pulse width are nearly identical.

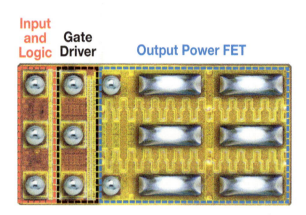

Figure 13.18 EPC22704 80 V, 175 A laser driver IC. Die measures 3.0 mm × 1.5 mm. *Source:* With permission of EPC.

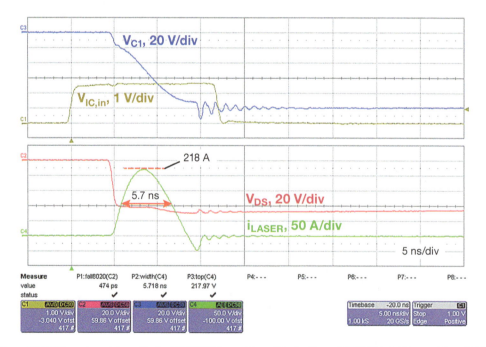

Figure 13.19 Experimental results for EPC91110 laser driver fitted with an EPC22704 GaN laser driver IC [39] with V_{IN} = 70 V and delivering a peak current I_{DLpk} = 218 A with t_w = 5.7 ns. The drain voltage exhibits a maximum dV_{ds}/dt = 108 V/ns and the drain current a maximum dI_d/dt = 111 A/ns. *Source:* Ref. [39]/Efficient Power Conversion Corporation.

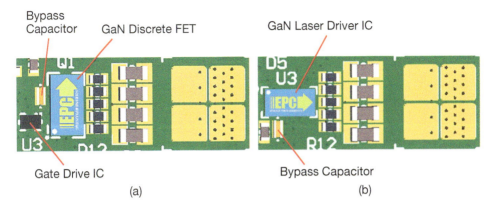

Figure 13.20 Power loop layout of the EPC9180 (a) using an EPC2218A and separate gate drive IC, and the comparable layout for the EPC91100 (b) using an EPC22704 monolithic GaN laser driver IC. *Source:* With permission of EPC.

There are a few details in Figure 13.19 which merit additional attention. First, note that the voltage (v_{DS}) fall time of the IC is nearly twice as fast as the discrete solution, yet there is no additional noise or ringing. This is a benefit of the reduced inductance of the gate loop due to integration of the driver. The IC input can also be controlled from a lower voltage logic level, as demonstrated by the 2.5 V applied input signal.

The EPC22704 is the first high-power GaN laser driver IC, and its performance is excellent. It eliminates the need for a high-performance gate driver IC, and this saves cost and PCB area. In addition, the shape of the laser driver IC is targeted toward lidar applications where it is relatively common to have multiple laser and driver channels. For optical design reasons, it is beneficial to have a high channels density, and the aspect ratio of the laser driver IC facilitates this. This can be seen in Figure 13.20, where the actual PCB layouts of the power loop for each of the EPC9180 and EPC91100 designs are shown at the same scale side by side.

13.5.4 High-Frequency Integrated GaN Laser Drivers

In recent years ITOF lidar technology has been advancing and applications are proliferating. As discussed in Section 13.2, ITOF applications tend to have lower peak currents, but operate at higher frequency and duty cycle. The lower peak current eases the requirements somewhat, but the high voltage spikes resulting from the hard turn-off generate substantial signal integrity challenges, and the high-frequency operation (more than 100 MHz in some cases) is near the limit of many single-ended logic families. These challenges are best met with fully integrated laser drive ICs.

One such driver is the EPC21603 [27], a fully integrated laser driver capable of 15 A, 40 V operation at 100 MHz or more (Figure 3.21a). This driver also has a low-voltage differential signaling (LVDS) input [40], so that it can directly accept the normal standard for high-frequency signaling in modern devices, without the need for an additional LVDS-to-CMOS receiver IC. The EPC9156 is a development board using the EPC21603, shown in Figure 13.21b [41].

The additional functionality of the LVDS shows the benefit of integration. Figure 13.22 shows two designs driven with LVDS input, one using the conventional approach (a), and the other using the EPC21603 monolithic GaN IC approach (b). On the left-hand side of each example is the schematic from the LVDS input to the laser, and on the right-hand side is a view

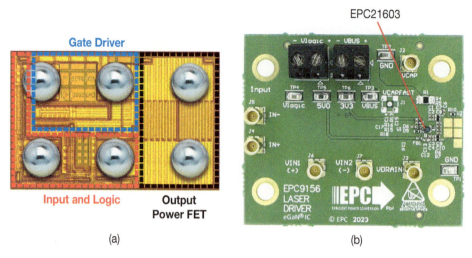

Figure 13.21 Die photograph of EPC21603 [27] high-frequency laser driver IC with LVDS input for experimental validation (a) and same IC mounted on EPC9156 [41] laser driver development board (b). The IC measures 1.5 mm × 1.0 mm. *Source:* Refs. [27] and [41]/Efficient Power Conversion Corporation.

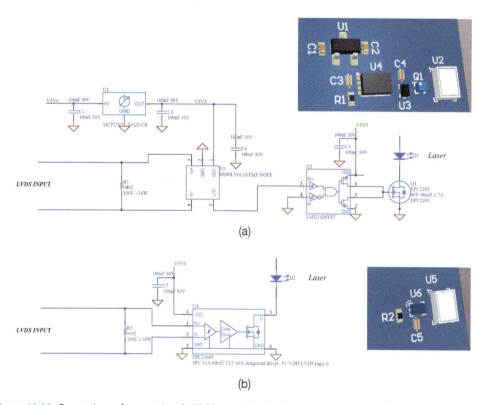

Figure 13.22 Comparison of conventional LVDS laser driver design (a) with GaN IC with integrated LVDS input (b). *Source:* With permission of EPC.

Table 13.4 Comparison of conventional LVDS laser driver design (a) with GaN IC with integrated LVDS input (b).

	Conventional design	**GaN IC design**
Necessary components in design	1× EPC2203 eGaN FET 1× LMG1020 gate driver IC 1× DS90LV012A LVDS receiver IC 1× MCP1703T 3.3V regulator IC 4× bypass caps 1× resistor	1× EPC21603 eToF laser driver IC 1× bypass caps 1× resistor
Total component count (excluding laser)	9	3
Max operating frequency	60 MHz (limited by gate driver IC)	100 MHz

of the required PCB layout for each design. Layouts are shown on the same scale. Clearly the integrated design greatly reduces both the number of components and the physical area of the driver design. Table 13.4 summarizes the differences.

Figures 13.23 and 13.24 show the measured behavior of the EPC21603 as it performs in the EPC9156 development board. Figure 13.23 shows the key waveforms for a 10-cycle burst at 100 MHz, with 19 V_{DC} measured at the bus and using a 2 Ω load. The output of the IC (v_{DS}) shows very clean switching of nearly the full voltage (18.4 V) with turn-on and

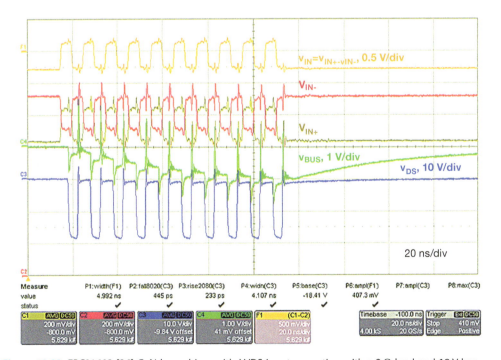

Figure 13.23 EPC21603 [26] GaN laser driver with LVDS input, operating with a 2 Ω load and 19 V bus. Waveforms show a 10-cycle burst at 100 MHz. *Source:* Adapted from Efficient Power Conversion Corporation [26].

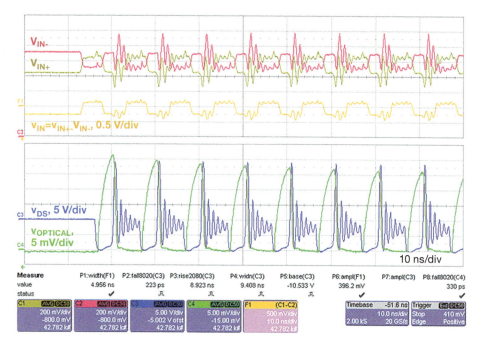

Figure 13.24 EPC21603 GaN laser driver with LVDS input [26] operating with a EGA2000-940-W laser load and 10 V bus. Waveforms show the beginning of a 10-cycle burst at 100 MHz. *Source:* Adapted from Efficient Power Conversion Corporation [26].

turn-off transition of 445 and 233 ps, respectively. This translates to a load current of 9.2 A. There is about 3 V of turn-off overshoot, but this is quickly damped by the load resistance. The extremely fast turn-off transition causes ground bounce which is evident on the input waveforms $V_{IN}+$ and $V_{IN}-$. Ground bounce is a common mode signal, so the differential input (top waveform) attenuates most of the effects and the IC functions as expected. Finally, note that the EPC9156 development board does not contain a current measurement shunt. The requirement for extremely fast transitions precludes the use of a shunt due to its added inductance.

Figure 13.24 shows the same IC and development board driving an ams OSRAM EGA2000 VCSEL, with the bus voltage reduced to 10 V. The optical waveform is nearly proportional to the current waveform, so it gives an idea of the current waveform. The optical waveform shows an exponential rise characteristic of an RL circuit, which indicates that it is the VCSEL inductance that limits the performance. This VCSEL is the same one shown in Figure 13.5d, and one can see that the manufacturer has made an effort to reduce its inductance. However, if more rectangular pulses are desired, there is still much work to be done.

During turn-off, the large overshoot of nearly 200% causes the IC output to reach 30 V, but this is easily tolerated due to the 40 V rating. This spike and subsequent ringing cause a large disruption to the input signals, but again the differential nature of the LVDS input attenuates most of the noise and the IC switches normally.

13.6 Additional Considerations for Laser Transmitter Design

There are several additional practical considerations that have a large impact on lidar driver design. The major ones are discussed in this section.

13.6.1 Bus Capacitors

Resonant capacitors should use NP0/C0G ceramic dielectric or other low loss, linear dielectric with stable properties, such as porcelain, glass, or mica. Class II dielectrics such as X, Y, or Z types, which all have very strong voltage-dependent capacitance and relatively high losses, should be avoided.

In non-resonant designs, the capacitors supplying the laser pulse require values at least one order of magnitude higher than with resonant designs. In most cases, this is not possible with NP0/C0G capacitors. Since the value is less critical, it is acceptable to use X-type dielectrics, but Y or Z types should still be avoided. Where applicable U2J dielectric capacitors can be used for higher capacitance values.

13.6.2 Resonant Capacitor Charging

In the examples shown, the resonant capacitor is charged through a linear charging resistor R_1 with a time constant τ_{chrg} given by Eq. (13.5). Since it takes $5\tau_{chrg}$ to charge the resonant cap to >99% of the final value, we can guarantee a full recharge between pulses by setting max pulse repetition frequency to $\text{PRF} = 1/5 \cdot \tau_{chrg}$. If a designer wants a higher PRF, they can reduce the value of R_1. Reducing R_1 will allow additional current to flow in Q_1 when on, and this may be acceptable for $5 \cdot \tau_{chrg} \gg t_w$.

For an ideal resonant system like that shown in Figure 13.7 and the associated waveforms in Figure 13.8, except for the very first time the capacitor C_1 is charged, the initial state of the capacitor is $V_{C1}(t_2) = V_{IN} - 2 \cdot V_{DLF}$. It is a good approximation that during recharge, all power dissipation occurs in R_1, and the energy dissipated in R_1 will be:

$$E_{R1chrg} = 2C_1(V_{IN} - V_{DLF})^2 \tag{13.12}$$

Since this is determined purely by conservation of charge and energy, it does not depend on the value or linearity of R_1, only that it is purely dissipative.

$$P_{R1chrg} = \text{PRF} \cdot E_{Rqchrg} \tag{13.13}$$

This power dissipation increases proportionally to PRF and adds to the power dissipation in the laser itself. If the power dissipation is too large, different recharge methods such as inductive charging should be considered. The simplest methods make use of a boost or flyback power stage operating in discontinuous conduction mode to directly charge the resonant capacitor. The design of such converters is well known.

13.6.3 Voltage Probing

To estimate a transition time with an error of less than 10%, we need to know the bandwidth the of the measurement system. Using the relationship between bandwidth (BW) and 10–90% transition time t_t for a single-pole analog filter $t_t = 0.35/\text{BW}$, we can write:

$$t_{t,meas}^2 = t_{probe}^2 + t_{scope}^2 + t_{L1}^2 \tag{13.14}$$

For a 10% error, $t_{t,meas} = 1.1\, t_t$. If we assume that $t_{probe} = t_{scope}$ as a starting point, then

$$t_{probe}^2 = \frac{1.21 \cdot t_t^2 - t_t^2}{2} \tag{13.15}$$

or

$$t_{\text{probe}} = 0.324 \cdot t_t \qquad (13.16)$$

This suggests that the scope and probe must have a minimum rise time at least three times faster than the waveform for usable results. If the designer wants to see any non-ideal behavior, a factor of at least 5–10 times faster is suggested. Thus, to accurately measure timing parameters such as transition times of less than 250 ps, combined oscilloscope and probe bandwidths of 6–12 GHz are appropriate, with 4 GHz being the absolute minimum. For basic troubleshooting and verification of operation, a 1 GHz oscilloscope with switchable 1 MΩ/50 Ω inputs and standard 10× probes is usable, but the corresponding 350–400 ps response means such an oscilloscope is not suitable for the accurate timing measurements necessary to verify performance.

Oscilloscopes with these bandwidths are readily available. While there are many probes available with this bandwidth for signal- or logic-level voltages, few are available for voltages approaching 100 V and above. One solution is the use of transmission line probes, which can have a nearly purely resistive input impedance and bandwidth of several GHz or more. However, the DC probe tip resistance is typically 500 Ω to 5 kΩ. To prevent probe failure and the DC loading effect, transmission line probes may be used with a DC block. The best approach is to build the probes into the PCB, which was done in all the development boards discussed in this chapter. This provides a near-ideal connection to the node of interest and is ground referenced, thus improving waveform fidelity and repeatability. The basic principles of such probes are discussed in [42].

13.6.4 Current Sensing

There are both pros and cons to current sensing in pulse laser drivers. The pros include verification of operation, timing determination of the laser pulse, and closed-loop optical power control for maximizing range while remaining eye-safe. However, current sensing has a long list of cons, including added inductance, increased power dissipation, poor waveform accuracy, cost, and reduced available voltage to overcome inductance.

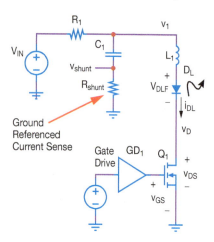

Figure 13.25 Simple capacitive discharge resonant laser driver showing implementation of current measurement shunt.

If pulse current measurement is a requirement, it can still be implemented with a resistive current shunt. However, a great deal of attention must be paid to all details. Presented is an example based on the EPC9179 demonstration board. The shunt resistor is placed between the resonant capacitor and ground so that a ground-referenced measurement can be made as shown in Figure 13.25. Typically, $R_{\text{shunt}} \ll R_1$, so the shunt current is nearly equal to the laser current. The shunt resistor comprises five parallel 0402 resistors to minimize added power loop inductance, as shown in Figure 13.11. These resistors are chosen such that the resistive element abuts the PCB surface to minimize shunt inductance. The shunt layout also follows the internal vertical layout methodology presented in Chapter 4, and the pickup point is centered to minimize the net magnetic field produced by the shunt current and its return current. It is connected to the opposite side of the PCB, and a 50 Ω transmission line brings it to the connector. The ground planes thus shield the low-voltage pickup from the laser driver loop. The low duty cycle of laser drivers allows the use of such physically small resistors even at very high currents.

For a given shunt, increasing the resistor value can help mitigate the effects due to the shunt inductance, so one should choose the largest value possible. In the EPC9179, a value of $R_{shunt} = 0.1$ Ω was used, which reduces the available voltage for overcoming the power loop inductance by about 10% at the peak device rating. The inductance of the shunt is approximately $L_{shunt} = 40$ pH, based on the method shown in [29]. This is still enough inductance to cause significant error for the time-scale of interest. The effect of the inductance is to differentiate part of the current signal, exaggerating the initial part and peak of the waveform. Furthermore, it introduces significant timing error as well. These errors worsen as the pulse gets shorter.

It is possible to compensate for the errors introduced by shunt inductance. When analyzed in the frequency domain, the shunt inductance L_{shunt} introduces a zero at $\omega_z = R_{shunt}/L_{shunt}$, and this can be cancelled by a pole ω_p at the same frequency. This is illustrated in Figure 13.26.

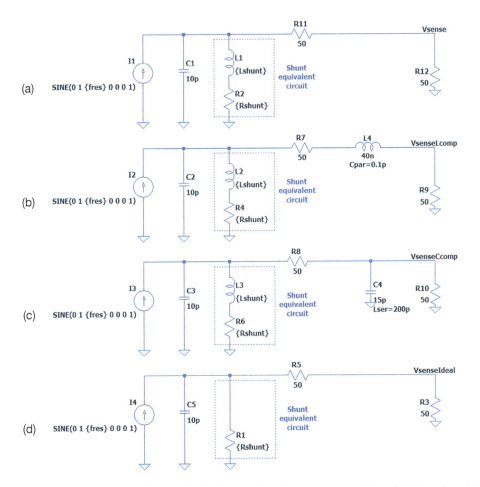

Figure 13.26 Simulation circuit diagrams for implementation of shunt compensation, with all versions driven with 2 ns FWHM pulses. An uncompensated shunt (a) is compared with an inductively compensated shunt (b), a capacitively compensated shunt (c), and ideal shunt (d).

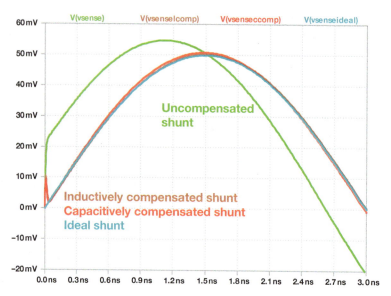

Figure 13.27 Results of the shunt compensation methods illustrated in Figure 13.26.

Figure 13.26a shows an uncompensated shunt with a 50 Ω termination driving a 50 Ω load, with the shunt output voltage denoted by v_{sense}. Figure 13.26b shows the use of an inductive pole to compensate the shunt zero, where the pole is formed by L4 and the terminating and load resistors. Figure 13.26c shows how to implement the same pole using a capacitor. Figure 13.26d shows an ideal shunt for comparison purposes. All these options are simulated, and the results are shown in Figure 13.27. The compensation corrects most of the errors, except for a very small spike of 20 ps duration at the beginning of the simulation. This spike is due to a combination of imperfect pole-zero cancellation, and the step change in the current stimulus derivative at the start of the simulation. This is strictly a simulation artifact as no known real system has a step change in the derivative of any state variable.

13.6.5 Dual-Edge Control

Resonant capacitive discharge laser drivers have compelling features. However, they do have a major limitation in that for a given power loop inductance, they enable good control over the pulse height via control of the bus voltage, but no control over the pulse width. Pulse width may also be used to control pulse energy, and it can be easier to control than pulse amplitude, especially if such control is desired for individual pulses. This is accomplished by tuning the transistor turn-off to force the laser current to zero. Since both the turn-on and turn-off times are used to control the laser current, this is referred to as dual-edge control.

The rectangular pulse driver of Figure 13.4b is an example of dual-edge control, but it is not the only one. Even with the capacitive discharge driver, the transistor can be turned off before the resonant half-cycle is complete. This means that the pulse can be made shorter than the length of the resonant half-cycle, not only controlling the pulse width, but also the pulse energy. For typical dual-edge control applications, the resonant capacitor and charging resistors will likely need to be changed. In the case where current must be limited, the charging resistor may be used for this purpose, since there is additional bus capacitance at the bus voltage input to the PCB. While dual-edge control appears straightforward, there are several component limitations that must be considered when working with high current pulses that have sub-nanosecond transition times and pulse widths under 10 ns. First, many gate drive ICs have minimum pulse width specifications. These are typically greater than 10 ns, although the Texas Instruments LMG1020 has been released with a nominal minimum pulse width of 1 ns. When dual-edge control is applied to the capacitive discharge driver, small changes in pulse width can result in large changes in amplitude because the amplitude is proportional to the product of the current slope and pulse width.

Finally, the current in the power loop inductance will be interrupted when the switch is turned off, and this will cause ringing and overshoot on the drain terminal of the transistor and the laser diode. This ringing will depend on the inductance, the current at the time of turn-off, and the capacitance of the laser, transistor, and PCB layout and stackup. This ringing may need to be clamped.

Finding a suitable clamp diode is very challenging. Most diodes have package inductance on the same order as the power loop inductance, and this limits response speed. Furthermore, if the clamp current is high, the larger diode to handle this will tend to have substantial capacitance that will contribute additional ringing, which in some cases can result in undesirable repeated optical pulses. One option is to integrate the clamp diode into GaN laser driver. This will help to reduce its parasitic inductance.

When examining dual-edge control applications, it is recommended that both careful simulation and experimentation are planned. The latter is especially important, as the available models may not accurately represent the behavior of the laser diodes for the extremely short transitions found in lidar applications.

13.7 Summary

Until recently, one of the main limitations in 3-D TOF lidar was the ability to achieve the necessary combination of power and speed in the laser pulses, and this limitation was due mainly to the semiconductor switch, the silicon MOSFET. Commercial GaN devices have enabled the development of cost-effective TOF lidar systems by making it possible to develop small, high-performance laser diode transmitters with very high peak power and pulse widths in the low single-digit nanosecond range. GaN technology has advanced to the point where it is now possible to monolithically integrate the gate driver and additional functions with the power switch, with demonstrated benefits in performance, size, and cost.

References

1 Glaser, J. (2017). How GaN power transistors drive high-performance lidar: generating ultrafast pulsed power with GaN FETs. *IEEE Power Electron. Mag.* 4 (1): 25–35. https://doi.org/10.1109/MPEL.2016.2643099.

2 McManamon, P. (2015). *Field Guide to Lidar*, 1e in SPIE field guides, no. FG36. SPIE https://doi.org/10.1117/3.2186106.

3 Ingram, K. (2012). Visualizing the forest. *ANR Blogs (UC Division of Agriculture and Natural Resources)*, June 13, 2012 https://ucanr.edu/blogs/blogcore/postdetail.cfm?postnum=7634. [Accessed: April 21, 2024].

4 McManamon, P.F. (2019). *LiDAR Technologies and Systems*. SPIE https://doi.org/10.1117/3.2518254.

5 Li, N., Chong, P.H., Xue, J. et al. (2022). A progress review on solid-state LiDAR and nanophotonics-based LiDAR sensors. *Laser Photon. Rev.* 16 (11): 2100511. https://doi.org/10.1002/lpor.202100511.

6 Siew, S.Y., Li, B., Zheng, H.Y. et al. (2021). Review of silicon photonics technology and platform development. *J. Lightwave Technol.* 39 (13): 4374–4389. https://doi.org/10.1109/JLT.2021.3066203.

7 Shekhar, S., Bogaerts, L., Chrostowski, L. et al. (2024). Roadmapping the next generation of silicon photonics. *Nat. Commun.* 15 (1): 751. https://doi.org/10.1038/s41467-024-44750-0.

8 Boulay, P. (2022). LIDAR for ADAS and AV applications: Emerging trends and developments, presented at the *Automotive LIDAR 2022 Conference and Expo*, Online (September 22, 2022). www.automotivelidar.com.

9 Hovanessian, S.A. (1984). *Radar System Design and Analysis*, 1e. Norwood: Artech House, Inc.

10 IEC 60825-1:2014. International Electrotechnical Commission (May 15, 2014). https://webstore.iec.ch/publication/3587. [Accessed: April 14, 2024].

11 O'Sullivan P., and Le Dortz, N. (2021), Time of flight system design–part 1: System overview. ADI: Analog Dialogue. https://www.analog.com/en/resources/analog-dialogue/articles/time-of-flight-system-design-part-1-system-overview.html. [Accessed: April 22, 2024].

12 Melexis, NV (2020). Time-of-flight (ToF) Basics, Application note AUH-2020, September 2020. https://media.melexis.com/-/media/files/documents/application-notes/time-of-flight-basics-application-note-melexis.pdf.

13 Payne, A.D., Dorrington, A.A., and Cree, M.J. (2011). *Illumination waveform optimization for time-of-flight range imaging cameras* (ed. F. Remondino and M.R. Shortis), 80850D. Munich, Germany: SPIE Optical Metrology https://doi.org/10.1117/12.889399.

14 Yoshida, J. (2020). Breaking down iPad pro 11's LiDAR scanner. *EE Times*, June 05, 2020. https://www.eetimes.com/breaking-down-ipad-pro-11s-lidar-scanner/. [Accessed: April 14, 2024].

15 OSRAM Opto Semiconductors, Inc. (2019). SPL S1L90A_3 A01 datasheet.

16 OSRAM Opto Semiconductors, Inc. (2019). SPL S4L90A_3 A01 datasheet.

17 ams AG (2021). EGA2000_940_W_short_datasheet.

18 Wilson, J. and Hawkes, J. (1989). *Optoelectronics, An Introduction*, 2e. Prentice Hall International (UK) Ltd.

19 Diehl, R.D. (ed.) (2000). High-power diode lasers: fundamentals, technology, applications, with contributions by numerous experts. In: *Topics in Applied Physics*, 78. Berlin; New York: Springer.

20 Aboujja, S., Bean, D. (2021). 1550 nm triple-junction laser diode excels for automotive lidar. *Laser Focus World*, (November 11, 2021). https://www.laserfocusworld.com/lasers-sources/article/14212074/1550-nm-triple-junction-laser-diode-excels-for-automotive-lidar. [Accessed: April 14, 2024].

21 Efficient Power Conversion Corporation (2023). EPC2252 – enhancement-mode GaN power transistor, EPC2252 datasheet. http://epc-co.com/epc/Portals/0/epc/documents/datasheets/epc2252_datasheet.pdf.

22 Infineon Technologies AG (2019). IAUZ40N08S5N100 datasheet. https://www.infineon.com/dgdl/Infineon-IAUZ40N08S5N100-DataSheet-v01_00-EN.pdf?fileId=5546d4626f1cf1c5016f1e6c94a801de.

23 Texas Instruments (2018). LMG1020 5-V, 7-A, 5-A low-side GaN and MOSFET driver for 1-ns pulse width. Application datasheet (Rev. B). https://www.ti.com/lit/gpn/lmg1020.

24 Texas Instruments (2020). LMG1025-Q1 Automotive low side GaN and MOSFET driver for high frequency and narrow pulse applications. https://www.ti.com/lit/ds/symlink/lmg1025-q1.pdf.

25 3PEAK Inc. (2022). TPM2025/TPM2025Q Automotive dual-channel ultra-highspeed GaN predriver. www.3peakic.com.cn.

26 Efficient Power Conversion Corporation (2022). EPC21601 – Enhancement-mode GaN power transistor, EPC21601 data sheet. December. 2022. https://epc-co.com/epc/Portals/0/epc/documents/datasheets/EPC21601_datasheet.pdf.

27 Efficient Power Conversion Corporation (2023). EPC21603 – Enhancement-mode GaN power transistor, EPC21603 datasheet. June 2023. https://epc-co.com/epc/Portals/0/epc/documents/datasheets/EPC21603_datasheet.pdf.

28 Efficient Power Conversion Corporation (2022). EPC21701 – enhancement-mode GaN power transistor, EPC21701 datasheet. December 2022. https://epc-co.com/epc/Portals/0/epc/documents/datasheets/EPC21601_datasheet.pdf.

29 Glaser, J. (2018). High power nanosecond pulse laser driver using an GaN FET. *Proceedings of the Renewable Energy and Energy Management PCIM Europe 2018. International Exhibition and conference for power electronics, intelligent motion*, June 2018: 1–8.

30 Glaser, J.S. (2019). eGaN FETs for lidar – getting the most out of the EPC9126 laser driver, Efficient Power Conversion Corporation. https://epc-co.com/epc/Portals/0/epc/documents/application-notes/AN027%20Getting-the-Most-out-of-eGaN-FETs.pdf. [Accessed: March 14, 2019].

31 Morgott, S. (2004). Range finding using pulse lasers. Application note. Osram Opto Semiconductors, Regensberg, Germany. September 2004. http://www.osram-os.com/Graphics/XPic2/00054201_0.pdf/Range%20Finding%20using%20Pulsed%20Laser%20Diodes.pdf.

32 Glaser, J. (2020). GaN-based solutions for cost effective direct and indirect time-of-flight lidar transmitters. *PSMA Power Technology Roadmap Webinar Series*, December 10, 2020.

33 IPC International, Inc. (2003). IPC-2221A generic standard on printed board design. May 2003.

34 IPC International, Inc. (2006). IPC-4761 Design guide for protection of printed board via structures. July 2006. https://shop.ipc.org/ipc-4761/ipc-4761-standard-only/Revision-0/english.

35 Efficient Power Conversion Corporation (2024). EPC9179 – 80 V, 75 A laser diode driver, evaluation board. https://epc-co.com/epc/products/evaluation-boards/epc9179. [Accessed: April 14, 2024].

36 Glaser, J.S. (2023). EPC9179, EPC9180, and EPC9181, quick start guide rev. 2. Efficient Power Conversion Corporation, 2023. https://epc-co.com/epc/documents/guides/EPC9179_qsg.pdf.

37 Efficient Power Conversion Corporation (2020). EPC2218A – enhancement-mode GaN power transistor, EPC2218A datasheet. May 2020. http://epc-co.com/epc/Portals/0/epc/documents/datasheets/epc2216_datasheet.pdf.

38 Efficient Power Conversion Corporation (2024). EPC9180 – 80 V, 230 A Laser diode driver, evaluation board. https://epc-co.com/epc/products/evaluation-boards/epc9180. [Accessed: April 14, 2024].

39 Efficient Power Conversion Corporation. EPC22704 – enhancement-mode GaN power transistor, EPC22704ENGR datasheet. December 2024.

40 Texas Instruments (2008). LVDS Owner's Manual Including High-Speed CML and Signal Conditioning. https://www.ti.com/lit/ug/snla187/snla187.pdf.

41 Efficient Power Conversion Corporation (2021). Development Board EPC9156, quick start guide. https://epc-co.com/epc/Portals/0/epc/documents/guides/EPC9156_qsg.pdf.

42 Weber, J. (1969). *Oscilloscope Probe Circuits*. Tektronix, Inc.

14

Motor Drives

14.1 Introduction

Today many electrical motors require variable-frequency drives to adapt the motor's speed and torque to the needs of the specific application operating point. Variable-frequency drives are implemented with a two-level voltage source inverter topology consisting of three phases because most motors are wound with three windings displaced at 120 electrical degrees. Windings are often star-connected with three terminals available, while the internal connection (neutral point) is not externally available.

14.2 Motor Types

The permanent magnet motor, also known as brushless DC motor (BLDC), is widely used and offers higher torque capability per cubic centimeter and higher dynamics compared to induction motors and brushed DC motors. Permanent magnet motors require less current to generate torque because the magnets generate the flux in the rotor. On the other hand, induction motors are widely used in the industrial world because of their superior reliability and offer an advantage if an overspeed constant-power field weakening mode needs to be used [1]. Although brushed DC motors are simple to use, they are more complex to build and are being phased out due to their lower performance and reliability compared to the other motor types.

In the last decade, the brushed DC motor phase-out and the subsequent advent of brushless motor adoption have been aided by motor control algorithms that allow a precise and controlled orientation of the magnetic field in the stator to obtain the optimum operating point at every speed and torque. With these algorithms, the torque and phase current have a linear relationship.

14.3 Inverter

Today silicon-based power devices have dominated in inverter power electronics, but their performance is nearing their theoretical limits [2]. There is an increasing need for higher power density in terms of the amount of power, volume, and weight the inverter can transfer to the motor from the DC source. Gallium nitride transistors and integrated circuits (ICs) have the best attributes to satisfy these needs. GaN's superior switching behavior helps to remove

---

GaN Power Devices for Efficient Power Conversion, Fourth Edition. Alex Lidow, Michael de Rooij, John Glaser, Alejandro Pozo, Shengke Zhang, Marco Palma, David Reusch, and Johan Strydom.
© 2025 John Wiley & Sons Ltd. Published 2025 by John Wiley & Sons Ltd.

dead time and increase PWM frequency to obtain near-perfect sinusoidal voltage and current waveforms for smoother, silent operation with higher system efficiency. Power density increases by substituting electrolytic capacitors in the input filter with smaller, cheaper, and more reliable ceramic capacitors.

14.4 Typical Applications

Electrical motors are used in many different applications, and where an important requirement is that the motor drive is integrated within the motor, GaN devices are the best fit. Moreover, higher total system efficiency is crucial for extended use of each single battery charge and longer battery life in battery-operated applications. Typical examples are industrial drones, humanoid and surgical robots, power tools, servo drives, e-bikes, reaction wheels in satellites, and e-scooters. In all these cases, GaN devices drive miniaturization, improve efficiency, and reduce heat generation and weight.

14.5 Voltage Source Inverters and Motor Control Basics

Figure 14.1 shows the two-level, three-phase voltage source inverter (VSI) topology. It consists of three half bridges connected to each motor phase and powered by a DC source. In motor control applications, a controller circuit consists of a digital processor that acquires phase currents and bus voltage data through analog-sensing circuits [3] and implements a specific motion control algorithm.

Given the mechanical load's specific requirement and operating point, the controller determines the proper frequency, voltage amplitude, and power direction. The power can flow from the DC source to the motor when the system is actively driving the mechanical load (i.e. accelerating or keeping constant mechanical speed) or can flow from the mechanical system back to the DC source when the system is actively braking, with the motor and the mechanical load reducing its rotational speed. In the latter case, the braking operation must always be controlled to avoid too much energy being fed back to the inverter source that can over-voltage the DC bus rails, especially if the input capacitor bank is made of electrolytic capacitors that are extremely sensitive to over-voltage [4–6] and may be unable to absorb all the energy.

In voltage source inverters, the controller generates the proper pulse-width modulation (PWM) commands so that the inverter can apply an amplified PWM at the DC bus voltage level to the motor terminals. The PWM fundamental harmonic and its amplitude determine the three sinusoidal voltages the motor circuit receives as input because the motor's R-L circuit

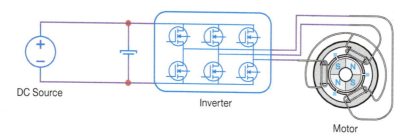

Figure 14.1 Overview of a brushless DC motor drive system.

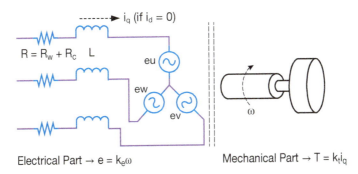

Figure 14.2 Electro-mechanical diagram of a permanent magnet synchronous motor (PMSM) with equivalent electrical schematic.

acts as a low-pass filter. The higher-order voltage harmonics of the applied PWM generate harmonic currents, which are responsible for increased winding ohmic losses and high-frequency alternating (or rotating) magnetic fields inside the iron lamination, which create additional losses in stator and rotor cores [7]. In principle, the motor losses are inversely proportional to the applied voltage waveform's total harmonic distortion (THD).

Figure 14.2 shows a simplified diagram of how mechanical and electrical worlds are linked in a motor. In the low-frequency electrical model of the motor, each phase can be modeled with a series of two resistors, an inductor, and a voltage generator representing the motor back electromotive "force" *bemf*. The equivalent resistor R and the inductor L determine the motor's electrical pole in frequency. The amplitude of the *bemf* is proportional to the motor mechanical speed, and k_e is called "*bemf* constant."

The model of the motor must include the losses due to the higher-order voltage harmonics. A methodology that characterizes motor impedance with frequency is extremely useful when dealing with GaN inverters that allow higher PWM frequencies than silicon-based counterparts. The high-frequency machine loss model can be derived from finite element simulation [7], or a direct experimental measurement as described in the next paragraphs. The finite element method provides good results only when all details of geometry and construction materials are known in advance. Direct measurement is more practical when all motor construction details are unknown.

14.5.1 High-Frequency Motor Loss Model

The motor losses can be modeled considering the three main components:

- P_W: winding losses that include ohmic losses in the stator's winding and, if present, in the rotor's winding. Considering the hth harmonic I_h of the current and referencing Figure 14.2, the losses are inversely proportional to the square of the frequency f_h as per Eq. (14.1) [8].

$$P_{W,h} \propto R_W I_h^2 \approx R_W \left(\frac{V_h}{2\pi f_h L}\right)^2 \propto \frac{1}{f_h^2} \tag{14.1}$$

- P_C: The flux density variation within the iron lamination generates core losses in the stator and rotor. These losses are based on eddy currents and the hysteresis loss mechanism. Given the h_{th} harmonic V_h of the applied voltage, the related flux density B_h is proportional

GaN Power Devices for Efficient Power Conversion

to V_h/f_h. The losses can be expressed as a function of the harmonic frequency f_h as per Eq. (14.2), where $\alpha \cong 2$ is the coefficient that defines the hysteresis loss dependency on flux.

$$\begin{cases} P_{c,\mathrm{hys},h} \propto f_h B_h^\alpha \propto f_h^{1-\alpha} V_h^\alpha \propto f_h^{1-\alpha} \approx \dfrac{1}{f_h} \\ P_{c,\mathrm{eddy},h} \propto f_h^2 B_h^2 \propto V_h^2 \propto \mathrm{const} \end{cases} \tag{14.2}$$

The eddy current losses are constant up to frequencies where the skin effect on core lamination becomes dominant. Above these frequencies, eddy losses decrease [9].

- P_M: The magnetic losses behave like the core losses, with the difference that magnets do not have lamination. Hence, the eddy current losses dominate.

Equations (14.1) and (14.2) indicate that eddy current and hysteresis losses dominate at higher frequencies. The motor impedance can then be described as:

$$Z = (Z_w) + (Z_c) = (R_w) + (R_c + j2\pi f L) = R + j2\pi f L \tag{14.3}$$

The equivalent circuit is shown in Figure 14.2, where the three components R_W, R_C, and L vary with frequency [10]. The high-frequency model of the motor should also include the capacitance between the phases, which depends on the motor construction. The model of Figure 14.2 is a good representation of the motor up to the winding L-C self-resonance.

A qualitative representation of the phase equivalent circuit can be obtained considering that there are three frequency regions, low, middle, and high, before L-C resonance:

1) **Winding losses are dominant:** the R_C is negligible in this region, and R_W and L are constant with frequency. Hence, the total impedance Z is proportional to the frequency.
2) **Core losses are dominant:** the eddy currents become the dominant loss contribution, the total resistance rises with the square of the frequency, and L is still constant with frequency. The total impedance Z is proportional to the frequency because the reactance is the major component [11].
3) **Lamination skin effect core losses dominant:** in this region, the inductance decreases inversely proportional to the square root of the frequency, while the resistance increases with the square root of the frequency. Thus, the total impedance Z continues rising proportional to the square root of the frequency [11].

Figure 14.3 shows the qualitative representation of the phase impedance variation with the frequency of an electrical motor.

By knowing R and the impedance Z as a function of the frequency, it is possible to derive the power loss P_h at the hth harmonic of the applied voltage V_h:

$$P_h = \frac{1}{2} R I_h^2 = \frac{1}{2} R \left(\frac{V_h}{|Z|}\right)^2 \tag{14.4}$$

By normalizing by V_h^2, it is possible to derive the motor's harmonic power loss factor LF:

$$\mathrm{LF}(f_h) = \frac{P_h}{V_h^2} = \frac{1}{2} \frac{R(f_h)}{|Z(f_h)|^2} \tag{14.5}$$

LF depends on the harmonic frequency and is independent of the applied voltage. From Eq. (14.5), considering the three regions introduced in the previous paragraph, in the region I, since R_W and L are constants, the losses are dominated by the ohmic component and are inversely proportional to the square of the frequency. In region II, the eddy current losses

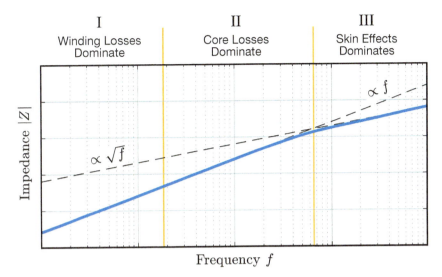

Figure 14.3 Qualitative representation of the phase impedance variation with frequency of an electrical motor.

become dominant, R is proportional to the square of the frequency, while L is constant; hence, the losses are constant. In region III, the losses decrease inversely proportional to the square root of the frequency [12, 13]. The qualitative behavior of the harmonic loss factor LF is represented in Figure 14.4.

From Figure 14.4, it is evident that higher PWM harmonic content corresponds to lower losses in the motor. On the other hand, higher PWM frequency means higher switching losses in the inverter. Hence, given a motor and the inverter, the designer must find the PWM frequency that reduces the total system losses.

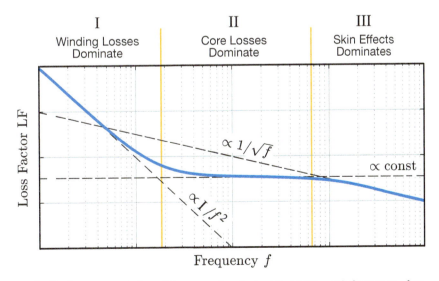

Figure 14.4 Qualitative representation of the power loss factor LF variation with frequency of an electrical motor.

GaN Power Devices for Efficient Power Conversion

A method to derive the LF is based on phase impedance direct measurement, with the motor in locked rotor and under a no-load condition. The rotor must be locked at the angle that allows maximum reluctance. The measurement can be done with an LCR meter that measures phase-to-phase terminals (keeping the third phase open) or with a setup that consists of a signal generator, an amplifier, and a power meter. The first setup is the simplest, but it is limited by the maximum current that can be generated by an LCR meter (usually 100 mA). The second setup permits higher accuracy at low frequencies, especially when the motor inductance is low [7].

14.6 Field-Oriented Control Basics

Thanks to the advent of high-performance digital controllers, modern motor control algorithms allow a precise and controlled orientation of the magnetic field in the stator to obtain the optimum motor operating point at every speed and torque. The torque T and phase current I have a linear relationship with these algorithms. Field-oriented control (FOC) is a widely used optimum control algorithm. The current, I, flowing in the motor phase can be decomposed into two orthogonal components, I_d and I_q. I_d is responsible for the flux generation (or weakening), and I_q is responsible for the generated mechanical torque. When operating with the FOC, the motor phase current, I, can be assumed to be equal to the "quadrature" current, I_q, while the "direct" current, I_d, can be considered null. Thus, the torque, T, is proportional to the I_q current via the torque constant, k_t. The motor mechanical speed, ω, determines a back electromotive "force" (*bemf*), which is a voltage e_u that opposes the input voltage command. The *bemf* e_u is proportional to the motor speed ω via the *bemf* constant k_e. By equating the electrical and mechanical power $I_q \cdot e_u \cong T \cdot \omega$, the constants k_e and k_t are proportional, so a motor that develops more torque per ampere (rms) also generates a higher back emf voltage that is acting against the inverter applied voltage. In modern motion control systems, the task of the controller is to apply a voltage to motor terminals so that the resulting current $I = \sqrt{I_q^2 + I_d^2}$ in each phase has the following characteristics:

- I_d component, which is generating a field aligned to the current rotor magnet position, is controlled to be zero
- I_q component, which generates a field at 90 electrical degrees to the current rotor magnet position, is controlled to generate the desired torque as per the speed loop output reference command

In other words, the main controller task is to apply a voltage so that the resulting sinusoidal current I is always in phase with the *bemf* sinusoidal voltage for the higher efficiency and maximum power transfer.

A field-oriented control scheme is commonly used to achieve this task. Figure 14.5 shows the field-oriented control (FOC) scheme block diagram. To achieve the optimum operating point at every condition, the controller acquires the DC bus voltage value and the phase current instant value per each phase and obtains the rotor angular position by using a sensor on the shaft or a sensorless algorithm. The sensorless rotor angle and speed observer description falls outside this chapter's scope.

By knowing the instantaneous rotor angular position and the currents in the three-phase stator *abc frame*, a coordinate rotation is performed so that in the Cartesian virtual *dq frame*, which is synchronous to the rotor, the motor appears as a conventional DC brushed motor.

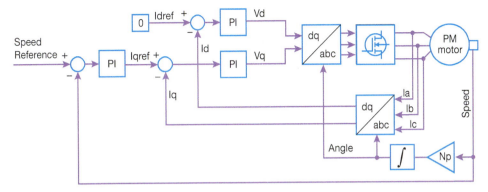

Figure 14.5 Field-oriented control (FOC) scheme.

The key to this transformation is that all sinusoidal values become DC values and can be controlled with regular proportional-integrative (PI) controllers.

Depending on the rotor structure, synchronous motors are divided into motors with a non-salient pole rotor and motors with a salient pole rotor. In non-salient pole motors, the magnets are evenly distributed on the rotor's surface, and the rotor's phase-to-phase inductance is not influenced by the rotor angle (Ld = Lq). For these motors, the speed controller generates the reference for the torque I_q controller, while the I_d controller imposes the I_d to be zero.

For salient pole motors, where the magnets are either buried within the rotor or are not evenly distributed on rotor's surface, the phase-to-phase inductance varies with the rotor angle (Ld ≠ Lq), and, to get the advantage of the reluctance torque due to the anisotropy, the speed controller generates the reference for both the I_q and the I_d controllers by imposing the I_d to be negative with an amplitude proportional to the required torque. The output of the controllers is the voltage $V = \sqrt{V_q^2 + V_d^2}$ to be applied to the motor in the *dq frame*. V, in the rotating *dq frame*, is a DC value. By executing an inverse coordinate rotation, the voltage is converted into three reference sinusoidal voltages in the *abc frame* for the PWM modulator, which converts the information into an instantaneous duty cycle for all three inverter legs.

At this point, the motor's electrical pole, determined by the phase resistance and inductance values, is essential in filtering out all unnecessary modulation harmonics. Hence, the PWM carrier frequency has an essential impact on the total system behavior. As described in the previous paragraph, the motor's electrical frequency response and the inverter losses must be considered to find the optimal PWM frequency for the system's efficient operation. A good design practice is to plot the motor and inverter losses as a function of frequency. In many cases, using GaN technology makes it possible to increase efficiency because GaN inverters exhibit lower losses than their Si counterparts, even at higher PWM frequency.

The motor estimated losses can be obtained from Eq. (14.5) by multiplying the measured loss factor, *LF*, per each harmonic of the applied voltage. As a valid approximation, the designer can use a square wave of amplitude V_{DCbus} for the applied voltage, 50% duty cycle, and rise and fall times, t_r and t_f, obtained considering the desired dv/dt. The harmonic amplitudes V_h can be obtained using a fast Fourier transform (FFT) function in a mathematical application program and then applying Eq. (14.6).

$$P_{\text{motor}} = \sum_h \text{LF}(f_h) \cdot V_h^2 \tag{14.6}$$

Table 14.1 Power tool prototype motor parameters.

Motor type	Poles	R (mΩ)	L (μH)	k_e (mV/s)
Surface permanent magnet motor for power tool	10	75	90	14

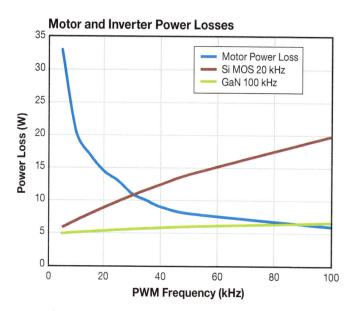

Figure 14.6 Estimated power losses for a 10-pole permanent magnet motor versus PWM frequency. In the same graph, there are the simulated losses for two different inverters operating at 20 V_{DC} and 10 A_{RMS} phase current.

The loss factor, *LF*, of a 10-pole motor used for a power tool application has been measured, and its power loss has been estimated considering 20 V_{DC} and 5 V/ns dv/dt. The motor parameters are shown in Table 14.1. Figure 14.6 shows an example of the estimated power loss obtained using Eq. (14.6). The estimated power losses of a 20 kHz MOS inverter and a 100 kHz GaN inverter are shown in the same figure. Combining the 100 kHz GaN inverter with the power tool motor under test leads to a lower power dissipation.

All current and voltage measurements and subsequent calculations must be performed at a rate at least double that of the required torque loop bandwidth. Modern motors are becoming more compact with less inductance, requiring higher bandwidth controllers and a higher PWM frequency.

14.7 Current Measurement Techniques

While the DC bus and voltage measurements of the phases are simple and obtained by a resistor divider technique, the current measurement can be done either in series with the motor terminals by using sensors that can extract the differential signal from the high common mode of the PWM voltage, or by reading the voltage drop in shunts placed in series to the source of

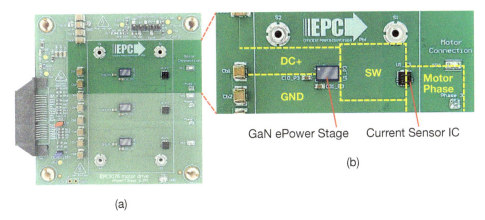

Figure 14.7 (a) EPC9176 motor drive inverter [14, 15] made of three EPC23102 ePower stage ICs [16]; (b) zoom of one inverter leg showing the ePower stage IC and the in-phase current sense integrated circuit.

the low side switches of the inverter. The leg shunts technique requires a cheaper and simpler analog interface circuit but poses a serious challenge in terms of power loop stray inductance.

Given today's demanding torque bandwidth and accuracy requirements, to make full use of GaN technology it is better to sense directly the phase currents at the motor terminals. In-phase current can be measured with shunt resistors in series to the motor terminals or by using contactless methods, such as Hall effect or tunnel magnetoresistance (TMR)-based sensors. The contactless method reduces power losses otherwise dissipated by the shunts, and thus increases the system's efficiency and reduces system size. In today's applications, both techniques are used, and the choice between the two is dictated by the specific application requirements and the designer's choice. Figure 14.7a shows a layout example of a GaN inverter consisting of three EPC23102 ePower™ Stage integrated circuits [16, 17]. Figure 14.7b shows the detailed view of one inverter leg with the in-phase current measure made with a Hall-based current sensor integrated circuit [18].

14.8 Power Dissipation in Motor and Inverter

Power dissipation and subsequent overall efficiency in an inverter plus motor system depend on the efficiency of the inverter and motor stages at the specific operating point.

In Figure 14.8, the inverter efficiency is defined as $\eta_{inverter} = \dfrac{P_{out}}{P_{in}}$, while the efficiency of the motor is defined as $\eta_{motor} = \dfrac{P_{mech}}{P_{out}}$, and the overall efficiency can be determined by the product of the two $\eta_{motor\ system} = \eta_{inverter} \cdot \eta_{motor}$
where,

- P_{in} is the input power and is calculated as $P_{in} = V_{dcbus,rms} \cdot I_{dcbus,rms}$.
- P_{out} is the output power; it is difficult to calculate P_{out} due to PWM modulation, and instead it is better to be measured with a three-phase power meter.
- P_{mech} is the mechanical power delivered to the load, calculated as $P_{mech} = T \cdot \omega$, where torque T and speed ω are measured by a torque/speed transducer.

As a result of the inverter control, the motor is generating a torque $T = k_t \cdot I_q$, where the I_q current is the controlled-phase RMS current when in field-oriented control operation,

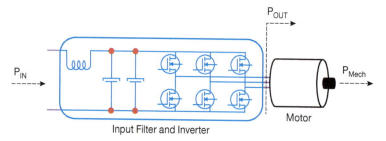

Figure 14.8 Inverter-motor system showing how the power flows from the DC input to the mechanical load.

and k_t is the torque constant. Given a torque T, the mechanical power depends on the load mechanical speed ω.

It is important to note that the inverter power dissipation is not directly dependent on mechanical power but instead depends directly on the DC bus voltage, the phase RMS current, and the PWM frequency. For this reason, it is always better to specify the inverter capability in terms of voltage and current at a given temperature.

14.9 Silicon Inverter Limitations

Inverter power dissipation consists of conduction losses and switching losses. Conduction losses are proportional to the switches' conduction resistance, $R_{DS(on)}$. Reducing the channel resistance helps to reduce conduction losses, but it can also increase switching losses. The relation between conduction losses and switching losses depends on the specific switch technology.

DC and battery-operated motor drive applications have a DC bus voltage spanning from 24 V_{DC} to 96 V_{DC}. Typically, when silicon MOSFETs are used in the inverter, the PWM frequency is kept below 40 kHz, and the dead time is kept within the range of 200–500 ns. In this scenario, most of the inverter power dissipation is due to conduction losses since PWM frequency is kept low to avoid the switching loss penalty that becomes quite high with Si MOSFET technology.

14.10 LC Filter Dissipation

Figure 14.9 shows an inverter integrated into the motor and connected to a DC source with cables. During its switching operation, the low PWM frequency silicon inverter generates a ripple in the voltage and the current on the cables, which become major sources of EMI. An LC filter must be inserted between the inverter and the DC source cables to reduce the interference. For example, an e-bike LC filter is made of a 2.7 µH inductor and two electrolytic capacitors with a total capacitance of 660 µF. The LC filter dissipation enters in the efficiency equation, $\eta_{inverter} = \dfrac{P_{out}}{P_{in}}$, and reduces the inverter and the overall system efficiency.

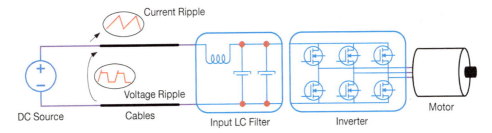

Figure 14.9 Cables from the DC source are the source of EMI and require the insertion of an LC filter at the inverter input.

14.11 Torque Sixth Harmonic Dissipation

Another drawback of silicon MOSFETs is the necessary insertion of dead time. Dead time is responsible for the sixth harmonic of the electrical frequency in the generated torque. Torque harmonics have no effect on the average torque applied to the load but decrease the motor efficiency by increasing the vibrations transmitted to the load and the windings' temperature.

14.12 GaN Advantage

GaN devices show lower switching losses than Si MOSFETs and do not have a body diode pn junction, so there is no associated reverse recovery in hard switching operation. These two factors help to eliminate the dead time and increase the PWM frequency so that the input filter made of one inductor and one or two electrolytic capacitors may be substituted with ceramic capacitors.

The higher PWM frequency results in a lower power loss in the motor because the loss factor LF decreases with frequency. The resultant advantage is quieter operation with a smaller and lighter inverter. The motor runs smoother at a lower temperature, is more efficient, and the sixth harmonic of the torque is completely removed. In addition, moving from electrolytic to ceramic capacitors reduces cost, increases system reliability, and reduces overall system size.

14.13 GaN Switching Behavior

Three factors contribute to the behavior of a device during the switching event:

- Reverse capacitance, C_{RSS}
- Body diode reverse recovery, Q_{RR}, and t_{RR}
- Power loop and common source inductance due to package and layout.

The factors were discussed in detail in Chapters 4 and 7, and the overall effect is that in a motor drive GaN inverter, the dead time can be reduced to tens of nanoseconds [19–22, 27].

14.14 Dead Time Elimination Effect

When using discrete GaN FETs or a GaN ePower^{TM} stage IC [16, 17] in an inverter for motor drive, the dead time can be reduced to tens of nanoseconds, allowing a smooth voltage waveform to be applied to the motor terminals. Figures 14.10 and 14.11 show the modulation voltage and phase current difference between two different dead time values. Eliminating the dead time improves the quality (in terms of total harmonic distortion [THD]) of the applied sinusoidal voltage that is, in turn, reflected in reduced distortion in the phase current, less vibration, less heat generation, and less acoustic noise generated by the motor.

Dead time insertion is responsible for the discontinuity of the voltage modulation during the zero crossing of the phase current, and the distortion is also reflected in the other motor phases. Six discontinuities in total for each electrical cycle appear as a sixth harmonic in the applied torque to the motor. The comparison of the torque signal spectrum is shown in Figures 14.12 and 14.13. With the GaN inverter (Figure 14.13), when the dead time is almost null, the applied torque is smoother, and the motor is more efficient because all current is converted into torque when applied to the load.

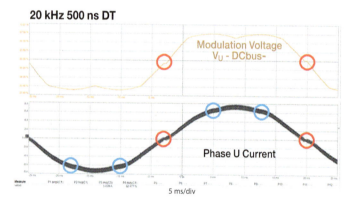

Figure 14.10 Measured voltage modulation and phase current at 20 kHz with 500 ns dead time of a motor operating from 36 V_{DC} and running at 400 RPM.

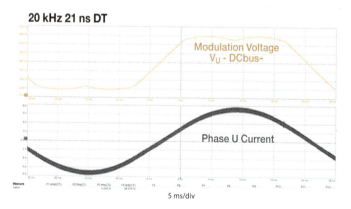

Figure 14.11 Voltage modulation and phase current at 20 kHz with 21 ns dead time of a motor operating from 36 V_{DC} and running at 400 RPM.

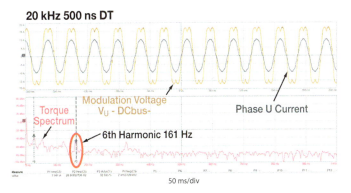

Figure 14.12 500 ns dead time effect on torque signal. In the motor under test at 400 RPM, the electrical frequency is 26.8 Hz, and the sixth harmonic is visible. The torque signal is obtained from a torque transducer.

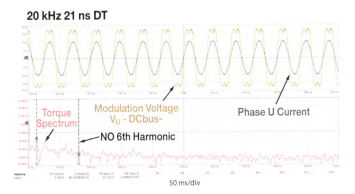

Figure 14.13 21 ns dead time effect on torque signal; in the motor under test at 402 RPM, the electrical frequency is 26.8 Hz and the sixth harmonic is null. The torque signal is obtained from a torque transducer.

14.15 PWM Frequency Increase Effect

14.15.1 Reduction of Input Filter

In motor applications, the dv/dt that can be applied at the motor terminals is limited by the winding insulation and by EMI requirements. Maximum dv/dt values span from 0.5 V/ns in automotive applications to 10 V/ns in industrial applications. A GaN inverter can easily be operated at 100 kHz PWM frequency thanks to its lower switching dissipation and smoother switching at the allowed dv/dt [19, 20]. Considering the worst case of a pure inductive load to simplify the equations, the input voltage ripple Δv_{pp} is a function of the inverter output peak current I_0, the input capacitance C_f value, and the PWM switching frequency f_{PWM} [23]:

$$\Delta v_{pp} \propto \frac{1}{4 f_{PWM}} \frac{I_0}{C_f} \tag{14.7}$$

As a first-order approximation, if the PWM frequency is increased from 20 to 100 kHz, the input capacitance can be reduced by at least a factor of five to preserve the same input voltage ripple. Capacitor technology plays a role here, so by increasing the PWM frequency, the reduction factor is larger than the theoretical. This will be explained in Section 14.15.2.

The input current ripple Δi_{pp} is inversely proportional to the PWM frequency. During a PWM cycle, each inverter leg can be considered equivalent to a buck converter with an inductor L, operated at duty D and voltage ΔV at constant current and similar formulas apply:

$$\Delta i_{pp} \cong \frac{\Delta V \cdot D}{f_{PWM} \cdot L} \propto \frac{1}{f_{PWM}} \tag{14.8}$$

Increasing the PWM frequency has the double effect of reducing both the input current ripple and input voltage ripple.

14.15.2 Capacitor Technology Effect on Input Voltage and Current Ripple

With low PWM frequencies ($\approx$10–50 kHz), the amount of input capacitance to reduce the input ripple current sufficiently requires polarized capacitors, such as electrolytic or tantalum because of their high capacitance density per unit volume. However, electrolytic capacitors impose limits on the magnitude of RMS current they can support, and it is common to see at least two connected in parallel for the input filter. Tantalum capacitors have lower capacitance density per volume than electrolytic counterparts and are expensive. Moreover, tantalum capacitors are prone to fail as a short circuit and cause issues with DC sources [24].

When the PWM frequency is increased, the minimum required capacitance decreases, allowing the exclusive use of ceramic capacitors. This brings several advantages to the design because ceramic capacitors exhibit lower series impedance with a minimum in the region between 100 and 200 kHz, are more stable as a function of temperature, and are more reliable. The equivalent series resistance (ESR) of a 100 V rated, 22 µF X7S capacitor [25] is shown in Figure 14.14. The result is a more compact and reliable inverter, given the same power dissipation and power output.

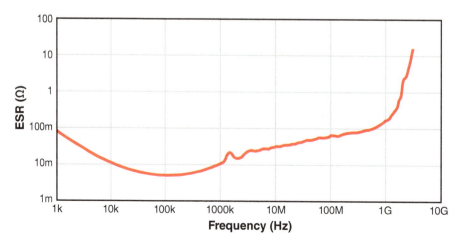

Figure 14.14 Equivalent series resistance of TDK 100 V 22 µF X7S capacitor CKG57NX7S2A226M500JH. *Source:* Adapted from TDK [25].

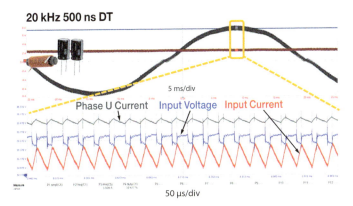

Figure 14.15 Conventional inverter with LC input filter at PWM = 20 kHz, DT = 500 ns, $L = 6\,\mu H$, and $C = 2 \times 330\,\mu F$ eCaps; phase U current 500 mA/div, input voltage 200 mV/div, input current 200 mA/div, and 50 μs/div zoom timescale. *Source:* With permission of EPC.

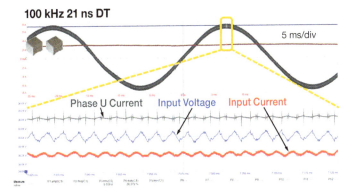

Figure 14.16 GaN inverter without input filter at PWM = 100 kHz, DT = 21 ns, and $C = 2 \times 22\,\mu F$ ceramic; phase U current 500 mA/div, input voltage 200 mV/div, input current 200 mA/div, and 10 μs/div zoom timescale.

Figures 14.15 and 14.16 show the input current ripple, the input voltage ripple, and the output current ripple when comparing two different setups consisting of:

- conventional inverter with LC input filter operating at PWM = 20 kHz, DT = 500 ns, $L = 6\,\mu H$, and $C = 2 \times 330\,\mu F$ (electrolytic).
- GaN inverter without input filter operating at PWM = 100 kHz, DT = 21 ns, and $C = 2 \times 22\,\mu F$ (ceramic).

Both the conventional inverter with an LC input filter and the GaN inverter without an input filter are powering an ebike motor source with a 36 V_{DC} battery voltage and delivering 5 A_{RMS} phase current. The input voltage and current ripple are similar, even though the solution at 100 kHz does not have an input filter. The GaN inverter shows a reduced output ripple, and the current in the motor has a better sinusoidal shape.

14.15.3 System Efficiency

When comparing the two setups of Figures 14.15 and 14.16 using a power meter system, the GaN inverter operating at 100 kHz PWM and 21 ns dead time shows higher system efficiency

GaN Power Devices for Efficient Power Conversion

Table 14.2 Efficiency comparison of Si and GaN inverter-based systems. Torque and speed are measured with a Magtrol TS107 transducer.

Setup	Inverter 20 kHz 500 ns dead time 400 RPM 5 A_{RMS}	GaN inverter 100 kHz 21 ns dead time 400 RPM 5 A_{RMS}
Input inductance	2.7 µH	None
Input capacitor	660 µF electrolytic	44 µF ceramic
P_{in}	121.3 W	113.3 W
P_{out}	119.6 W	111.3 W
$\eta_{inverter}$	98.5%	98.2%
Speed	42.25 rad/s	41.94 rad/s
Torque	1.876 N	1.940 N
P_{mech}	79.3 W	81.36 W
η_{motor}	66.3%	73.1%
η_{total}	65.3%	71.8%

than the conventional inverter running at 20 kHz frequency, 500 ns, and with an input LC filter.

Table 14.2 shows that transitioning from a silicon-based 20 kHz inverter to a GaN-based 100 kHz inverter with almost no dead time reduces the input filter's size, weight, and cost, and increases the total system efficiency at the specific operating point by 6.5%.

14.15.4 Effect of PWM Frequency Increase on the Output Voltage Spectrum

Moving from 20 to 100 kHz PWM frequency also brings other effects that need to be considered. Figure 14.17 shows the theoretical spectrum envelope [23] comparison between lower frequency and slower dv/dt, and higher frequency and higher dv/dt PWM.

In Figure 14.17, t_{rf} indicates the rise and fall times, i.e. at 5 V/ns the switching node requires 7.2 ns to rise from 0 to 36 V and to fall from 36 to 0 V. Increasing the PWM frequency shifts the spectrum content to higher frequencies, so to get the full advantage of GaN-based inverter technology, integrating the inverter with the motor would help. The power cables to the motor may also be sources of EMI, but they are reduced to a minimum when the inverter is integrated into the motor.

Besides the theoretical spectrum envelope, the designer must consider that EMI emissions result from many factors, and it is difficult to predict the radiated spectrum just by examining the dv/dt, the PWM frequency, and the rise and fall times. Figure 14.18 compares the radiated emissions measured inside a fully anechoic chamber of three inverters powering the same motor and load, using the same input DC power supply. The radiated emission results are obtained from the inverters described in Table 14.3.

A 10-pole BLDC motor with 0.55 Nm nominal torque and 6000 RPM nominal speed was used during the EMI tests. The load and setup configuration was kept the same. One would expect lower emissions from the MOSFET inverter, especially at higher frequencies, because of very slow dv/dt and low PWM frequency. Figure 14.18 shows that the MOSFET inverter has

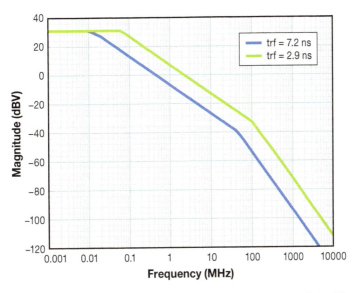

Figure 14.17 Inverter output voltage spectrum envelope, where the blue line is 20 kHz 5 V/ns and the red line is 100 kHz 12 V/ns. The DC source is 36 V.

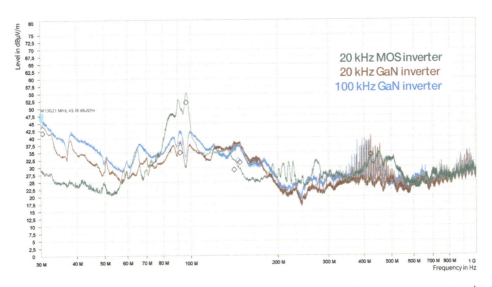

Figure 14.18 Measured radiated EMI spectra of the unfiltered inverters powering a gate opener motor as load. Measurements taken in a fully anechoic chamber using the same setup and load conditions.

lower emissions below 55 MHz, higher emissions between 80 and 100 MHz, and similar emissions in the other spectrum regions. It is also possible to discern a small difference between the 20 kHz GaN inverter and the 100 kHz GaN inverter, but it is definitely lower than theoretically predicted in Figure 14.17.

Table 14.3 Inverters tested in the fully anechoic chamber for radiated EMI comparison test.

Inverter	DC$_{Bus}$ voltage (V)	Motor phase current (A_{RMS})	PWM frequency (kHz)	Dead time (ns)	dv/dt (V/ns)
MOS	36	5.5	20	500	0.75
GaN EPC9194	36	5.5	20	100	5.5
GaN EPC9194	36	5.5	100	100	6.1

14.16 Layout Considerations for Motor Drives

In Chapter 4, the PCB layout best practices for GaN devices were presented. When dealing with motor applications, a designer must consider that motor drive inverters typically switch slower than other converters and that the phase currents may be higher. The acceptable dv/dt and the di/dt in motor applications are limited by the motor design, the winding insulation type, the rotor bearings, and the parasitic capacitances. Excessive dv/dt and di/dt impact the motor's reliability and are responsible for radiated and conducted emissions, whose limits vary by application and as imposed by international standards.

The load current is also a differentiating factor between the applications; for example, in a DC–DC converter, the designer may add more phases to reduce or limit the current per phase. This is impossible in a motor inverter because the number of phases is defined by the motor design, and most motors are three phase. The motor phase current of an inverter leg can be higher than the current in one phase of a DC–DC converter. This pushes the PCB design to the extremes for both current conduction capability and heat extraction, and the designer must trade off these characteristics with power loop inductance.

14.16.1 High-Frequency Decoupling Capacitors

Due to the motor drive inverters' slower switching nature, the internal-vertical layout design described in Chapter 4 is not required for GaN FET motor inverter design. A typical motor inverter design does not need high-frequency capacitors on the same copper layer where the switching cell devices are mounted. Therefore, the ground return plane, as required in the internal-vertical layout, can be put on a different (e.g. deeper) inner layer.

The most common layout technique for motor inverters is the external-vertical loop, as described in Figure 4.7. The power loop inductance is slightly increased in favor of a reduced PCB resistance and better thermal conductivity. In fact, the designer can increase the copper thickness of the layers connecting to the GaN devices to reduce PCB contact and conduction resistance and provide a good lateral path to remove the heat generated in the PCB traces.

14.16.2 Paralleling GaN Devices in Motor Drives

An inverter leg in a motor drive application may require several devices in parallel, posing serious challenges if the designer wants to achieve complete symmetry, as described in Chapter 4, Figure 4.16. Moreover, paralleling the half bridges in motor drives, as described in Figures 4.18 and 4.20, is impractical.

Thanks to the slower nature of the motor inverter, the designer can trade off the complete symmetry with a simpler layout by putting extreme care on the gate signals distribution, which, even if it is not symmetrical, must be shielded and properly connected to avoid the detrimental effects of the common source inductance. The EPC9186 [26] experimental example shows how to properly design a parallel layout, in particular the gate signals and their returns.

14.17 GaN Devices for Motor Applications

The motor drive inverter can be designed based on several options, including a large selection of discrete GaN FETs and ePower™ power stage ICs that integrate gate drivers with power switches [28]. The motor phase RMS current and the boundary conditions for the heat management determine the switches' current rating (for a detailed discussion of thermal management using GaN devices, see Chapter 6). The DC bus voltage and the chosen topology (i.e. 2-level inverter or multilevel inverter) determine the maximum V_{DS} rating.

Figure 14.19 shows the most popular GaN devices used in motor drive applications. The discrete GaN transistors span from 80 V to 200 $V_{DS,max}$. The fully integrated ePower Stage ICs are rated at 80 and 100 V.

14.18 Application Examples

14.18.1 Humanoid Robots' Motor Joints

Humanoid robots are complex mechatronics structures and can be divided into two main classes: (i) those dedicated to industrial applications, and (ii) those dedicated to human-robot interaction. From a design point of view, there are at least two points to consider: (i) the mass distribution and (ii) the undesirable mechanical resonances. The latter implies suppressing resonance at the level of the joints and the links. For this reason, most humanoid robots are made rigid to achieve high-precision control. However, with increased physical interactions with

Figure 14.19 EPC discrete GaN FETs and monolithic ePower Stage IC for motor drive applications. *Source:* Ref. [28]/Efficient Power Conversion Corporation.

humans, the robot actuators must be able to safely reduce the power and introduce a level of perceived mechanical flexibility [29].

The emergence of a new generation of high-speed biped robots marks a significant advancement in the field. These robots require motors capable of delivering increased torque at higher speeds. The mass distribution depends mainly on the actuators. To limit their inertial effects on the robot dynamic, it is mandatory to bring the actuator as close as possible to the root of the link on which it is fixed.

The approach consists of servomotors, which allow the torque and the speed controller loop gains to be lowered dynamically and the actuator to comply with safety rules. While servomotors can meet these performance demands, they necessitate a more complex control system and command a high-performance power supply. A torque sensor is required on each joint to reject perturbations through torque control, and the motor must generate the least amount of heat possible because of the limited passive cooling system and the increased interaction with humans, which requires surface temperatures not to exceed 40 °C. GaN integrated circuits permit inverter miniaturization so that it can be integrated within the motor. Furthermore, as seen in previous paragraphs, the GaN inverter allows for more accurate and efficient motor control, reflected in cooler motor operation.

Figure 14.20 shows a motor joint that integrates the inverter and the control. The motor has an embedded planetary gearbox to reduce the motor speed and increase the torque. From the inverter's control perspective, it is a high-speed motor. These motors operate with battery voltage between 36 and 64 V and have high torque constant thanks to the gearbox. The motor in Figure 14.20 has a maximum torque of 23.7 Nm and a maximum speed of 30 rad/s with a reduction ratio of 6.33. The torque constant is approximately k_t = 0.6 Nm/Arms, which means that the inverter must be able to deliver 39.5 Arms during the peak operation.

Figure 14.7 shows the EPC9176 inverter reference design, which has been developed to address this type of application. The GaN integrated circuit, ECP23102, includes the gate drivers, the power switches, the level shifters, and the bootstrap circuit within an FCQFN package of 3.5 × 5 mm.

Figure 14.21 shows the inverter block diagram, while Figure 14.7 shows a zoom of the switching cell where it is possible to see that the external vertical layout has been used because there are no high-frequency capacitors adjacent to the EPC23102 integrated circuit. The board is made of eight layers with different thicknesses to satisfy the layout clearance and creepage

Figure 14.20 Unitree A1 – For robotic motor joints. Dimensions: diameter 9 cm, height 4.5 cm. *Source:* Unitree Robotics [30]/HangZhou YuShu TECHNOLOGY CO., LTD.

Motor Drives | 423

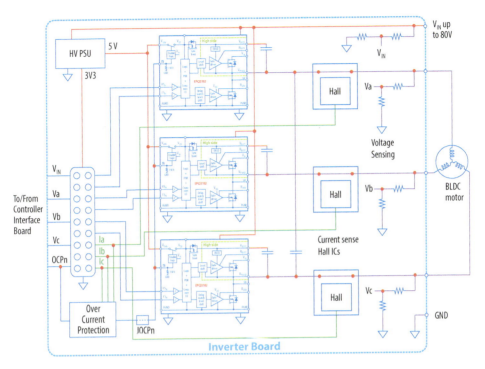

Figure 14.21 EPC9176 functional block diagram.

rules, which depend on the thickness of the copper. The top and bottom copper layers are 70 μm thick, while the internal layers are 140 μm thick. Each internal layer has the same layout as depicted in Figure 14.7 to reduce the copper resistance and increase the thermal conductivity across the layers. The FR4 dielectric has poor thermal conductivity, so the design includes several vias between the layers.

Figure 14.22 shows that the GaN integrated circuit with top-side cooling and no air convection can deliver up to $20\,A_{RMS}$ in continuous operation. If the motor's thermal mass is

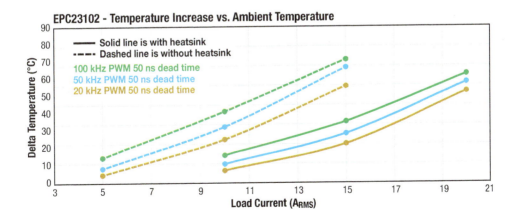

Figure 14.22 EPC9176 steady-state operation at 48 V_{DC} 500 RPM. Temperature increase versus the ambient at various PWM frequencies, and with or without top-side cooling.

adequately designed, the peak operation of up to 40 A_{RMS} can be reached if the total RMS current in the mission profile is lower than the current shown in Figure 14.22.

14.18.2 Food Delivery and Agricultural Drones

Battery-powered unmanned aerial vehicles (UAV) are used today for food and drug delivery in areas that are difficult to reach with conventional vehicles and in agriculture, where they are used to spread seeds, spray fertilizers, and more. These drones use motor drives that need to be lightweight and efficient to maximize working battery life and extend flight time between recharging.

The most reliable UAV is multi-rotor because it can perform demanding maneuvers. Numerous configurations of multi-rotor UAVs exist, such as Quadcopter, Hexacopter, and Octocopter. The Quadcopter is the simplest form and offers the minimum stability required, thanks to the equal distribution of rotor forces during hover. Two counter-rotating propellers exist for each side's pitch and roll axis. A pair of the rotors turns clockwise (CW), and the other pair rotates counterclockwise (CCW). Hexacopter and Octocopter UAVs are also widely used in operations when a higher payload must be transported. The permanent magnet motor with the propeller is a fundamental building block whose efficiency, dimensions, and weight determine the drone weight, payload, and run time between two subsequent battery charge cycles.

Figure 14.23 shows an example of a motor for a drone with the inverter positioned next to the motor under the airflow generated by the rotating propeller. Compared to other motor drive applications that generally do not have forced air cooling, the drone has better cooling because the inverter is positioned under the propeller. It is important to have a lightweight inverter because the mass must be concentrated in the drone's central body for better controllability during maneuvers.

Si MOSFET-based inverters switch at 16 kHz PMW using trapezoidal sensorless modulation or sinusoidal-field-oriented control (FOC) sensorless modulation techniques. However, due to their construction and high-frequency operation, the motors usually have a power loss factor, LF, that is consistently improved at frequencies above 50 kHz. The motor's steady output power for this application spans from 1.5 kW to 3 kW and can reach up to 6 kW for some

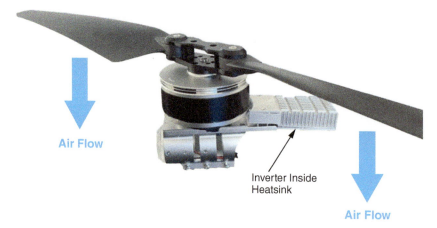

Figure 14.23 Eagle Power [31] motor with inverter and propeller. 75 V_{DC} 120 A_{RMS}, 1.20 m propeller diameter. *Source:* Zhongshan EAGLEPOWER Technology [31]/EAGLEPOWER Technology Co., Ltd.

seconds during the takeoff phase. Depending on the payload, the drones can be divided into three categories:

- Light load: up to 500 g, used for recreational and audio/video activities
- Medium load: up to 9 kg, used for food and drug delivery
- Heavy load: up to 30 kg, used in agriculture.

A GaN inverter reference design EPC9194 [32] has been tested on two different motors for medium and heavy loads. The EPC9194 GaN is a three-phase, two-level inverter comprising six 100 V rated, 1.8 mΩ $R_{DS(on)}$ EPC2302 GaN transistors in a PQFN package [33]. The EPC9194 inverter board is displayed in Figure 14.24. From Figure 14.24a, it is possible to see that the GaN FETs and their gate resistors are on the top side with four leg shunts in parallel per each switching cell and that there are no high-frequency capacitors. In this reference design, the external vertical layout is used. The half-bridge gate drivers, one per each inverter's phase, are mounted on the opposite side of the board.

Figure 14.25 shows the block diagram of EPC9194. It is essentially same as the block diagram in Figure 14.21. The current sensing scheme is different because the EPC9194 uses shunts (on the phase or on the legs, selectable by the user) and fast amplifiers. Moreover, due to the phase current required to drive a drone motor, the EPC9194 is based on discrete transistors with low $R_{DS(on)}$ of 1.8 mΩ.

The EPC9194 was used to power a 2.5 kW Eagle Power [31] trapezoidal motor and a 1.5 kW Tohan sinusoidal motor. In both cases, the GaN inverter running at higher PWM frequencies increases system efficiency; in fact, it requires less input power to run at the same speed.

According to the diagram displayed in Figure 14.26, the Eagle Power motor driven by the GaN inverter shows a significant reduction in input power. This is due to two main factors that contribute to the GaN inverter's higher efficiency: the use of a sinusoidal FOC (compared to the trapezoidal scheme) and a higher PWM frequency. Figure 14.27 shows a lower reduction in the Tohan sinusoidal motor, which is not negligible, considering that the input power reduction must be multiplied by the number of motors in a drone.

14.18.3 Servo Drives for Forklifts and Power Tools

Battery-powered industrial vehicles such as forklifts, power tools, manual handlers, or warehouse automatic vehicles require high-current inverters to drive the electric motors. Gallium nitride technology helps to increase power capability and simplifies the inverter design in these applications. An inverter for forklift applications is powered with a DC voltage between 24 and 120 V and can source up to 900 A_{RMS} motor phase current. Generally, each industrial vehicle producer has a platform approach and sells product families divided by voltage range, where the inverters are sized on the maximum current that can be achieved for a transient period (e.g. two minutes). A typical inverter for these applications is contained in an IP65-rated enclosure (150 mm × 120 mm × 60 mm) with a thick aluminum baseplate. Inside the enclosure, the power transistors are soldered to an Insulated Metal Substrate (IMS) board thermally and mechanically connected to the aluminum baseplate.

Above the IMS board is a super-dense PCB with gate drivers, analog signal conditioning, power supplies, and at least two microprocessors, one dedicated to functionality and the other to safety. A certain number of transistors in parallel are required to process the current and the heat generated by the conduction and switching dissipation.

Solutions based on silicon MOS technology are subject to constraints on the maximum number of devices that can be used in parallel, the maximum PWM switching frequency, and the dead time between complementary switches. The first constraint limits the maximum

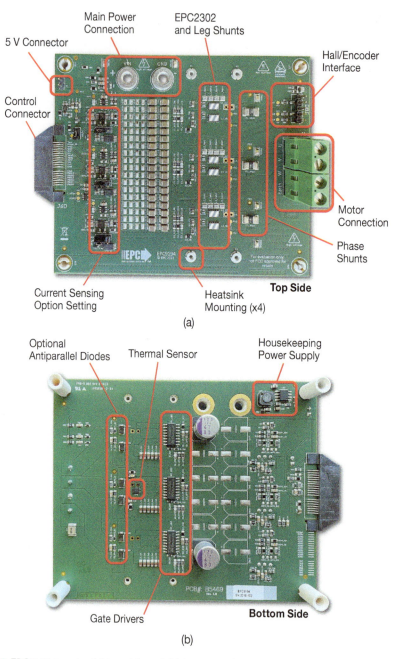

Figure 14.24 EPC9194 inverter (a) top side and (b) bottom side with functional circuits highlighted. *Source:* With permission of EPC.

current, while the other two degrade motor efficiency. With GaN technology, the GaN transistors allow more compact and more efficient drives due to the device dimensions and higher PWM with smaller dead time operation.

Thermal management remains the first problem to be addressed. Techniques such as those described in Chapter 6 must be used to exploit the full advantage of the new technology.

Motor Drives | 427

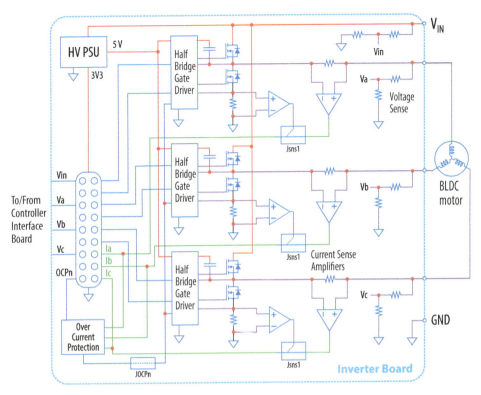

Figure 14.25 EPC9194 inverter functional block diagram.

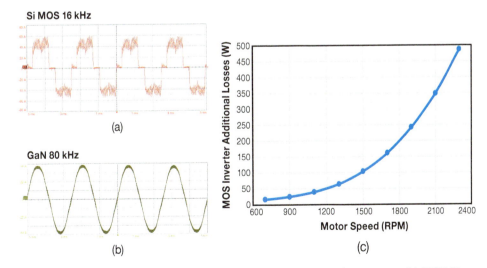

Figure 14.26 (a) Original Eagle Power PM120 trapezoidal motor inverter phase current; (b) EPC9194 motor phase current; (c) additional input power required by Si MOS inverter with respect to GaN inverter operating the motor at same propeller speed.

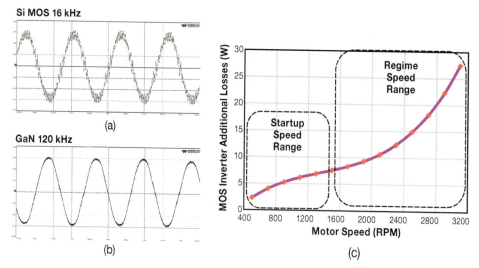

Figure 14.27 (a) Original Tohan T15S 18S-120A sinusoidal motor inverter phase current; (b) EPC9194 motor phase current; (c) additional input power required by Si MOS inverter with respect to GaN inverter operating the motor at same propeller speed.

Paralleling the GaN devices must be done carefully; however, considering that it is a motor drive application and that di/dt and dv/dt are slower than in other converters, the paralleling rules can be relaxed respect to what it has been described in Chapter 4.

Figure 14.28 shows the EPC9186 inverter reference design [26], with a closeup image of the switching cell. The external vertical layout has also been used in this case because there are no

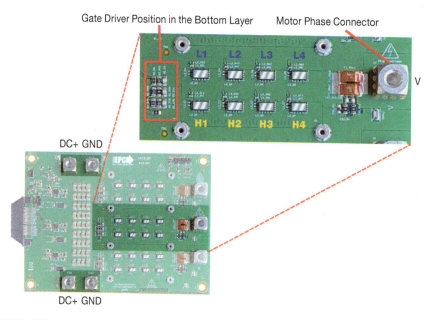

Figure 14.28 EPC9186 [26] inverter reference design with a zoomed section of the phase V switching cell. *Source:* Ref. [26]/Efficient Power Conversion Corporation.

high-frequency capacitors on the top side of the PCB where the GaN FETs EPC2302 are mounted. The gate driver is mounted on the bottom side, as indicated in the picture by the dashed red box on the left side. The high-side transistors are indicated with the letters H1 to H4, and the low-side transistors are indicated with the letters L1 to L4. The half bridge, composed of H1 and L1, is nearer to the gate driver than the half bridge H4-L4. This approach trades off layout symmetry for simplicity and scalability. If the gate signals are properly routed and shielded with the source Kelvin return signals, the common source inductance and the inductance mismatch effects are reduced and negligible.

Figure 14.29 shows the gate voltages measured at the four corner transistors, H1, L1, H4, and L4, at positive and negative currents. A mismatch in the slopes of the rising V_{GS} signals, between 0 and 2 V, would indicate a mismatch of the CSI inductances. From Figure 14.29, it is possible to see that the waveforms overlap perfectly.

Figure 14.30 shows that in the EPC9186 PCB, which has eight copper layers, the gate signals are distributed between layers six, seven, and eight, which are distant from the first layer where the GaN FETs are mounted. In layers six and eight, two shielding planes are connected to each transistor source with a single connection. Decoupling the source shielding planes from the source power planes that carry the current is important. The current-carrying planes are positioned from the first to the fifth layer to efficiently use the copper next to the GaN devices.

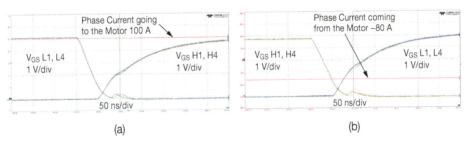

Figure 14.29 EPC9186 gate–source voltages at the GaN FETs H1, L1, H4, L4. (a) Positive current (b) negative current. Current 50 A/div, V_{GS} 1 V/div, timescale 50 ns/div.

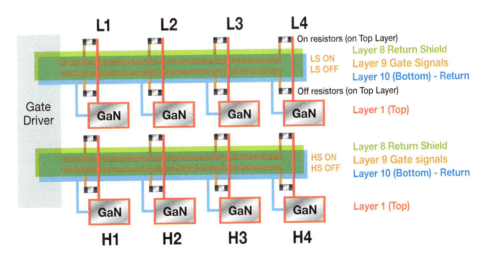

Figure 14.30 EPC9186 eight layers PCB. Example of gate signals and Kelvin source returns routing.

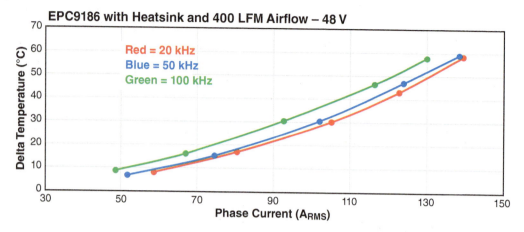

Figure 14.31 EPC9186 continuous operation at 48 V_{DC} with small heat sink and 400 LFM forced air convection.

Figure 14.31 shows the continuous operation capability of the EPC9186 inverter equipped with a small heat sink under 400 LFM of forced air convection. The inverter operates at 100 kHz with 75 ns dead time and carries up to 150 Arms phase current switching without any ringing.

14.19 Summary

This chapter discusses the benefits that GaN technology brings to motor drive applications. The motor power loss mechanisms analysis shows that the motor phase impedance characterization with frequency is of fundamental importance when optimizing system efficiency. The motor loss factor, LF, decreases with frequency, indicating that many motors run more efficiently if driven at higher PWM frequencies. Moreover, GaN transistors allow for reducing the dead time, the torque harmonics, and, hence, vibrations and losses.

The next chapter will show further benefits of GaN technology that are crucial in space applications due to the inherent radiation tolerance capability of GaN devices.

References

1 Murray, A., Palma, M., and Husain A. (2008). Performance comparison of permanent magnet synchronous motors and controlled induction motors in washing machine applications using sensorless field oriented control. *Conference Record – IAS Annual Meeting*, November 2008.
2 Baliga, B.J. (1989). Power semiconductor device figure of merit for high frequency applications. *IEEE Electron Device Lett.* 10: 455.
3 Efficient Power Conversion Corporation (2023). EPC9194 Schematic. 8–10 https://epc-co.com/epc/Portals/0/epc/documents/schematics/EPC9194_Schematic.pdf.
4 Albertsen A. Electrolytic capacitor lifetime estimation. Jianghai Europe Electronic Components GmbH. https://www.yumpu.com/en/document/read/6654541/electrolytic-capacitor-lifetime-estimation-jianghai-europe.

5 Kirisken, B. and Ugurdag, H.F. (2014). Cost-benefit approach to degradation of electrolytic capacitors. *2014 Reliability and Maintainability Symposium,* Colorado Springs, CO.

6 Shrivastava, A., Azarian, M.H., and Pecht, M. (2017). Failure of polymer aluminum electrolytic capacitors under elevated temperature humidity environments. *IEEE Trans. on Comp., Packag., and Manufact. Technol.* 7 (5): 745–750.

7 Cittanti, D., Mallemaci, V., Mandrile, F. et al. (2021). PWM-induced losses in electrical machines: an impedance-based estimation method. *24th International Conference on Electrical Machines and Systems (ICEMS)*, Gyeongju, Republic of Korea. 548–553. https://doi.org/10.23919/ICEMS52562.2021.9634438.

8 Schweizer, M. (2012). System-oriented efficiency optimization of variable speed drives. Ph.D. dissertation. ETH Zurich, Zurich, Switzerland. https://www.google.com/url?sa=t&source=web&rct=j&opi=89978449&url=https://www.pes-publications.ee.ethz.ch/uploads/tx_ethpublications/PhD_MSC_web.pdf&ved=2ahUKEwi18-mqj4KGAxXwLkQIHYr-CqsQFnoECBIQAQ&usg=AOvVaw2x1ogusNRFxLPHD82FPT_V.

9 Boglietti, A., Ferraris, P., Lazzari, M., and Pastorelli, M. (1995). Change of the iron losses with the switching supply frequency in soft magnetic materials supplied by PWM inverter. *IEEE Trans. Magnetics* 31 (6): 4250–4252.

10 Polinder, H. (1998). On the losses in a high-speed permanent-magnet generator with rectifier. Ph.D. dissertation. TU Delft, Delft, Netherlands.

11 Stoll, R.L. (1974). *Analysis of Eddy Currents*. Oxford: Oxford University Press.

12 de Buck, F.G.G., Giustelinck, P., and de Backer, D. (1984). A simple but reliable loss model for inverter-supplied induction motors. *IEEE Trans. on Indust. Appl.* IA-20 (1): 190–202.

13 Cittanti, D. (2018). System level efficiency optimization of an electric vehicle traction drive. Master's thesis. Politecnico di Torino, Turin, Italy.

14 Efficient Power Conversion Corporation (2022). EPC9176 Development board. EPC9176 20 A_{RMS} 3-Phase BLDC motor drive inverter reference design board.

15 Efficient Power Conversion Corporation (2022). EPC9176 Quick start guide. https://epc-co.com/epc/documents/guides/EPC9176_qsg.pdf.

16 Efficient Power Conversion Corporation (2024). EPC23102 – 100 V, 35 A ePower™ Stage IC. EPC23102 datasheet. https://epc-co.com/epc/Portals/0/epc/documents/datasheets/EPC23102_datasheet.pdf.

17 Efficient Power Conversion Corporation (2021). EPC2152 – 80 V, 15 A ePower™ Stage. EPC2152 datasheet. [Rev. 2.0 March 2021]. https://epc-co.com/epc/Portals/0/epc/documents/datasheets/EPC2152_datasheet.pdf.

18 Allegro Microsystems. ACS71240 Cost-effective shunt alternative Hall-based current sensor. ACS71240 datasheet. https://www.allegromicro.com/en/products/sense/current-sensor-ics/zero-to-fifty-amp-integrated-conductor-sensor-ics/acs71240.

19 Barba, V., Musumeci, S., Palma, M. and Bojoi, R. (2023). Dead time reduction strategy for GaN-based low-voltage inverter in motor drive system. *IEEE Applied Power Electronics Conference and Exposition (APEC)*, Orlando, FL, USA, (March 19–23, 2023), 2385–2390. https://doi.org/10.1109/APEC43580.2023.10131652.

20 Mandrile, F., Musumeci, S. and Palma, M. (2021). Dead time management in GaN based three-phase motor drives. *23rd European Conference on Power Electronics and Applications (EPE'21 ECCE Europe)*, Ghent, Belgium, (September 6–10, 2021), 1–10. https://doi.org/10.23919/EPE21ECCEEurope50061.2021.9570665.

21 Barba, V., Stella, F., Musumeci, S., et al. (2023). Optimal dead time selection in GaN FET switching leg via thermal analysis. *IEEE Energy Conversion Congress and Exposition (ECCE)*, Nashville, TN, USA, (October 29–November 2, 2023), 5392–5397. https://doi.org/10.1109/ECCE53617.2023.10362795.

22 Glaser, J., Reusch, D. (2016). Comparison of deadtime effects on the performance of DC-DC converters with GaN FETs and silicon MOSFETs. *IEEE Energy Conversion Congress and Exposition (ECCE)*, Milwaukee, WI, USA, (September 18–22, 2016). https://doi.org/10.1109/ECCE.2016.7854939.

23 Vujacic, M., Hammami, M., Srndovic, M. et al. (2018). Analysis of dc-link voltage switching ripple in three-phase PWM inverters. *Energies* 11 (2): 471. https://doi.org/10.3390/en11020471.

24 Fresia, E.J. and Eckfeldt, J.M. (1963). Failure modes and mechanisms in solid tantalum capacitors. *Second Annual Symposium on the Physics of Failure in Electronics*, Chicago, IL, USA, (September 25–26, 1963). 483–497. https://doi.org/10.1109/IRPS.1963.362263.

25 TDK. CKG57NX7S2A226M500JH 100 V 22 μF X7S Multilayer ceramic capacitor. CKG57NX7S2A226M500JH datasheet. https://product.tdk.com/en/search/capacitor/ceramic/mlcc/info?part_no=CKG57NX7S2A226M500JH.

26 Efficient Power Conversion Corporation (2023). EPC9186 Reference design. EPC9186 quick start guide. https://epc-co.com/epc/products/evaluation-boards/epc9186.

27 Barba, V., Musumeci, S., Stella, F. and Palma, M. (2023). Dead time constraints in gallium nitride devices for inverter applications. *25th European Conference on Power Electronics and Applications (EPE'23 ECCE Europe)*, Aalborg, Denmark, (September 4–8, 2023), 1–8, https://doi.org/10.23919/EPE23ECCEEurope58414.2023.10264571.

28 Efficient Power Conversion Corporation. GaN products for motor drives. https://epc-co.com/epc/applications/gan-for-motor-drives.

29 Stasse, O. and Flayols, T. (2019). An overview of humanoid robots technologies. in Venture, G., Laumond, J-P, and Watier, Bruno (editors), *Biomechanics of Anthropomorphic Systems*, New York: Springer, 281–310. https://laas.hal.science/hal-01759061v1/document.

30 Unitree Robotics. Joint A1 Motor. https://www.unitree.com/a1motor.

31 Zhongshan EAGLEPOWER Technology. PM120 Drone motor. PM120 datasheet. http://en.rc-eaglepower.com/products/show_4.html.

32 Efficient Power Conversion Corporation (2023). EPC9194 Reference Design. EPC9194 quick start guide. [Revised 1.1]. https://epc-co.com/epc/products/evaluation-boards/epc9194.

33 Efficient Power Conversion Corporation (2024). EPC2302 – Enhancement mode power transistor. EPC2302 datasheet. [Revised March 2024]. https://epc-co.com/epc/products/gan-fets-and-ics/epc2302.

15

GaN Transistors and Integrated Circuits for Space Applications

15.1 Introduction

Radiation in space is generated by many sources within and outside of our solar system. This radiation comes in the form of gamma rays, energetic electrons, protons, and heavier ions that are all known to cause damage in semiconductors. Over the years, silicon-based devices have been well characterized under various radiation conditions with vulnerabilities being identified and, to some extent, mitigated through design and process improvements. NASA has published guidelines to help designers of satellite systems consistently design for the different environments encountered in various earth orbits [1].

In this chapter, research on GaN capabilities under exposure to different types of radiation, with a particular focus on GaN transistors and integrated circuits used in power conversion, will be examined. In turn, actual measurements of enhancement-mode GaN transistors and integrated circuits and their capability will be compared to radiation-resistant silicon power MOSFETs and ICs.

15.2 Failure Mechanisms in Electronic Components Used in Space Applications

An energetic particle or photon can cause damage to a semiconductor device in two primary ways: (i) physical damage to the crystal (displacement damage); and (ii) ionization damage that results from the cloud of electron–hole pairs that are created by the particle [2]. The ionization, in turn, leads to undesired trapping of electrons or holes in non-conducting layers that can alter the electrical characteristics of the device. The ionization cloud can also cause the device to momentary conduct along the particle track, leading to various catastrophic failure modes that will be discussed in the next section.

Power MOSFETs, in particular, are vulnerable to radiation in two ways. First, gamma (electron) radiation can cause positively charged traps (holes) to develop in the thin gate oxide [3, 4]. The addition of positive charges between the gate electrode and the channel reduces the threshold voltage of the n-channel power MOSFET. The second major vulnerability stems from energetic particles that can fully penetrate the semiconductor device. These particles cause single-event effects (SEE), such as single-event gate rupture (SEGR) and single-event burn-out (SEB).

GaN Power Devices for Efficient Power Conversion, Fourth Edition. Alex Lidow, Michael de Rooij, John Glaser, Alejandro Pozo, Shengke Zhang, Marco Palma, David Reusch, and Johan Strydom.
© 2025 John Wiley & Sons Ltd. Published 2025 by John Wiley & Sons Ltd.

434 | *GaN Power Devices for Efficient Power Conversion*

To understand how these same radiation effects impact the electrical performance of GaN transistors, Schottky-based depletion-mode devices (Figure 1.6a in Chapter 1), Metal Insulator Semiconductor (MIS)-based depletion-mode devices (Figure 1.6b in Chapter 1), and pGaN-based enhancement-mode devices (Figure 1.10 in Chapter 1) need to be considered. All three of these structures use an AlGaN/GaN barrier to generate the 2DEG, but have different gate structures that can lead to different vulnerabilities when exposed to radiation. Cascode-configured GaN transistors will not be considered in this chapter because their behavior would be limited by the associated silicon MOSFET's radiation tolerance.

15.3 Standards for Radiation Exposure and Tolerance

The NASA guidelines for transistors [1] established the standard set for a radiation-hardened (rad-hard) device as having the ability to withstand a total incident dose (TID) between 200 kRad(Si) and 1 MRad(Si), and a single-event upset (SEU) threshold linear energy transfer (LET) of 80–150 MeV/mg/cm^2. A "Rad" is defined as the mean energy absorbed per unit mass of irradiated material at the point of interest [4]. It is also common to see mean energy absorption quantified in Grays (Gy), a unit derived from standard international units.

1 Rad = 100 ergs/g
1 Gy = 1 J/kg
100 Rad = 1 Gy

A Rad(Si) or Gy(Si) would be a measure of absorption in a silicon crystal. In order to maintain comparisons with silicon power MOSFETs, most researchers have elected to use Rad(Si) to report their results for GaN devices.

Linear energy transfer (LET) is a measure of the energy deposited by an incident particle (the "stopping power") per unit of track length. A heavier ion, such as Au, would have a higher LET for the same energy than a lighter ion, such as a proton.

15.4 Gamma Radiation

Gamma radiation consists of high-energy photons that interact with electrons, where the resultant electrons can be trapped in energetic traps contained within the device structure. The common approach to measuring device tolerance to gamma radiation is to use the decay radiation from a Cobalt-60 (^{60}Co) source. Using this method, testing of depletion-mode GaN Schottky gate transistors designed for RF applications has been reported in references [5, 6], where the results showed no degradation when exposed up to 600 MRads(Si) and are consistent with results for silicon JFETs [3]. This lack of degradation is not surprising because there are no oxides in the gate structure that can trap charge, and thereby modulate, the threshold voltage.

In a silicon-based MOSFET, the gamma radiation causes the generation of hole–electron pairs in the vicinity of the gate. Holes become trapped within the gate oxide leading to voltage shifts [2–4]. The positive charge reduces the threshold voltage of the device until the transistor goes from normally off – or enhancement mode – to normally on, which is a depletion-mode state. At this point the system will need a negative voltage to turn the MOSFET off. Typical ratings for radiation-hardened (rad-hard) devices range from 100 kRads to 300 kRads. In some cases, devices can be made to go up to 1 MRad, but these tend to be very expensive.

GaN Transistors and Integrated Circuits for Space Applications | 435

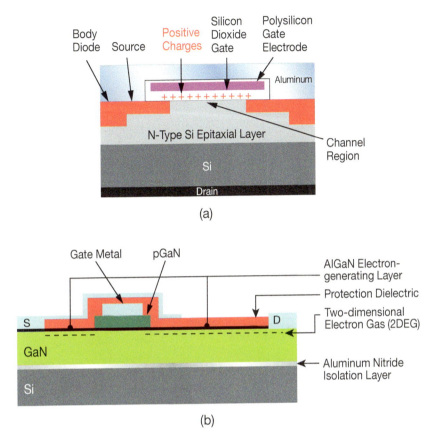

Figure 15.1 (a) Cross section of a typical silicon MOSFET (b) Cross section of a typical enhancement-mode GaN (eGaN®) device.

Figure 15.1a is a cross section of a typical silicon MOSFET. It is a vertical device with the source and gate on the top surface and the drain on the bottom surface. The gate electrode is separated from the channel region by a thin silicon dioxide layer. As illustrated, gamma radiation can induce positive charges in the gate oxide layer which can cause the threshold voltage to fall below zero volts.

Enhancement-mode GaN devices that use a pGaN gate such as described in Chapter 1 are built very differently from silicon MOSFETs. As shown in Figure 15.1b, all three terminals, gate, source, and drain, are located on the top surface. As in a silicon MOSFET, conduction between source and drain is modulated by biasing the gate electrode from zero volts to a positive voltage – usually 5 V. In enhancement-mode GaN devices the gate metal is separated from the underlying channel by a semi-insulating pGaN region as well as an aluminum gallium nitride layer. These layers do not accumulate charge when subjected to gamma radiation – a key benefit compared to MOSFETs.

Enhancement-mode pGaN HEMT devices have been tested extensively [7, 8] with favorable results. In these tests, the devices had either drain-to-source bias during testing, or gate-to-source bias. By exposing devices to gamma rays while applying bias on each of the transistor terminals, the in-circuit performance in a high-radiation environment can be projected. Figure 15.2 is a graph showing the progression of drain-to-source and gate-to-source leakage

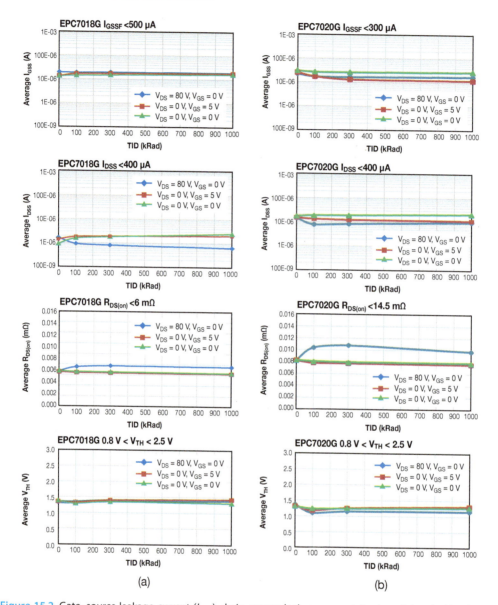

Figure 15.2 Gate–source leakage current (I_{GSS}), drain–source leakage current (I_{DSS}), and threshold voltage (V_{TH}) responses after exposure to different doses of total ionizing dose (TID) from a ^{60}Co source under different biasing conditions on the gate and drain as indicated for, (a) a 100 V HEMT platform and, (b) a 200 V HEMT platform. *Source:* unpublished data from Efficient Power Conversion.

currents along with threshold voltage of GaN HEMT devices from 100 to 200 V platforms tested up to 1 MRad(Si).

There are many failure mechanisms for silicon-based power MOSFET devices under various types of radiation conditions, and each of these failure mechanisms has led to design and performance compromises. For example, under many forms of radiation-trapped charges in the silicon dioxide gate of a power MOSFET has the effect of lowering the threshold voltage of an n-channel device. With enough trapped charges, the device will go from enhancement mode to depletion mode (negative threshold voltage with respect to the source voltage). The trapped charges will cause system failure unless a negative voltage is applied to turn the device off. Reference [11] has an excellent summary of the impact of various forms of heavy ion bombardment on the threshold voltage of a modern commercial trench-gate power MOSFET showing significant degradation of threshold voltage after a few hundred kRad(Si). Enhancement-mode GaN transistors with pGaN gate electrodes therefore offer a distinct performance advantage over silicon MOSFETs when subjected to gamma radiation.

15.5 Neutron Radiation (Displacement Damage)

Displacement damage (DD) is a cumulative effect due to the long-term, non-ionizing damage from neutrons. Displacement damage is the result of nuclear interactions, typically scattering, causing lattice defects. It causes permanent degradation of electrical parameters similar to the total ionizing dose. In a silicon MOSFET, the resulting material changes increase the junction leakage currents, reduce the minority carrier lifetime, or change the resistivity of the material and hence parameter degradation [18].

Radiation-hardened silicon power MOSFETs have shown stable leakages under neutron radiation, but also have exhibited significant threshold voltage shifts due to the charge build-up in the gate oxide similar to the gamma radiation effects. Significant $R_{DS(on)}$ shift can also be observed by displacement damages for neutron fluences greater than about 1×10^{14} neutrons/cm^2 [19].

Measuring displacement threshold energy, E_d (eV), of different materials [20] will predict how enhancement-mode GaN transistors respond to displacement damage. Figure 15.3 shows the difference in E_d (GaN) compared to that of E_d (Si) plotted as a function of the inverse of the lattice constant. There is an approximate linear relationship between the inverse of the lattice constant and the displacement threshold energy. The higher these values for a given material provides an indication of how strongly bonded the component atoms are, thus making it more

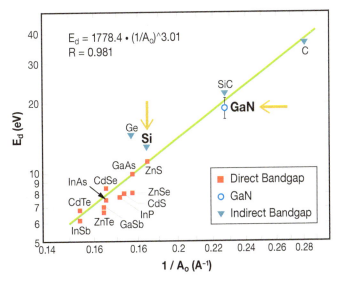

Figure 15.3 Graph of displacement threshold energy versus the inverse of the lattice constant for different materials. *Source:* Adapted from Efficient Power Conversion Space Division [19].

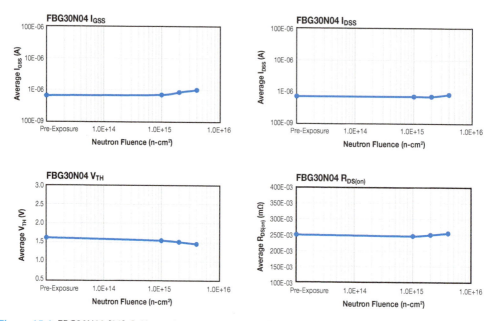

Figure 15.4 FBG30N08 [20] GaN transistor parameters under neutron irradiation [21]. *Source:* Adapted from Refs. [20] and [22].

difficult to cause lattice damage and the subsequent creation of point defects during irradiation. It is apparent that GaN is significantly more resistant to damage than Si.

The results of neutron irradiation on an FBG30N04 [20] GaN transistor, shown in Figure 15.4, demonstrate the stability of the GaN technology across all device parameters up to 4×10^{15} neutrons/cm^2. Similarly, other researchers have shown the tolerance of GaN transistors to neutron irradiation-induced displacement damage [21]. As of this writing, devices have not been induced to fail during neutron bombardment, suggesting they can endure significantly higher levels of bombardment.

15.6 Single-Event Effects (SEE) Testing

As with gamma and neutron testing, there has been significant research on SEE in Schottky gate RF HEMT transistors [8, 12], but no MIS-gate HEMT device testing has been reported. In the case of a cascode GaN transistor discussed in Chapter 1, the reaction to single-event radiation would be determined by the combination of the silicon MOSFET and the GaN transistor in their series configuration. Enhancement-mode pGaN-gated devices have also had extensive testing [7, 13–15] with differing levels of success for SEE hardness. Whereas gamma or neutron radiation tolerance is a by-product of the materials properties of GaN and the adoption of a pGaN gate electrode, SEE sensitivity requires special structural design and processing.

SEE are caused by heavy ions generated by the impact of galactic cosmic rays, solar particles, or energetic neutrons and protons. This can be simulated terrestrially using a cyclotron [16] to create beams of different ions. Two of the most common ions used to evaluate radiation tolerance of electronics components are xenon (Xe), with a LET of about 50 MeV cm^2/mg, and Gold (Au), with an LET of about 85 MeV cm^2/mg.

In a silicon MOSFET there are two primary failure mechanisms caused by these heavy ions, single-event gate rupture (SEGR) and single-event burnout (SEB). SEGR is caused by the energetic atom creating such a high transient electric field across the gate oxide that the gate oxide ruptures, as illustrated in the upper cross section in Figure 15.5. Whereas SEB is caused when the energetic particle transverses the drift region of the device where there are relatively high electric fields. The energetic particle loses its energy while generating a large number of hole electron pairs and a momentary short circuit through the device.

The beam flux, measured in number of ions passing through a given area in a given time, (with units of ions/cm^2/s) is established and the equivalent dose density, or fluence, is calculated by multiplying the flux by the amount of time the devices are exposed to the beam (with units of ion/cm^2). Depending on the mission requirements, fluences of between 1×10^5 and 1×10^7 ions/cm^2 are required. The amount of hits by an ion on the active area of the device can be calculated by multiplying the fluence by the active area of the device exposed to the beam. In Si MOSFETs, these ions create physical damage to the crystal that can cause Single-Event Gate Rupture (SEGR) or Single-Event Burnout (SEB) failures in a device.

The Schottky gate and pGaN gate devices show remarkable resistance to SEE radiation. In reference [12], Schottky gate devices showed slow degradation of I_{DSS} when irradiated with bromine ions with an LET of 38 MeV cm^2/mg. Abrupt failures (SEB) were observed in these devices with 70 V drain-to-source bias at an LET of 60 MeV cm^2/mg. Schottky gate transistors in reference [8] also showed no degradation when exposed to lower levels of SEE radiation from FE, O, and C atoms. Reference [14] also showed excellent SEE tolerance with the main failure mechanism being the increase of I_{DSS} leakage, the rate of which scaled with the LET and drain-to-source voltage.

High levels of SEE capability for GaN HEMTs were reported in references [13, 17]. Tests showed eGaN® FET devices capable of withstanding 1×10^7 cm^{-2} Au ions at a LET of 87.2 MeV cm^2/mg with full-rated voltage applied to 40 V or 100 V rated transistors. An example of the stability of device characteristics that is achievable under Au bombardment in a correctly designed and processed device is shown in Figure 15.6. The devices shown in Figure 15.6 had received more than 500,000 hits by Au ions by the time the fluence had reached 1×10^7 ions/cm^2.

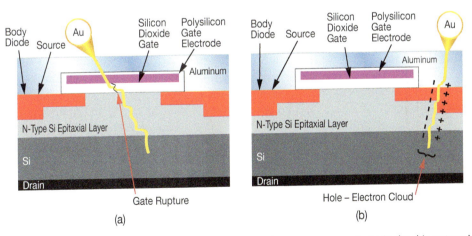

Figure 15.5 (a) In Si MOSFETs, heavy ions impacting gate regions can cause ruptures in the thin gate oxide region. This phenomenon is called single-event gate rupture (SEGR). (b) Heavy ions penetrating drain and source regions can cause clouds of hole–electron pairs to be generated that appear as a momentary short circuit. This phenomenon is called single-event burnout (SEB).

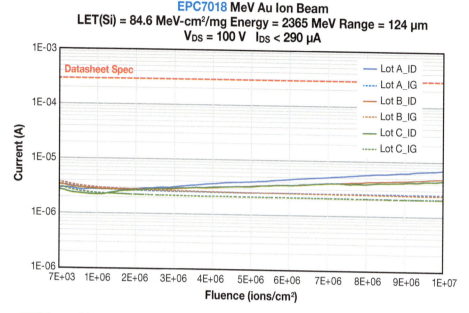

Figure 15.6 I_{DSS} and I_{GSS} progression for EPC7018G [9] GaN transistor during single-event testing up to a fluence of 1×10^7 ions/cm² Au at 84.6 LET and biased at 100 V_{DS}. *Source:* Adapted from Efficient Power Conversion Space Division [9].

15.7 Performance Comparison Between GaN Transistors and Rad-Hard Si MOSFETs

Schottky gate and pGaN gate HEMT transistors can tolerate high doses of radiation without significant performance degradation. In the case of the enhancement-mode GaN devices, as shown in Chapters 3–14, they significantly outperform state-of-the-art commercial power MOSFETs. Unlike GaN transistors, power MOSFETs designed to operate in high-radiation environments do not have dynamic switching performance comparable to commercial MOSFETs. Significant compromises in device geometries and fabrication processes are needed to harden the final product [4].

Illustrating these significant compromises to performance required of rad-hard MOSFET users, Table 15.1 compares a state-of-the art commercial power MOSFET with a comparable radiation-tolerant version [23, 24]. The last two rows in the table compare the soft-switching figure of merit (FOM) and the hard-switching FOM first discussed in Chapters 6 and 7, respectively. The latest commercial MOSFET are four to five times superior to the rad-hard MOSFET in these two key measures of device performance in switching converters.

Moving to radiation-tolerant GaN transistors, Table 15.2 compares the performance of a 200 V enhancement-mode GaN transistor [10, 25, 26] and a 200 V rad-hard MOSFET [23, 24]. The GaN transistors have comparable SEE capability, five times the gamma radiation tolerance (TID), a hard-switching FOM ($R_{DS(on)} \cdot Q_{GD}$) 50 times superior, and a soft-switching FOM ($R_{DS(on)} \cdot (Q_G + Q_{OSS})$) 10 times superior to the comparable rad-hard MOSFET. Any power conversion system would operate with significantly lower losses and higher gamma radiation tolerance using these, or similar enhancement-mode GaN transistors.

Table 15.1 Comparison of key electrical parameters for 200 V rated commercial MOSFET (IPB107N20N3 G) [23] and a comparably rated rad-hard MOSFET (IRHN57250SE) [24].

Parameter	IPB107N20N3 G	IRHN57250SE	Unit	Performance ratio
Rad tolerant	No	Yes		
BV_{DSS}	200	200	V	1.0
$R_{DS(on)}$	0.011	0.06	Ω	5.5
Q_G	161	132	nC	0.8
Q_{GS}	23	45	nC	2.0
Q_{GD}	8	60	nC	7.5
$Q_G \cdot R_{DS(on)}$	1.8	7.9	nC $\cdot \Omega$	4.5
$(Q_{OSS} + Q_G) \cdot R_{DS(on)}$	2.6	11.2	nC $\cdot \Omega$	4.3
$Q_{GD} \cdot R_{DS(on)}$	0.09	3.6	nC $\cdot \Omega$	41

Source: Adapted from Infineon [23] and [24].

15.8 GaN Integrated Circuits

Enhancement-mode GaN power devices use lateral conduction structure between drain and source terminals with the current modulated by the gate terminal. The construction of the device relies on multiple metal layers acting as interconnects to connect the thousands of unit cells in parallel to form the single power device. Alternatively, each lateral GaN unit cell can be scaled to different voltages, currents, switching speeds, and parametric matching requirements as needed by the circuit design. Then the different devices can be interconnected into an integrated circuit (IC) with additional device elements such as integrated resistors and capacitors as discussed in Chapter 1.

One inherent advantageous attribute of the GaN IC structure is the built-in radiation hardness of all its integrated devices. All the radiation-hardness properties of the GaN devices, as described earlier in the chapter, are still present in their integrated forms. Understanding the radiation effects allows circuit designers to design circuits and products that have superior radiation-hardness specifications compared to Si integrated circuits.

Using gallium nitride IC technology, circuit designers gain increased integration density for size reduction and can also benefit from the inherent advantages of the higher performance levels of monolithic integration not easily achieved with discrete or hybrid implementation.

One such example is the integrated half-bridge power stage IC as shown in block diagram form in Figure 15.7 and discussed in detail in Chapters 1, 3, 10, and 14. This monolithic IC includes the output high-side, and low-side, GaN power devices configured as half bridges, the floating high-side gate driver, the fixed referenced low-side gate driver, the level-shifting circuits, the synchronous bootstrap circuit and the input logic circuit. A radiation-hardened version of the commercial GaN IC, rated at a maximum input voltage of 50 V, is also available [28, 29].

One of the recurring issues with silicon MOSFETs in a half-bridge configuration under single-event testing is a failure mechanism known as single-event transient (SET). This is when the heavy ion causes both the upper and lower transistors in the half bridge to turn on simultaneously. This then leads to a momentary unwanted shoot-through condition that can fail the system.

Table 15.2 Comparison of key electrical parameters and radiation tolerance between a commercial-grade 200 V rated enhancement-mode GaN transistor (EPC2215) [25], a radiation-hardened (TID, SEE, & neutrons) enhancement-mode GaN transistor (EPC7020G) [10, 26], and a comparably rated rad-hard MOSFET (IRHN57250SE) [24].

	EPC2215 Commercial GaN transistor	EPC7020G Rad-Hard GaN transistor	IRHN57250SE Rad-Hard MOSFET	Units	Performance Ratio commercial GaN versus Rad-Hard MOSFET	Performance ratio Rad-Hard GaN versus Rad-Hard MOSFET	Method
BV_{DS}	200	200	200	V	1:1	1:1	
$R_{DS(on)}$	8	14.5	6	mΩ	1.3:1	2.4:1	
Q_G	17.7	13.5	132	nC	7:1	10:1	
Q_{GS}	3.3	3.8	45	nC	14:1	12:1	
Q_{GD}	2.0	2.5	60	nC	30:1	24:1	
$Q_G \cdot R_{DS(on)}$	0.142	0.196	7.9	nC $\cdot$ Ω	56:1	40:1	
$(Q_{OSS} + Q_G) \cdot R_{DS(on)}$	0.97	1.35	11.2	nC $\cdot$ Ω	12:1	8:1	
$Q_{GD} \cdot R_{DS(on)}$	0.016	0.036	3.6	nC $\cdot$ Ω	225:1	100:1	
Guaranteed SEE SOA at $\sim$84 LET ($V_{GS} = 0$ V)	N/A	200	200	V	N/A	1:1	MIL-STD-750E Method 1080
Guaranteed TID capability	N/A	1000	100	kRad(Si)	N/A	10:1	MIL-STD-750 METHOD 1019

Source: Adapted from [10, 24–26].

GaN Transistors and Integrated Circuits for Space Applications | 443

Table 15.3 83 MeV cm^2/mg LET SEE Testing of EPC7011 at rated BV of 50 V and above rated BV. Failure only occurred above the maximum rated voltage.

V_{DD}(V)	V_{BIAS}(V)	Dead time(s)	Frequency (Hz)	Edge slew rate(s)	Pulse triggering	Pass/Fail	Number of events	Comments
50	12	20 n	100K	20 ns	±500 ns	Pass	0	
60	12	20 n	100K	20 ns	±500 ns	Fail	0	SEB

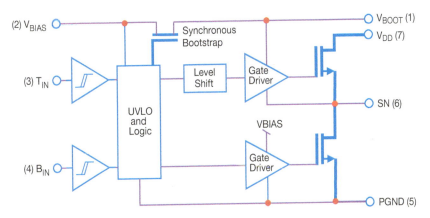

Figure 15.7 EPC7011 Simplified block diagram.

15.8.1 Single-Event Transient (SET) Testing

For single-event transient (SET) testing, the bias voltage (V_{BIAS}) was set to a nominal 12 V, and the input was set to a 50% duty cycle with a 20 ns dead time. The switching frequency was 100 kHz. The V_{DD} voltage was set to the worse-case rated voltage of 50 V for testing at LET of 63 and 84 MeV cm^2/mg. The switch-node output of the EPC7011 was monitored with an oscilloscope and the trigger was set to detect any pulse width change.

The switch-node waveforms are shown in Figure 15.8 for 63 MeV cm^2/mg LET and in Figure 15.9 for 84 MeV cm^2/mg LET. Whereas there are no transient effects at 63 LET, there

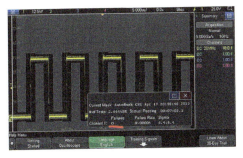 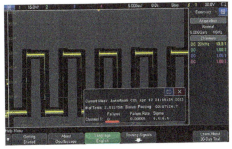

Figure 15.8 Switch-node waveforms for the EPC7011 at 100 kHz and 50% duty cycle. No transients can be seen at 63 LET.

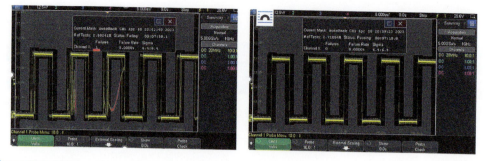

Figure 15.9 Switch-node waveforms for the EPC7011 at 100 kHz and 50% duty cycle and 84 LET. One non-destructive transient can be seen in the left oscillogram. The oscillogram on the right shows no transient effects.

is one non-destructive, yet abnormal, waveform shown on the left side of Figure 15.8 for a sample at tested 84 LET. The figure on the right side showed no transient effects.

Monolithic GaN-based driver ICs have also become available as of 2024 [30] with the same radiation resistant characteristics of GaN-based EPC7011 IC discussed above. A simplified block diagram of the first of these driver ICs is shown in Figure 15.10.

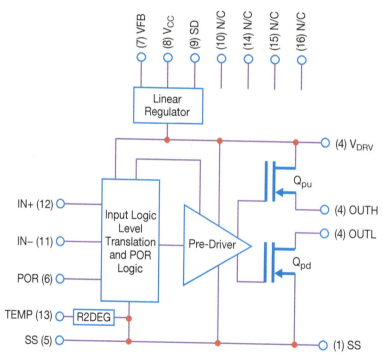

Figure 15.10 EPC7009 Simplified block diagram.

15.9 Summary

GaN transistors and integrated circuits have been tested under gamma irradiation, neutron irradiation, and heavy ion bombardment. These devices, when designed and processed correctly, demonstrate readiness for use in the most stringent of radiation environments and far exceed the capabilities of silicon power MOSFETs and ICs.

The problem designers encounter with silicon MOSFETs is that they must choose between radiation tolerance and electrical performance. Commercial MOSFETs have thick gate oxides and trap a lot of charge, resulting in large shifts in the threshold voltage and eventual failure at relatively low total-dose exposure. The radiation-hardened MOSFETs available have FOMs several times worse than their commercial counterparts, leading to either low efficiency or large size (due to the low switching frequency). Significant radiation limitations are applicable to the various radiation-resistant integrated circuits made with silicon IC technologies [31].

Correctly designed and processed enhancement-mode GaN transistors and integrated circuits give designers a new capability with electrical performance superior to cutting-edge Si MOSFETs and Si driver ICs, and radiation tolerance that exceeds the best radiation-hardened silicon devices available. GaN devices bring a combination of electrical and radiation performance that establishes a new state-of-the-art for transistor and IC space applications.

Factors that affect the rate of conversion from the aging power MOSFET to GaN transistors and integrated circuits will be discussed in the final chapter of this book.

References

1 National Aeronautics and Space Administration. NASA Practice No. PD-ED-1258. *Space Radiation Effects on Electronic Components in Low-Earth Orbit.* February 1999. https://llis.nasa.gov/lesson/824.

2 Holmes-Siedle, A. and Adams, L. (2002). *Handbook of Radiation Effects*, 2e. Oxford, UK: Oxford University Press.

3 Messenger, G.C. and Ash, M.S. (1986). *The Effects of Radiation on Electronic Systems.* New York, NY: Van Nostrand Reinhold Company.

4 Ma, T.P. and Dressendorfer, P.V. (1989). *Ionizing Radiation Effects in MOS Devices and Circuits.* New York: John Wiley and Sons, Inc.

5 Aktas, O., Kuliev, A., Kumar, V. et al. (2004). 60Co gamma radiation effects on DC, RF, and pulsed I-V characteristics of AlGaN/GaN HEMTs. *Solid-State Electron.* 48: 471–475.

6 McClory, J.W. (2008). The effect of radiation on the electrical properties of aluminum gallium nitride/gallium nitride heterostructures. Ph.D. dissertation. The Air Force Institute of Technology, Wright Patterson Air Force Base, Ohio.

7 Lidow, A., Witcher, J.B., and Smalley, K. (2011). Enhancement mode gallium nitride (eGaN®) FET characteristics under long-term stress. *GOMAC Tech Conference*, Orlando Florida (March 2011).

8 Sonia, G., Brunner, F., Denker, A. et al. (2006). Proton and heavy ion irradiation effects on AlGaN/GaN HFET devices. *IEEE Tran. Nucl. Sci.* 53 (6): 3661–3666.

9 Efficient Power Conversion Space Division (2023). EPC7018G – rad hard eGaN® 100V, 90A, 6mΩ surface mount (FSMD-G). EPC7018G datasheet. [Revised May 2023 Rev. Q1]. https://epc.space/documents/datasheets/EPC7018G-datasheet.pdf.

10 Efficient Power Conversion Space Division (2023). EPC7020G – rad hard eGaN® 200V, 80A, 14.5mΩ surface mount. EPC7020G datasheet. [Revised May 2023 Rev. Q1]. https://epc.space/documents/datasheets/EPC7020G-datasheet.pdf.

11 Felix, J.A., Shaneyfelt, M.R., Schwank, J.R. et al. (2007). Enhanced degradation in power MOSFET devices due to heavy ion irradiation. *IEEE Tran. Nucl. Sci.* 54 (6): 2181–2189.

12 Bazzoli, S., Girard, S., Ferlet-Cavrois, V. et al. (2007). SEE sensitivity of a COTS GaN transistor and silicon MOSFETs. *9th European Conference on Radiation and Its Effects on Components and Systems*, RADECS, 2007.

13 Lidow, A. and Smalley, K. (2012). Radiation tolerant enhancement mode gallium nitride (eGaN®) FET characteristics. *GOMAC Tech Conference*, Las Vegas, Nevada (March 2012).

14 Lidow, A., Strydom, J., and Rearwin, M. (2014). Radiation tolerant enhancement mode gallium nitride (eGaN®) FETs for high-frequency DC-DC conversion. *GOMAC Tech Conference*, Charleston, South Carolina (April 2014).

15 Scheick, L.Z. (2016). Recent gallium nitride power HEMT single event testing results. Poster W-6, *IEEE Nuclear and Space Radiation Effects Conference (NSREC)*, Portland, United States (2016). https://ieeexplore.ieee.org/abstract/document/7891731

16 Texas A&M University Cyclotron Institute. https://cyclotron.tamu.edu

17 Kuboyama, S., Maru, A., Shindou, H. et al. (2011). Single-event damages caused by heavy ions observed in AlGaN/GaN HEMTs. *IEEE Tran. Nucl. Sci.* 58 (6): 2734–2738.

18 Gillberg, J.E., Burton, D. I, Titus, J.L. et al. (2001). Response of radiation hardened MOSFET to neutrons. *IEEE Nuclear and Space Radiation Effects Conference (NSREC)*, Vancouver, Canada (2001).

19 Pearton, S.J., Ren, F., Patrick, E. et al. (2016). Review – ionizing radiation damage effects on GaN devices. *ECS J. Solid State Sci. Technol.* 5 (2): Q35–Q60.

20 Efficient Power Conversion Space Division (2023). FBG30N04 – rad hard eGaN® 300V, 4A, 400mΩ Surface Mount (FSMD-C). FBG30N04 datasheet, February 2023. https://epc.space/products/gan-discretes/FBG30N04C

21 Ling, L., Zhang, J.C., Xue, J.S. et al. (2012). Neutron irradiation effects on AlGaN/GaN high electron mobility transistors. *Chin. Phys. B* 21 (3): 037104.

22 Lidow, A., Nakata, A., Rearwin, M. et al. (2014). Single-event and radiation effect on enhancement mode gallium nitride FETs. *IEEE Nuclear and Space Radiation Effects Conference (NSREC)*, Paris, France (2014).

23 Infineon (2011). OptiMOS™3 power transistor. IPB107N20N3-G datasheet. [July 2011 Rev. 2.3]. https://www.infineon.com/cms/en/product/power/mosfet/n-channel/ipb107n20n3-g.

24 Infineon (2011). IRHN57250SE – Radiation hardened power MOSFET surface mount (SMD-1). IRHN57250SE datasheet, December 2011. https://www.infineon.com/cms/en/product/high-reliability/space/power/rad-hard-mosfets/n-channel-rad-hard-power-mosfets/irhn57250se/

25 Efficient Power Conversion Corporation (2023). EPC2215 – Enhancement-mode power transistor. EPC2215 datasheet, June 2023. http://epc-co.com/epc/documents/datasheets/EPC2215_datasheet.pdf.

26 Zafrani, M., Brandt, J., Strittmatter, R. et al. (2022). Radiation results for modern GaN-on-Si power transistors, *IEEE Radiation Effects Workshop (REDW)*, in conjunction with NSREC, Provo Utah. (2022).

27 Strydom, J., Lidow, A., and Goti, T. (2013). Radiation tolerant enhancement mode gallium nitride (eGaN®) FETs in DC-DC converters. *GOMAC Tech Conference*, Las Vegas Nevada, (March 2013).

28 Lidow, A. (ed.) (2022). *GaN Power Devices and Applications*. El Segundo, CA: Power Conversion Publication.

29 Efficient Power Conversion Space Division (2024). EPC7011L7SH – Space level radiation-hardened GaN power stage. EPC7011L7SH datasheet February 2024. https://epc.space/documents/datasheets/EPC7011L7SH-datasheet.pdf

30 Efficient Power Conversion Space Division (2024). EPC7009 – Radiation-hardened single output eGaN® Gate driver integrated circuit. EPC7009 datasheet April 2024. https://epc.space/products/rad-hard-gan-drivers-and-power-stages-ics/EPC7009L16SH/.

31 Renesas (2021). ISL70040SEH – Radiation hardened low-side GaN FET driver. ISL70040SEH datasheet February 2021 [Rev.9.00]. https://www.renesas.com/us/en/document/dst/isl70040seh-isl73040seh-datasheet.

16

Replacing Silicon Power MOSFETs

16.1 Introduction: GaN, Rapid Growth/Great Future

A goal of this book is to lay the foundation for understanding GaN technology, an emerging power device technology moving rapidly to replace the use of MOSFETs in traditional applications and opening new markets with its higher performance, smaller size, and lower cost. Early chapters presented the fundamentals of GaN technology and the challenges that must be overcome to be widely adopted into the design of power systems.

The first six chapters were followed by analyses of application examples where GaN devices have made inroads and have successfully been used to increase overall power system performance. This final chapter will review how GaN has emerged and the promising future for GaN's expanding adoption into power system designs.

16.1.1 What Controls the Rate of Adoption of a New Technology?

Many technologies have undertaken the challenge to displace existing technologies. In the world of semiconductors, the silicon power MOSFET journey, spanning more than 40 years, taught us that there are four key variables controlling the adoption rate of a disruptive power management technology [1].

1) Does it enable significant new capabilities?
2) Is it easy to use?
3) Is it VERY cost effective to the user?
4) Is it reliable?

GaN's journey replacing the silicon power MOSFET is well underway, having been initiated about 25 years ago. In this chapter, the growth and adoption of GaN power devices through the prism of these four fundamental criteria will be examined. In addition, the expansion of GaN technology for the design and manufacturing of integrated circuits to replace discrete power transistors will be explored.

16.2 New Capabilities Enabled by GaN Devices

The most significant new capabilities enabled by GaN transistors stem from their demonstrated disruptive improvement in power density, switching speed, and size. And this is only the beginning, since theoretically GaN technology performance curves – switching speed, efficiency, size, and cost – are merely at the beginning of an exponential improvement.

GaN Power Devices for Efficient Power Conversion, Fourth Edition. Alex Lidow, Michael de Rooij, John Glaser, Alejandro Pozo, Shengke Zhang, Marco Palma, David Reusch, and Johan Strydom.
© 2025 John Wiley & Sons Ltd. Published 2025 by John Wiley & Sons Ltd.

As discussed in Chapter 1, GaN transistors have a much higher critical electric field than silicon, which enables this new class of devices to withstand much greater drain-to-source voltage with much less penalty in on-resistance, capacitance, and size. This capability, coupled with higher electron mobility and innovative device packaging, has created a class of devices that are significantly smaller and faster than their silicon predecessors. This has inspired many new applications and capabilities that are changing our world – lidar, robotics, automotive, and satellites to name a few.

One leading-edge example of GaN transistors' capabilities is a system that synthesizes the transient currents and voltages experienced in artificial intelligence and high-speed supercomputing. Figure 16.1 shows an example of a load transient system from Picotest [2] where an array of GaN transistors is switching 2000 A with transition times of approximately one nanosecond, something impossible to achieve with silicon MOSFETs.

An early application to take advantage of the speed and high current capabilities of GaN devices was lidar. Lidar systems started using GaN transistors as early as 2011 and were quickly integrated into the development of autonomous vehicles such as the robotaxi shown in Figure 16.2. Robotaxis are now commonplace in several U.S. and Chinese cities, and lidar applications have spread beyond passenger cars to drones, robots, and sensors used for crowd and traffic control.

In Chapter 14, the impact of GaN devices on motor drives was demonstrated. Applications using GaN in motor drives include e-bikes and drones, where the benefits of high efficiency and small size are of great value. As humanoid and other untethered robots come onto the world stage, the advantages of GaN transistors and ICs in these demanding applications are extremely compelling, not just in the motors that drive the limbs and joints, but also as the primary sensor using lidar, giving them "vision" capability. Figure 16.3 shows the various motors in a typical humanoid robot.

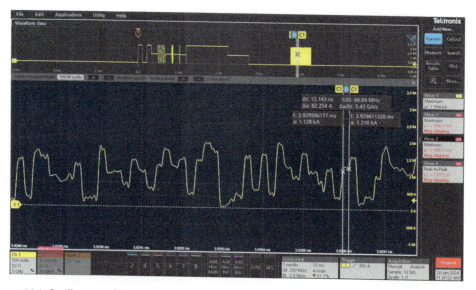

Figure 16.1 Oscillogram of a load transient system turning on and off an array of GaN transistors. Peak currents are 2000 A and transition times are approximately one nanosecond. (Image courtesy of Picotest).

Replacing Silicon Power MOSFETs | 451

Figure 16.2 Autonomous vehicles rely on lidar systems, utilizing GaN transistors, to create precise and rapid 3-D images of the surroundings (see Chapter 13 for a more detailed discussion of lidar).

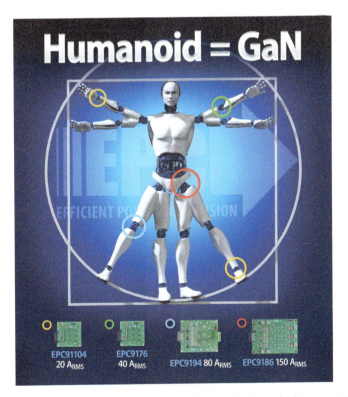

Figure 16.3 Humanoid robot with illustrations showing where the limbs and joints are powered by electric motors driven by GaN-based controllers (see Chapter 14 for a more detailed discussion of motor drives).

Figure 16.4 Size comparison between a GaN transistor (EPC2218) and a MOSFET (ISC027N10NM6). Both devices have approximately the same $R_{DS(on)}$, yet the GaN device is one-fourth the area.

Figure 16.4 shows the physical-size comparison between a GaN transistor and equivalent-rated MOSFET. An example of an application that puts a high value on small size, and therefore power density, is the artificial intelligence (AI) server. The cost of these server boards is very high, and they consume between three and eight kilowatts per board. Thus, there is a large motivation to use as little space as possible for power delivery systems leading to the distribution of power through a 48 V bus and have DC–DC converters that take up as little of the precious board space as possible.

An example of a DC–DC converter benefiting from the smaller size of GaN transistors is shown in Figure 16.5 [3]. Several of these types of 1 kW 48–12 V LLC converters are on each of the AI server boards. For a more detailed discussion of these types of DC–DC converters, see Chapter 10.

With GaN transistors and ICs gaining wider adoption, their role as the successor to the power MOSFET has become clear. Designers have been able to leverage the increased switching performance to improve power conversion system efficiency, size, and cost.

Examined in Chapters 7–11 were the advantages of GaN transistors in hard- and resonant-switching topologies from hundreds of kHz up to hundreds of MHz and even in RF amplifiers requiring the multi-GHz range. In Chapters 12–14, applications such as class D audio, lidar, and motor drives were described in detail and compared against state-of-the-art power

Figure 16.5 The EPC9159 is a 1 kW 48 V – 12 V DC–DC converter like the ones used in artificial intelligence servers. It measures 23 mm × 18 mm and has a power density of about 5100 W/in^3. The high-power density of the system stems from the small size and fast switching speed of the GaN devices [3].

MOSFETs in similar circuit topologies. Chapter 15 explored the exceptional capabilities of enhancement-mode GaN transistors and integrated circuits to withstand large amounts of radiation exposure.

In all these application examples, GaN technology shrank overall system size, enhanced system efficiency or performance, and increased power density. GaN technology does not only promise theoretically significant new capabilities, GaN-based devices are increasingly being demonstrated in actual, real-world applications. And, in many cases, GaN also reduced the direct cost of materials compared with earlier, lower-performance systems.

16.3 GaN Devices Are Easy to Use

How easy a device is to use depends on the skill of the user, the degree of difficulty of the circuit under development, how different the device is compared with devices within the experience of the user, and the tools available to help the user apply the device.

GaN transistor operating behavior is very similar compared to the ubiquitous power MOSFET and, therefore, designers can greatly leverage their past design experience. One key difference being the relatively higher frequency response, that is both a step function improvement over any prior silicon device, and becomes an added consideration for the user when designing a circuit.

As described in Chapters 3, 4, and 7, small amounts of stray parasitic inductance can cause increased power losses and a large voltage overshoot across the gate-to-source terminal that could potentially damage devices. The small size of GaN devices compared with silicon adds to the need for efficient heat extraction in many of the newest applications. Chapter 6 has tools needed to extract the heat efficiently from systems using GaN devices.

User-friendly tools can make understanding GaN device behavior easier. SPICE device models and interactive thermal models are widely available for download [4, 5]. Pre-assembled circuit kits are available from GaN device manufacturers, such as Efficient Power Conversion Corporation, Texas Instruments, Transphorm, Microchip, Analog Devices, Navitas, and Innoscience. These evaluation kits greatly simplify the first-step evaluation of the GaN devices, making it easy to directly compare the performance with traditional MOSFET circuits, if needed.

Furthermore, the GaN eco-system continues to grow with ICs specifically designed to drive GaN transistors and control GaN-based systems, making the designer's job easier by compressing designs and allowing the designer to focus more effort on previously regarded low-priority tasks such as reducing common source and power loop inductances. Finally, with the expanded user base of GaN, there are more and more experienced designers able to turn ideas into products quickly using the state-of-the-art GaN devices.

Like the increase in switching speed, the dramatic size reduction of GaN devices compared to MOSFETs can stretch the experience of manufacturers of power conversion systems. Fortunately, over time, manufacturing skills have improved to meet requirements initially set forth by cell phone manufacturers, automotive electronics manufacturers, satellite makers, and high-density computing equipment manufacturers. The size advantages and greater thermal efficiency of chip-scale packages are compelling, and the barriers to adoption have been reduced thanks to the overall trends toward higher-density electronics. GaN devices are also available in modern packages designed to minimize parasitic inductance as well as increased size.

The emergence of power transistors with integrated drivers, as well as fully monolithic half-bridge ICs with integrated level-shifting gate driver and protection circuits, further reduces design time and risk while lowering system losses, size, and cost, as shown in Chapters 1, 10, 13, 14, and 15. This makes it easier for designers to fully harvest the advantages of GaN technology with lower risk of product launch delays.

16.4 GaN Cost Reduction over Time

The first time GaN transistors cost less than a silicon power MOSFET with equivalent voltage and on-resistance was in 2015 [6]. Since then, GaN transistors have continued to shrink in size and increase in volume. In 2024, 100 V rated GaN devices are about five times smaller in die area than equivalent silicon MOSFETs. This translates into more devices per wafer manufactured.

Since the GaN manufacturing process can be performed side-by-side silicon MOSFETs, and generally have fewer manufacturing steps and equivalent manufacturing yields, the costs are comparable, except for the cost of growing the thin GaN/AlGaN heterostructure discussed in Chapter 1. However, with the advent of new-generation metal organic chemical vapor deposition (MOCVD) reactors, the cost of this unique step has also dramatically declined.

Currently, for devices with voltage ratings 80 V and higher, the cost of manufacturing, and the price to purchase are consistently lower than modern silicon MOSFETs. Expect the cost of GaN transistors to continue to decline as process technology evolves and the devices shrink further.

16.5 GaN Devices Are Reliable

The cumulative reliability information available for silicon power MOSFETs is staggering. Many years of work have been devoted to understanding failure mechanisms, controlling and refining processes, and designing products that have distinguished themselves as the highly reliable backbone of any power conversion system. GaN devices are now several years into this journey with a growing base of exemplary reliability data. Manufacturers have published results from their qualification tests [7–9], and devices have been applied successfully to many highly demanding applications, such as automotive [10], space [11], and power applications [12, 13], with good results. In addition, field information covering hundreds of millions of units in use for hundreds of billions of hours in real-world applications demonstrates that GaN technology is robust under a wide range of actual and accelerated life-test conditions.

In recent years there has been much progress in understanding the fundamental physical mechanisms governing the wear-out of GaN devices under a variety of conditions. These physics of failure studies have led to the ability to predict device lifetimes over a wide variety of actual mission profiles including lidar, motor drives, solar panel optimizers, and DC–DC converters [14]. Chapter 5, Table 5.1, gives a summary of the primary wear-out mechanisms for each of these applications. Further, Table 5.2 shows the general mitigation strategies for each of these mechanisms.

In all cases explored, GaN devices are proving to be more reliable than the silicon MOSFET operating under similar stress conditions. This is mostly due to the intrinsic material properties of GaN itself and provides support for further adoption of GaN devices.

16.6 Future Direction of GaN Devices

The GaN technology journey is still in its early years. There are profound improvements that can be made in basic device performance as measured by figures of merit, including $R_{DS(ON)}$ times Area, $R_{DS(ON)}$ times Q (input and output), and cost. Today's GaN transistor performance is still more than two orders of magnitude away from the theoretical limits, and it is therefore quite reasonable to expect the pace to continue with improvement in the key figures of merit as illustrated in Figure 16.6.

Perhaps the greatest opportunity for GaN technology to impact the performance and cost of power conversion systems comes from the intrinsic ability to integrate both power-level and signal-level devices on the same substrate. Chapter 1 showed a monolithic power stage including power devices, drivers, and level-shifting circuits on a single chip.

Monolithic lidar ICs, as well as ICs designed for motor drives, have also appeared on the market as shown in Figures 16.7 and 16.8. These integrated circuits are all precursors to what will become a full power systems-on-a-chip. ICs for applications including motor drives, buck converters, synchronous rectifiers, audio amplifiers, wireless power transmit amplifiers and

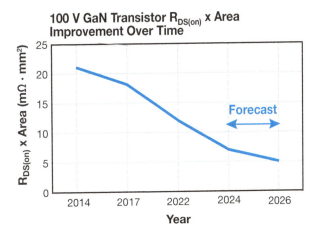

Figure 16.6 Projected improvement in $R_{DS(on)} \times$ Area figure of merit for 100 V GaN transistors over time. This figure of merit is a key indicator of cost and performance.

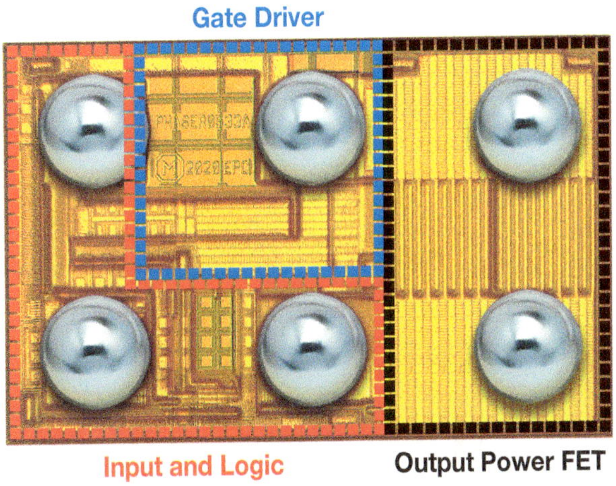

Figure 16.7 EPC21603 is a monolithic laser driver and control IC for lidar systems discussed in Chapter 13. Die dimensions are 1 mm × 1.5 mm.

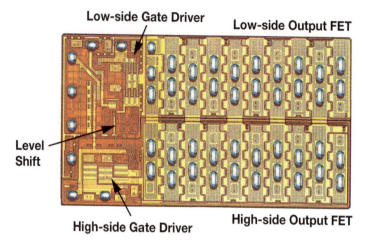

Figure 16.8 The EPC23102 [15] has integrated high-side and low-side GaN transistors with internal gate drivers and level shifter. It can handle 35 A at 1 MHz and has a maximum input voltage of 100 V. Die dimensions are 4.5 mm × 2.6 mm.

receivers, solar inverters, multilevel converters, power factor correction circuits, and a whole host of power systems, yet to be implemented, will benefit from the lower cost and smaller size of GaN-on-Si integration.

16.7 Summary

In the late 1970s, power MOSFET pioneers believed they had a technology that would displace bipolar transistors completely. Forty years later, plenty of applications remain that prefer bipolar transistors over power MOSFETs, but the size of the power MOSFET market is many times larger than the bipolar market. This is due to all the new applications and new markets enabled by that breakthrough technology. Today, GaN technology is at that same precipice. Like the early-power MOSFET, from the mid-1970s to 1980s, GaN manufacturers are at the early stages of an exciting journey with new products and breakthrough capabilities emerging almost monthly.

The power MOSFET is not dead, but is nearing the end of the road for major improvements in performance and cost. As more and more GaN transistor-based designs come to market and GaN-on-silicon manufacturing volumes grow, and integration leads to full power systems-on-a-chip, this newer technology will most probably become dominant due to its expanding advantages in integration, performance, size, and cost.

References

1 Lidow, A. (2010). Is it the end of the road for silicon in power management? *CIPS 2010 Conference*, Nuremburg, Germany (March 2010).
2 Sandler, S., Dannan, B., Barnes, H., et al. (2024). Design, simulation, and validation challenges of a scalable 2000 amp core power rail. *DesignCon 2024*, Santa Clara, California (January 2024).

3 Efficient Power Conversion Corporation (2023). EPC9159: 1 kW, 48V/12V LLC Converter, EPC9159 quick start guide, 2023. https://epc-co.com/epc/products/evaluation-boards/epc9159.

4 Efficient Power Conversion Corporation. eGaN FET SPICE Models. http://epc-co.com/epc/DesignSupportbr/DeviceModels.aspx.

5 Infineon (2022). Reliability and qualification of high-voltage CoolGaNTM GIT HEMTs, White paper. www.infineon.com/dgdl/Infineon-2_WhitePaper_Reliability_and_qualification_of_CoolGaN_EN-Whitepaper-v01_00-EN.pdf?fileId=5546d46266a498f50166c9bebef322aa&da=t.

6 Schiro, A. Application-specific reliability methods drive successful Navitas GaNFastTM IC field results in high-volume production, White paper. https://navitassemi.com/wp-content/uploads/2023/02/Application-Specific-Reliability-Methods-Drive-Successful-Navitas-GaNFast%E2%84%A2-IC-Field-Results-in-High-Volume-Production.pdf.

7 Marcon, D. (2024). Debunking GaN cost and reliability myths. *EEPower*, 17 January 2024. https://eepower.com/industry-articles/gan-power-proven-reliability-competitively-priced/#.

8 Ohnsman, A. (2016). Velodyne unveils low-cost LiDAR in race for robo-car vision leadership. *Forbes*, 13 December 2016. https://www.forbes.com/sites/alanohnsman/2016/12/13/velodyne-unveils-lower-cost-lidar-in-race-for-robo-car-vision-leadship/?sh=339904e474c4.

9 Whitock, P. (2018). Will GaN and the Tesla SpaceX car survive space radiation? Yes and no. *ElectroPages*, February 2018. https://www.electropages.com/2018/02/will-gan-and-the-tesla-spacex-car-survive-space-radiation/

10 Takahashi, D. (2018). Gaming laptops will have smaller power supplies with EPC's gallium nitride chips. *Venture Beat*, 27 February 2018. https://venturebeat.com/2018/02/27/gaming-laptops-will-have-smaller-power-supplies-with-epcs-gallium-nitride-chips/.

11 Sverdlik, Y. (2017). GaN is eyeing silicon's data center lunch. *Data Center Knowledge*, 9 February 2017. https://www.datacenterknowledge.com/archives/2017/02/09/gan-is-eyeing-silicons-data-center-lunch

12 Gajare, S., Li, D., Garcia, R. et al. (2024). GaN reliability and lifetime projections: phase 16. EPC El Segundo, CA (2024). https://epc-co.com/epc/design-support/gan-fet-reliability/reliability-report-phase-16.

13 Witham, J. (2022). "Beating the heat: How GaN keeps the data center cool." Data Center Dynamics, July 2022.

14 Zhang, S., Gajare, S., and Garcia, R. (2023). "Using Test-to-Fail Methodology to Predict How GaN Devices Can Last More than 25 Years in Solar Applications," *PCIM Asia 2023; International Exhibition and Conference for Power Electronics, Intelligent Motion, Renewable Energy and Energy Management*, Shanghai, China.

15 Efficient Power Conversion Corporation (2024). EPC23102 – ePower StageTM IC. EPC23102 datasheet. https://epc-co.com/epc/documents/datasheets/EPC23102_datasheet.pdf.

Appendix

Glossary of Terms
4th Edition

Term	Symbol	Units	Definition	Chapter
abc frame			The reference frame based on the three electrical phases of the stator of a motor. This frame is not Cartesian; there are three directions for a plane. This reference frame is static with the stator	14
bemf constant	k_e	volts/krpm	The bemf constant defines the voltage back a motor generates when spinning at a given speed. It is directly proportional to the torque constant	14
Back electromotive force – bemf	e_u	volts	It is a voltage that opposes the change in current that induced it, as Lenz's law describes. In a motor, the changing magnetic field, or change in current, is caused by a relative movement of the magnetic poles with respect to its stator winding coils. The bemf is a voltage that opposes the applied inverter's voltage, and it is proportional to the motor speed	14
Bandgap energy	E_G	eV	A bandgap, also called "an energy gap" or "bandgap energy," is an energy range in a solid where no electron state can exist. The bandgap is a major factor determining the electrical conductivity of a solid. Substances with large bandgaps are generally insulators; those with smaller bandgaps are semiconductors, while conductors either have very small bandgaps or none, because the valence and conduction bands overlap	1

(*Continued*)

GaN Power Devices for Efficient Power Conversion, Fourth Edition. Alex Lidow, Michael de Rooij, John Glaser, Alejandro Pozo, Shengke Zhang, Marco Palma, David Reusch, and Johan Strydom.
© 2025 John Wiley & Sons Ltd. Published 2025 by John Wiley & Sons Ltd.

Term	Symbol	Units	Definition	Chapter
Bootstrap circuit			The bootstrap circuit is used to charge the floating capacitor that supplies the charge to a circuit that is not referenced to the ground reference voltage. It is usually made of a resistor, a diode, and a floating capacitor	11
Breakdown voltage	BV_{DSS}	volts	The maximum voltage that can be applied in steady state from drain-to-source terminals in a GaN transistor or MOSFET	2, 5, 13
Brushless DC motor	BLDC		The permanent magnet synchronous motor (PMSM), also known as DC brushless motor (BLDC), is widely used and offers higher torque capability per cubic inch and higher dynamics when compared to induction motors and DC brushed motors	14
Capacitance (drain–source)	C_{DS}	farads	The capacitance between the drain terminal and the source terminal	2, 3, 4
Capacitance (gate–drain)	C_{GD}	farads	The capacitance between the gate terminal and the drain terminal	2, 3, 4
Capacitance (reverse transfer)	C_{RSS}	farads	The capacitance between the gate terminal and the drain terminal	2, 3, 4
Capacitance (gate–source)	C_{GS}	farads	The capacitance between the gate terminal and the source terminal	2, 3, 4, 7, 8, 10
Capacitance (input)	C_{ISS}	farads	The input capacitance of the device is the sum of the C_{GD} and C_{GS}	2, 3, 4, 7, 8, 10
Capacitance (output)	C_{OSS}	farads	The output capacitance of the device is the sum of the C_{GD} and C_{Ds}	2, 3, 4, 7, 8, 9, 10
Carrier mobility	μ	cm^2/V·s	The electron mobility characterizes how quickly an electron can move through a metal or semiconductor, when pulled by an electric field	1, 2, 7, 10
Cascaded full-bridge converter	CHB		The cascaded full-bridge converter is a topology made of cascaded full bridges (also known as H-bridges). As the MMC, it requires separated and isolated DC voltage sources	11

Term	Symbol	Units	Definition	Chapter
Charge stored in flying capacitor	Q_{CFLY}	Coulomb	The charge stored in the flying capacitors in the intermediate levels	11
Chip-scale package	CSP		A type of packaging technology where the solder bumps are directly mounted on the die	5
Clarke transforms and Park transforms			In the context of electrical engineering of three-phase circuits, the Clarke transforms and Park transforms and their inverses are used to rotate the reference frames of AC waveforms such that they become DC signals	14
Coefficient of thermal expansion	CTE	ppm/°C	Material's expansion and compression during temperature variation	5
Common-mode transient immunity	CMTI	V/ns	Maximum tolerable slew rate (dv/dt) of the common-mode voltage applied between two isolated circuits	3, 4
Common-source inductance	CSI	henrys	Inductance shared by the drain-to-source power current path and gate driver	3, 4, 7, 10, 13, 14
Converter efficiency	P_{OUT}/P_{IN}		The ratio between the output and the input power in a converter. It is a number expressed in percentage. The closer to 100%, the better	11
Converter inductance	L	henrys	The inductance used to store energy in the form of magnetic energy in buck, boost, and PFC converters	11
Converter input power	P_{IN}	watts	The converter input power	11
Converter input voltage	V_{IN}	volts	Input voltage applied to a converter	11
Converter output current	i_{load}	amperes	The current supplied to the load by a converter	11
Converter output power	P_{OUT}	watts	The converter output power	11
Converter output voltage	V_{OUT}	volts	The converter's output voltage	11
Crest factor	CF	dB	The ratio, expressed in dB, between the peak voltage/current and the RMS voltage/current.	12

(*Continued*)

Term	Symbol	Units	Definition	Chapter
Critical electric field	E_{crit}	MV/cm	When the electric field gradient exceeds the critical electric field, valence bonds between atoms are ruptured and current flows	1, 2, 16
Cubic power density	W/in^3	watts/inch^3	An indicative factor that represents the amount of power processing capability of a converter. In general, it is the maximum power that a converter can process divided by its volume	11
Dead time	dt	seconds	The time between the conduction cycles of the low-side and high-side transistor in a half bridge. Dead time prevents the simultaneous conduction of both high-side and low-side transistors. This conduction is called shoot-through	14
Device junction temperature	T_J	°C	The temperature of the device junction during operation	2, 5
Diode reverse recovery charge	Q_{RR}	coulombs	The amount of charge that needs to be pulled out of the body diode of a FET in order to turn the device OFF. Enhancement-mode GaN transistors with pGaN gates have zero Q_{RR}	2, 7, 10, 11, 14
Direct current	I_d	amperes	Component of the current that generates a flux that is aligned to the magnet direction. The I_d current, also known as flux component, when different from zero, can increase or decrease the flux of the magnet	14
Direct time of flight	DTOF		Used to describe a TOF lidar where the time between the transmitted and received optical pulses is directly measured	13
dq frame			The Cartesian reference frame static with the rotor. The d direction is aligned with the magnet poles. The q direction is orthogonal to the magnet	14
Drain-to-source leakage current	I_{DSS}	amperes	The current that flows in a GaN transistor or MOSFET from source to drain when a positive voltage is applied from drain to source.	2, 5
Drain current at the quiescent operating point	I_{DQ}	amperes	Drain current in the transistor at the quiescent operating point in a linear amplifier	9

Term	Symbol	Units	Definition	Chapter
Drain efficiency	η_D	percentage	Ratio of P_{RFOUT}/P_{DC}	9
Duty cycle	D		The duty cycle is the fraction of one PWM period in which the power cell complementary high-side transistor is active. The duty cycle is commonly expressed as a percentage or a ratio	11
Edge-emitting laser	EEL		Semiconductor diode laser that emits light from the edge, parallel to the wafer plane	13
Effective dead time	t_{eff}	seconds	The time from when the gate voltage of the device reaches the turn-off plateau voltage to when the other device's load current commutates from the diode	3, 7, 8, 10, 16
eGaN®FET	eGaN FET		Trademarked symbol for enhancement-mode gallium nitride on silicon FET	1, 5, 7, 9, 10, 13, 15, 16
Electromigration	EM		A failure mode that is caused by the movement of atoms and void formation	5
Electromagnetic interference	EMI		EMI is a disturbance generated by an external source that affects an electrical circuit by electromagnetic induction, electrostatic coupling, or conduction	11
External vertical power loop			A layout of a half-bridge circuit that utilizes the opposite layer of the PCB as the power loop return path	14
Failure in time	FIT		Number of failures in 10^9 (1 billion) device hours	5
Field-oriented control	FOC		An algorithm to control the BLDC motor that maximizes the efficiency of the motor and inverter by applying a rotating stator field to obtain the maximum torque from the magnetic rotor	14
Figure of merit	FOM	$m\Omega \cdot nC$	The FOM is a way to compare different device technologies. The most common FOM is calculated by multiplying the $R_{DS(on)}$ or a given device times the Q_G	1, 3, 4, 7, 8, 9, 10, 11, 13, 16

(Continued)

Term	Symbol	Units	Definition	Chapter
Figure of merit for multilevel flying capacitors	FOM_{ML}		The multilevel flying capacitors FOM is a way to compare different numbers of levels and device technologies within the flying capacitor topology	11
Figure of Merit – hard switching	FOM_{HS}	$m\Omega \cdot nC$	The hard-switching FOM is a way to compare different technologies in hard-switching applications such as buck converters. $FOM_{HS} = (Q_{GD} + Q_{GS2}) \cdot R_{DS(on)}$	7, 10
Figure of Merit – soft switching	FOM_{ss}	$m\Omega \cdot nC$	The soft-switching FOM is a way to compare different technologies in soft-switching and resonant applications. $FOM_{SS} = (Q_{GD} + Q_{G}) \cdot R_{DS(on)}$	8, 10
Flying capacitance	C_{FLY}	farads	In FCML converters it is the capacitance of the flying capacitors in intermediate levels	11
Flying capacitor multilevel converter	FCML		The flying capacitor multilevel is a topology made of stacked transistors and interposed capacitors. A N-levels FCML allows to generate a voltage output waveform made of N steps.	11
Frequency-modulated continuous wave	FMCW		Used to describe a form of lidar that modulates the optical frequency	13
Gallium Nitride	GaN		Wide bandgap semiconductor material used in light-emitting diodes (LED), as well as RF and power transistors and integrated circuits.	1, 11, 13
Gate charge required to increase gate voltage to threshold voltage	Q_{GS1}	coulombs	Charge required to increase gate voltage from zero to the stated threshold voltage of the device	3, 4, 7
Gate charge for the current transition interval	Q_{GS2}	coulombs	Charge required to increase gate voltage from the stated threshold voltage of the device to the plateau voltage (current conduction interval)	3, 4, 7
Gate charge from zero to the onset of the plateau voltage	Q_{GS}	coulombs	Charge required to increase gate voltage to the plateau voltage	3, 4, 7
Gate charge (total)	Q_{G}	coulombs	Total gate charge required to drive a device from zero to rated gate voltage (fully enhanced)	2, 3, 4, 7, 8, 10, 11, 13

Term	Symbol	Units	Definition	Chapter
Gate–drain charge	Q_{GD}	coulombs	The charge transferred between the gate terminal and the drain terminal when the drain voltage changes	2, 3, 4
Gate driver output voltage	V_{DR}	volts	On-state output voltage of gate driver	3, 4, 7, 8
Gate plateau voltage	V_{pl}	volts	Gate voltage at which the drain–source voltage transition occurs during a hard-switching event	3, 4, 7
Gate threshold voltage	V_{th}	volts	Gate voltage that needs to be applied to initiate drain–source conduction	2, 3, 4, 5, 7
Half-maximum full-width	HMFW	ns	Pulse-width metric where starting and ending thresholds are half the maximum value of the pulse	13
High electron mobility transistor	HEMT		High electron mobility transistor (HEMT), also known as heterostructure FET (HFET) or modulation-doped FET (MODFET), is a field-effect transistor incorporating a junction between two materials with different bandgaps (i.e. a heterojunction) as the channel instead of a doped region, as is generally the case for MOSFET	1, 2, 3, 4, 5, 7, 9, 10, 15
High-Temperature Gate Bias	HTGB		A reliability test in which the devices are biased at positive gate bias and high temperature (150 °C)	5
High-Temperature Reverse Bias	HTRB		A reliability test in which the devices are biased at positive drain bias and high temperature (150 °C)	5
Indirect time of flight	ITOF		Used to describe a TOF lidar where the time between the transmitted and received optical pulses is determined by measuring the phase shift of the envelope modulation of a received optical signal with respect to that of the transmitted signal and computing the time difference	13
Inductor current ripple	Δi_{pp}	amperes	It is the peak-to-peak current ripple in the inductor in those converters that have an inductor as the energy storage device	11, 14
Input voltage ripple	Δv_{pp}	volts	Peak-to-peak voltage ripple at the input of the motor inverter	14

(Continued)

Term	Symbol	Units	Definition	Chapter
Intermittent Operating Life	IOL		The devices are subjected to intermittent temperature variation due to device self-heating	5
Internal gate resistance	R_G	ohms	The resistance inside the transistor gate that limits how fast charge can be pulled out of, or pushed into, the gate electrode	3, 4, 8, 13
Internal vertical power loop			A layout of a half-bridge circuit that utilizes the first inner layer of the PCB as the power loop return path	14
Insulated Metal Substrate board	IMS		A single (or multi) layer circuit made of copper layers interposed with insulating layers with high thermal conductivity	14
Inverter efficiency	$\eta_{inverter}$		The efficiency of an inverter, defined as the ratio between the output to the input power at the inverter terminals	14
Joule heating	P_{JH}	watts	The heat produced when current passes through a conductor.	6
Junction temperature	T_J	°C	The maximum temperature of the active region of the device during operation	6
Land grid array package	LGA		The land grid array (LGA) is a type of surface-mount packaging for integrated circuits (ICs) and eGaN FETs. An LGA can be electrically connected to a printed circuit board (PCB) by soldering directly to the board	1, 2, 3, 4, 5, 6, 7, 8, 10, 13, 16
Lidar			Light detection and ranging sensor is used in many autonomous vehicle and mapping functions.	5, 13
Linear energy transfer	LET	Me V · cm²/ mg	Linear energy transfer (LET) is a measure of the energy transferred to material as an ionizing particle travels through it. Typically, this measure is used to quantify the effects of ionizing radiation on electronic devices	15
Linear feet per minute	LFM	LFM	Unit to measure air velocity	6, 10, 11, 14
Mean Time to Fail	MTTF		The population reaches 50% failure rate	5

Term	Symbol	Units	Definition	Chapter
Metal oxide chemical vapor deposition	MOCVD		MOCVD is a chemical vapor deposition method of epitaxial growth of materials, especially compound semiconductors from the surface reaction of organic compounds or metalorganics and metal hydrides containing the required chemical elements	1, 15
Miller ratio	Q_{GD}/Q_{GS}		The Miller ratio is a gauge of how much the gate terminal may be susceptible to false turn-on when a high dv/dv is applied at the drain terminal	2, 3, 4, 10, 16
Modular multilevel converter	MMC		The modular multilevel converter is a topology made of cascaded half bridges. It requires separated and isolated DC voltage sources	11
Motor electrical pole	τ_e	seconds	The electrical pole of a motor is determined by the phase inductance L and the phase resistance R. The electrical pole is used to tune the current controllers in the FOC algorithm	14
Motor efficiency	η_{motor}		The efficiency of a motor, defined as the ratio between the mechanical output power at the shaft to the electrical input power at the motor terminals	14
Motor impedance	Z	ohms	The phase-to-phase impedance vs. frequency of a motor with locked rotor. Measured with an LCR meter	14
Motor mechanical pole	τ_m	seconds	The mechanical pole of a motor is determined by the inertia J and the viscous friction β. The mechanical pole is used to tune the speed controller in the FOC algorithm	14
Motor mechanical speed	ω	rpm	The mechanical speed of the motor shaft. Multiplied by the k_e bemf constant, it gives the bemf back generated voltage that opposes to the applied inverter's voltage	14
Motor Power Loss Factor	LF	watts/ volts^2	The power loss factor is obtained by normalizing the power loss P_h by the applied voltage V_h	14

(*Continued*)

Term	Symbol	Units	Definition	Chapter
Motor shaft moment of inertia	J	kg m^2	The moment of inertia of the shaft of the motor	14
Motor shaft viscous friction constant	β	N s/m^2	The viscous friction factor at the shaft of a motor. The resisting torque is proportional to the speed and to the viscous friction constant	14
Multilevel converter			A switching converter that uses multiple stacked transistors that breaks the supply voltage into multiple levels for the output.	11
Multilevel effective duty cycle	D_{eff}		In case of FCML converters, it is the effective duty applied to the load due to the status of two adjacent switching cells	11
On-resistance	$R_{DS(on)}$	ohms	When the eGaN FET is in the on-state, it exhibits a resistive behavior between the drain and source terminals. This resistance is called $R_{DS(on)}$ for "drain-to-source resistance in on-state" and is the sum of many elementary contributions	1, 2, 3, 4, 5, 8, 9, 10, 11, 15, 16
Output Charge	Q_{oss}	coulombs	The charge required to be supplied to the drain terminal to achieve a certain voltage on the drain relative to the source	3, 4, 7, 8, 10, 11, 15
PCB FR4 dielectric	FR4		FR4 is a type of printed circuit board base material made from a flame-retardant epoxy resin and glass fabric composite. Dielectric layers in FR4 have a poor thermal conductivity	14
Permanent magnet synchronous motor	PMSM		See Brushless DC motor	14
Phase-shifted PWM	PSPWM		In FCML converters it is a technique to generate complementary PWM command signals for N levels. The original computed two-level signal is shifted by 360 / (N−1) electrical degrees per each subsequent level	11
Phase shift time	Δt_{PS}	seconds	In FCML converters using PSPWM modulation, it is the displacement in time between one command signal and the adjacent cell command signal. It is equal to the PWM period divided by (N−1)	11

Term	Symbol	Units	Definition	Chapter
PI controller	PI		Proportional integral controller. Widely used in motor control. The FOC algorithm transforms AC values into DC values so that simple PI controller can be used	14
Poles	p		Number of poles of a motor. It can be only an even number. If $p = 2$, the modulating frequency equals the mechanical speed. Otherwise, the electrical frequency is a higher multiple of the mechanical speed	14
Pole pairs	pp		The number of poles (p) divided by 2	14
Power (DC power to the RF transistor)	P_{DC}	watts	DC power delivered to the RF transistor	9
Power Factor Correction	PFC		A PFC circuit is used to correct the power factor of a load circuit connected to an AC source so that the load circuit appears as a resistive load to the AC source.	11
Power loop inductance	L_{Loop}	henrys	Total parasitic in the power loop. In a half bridge the power loop is formed by the bypass capacitors, high-side and low-side power FETs and connections between them	3, 4, 6, 14
Power losses (turn-off)	P_{off}	watts	Power losses due to the turn-off switching transition	7
Power losses (turn-on)	P_{on}	watts	Power losses due to the turn-on switching transition	7
Power loss at the hth harmonic	P_h	watts	The power loss in a motor when a sinusoid at frequency f_h and voltage V_h is applied to the motor with locked rotor	14
Power (output RF)	P_{RFout}	watts or dBm	RF power of the amplifier	9
Power Quad Flat No-lead	PQFN		The Power Quad Flat No-lead is a type of surface-mount packaging for electronic devices. In it the device is mounted on a lead frame and encapsulated in mold compound. On the bottom surface of the package, parts of the lead frame are exposed for soldering to a PCB	4, 5, 6, 10

(Continued)

Term	Symbol	Units	Definition	Chapter
Pulse repetition frequency	PRF	Hz	Pulse frequency or rate of pulse repetition, frequently used in the context of lidar or radar	13
Pulse-width modulation	PWM		Type of modulation where a low-frequency signal (the modulating signal) is compared to a high-frequency triangular signal (the carrier) to produce a square wave with varying duty cycle	11, 14
PWM frequency	f_{PWM}	hertz	In the pulse-width modulation technique, it is the frequency of the carrier signal that is modulated by a lower-frequency signal. It is the inverse of the PWM period	11, 14
PWM period	T_{PWM}	seconds	In the pulse-width modulation technique, it is the period of the carrier signal that is modulated by a lower-frequency signal. It is the inverse of the PWM frequency	11
Quadrature and direct inductances	L_q, L_d		The inductance measured across the magnetic circuit in the q and the d directions, respectively. In isotropic motors, $L_q = L_d$	14
Quadrature current	I_q	amperes	Component of the current that generates a flux that is 90° to the magnet direction. The I_q current, also known as torque component, when I_d is zero, generates the motor torque via the k_t constant	14
Reflection coefficient (input port)	s_{11}	percentage	The percentage of the incident wave that is reflected back from the input port	9
Reflection coefficient (input power forward gain)	s_{21}	percentage	The percentage of the input port incident wave that is reflected to the output port	9
Reflection coefficient (output port reverse gain)	s_{12}	percentage	The percentage of the output port incident wave that is reflected to the input port	9
Reflection coefficient (output port)	s_{22}	percentage	The percentage of the incident wave that is reflected back from the output port	9
Reverse conduction			When current flows into the source terminal of a transistor. See also Diode Reverse Recovery	1, 2, 3. 7, 11, 12, 14

Term	Symbol	Units	Definition	Chapter
Rise and fall times	t_{rf}	seconds	Indicates the rise and fall times of the switching node	14
Rollett stability factor	K		Rollett stability factor is a test for unconditional stability	9
Safe Operating Area	SOA		A test condition where high current and high drain–source voltage are applied to the device simultaneously	5
Signal-to-noise ratio	SNR	dB	The ratio of the total signal power to the total noise power in a signal processing system	13
Silicon Carbide	SiC		Wide bandgap semiconductor made of silicon and carbon compound	1, 11
Single-event burnout	SEB		Single-event burnout (SEB) is typically a latch-up event that is possible with power eGaN FETs. It may be triggered by the passage of a heavy ion. If unmitigated, SEB can be destructive. SEB can be prevented by limiting drain current or switching on the eGaN FET when SEB is detected	15
Single-event effects	SEE		Single-event effects (SEE) occur when a high-energy particle travels through a semiconductor; it leaves an ionized track behind. This ionization may cause a highly localized effect similar to the transient dose one – a benign glitch in output, a less benign bit flip in memory or a register, or, especially in high-power transistors, a destructive latch up and burnout. Single-event effects have importance for electronics in satellites, aircraft, and other civilian and military aerospace applications	15
Single-event gate rupture	SEGR		Single-event gate rupture (SEGR) is a destructive event that can occur when an energetic ion passes through the gate of a power transistor. A high field is created across the oxide causing permanent failure. SEGR cannot be mitigated using circuit-level approaches	15
Sinusoidal modulation			A type of PWM modulation technique that applies a sinusoidal modulating voltage	14

(Continued)

Term	Symbol	Units	Definition	Chapter
Smith chart			The Smith chart is a graphical aid designed for electrical engineers specializing in radio frequency (RF) engineering to assist in solving problems with transmission lines and matching circuits. The Smith chart can be used to represent many parameters including impedances, admittances, and reflection coefficients	9
Switching energy	E_{ON}, E_{OFF}, E_{rec}	joules	The energy dissipated during the change of state of a switching cell. E_{rec} is present only when there are bipolar diodes in the switching cell	11
Switching frequency	f_{sw}	hertz	Operating frequency of the transistor	7, 8, 10
Switching-node current variation rate	di/dt	amperes/ second	The rate at which the path of the current changes in a converter-switching cell when it changes state	11
Switching-node voltage variation rate	dv/dt	volts/ second	The rate at which the switching node in a converter changes state from one level to the next	11
Switching power cell			The basic topology within a converter, made of two complementary transistors	11
Target voltage ripple percentage	α_{ripple}		In FCML converters, the desired maximum voltage ripple of the flying capacitors in all operating conditions	11
Temperature Cycling	TC		An environmental reliability testing where the temperature swings from T_{Max} and T_{Min} and vice versa	5
Thermal impedance-junction-to-board	$Z_{\theta JB}$	°C/W	Thermal impedance from the device junction to the bottom of the solder bumps, without consideration of the type or size of the mounting circuit board. The impedance is typically used when the device is not in steady state	2
Thermal resistance	R_θ	°C/W	The opposition to heat flow within a material or system	6
Thermal resistance junction-to-ambient	$R_{\theta JA}$ or R_{THJA}	°C/W	Thermal resistance junction-to-ambient is measured with the DUT. It is mounted onto a single-sided 2-ounce FR-4 circuit board whose area is 1 in.^{2} (645.16 mm^2)	2, 6

Term	Symbol	Units	Definition	Chapter
Thermal resistance junction-to-board	$R_{\theta JB}$ or R_{THJA}	°C/W	Thermal resistance from the device junction to the bottom of the solder bumps, without consideration of the type or size of the mounting circuit board	2, 6
Thermal resistance junction-to-case	$R_{\theta JC}$ or R_{THJC}	°C/W	Thermal resistance from the device junction to the silicon backside of the transistor	2, 6
Thermal resistance (effective)	$R_{\theta(Effective)}$	°C/W	$R_{\theta(Effective)}$ is the effective thermal resistance when all resistance paths are added together	2, 6
Threshold voltage	$V_{GS(TH)}$ or V_{TH}	volts	In a GaN transistor the threshold voltage is the voltage applied between gate and source that enhances the 2DEG under the gate enough to begin conducting current from drain to source	1, 2, 3, 4, 7, 15, 16
Time of flight	TOF		Used to describe a form of lidar that measures distance to one or more targets by measuring the time it takes for light to travel a distance.	13
Total Harmonic Distortion	THD		The total harmonic distortion of a signal is defined as the ratio of the sum of the powers of all harmonic components to the power of the signal's fundamental frequency	11, 12
Total Incident Dose	TID	Rad(si)	The NASA guideline for a "Rad Hard" device is the device with the ability to withstand a total incident dose (TID) between 200 kRad(Si) and 1 MRad(Si)	15
Torque	T	Nm	The torque generated at the motor shaft as $k_t \times I_q$ (when $I_d = 0$)	14
Torque constant	k_t	Nm/A_{RMS}	The torque constant defines the torque generated at the motor's shaft at a given phase current in conditions of optimum control (see FOC)	14
Totem Pole Boost PFC converter			A topology of a PFC circuit made of one low-frequency switching leg and a high-frequency switching leg	11
Transient thermal impedance	$Z_{\theta JB}$ or $Z_{\theta JC}$	K/W	Thermal impedance from the device junction to the bottom of the solder bumps or back of the die, without consideration of the type or size of the mounting circuit board. The impedance is used when the device is not in steady state	6

(Continued)

Term	Symbol	Units	Definition	Chapter
Transconductance	g_m	siemens	Transconductance is the ratio of the current change at the output port to the voltage change at the input port	3, 4, 7
Trapezoidal modulation			A type of PWM modulation technique, also called six steps, that applies a trapezoidal modulating voltage	14
Turn-off current fall time	t_{CF}	seconds	Hard-switching fall time for current in the transistor	3, 4, 7
Turn-on voltage fall time	t_{VF}	seconds	Hard-switching fall time for voltage across the transistor	3, 4, 7
Turn-on current rise time	t_{CR}	seconds	Hard-switching rise time for current in the transistor	3, 4, 7
Turn-off voltage rise time	t_{VR}	seconds	Hard-switching rise time for voltage across the transistor	3, 4, 7
Two-dimensional electron gas	2DEG		A two-dimensional electron gas (2DEG) is a gas of electrons free to move in two dimensions, but tightly confined in the third. This tight confinement leads to quantized energy levels for motion in that direction, which can then be ignored for most problems. Thus, the electrons appear to be a 2D sheet embedded in a 3D world	1, 2, 5, 15
Under-Voltage Lockout	UVLO	volts	The under-voltage lockout is a circuit that turns off the gate driver when the supply voltage falls below a minimum value to avoid excessive power consumption of the power transistor due to low V_{GS} voltage applied to the gate	11
Unmanned aerial vehicles	UAV		Type of aerial vehicles which are autonomous and do not require a human aboard	14
Vertical cavity surface-emitting laser	VCSEL		Semiconductor diode laser that emits light from the edge, perpendicular to the wafer plane	13
Voltage source inverter	VSI		Type of inverter that imposes a defined voltage waveform to the load (e.g. a motor). It requires a unidirectional voltage blocking – bidirectional conducting switch	14

Term	Symbol	Units	Definition	Chapter
Wavelength	λ	nm	The optical wavelength of light in a system, e.g. a lidar	13
Zero-voltage switching time	t_{zvs}	seconds	The time required to discharge a transistors output capacitance to achieve a zero-voltage switching transition	8
Zero-Current Switching	ZCS		Switching event that occurs with zero current flowing in the transistor power terminals	3, 4, 7, 8, 10
Zero-Voltage Switching	ZVS		Switching event that occurs with zero voltage across in the transistor power terminals	3, 4, 7, 8, 10, 14

Index

a

Adoption rate   449
Aluminum gallium nitride (AlGaN)   2, 5–8, 10–13, 27, 31, 34, 434–435, 454
Aluminum nitride (AlN)   12–13, 435
Artificial intelligence   450, 452
ASIC controller   295, 299–301
Autonomous navigation   367

b

Bandwidth   273, 292, 353, 358, 384, 388, 395–396, 410–411
Body diode   38–39, 54–56, 197, 214, 216–217, 263, 337, 413, 435, 439, 462
Bootstrap   55–60, 66–69, 84, 240, 243, 332–338, 340, 346, 422, 441, 443, 460
Bottom-side cooling   156, 175–176
Breakdown voltage (BV$_{DSS}$)   1, 3–4, 9, 19, 25, 27–30, 32, 37, 65, 172, 174, 322, 379, 460
Brushless motors   187, 403–404, 460
Buck converter   18–19, 139–142, 181, 187, 196, 203–204, 206, 208, 211–213, 215–216, 224–226, 231, 237, 240–242, 244, 253, 262, 295, 297–298, 300–303, 306–307, 329–330, 334, 336, 338–339, 341, 351, 353, 416, 455
Bus converter   139, 249, 261–264, 266–270, 308

c

Capacitance   25, 35–39, 42, 47, 51, 53, 59–62, 65, 67, 70, 75, 95, 114, 139, 174, 196–197, 199, 201–205, 207, 209–214, 216–217, 220, 222, 225–226, 229, 236, 238, 240–245, 249–251, 254–258, 262–263, 273, 292, 312, 314, 324, 327–329, 344, 353, 361, 382, 384, 386, 395, 398–399, 406, 412–413, 415–416, 420, 450, 460, 464, 475
Capacitive discharge   378–381, 396, 398–399
Cascaded bootstrap   332–336, 338
Cascode transistor   9
Charge trapping   434, 445
Chip-scale package (CSP)   78, 96, 122, 133, 142, 144, 155, 157, 353, 376, 453, 461
Class A amplifier   275, 278–279, 351
Class D amplifier   12, 348, 351–353, 355–364, 452
Closed-loop amplifier   162, 355, 396
Common mode rejection ratio (CMRR)   60–61, 296, 324, 331, 394, 410
Common mode transient immunity (CMTI)   61, 461
Common-source inductance (CSI)   53–54, 59, 62, 75, 77, 80, 86–87, 90, 93, 217–221, 237–238, 296, 306, 353, 386, 413, 421, 429, 461
Conduction losses   31, 81, 139, 196–197, 205, 223–224, 231, 233, 236, 239, 257, 266–267, 308, 314, 351, 361, 412
Cost   1–3, 11–12, 15, 22, 134, 139, 144, 153, 161, 164–165, 167–168, 172, 174, 176–177, 186, 191, 196, 224, 262, 273, 292, 324, 354, 367, 369, 371, 373, 375–378, 382, 391, 396, 399, 413, 418, 449, 452–456
C_{RSS} (reverse transfer capacitance)   37–38, 199, 229–230, 255, 413, 460

d

Dead time   10, 52, 56–58, 67–69, 197, 206–210, 214–216, 225–226, 231–233, 235, 240, 266–267, 303, 310, 314, 333,

---

GaN Power Devices for Efficient Power Conversion, Fourth Edition. Alex Lidow, Michael de Rooij, John Glaser, Alejandro Pozo, Shengke Zhang, Marco Palma, David Reusch, and Johan Strydom.
© 2025 John Wiley & Sons Ltd. Published 2025 by John Wiley & Sons Ltd.

337, 353–354, 361, 404, 412–415, 418, 425–426, 430, 443, 462

di/dt 19, 53–55, 69, 331, 420, 428, 472

Dielectric breakdown   100, 174

Digital controller   408

Digital isolator   313

Direct-drive devices   47, 48, 51, 53, 65–66

Direct Time of Flight (DTOF)   369–374, 388, 462

Drain efficiency (η_D)   274–275, 277, 290–292, 463

Drain-source leakage current (I_{DSS})   27–29, 436, 439, 462

Drones   95, 144, 404, 424–425, 450

Duty cycle   30, 100, 103–104, 109, 112–113, 146, 149, 159–160, 226, 236, 241, 266–268, 326, 338, 340, 354, 367, 373, 380, 384, 391, 397, 409, 443–444, 463

dv/dt 50–53, 55, 60–61, 67, 69–70, 141, 174, 183, 331, 409–410, 415, 418, 420, 428, 472

Dynamic losses   225, 235

Dynamic $R_{DS(on)}$ 104, 109, 111–113, 136, 141, 143

e

Edge-emitting laser (EEL)   373, 463

Efficient Power Conversion Corporation (EPC)   2, 15, 28, 64, 453

Electromagnetic interference   221, 331, 463

Electron trapping   105, 140, 143

Electromigration   96–97, 113, 119–122, 463

Enhancement-mode GaN (eGaN)   2, 6–10, 12, 14–19, 24, 26, 28, 32–35, 37–39, 41, 48, 61, 64, 66, 186, 197, 204, 214–217, 233, 235, 254, 261–262, 269, 276, 285, 290, 292, 363, 393, 433, 435, 437, 439–442, 445, 462–463, 466, 468, 471

Epitaxial growth   467

f

Failure in Time (FIT)   98–99, 463

Fall time   70, 216, 231–232, 388, 391, 409, 418, 471, 474

Field oriented control (FOC)   408–409, 411, 424–425, 463

Figure of merit (FOM)   1, 45, 47, 139, 217, 257–261, 266, 269–270, 274, 283, 295, 321–322, 325–326, 348, 376–377, 440, 455, 463

Filter   59–60, 69, 181, 196, 224, 243, 251, 288–289, 297–298, 310, 329–331, 345, 351, 355, 357, 359, 364, 395, 404–405, 409, 412–413, 415–418

Flying capacitor   322–328, 333, 343–345, 347, 464, 472

Forced air convection   160, 430

Full bridge   20, 310, 313–314, 322–323, 351, 460

g

Gain   1–2, 4, 101, 257, 274–275, 277–278, 283–287, 289–292, 296, 308, 355, 357–358, 361, 422, 441, 470

Gallium nitride (GaN)   2, 5, 11, 21, 29, 95, 324, 364, 403, 425, 435, 441, 464

Gamma radiation   98, 434–437, 440

GaN FETs   2, 78, 139, 142–144, 146–148, 158, 171–173, 184, 188–189, 205, 295, 299–303, 314, 322, 325, 340, 342, 345–346, 351, 356, 361, 384, 414, 421, 425, 429, 463, 468, 471

GaN integrated circuits   15, 179, 422–423, 441–444

GaN reliability   95–150

Gate charge (Q_G)   36–38, 45–47, 50, 139, 196–201, 205, 216, 228–231, 254–255, 258–260, 264, 269, 271, 314, 321, 324–325, 353, 361, 376–377, 440, 463–464, 467

Gate-drain charge (Q_{GD})   37–38, 46, 52, 196, 198–199, 202, 205, 216–217, 228–230, 235, 239, 245, 254–255, 259, 264–265, 269, 440–442, 464–465

Gate drive   9, 37–39, 47–53, 55–56, 58–60, 65, 70, 75–76, 87, 89–90, 92, 99, 116, 139, 197, 216–217, 225, 239, 258–260, 265–266, 268, 317, 333–337, 348, 361, 376–380, 383, 385, 387, 391, 396, 399

Gate loop   48–53, 55, 64, 66, 68–69, 75, 82–83, 87, 92–93, 101, 301, 303, 313, 378, 386–388, 391

Gate plateau voltage (V_{Pl})   45, 200–202, 206–211, 219–220, 228–230, 232–235, 238–240, 259, 465

Gate resistance (R_G)   37–38, 49–51, 69, 75, 85, 197, 220, 225, 236–237, 239, 258, 352, 376–377

Gate-source charge (Q_{GS})   37–38, 46, 196, 200–201, 228–230, 238, 255, 259, 441, 464

h

Half bridge   16, 18–21, 52, 55–56, 58–62, 64, 67, 75–76, 78–79, 84–85, 89–93, 164, 168, 172, 176–184, 196, 202–203, 205–206, 208, 211–212, 217, 219, 227–228, 296, 303, 307, 313, 322, 337, 351, 353, 355–356, 360–361, 404, 420, 425, 427, 429, 441, 462–463, 466–467, 469

Hard-switching figure of merit   217, 322

Heatsink   41, 114, 155–163, 165, 169–172, 174–183, 185–191, 287, 298–300, 303, 305, 316–317, 340, 345, 357, 362–364, 423–424, 426, 430

Heat-spreading   160, 163–166, 168, 170, 185, 189

Heteroepitaxy   11–13

High electron mobility transistors (HEMT)   2, 6–18, 20, 64–65, 95, 273, 277, 435–436, 438, 440, 465

Humanoid robot   95, 421–424, 450–451

i

Idle losses   357, 360–361

I_{DSS} (drain-source leakage current)   27–29, 436, 439–440, 462

Impact ionization   3, 101–102, 148

Impedance   8, 45, 52, 66–67, 69, 86, 156, 159–160, 171–172, 191, 198–199, 203, 238, 243, 252, 273, 277, 281–282, 287, 289, 292, 357–358, 368, 377–378, 381, 383, 396, 405–408, 416, 430, 467, 472–473

Indirect Time of Flight (ITOF)   369–370, 372–374, 384, 388, 391, 465

Inductance   15–16, 19, 48–49, 51–54, 59, 62, 64–66, 69–70, 75–82, 84–87, 89–90, 92–93, 101, 135, 140, 144, 199, 214, 217–222, 224–225, 237–238, 242–245, 250, 262–264, 296–297, 301, 303, 306–307, 313–314, 329–330, 336, 338, 340, 346, 352–353, 361, 369, 376–389, 391, 394, 396–399, 406, 408–411, 413, 418, 420–421, 429, 453, 461, 467, 469–470

Interleaving   81

Intermediate bus converter (IBC)   139, 249, 262, 308

Intermodulation distortion (DIM)   355, 360

Internal-vertical power loop   79, 81–83, 85–86, 466

Isolated DC–DC converter   263, 333

j

Joule heating   120, 169–170, 243, 466

Junction temperature (Tj)   32, 34, 42, 111, 113–116, 119–120, 122, 146–147, 160, 180–182, 185, 466

Junction-to-ambient thermal resistance ($R_{\theta JA}$)   41–42, 156–159, 163, 177, 185, 472

Junction-to-board thermal resistance ($R_{\theta JB}$)   40–42, 157–160, 163, 169–171, 174–176, 180–181, 473

Junction-to-case thermal resistance ($R_{\theta JC}$)   41–42, 156–158, 160, 163, 170, 174–175, 181, 473

k

K (Rollett stability factor)   274, 281, 286, 471

l

Land grid array (LGA)   17, 466

Laser   144, 146–150, 367–369, 372–399, 455, 463, 474

Laterally diffused metal-oxide-semiconductor (LDMOS)   64, 273, 291–292

Layout   16, 20, 41–42, 45, 49, 53, 55, 59, 61, 65, 68–70, 75–83, 85–87, 89–93, 102, 158, 163, 165–166, 168, 176–177, 184–185, 214, 217, 220, 237, 244, 263, 292, 296–297, 299–304, 306–307, 309–310, 312, 314, 317, 331, 333, 336, 362, 369, 378, 381–382, 384, 386–387, 390–391, 393, 396, 399, 411, 413, 420–423, 425, 428–429, 463, 466

Lidar   20, 69, 95–97, 133, 143–145, 147–150, 367–380, 386, 389–391, 394, 399, 450–452, 454–455, 466

Lifetime projection   129, 138, 148

Linear energy transfer (LET)   434, 438–440, 443–444, 466

LLC resonant converter   139, 251, 308, 310–314, 357, 452

m

Magnetic field self-cancellation   76, 80–82, 89

Magnetics   223–224

Mapping   367, 466

Matching network   274, 281–282, 284, 287–290

Maximum likelihood estimation (MLE)   123, 147

Mean time to failure (MTTF)   98, 101–102, 123–126, 129–130, 134, 139, 148, 466

Miller charge   52, 198–200, 264

Miller induced turn-on   51, 66

Miller ratio   37, 52–53, 467

Mission-specific reliability   133

Monolithic GaN power-stage IC   68, 85–86, 144, 182, 185, 191, 296–297, 378, 388, 391, 444

Monolithic half bridge   16, 19, 84–85, 180

Monolithic integration   84–85, 179, 296, 441

Monolithic power stage   84, 455

MOSFET   1–4, 6, 8–10, 15, 19, 21–22, 25, 28–30, 32, 34–35, 38–40, 45, 47–48, 50, 52–53, 56, 61–62, 64–70, 75, 89, 97, 101, 114–115, 139, 155, 157–159, 196–198, 210, 214–218, 220, 223, 238, 249, 254–269, 295, 306, 314, 317, 321–322, 325–326, 342, 345, 352–353, 361, 376–377, 384, 399, 412–413, 418, 424, 433–442, 445, 449–450, 452–454, 456, 460, 462, 465

Motor loss factor (LF)   409–410, 413, 424, 430, 467

Multilevel   317, 321–327, 329–333, 336, 338–339, 342, 345, 348, 421, 456, 464, 467–468

Multiphase   84

n

NASA   433–434, 473

Natural convection   160–162, 166, 184, 187, 189–190

η_D (drain efficiency)   274–275, 277, 290–292, 463

Neutron radiation   98, 437–438

Noise floor   358–360

o

On-resistance ($R_{DS(on)}$)   1, 3–4, 9–10, 19, 22, 25, 27, 31–33, 38–40, 42, 45, 47–48, 64, 66–67, 75, 89, 95, 97, 104–115, 123, 125, 133, 135–136, 138–143, 145, 216–218, 255, 257–258, 260–261, 263–265, 270, 306, 310, 314, 321–322, 324–326, 352, 376–377, 412, 425, 436–438, 440–442, 450, 452, 454–455, 463–464, 468

Operating temperature   3, 40, 84, 130, 135, 137, 163, 299

Output capacitance   38, 59, 62, 65, 196–197, 202–205, 207, 209–214, 216, 222, 225–226, 236, 241–243, 249–250, 254–256, 262–263, 273, 314, 384, 460, 475

Output charge (Q_{OSS})   38, 203–204, 207, 209–213, 226, 231–235, 249, 252, 254–258, 260, 264–265, 268, 270–271, 314, 321–322, 324–325, 361, 440, 468

Overshoot   29–30, 37, 48–49, 62, 75, 77, 85, 103, 109, 112, 135, 142, 149, 221–222, 298, 352, 377, 384, 394, 399, 453

Overvoltage   28–30, 57, 96, 100–104, 107–113, 140–142, 146, 149

p

Paralleling   85–93, 176, 295, 303–308, 314, 420–421, 428

Parasitic inductance   16, 19, 64–65, 70, 75–78, 80–81, 84, 92–93, 135, 140, 144, 214, 220, 225, 263, 303, 306–307, 369, 376, 399, 453

Partial power   310–311, 314–316

Permanent magnet motors   403, 405, 410, 424, 468

pGaN   7–8, 10–14, 31, 34, 48, 434–435, 437–440, 462

Photovoltaic   134

Plateau voltage (V_{pl})   45–46, 50–51, 89, 198, 200–202, 207–211, 220, 228–230, 232–235, 238–240, 463–465

Power loop   15, 19, 48–49, 53–55, 64, 69, 75–87, 89–93, 176, 217, 219–220, 222, 242–243, 297, 301, 304, 306–309, 312, 378, 381–382, 384–388, 391, 396–399, 411, 413, 420, 453, 463, 466, 469

Power loop inductance (L_{Loop})   19, 69, 75–77, 79–85, 93, 217, 219–222, 242–243, 297, 306, 382, 384–386, 388, 396–399, 420, 453, 469

Power quad flat no lead (PQFN)   15, 17–18, 41, 78–79, 85, 96, 111, 113, 119, 122, 124, 126, 155, 157, 165–166, 191, 296, 353, 356, 425, 469

Power tools   404, 410, 425–430

Pulsed laser driver   368–378

Pulse width modulation (PWM)   69, 195–196, 253–254, 296, 324, 326–332, 338, 353–357, 404–405, 407, 409–413, 415–420, 423, 425–426, 430, 463, 468, 470–471, 474

q

Q_G (total gate charge)   36–38, 45–47, 50, 139, 196–202, 205, 216, 228–231, 254–255, 258–260, 264, 266–267, 269, 271, 314, 321–322, 324–325, 353, 361, 376–377, 440, 463–464

Q_{GD} (charge required into the gate to bring the drain voltage down from stated drain voltage to near zero volts) 37–38, 46, 52, 196, 198–199, 202, 205, 216–217, 228–230, 235, 239, 245, 254–255, 259, 264–265, 269, 440–442, 464–465, 467

Q_{GS} (charge required to increase gate voltage to the plateau voltage) 37–38, 46, 196, 200–201, 228–230, 238, 255, 259, 441, 464

Q_{GS1} or $Q_{GS(th)}$ (charge required to increase gate voltage from zero to the stated threshold voltage) 46, 52, 198, 200–201, 229, 259, 464

Q_{GS2} (charge required to increase gate voltage from zero to the plateau voltage) 46, 198, 200–202, 207, 209–210, 217–218, 220, 228–229, 232, 234, 237, 239–240, 254, 259, 269, 464

Q_{OSS} (output charge) 38, 203–204, 207, 209–213, 226, 231–235, 249, 252, 254–258, 260, 264–265, 268, 270–271, 314, 321–322, 324–325, 361, 440, 468

Qromis substrate technology (QST) 11–12

Q_{RR} (reverse recovery charge) 38–40, 65, 214, 216, 241, 243–244, 321–325, 413, 462

r

Rad-hard 434, 440–442

Rad-hard Si MOSFET 440

Radiation resistant 443–445

$R_{DS(on)}$ (on-resistance) 1, 4, 9, 31–33, 39–40, 42, 47–48, 64, 66–67, 97, 104–115, 123, 125, 133, 135–136, 138–143, 145, 216–218, 255, 258, 260–261, 264–265, 270, 310, 314, 321–322, 324–326, 352, 376–377, 412, 425, 436–438, 440–442, 452, 455, 468

Rectifier 202, 217, 312, 321, 324, 331, 351, 353, 455

  center-tapped 310, 314

  secondary side 139, 263, 310

  synchronous 10, 63, 81, 204–209, 215, 217, 221–222, 224–228, 230–233, 236–240, 242, 302–303, 307–310, 314, 455

  totem boost and telecom 331, 342, 344–346, 348, 473

Reliability 2, 15, 22, 64, 69, 95–153, 167–168, 222, 321, 324, 403, 413, 420, 454, 465, 472

Resonant converter 250, 252–254, 261–262, 266, 268, 308, 310

Resonant pulse 378, 380, 384, 388

Reverse conduction 10, 25, 38–40, 42, 56, 67–68, 196–197, 205–209, 214–215, 217, 226–228, 231–235, 239–240, 244, 333, 353, 470

Reverse conduction losses 196–197, 205, 231, 233, 239

Reverse recovery charge (Q_{RR}) 38–40, 59, 65, 214, 216, 241, 243, 321–322, 324–325, 352, 413, 462

Reverse transfer capacitance (C_{RSS}) 37–38, 199, 229–230, 255, 413, 460

RF transistor 2, 273–278, 281, 469

R_G (gate resistance) 37–38, 49–51, 69, 75, 85, 197, 220, 225, 236–237, 239, 258, 352, 376–377, 466

Ripple 224–226, 228, 240, 326, 328–331, 338, 345–346, 353, 357, 360–361, 412–413, 416–417, 465, 472

Rise time 37, 53, 69, 396, 474

Rollett stability factor (K) 274, 281, 286, 471

R_θ (thermal resistance) 40–42, 84, 93, 155–161, 163, 165, 167–171, 173–175, 177–185, 191, 363, 472–473

$R_{\theta JA}$ (junction-to-ambient thermal resistance) 41–42, 156–159, 163, 177, 185, 472

$R_{\theta JB}$ (junction-to-board thermal resistance) 40–42, 157–160, 163, 169–171, 174–176, 180–181, 473

$R_{\theta JC}$ (junction–to–case thermal resistance) 41–42, 156–158, 160, 163, 170, 174–175, 181, 473

s

Saturation current 115–116, 379, 384

Schottky 6–8, 18, 31, 34, 56–57, 59, 67–68, 214–216, 231, 333, 434, 438–440

Short circuit 6, 97, 113, 115–116, 118–119, 382, 416, 439

Shunt compensation 397–398

Shunt inductance 396–397

Silicon carbide 2–5, 11–14, 322, 325, 437, 471

Single-event effects (SEE) 433, 438–440, 442–443, 471

Single event gate rupture (SEGR) 433, 439, 471

Single event transient (SET) 441, 443–444

Solar 95–96, 133–138, 295, 375, 433, 438, 454, 456

Soft-switching 51, 249–271, 295, 440, 464
Soft-switching figure of merit 260–261, 264, 266, 269, 440, 464
Solder joint 120, 122–128, 130–133, 137, 143, 157, 163, 167–168
s-parameters 276–278, 281–285
Stability 95, 144–145, 189, 274, 278, 281–282, 285–287, 424, 438–439, 471
Stressor 97–99, 120, 134–135, 138–141, 143–144, 150
Switches 104, 195, 228, 243, 263, 310, 314, 322–325, 331–334, 337, 343, 351, 376, 394, 411–412, 421–422, 425
Switching losses 19, 53, 75, 139, 196–198, 200, 202, 216–217, 220, 228, 236–237, 245, 254, 264, 310, 314, 317, 351, 364, 407, 412–413
Synchronous bootstrap 333–336, 338, 441, 443
Synchronous rectifier 10, 63, 81, 204–209, 215, 221–222, 224–228, 230–231, 233, 236–237, 239–240, 242, 302–303, 307–310, 314, 455

t

Temperature coefficient 32, 89, 384
Temperature cycling 122–134, 136–139, 142–143, 168, 472
Temperature measurement 155, 182–185, 191
Test-to-fail 95–99, 104, 132, 134–135, 138–139, 144
Thermal coupling 180–181
Thermal design 155, 160–161, 163, 170, 172, 174, 191, 292
Thermal interface material (TIM) 156, 160, 162, 170–176, 178, 181–182, 186–189, 191
Thermal resistance (R_θ) 40–42, 84, 93, 155–161, 163, 165, 167–171, 173–175, 177–185, 191, 363, 472–473
T_J (junction temperature) 32, 34, 42, 111, 113–116, 119–120, 122, 146–147, 160, 180–182, 185, 466
Thermo-mechanical wear-out 95, 122–133
Threshold voltage (V_{TH}) 25, 34–35, 42, 46, 51, 55, 89, 200, 433–437, 445, 473
Through power 310–311, 314–316
Time of flight (TOF) 20, 367–369, 373–374, 399, 473
Time to failure (TTF) 102, 116, 119, 122, 142, 148
Top-side cooling 171, 176, 178–179, 181, 185–187, 423

Total harmonic distortion (THD) 331, 346, 352–355, 357–358, 361, 364, 405, 414, 473
Total incident dose (TID) 434, 436, 440, 442, 473
Trapping 104–106, 140, 143, 433
Transconductance 51, 89, 220, 236, 238, 474
Transient load 28–30, 37, 48, 59–61, 75, 92, 100–101, 103, 109, 113, 115, 139, 141, 149, 222, 225, 231–233, 238, 295, 303, 355, 386, 425, 439, 441, 443–444, 450, 461
Transient thermal impedance 159–160, 191, 473
Two-dimensional electron gas (2DEG) 2, 5–8, 10, 12–13, 18, 20, 27, 31–34, 36, 38–39, 101, 104–105, 107, 140, 434–435, 473–474
Two-phase buck converter 300–301, 341
Two-port network 176, 276–277

u

Underfill 96, 122, 130–133, 137–138, 142–143

v

Vertical cavity surface emitting lasers (VCSEL) 373–375, 394, 474
Vertical GaN 10–11
Voltage measurement 410
Voltage overshoot 29–30, 49, 75, 77, 85, 109, 135, 221–222, 352, 377, 453
Voltage probes 388, 396
Voltage ringing 50, 113, 140–141, 146
Voltage source inverters 321, 404
V_{pl} (gate plateau voltage) 45, 200–202, 206–211, 219–220, 228–230, 232–235, 238–240, 259
V_{TH} (threshold voltage) 25, 34–35, 42, 46, 51, 55, 89, 114, 200, 229, 433–437, 445, 473

w

Wafer fabrication 13–17, 454
Wafer-level chip-scale package (WLCSP) 376
Wear-out mechanism 95–101, 104, 113, 119, 122, 134–135, 139, 142–143, 149–150, 454
Weibull distribution 123, 125–126, 143, 147

z

Zero-current switching (ZCS) 249–254, 257, 259, 262–263, 265, 302, 310, 475
Zero-voltage switching (ZVS) 206, 208–211, 231, 233, 249–254, 257, 259, 262–263, 265–268, 302, 310, 360–361, 475